Agriculture, Business and Heritage Property Relief

Seventh Edition

Agriculture, Business and Heritage Property Relief

Seventh Edition

Chris Erwood CTA, ATT, TEP

Bloomsbury Professional

Bloomsbury Professional
An imprint of Bloomsbury Publishing Plc

Bloomsbury Professional Ltd
41–43 Boltro Road
Haywards Heath
RH16 1BJ
UK

Bloomsbury Publishing Plc
50 Bedford Square
London
WC1B 3DP
UK

www.bloomsbury.com

BLOOMSBURY and the Diana logo are trademarks of Bloomsbury Publishing Plc

British Library Cataloguing-in-Publication Data

A catalogue record for this book is available from the British

ISBN: PB: 978 1 52650 376 3

Typeset by Phoenix Photosetting, Chatham, Kent
Printed and bound in Great Britain by CPI Group (UK) Ltd, Croydon, CR0 4YY

Preface

This book is now in its 7th edition and reflects the law as it stands at January 2018. It thus takes into account all pertinent changes in the legislation since the last edition published in April 2014, has been fully updated for interim case law development, contains detailed reference to current HMRC practice and provides cross reference to other professional commentary. In examining legislation the author has, where appropriate, paraphrased the text, repeated words to help explain the meaning or following the example of Blackburne J in Starke, broken it up into 'bite-sized' portions. All pertinent references have now been amended to refer to 'HMRC' which term should be taken to include their predecessors.

The book is comprehensive, wide reaching and tackles head-on the complexity of the four principal reliefs. Intended as much more than a mere 'aide memoire', this work is heavy on practical detail and is specifically aimed at the busy practitioner rather than the mere academic – it is a solid reference work. Each chapter commences with a 'Quick guide' to summarise the contents and point the reader to the relevant section. In reviewing contentious areas the author has quoted at length from the key decided cases so that the practitioner with a limited library and time has sufficient detail to decide whether it is necessary to read the actual report before proceeding. Many of the worked examples are indeed drawn from life but adapted to better illustrate specific points in the legislation.

The illustrative examples

The book contains in excess of 100 detailed examples which serve to illustrate the technical points drawn out in the various chapters. The author has introduced fictitious advisers into the examples whose relative skill in tax matters is varied and sometimes a little off target as will quickly be apparent to the reader. Cerys is an accountant and chartered tax advisor (as well as a Trust and Estate Practitioner who qualified under the qualified practitioner route) route, Jack is a solicitor and Leslie, his assistant, is a Trust and Estate Practitioner (who qualified under the Diploma route). It is a hard and rather sobering fact that more can be learned from cases that go horribly wrong than from those which mirror textbook approach. In the examples, as in real life, there may be details that are not perceived as absolutely essential to the points in issue but which nonetheless have an impact on the eventual outcome. Many readers may find it challenging to work through the examples so, to better to explain a particularly awkward corner of the law or simply to add interest, the examples also occasionally illustrate how simply things can 'go wrong'.

Team work

The author gratefully acknowledges the editorial team at Bloomsbury Professional for their encouragement, patience and hard work in bringing this updated work to fruition but importantly Toby Harris, with whom the author collaborated on the previous 6th edition of this work. Toby has been an immense inspiration to the author and has heavily contributed to the footprint of much of the technical text and interpretation still reflected within the current work.

Chris Erwood CTA ATT TEP
North Devon,
January 2018

Contents

Preface v
Table of Statutes xxi
Table of Statutory Instruments xxxi
Table of Cases xxxiii
Table of Examples xli

Chapter 1: Introduction **1**
The place of BPR/APR in inheritance tax (IHT) 1
History of BPR/APR 1
The need for heritage property relief 2
Outline of the scope of this book 3
Territoriality of APR, BPR and heritage relief 4
Interaction of APR and BPR with other reliefs 4
How APR and BPR apply computationally 6
Interaction with IHT instalment relief 7
The distinction between BPR and APR 8
Is there a need for both APR and BPR? 8

Chapter 2: Agricultural property: qualifying property **10**
Quick guide 10
Agricultural land or pasture 11
Woodland 11
Buildings used for intensive rearing of livestock or fish 11
Ancillary occupation 15
Farm cottages 15
Farmhouses 17
Tests of a farmhouse: historical background to the current law 17
The historical context of the law and of the subject dwelling 18
Big farmhouses 19
Tests applied in earlier cases 19
Higginson: an extreme case 20
Antrobus No 1: lifestyle farming 21
The Antrobus checklist 21
The 'elephant' test 22
Review of s 115(2) in Starke 23
Rosser v IRC 24
HMRC v Hanson 25
The leading case: McKenna 26
The meaning of 'farmhouse' in Hanson 29
The fine house with 'a bit of land' 30
Land and farmhouse not in identical ownership 31
Farmers who fall ill 32

Horses 32
Short rotation coppicing 33
Milk quota 34
Land within conservation and similar schemes 34
Farming companies 35
Territoriality 36
What does not qualify for APR? 37

Chapter 3: Agricultural property: agricultural value, the rates of relief, occupation and other aspects 39
Quick guide 39
Agricultural value 39
The rates of relief 41
- 100% APR 41
- The ownership or occupation test 41
- The test determinative of the rate of relief 42

Farming and trusts 43
- Must the holder of an interest in possession himself be the farmer? 44

Working farmer relief ('double discount cases') 45
- 'Wholly or mainly engaged' 46
- The occupation test for working farmers 46
- Other difficulties and restrictions relating to working farmer relief 47
- Working farmer relief and companies 48
- Limitation of working farmer relief by value or by acreage 48
- Working farmer relief: successions 50
- Restrictions on working farmer relief 51

Land let on new tenancies 51
'Renewing' the tenancy to secure 100% relief 53
Occupation for the purposes of agriculture 55
- Agricultural purposes 55
- Diversification 57
- The Basic Farm Payment (BFP) 57
- Who is the occupier? 58
- Occupation in partnership situations 59
- Occupation of the farmhouse for agriculture 60
- The farmer who falls ill 61
- Atkinson: what constitutes 'occupation for the purposes of agriculture'? 62
- Occupation of dwellings for agricultural purposes 62
- Farm cottages: the rate of relief 64
- Share farming and contract farming 65

50% APR 65

Chapter 4: Agricultural property: areas of particular concern 67
Quick guide 67
No contract for sale 68

Joint ownership	69
Partnerships	69
Trusts	71
Occupation by trust beneficiaries: Woodhall v IRC and Watkins v IRC	72
Replacement property	73
Test of replacement property as to occupation	73
Test for replacement property as to ownership	73
Controlled companies	75
Successive transfers and inherited ownership	75
Successions	75
Successive transfers	77
Replacement property	78
Problems arising out of successive *Gladstone v Bowers* and grazing agreements	79
Valuation of agricultural property	79
General considerations	79
Tenancies: Baird's Executors v IRC	82
Gray (surviving executor of Lady Fox) v IRC	83
Alexander v IRC	85
Lotting for sale: 'natural units'	85
Other aspects of valuation	87
Planning permission	87
Joint property: Arkwright v IRC	89
Price (Executor of Price, deceased) v HMRC	91
The valuation of farmhouses in the light of *Antrobus No 2*	92
Planning exercises relating to tenancy valuation	95
Sales of quoted shares within 12 months of death	96
Land sold within four years after death	97
Sales of related property within three years after death	99
Vacant possession of land held outside a partnership	100
Chapter 5: Business property: property qualifying for the relief	**102**
Quick guide	102
Introduction: the concepts	102
Relevant business property: the six categories	102
Section 105(1)(a): Business or an interest in a business	103
Activities constituting a business	104
Non-investment businesses	104
Business, but of the wrong kind for relief: McCall v HMRC	105
Corporate v non-corporate	107
The restrictive effect of s 105(3)	108
Woodlands	109
Sole proprietor's business	109
Authors, artists, sculptors and musicians	109
Partnership interest	110

The 'mere asset' exclusion: review of the early cases 110
Nelson Dance 111
Disposal of part of a partnership share 112
One business or several? What is the interest? 113
Lloyd's underwriters: the assets that qualify for BPR 114
Lloyd's Name losses 115
Assets supporting the underwriting business 116
Succession 118
Limited partnerships and limited liability partnerships (LLPs) 119
LLP accounts and formalities 120
Jersey LLPs 120
Transition to LLP 120
Section 105(1)(b): unquoted securities giving control 121
The meaning of 'control' 122
Section 105(1)(bb): unquoted shares 124
Section 105(1)(cc): controlling holdings in a quoted company 124
Section 105(2): restriction on relief 125
Preference shares: effect on control 125
Clearances for BPR 125
Situations not covered by the clearance procedure 127
Section 105(1)(d): land and buildings etc used in the business 128
Machinery and plant 129
Animals 130
Intellectual 'tools' and property 130
Equipment, not 'the setting' 131
New categories of 'property' 131
Bookmakers' pitches 132
Section 105(1)(e): land and buildings used in the tenant for life's business 133
The meaning of 'interest in possession' 133
History of relief under s 105(1)(d) 134
Timing 134
The meaning of 'assets used in the business' 135
The applicability of the lower rate of relief 136
Other conditions 136

Chapter 6: Business property: the rates of relief, limitations to the relief 138
Quick guide 138
The rates: manner of application 138
Interaction with quick succession relief 138
Interaction with other reliefs 139
Grossing up 140
The rates of relief 141
Period of ownership 141
Modification of the two-year rule 143

Replacement property	143
Period of ownership of unquoted shares: capital reconstruction	146
Extending the period of ownership	147
Period of ownership: detailed rules	148
Definitions for s 91	149
Transfers after 6 April 1996: special rules	149
No contract for sale	149
What is 'a binding contract for sale'?	150
Exception to the rule in s 113: underlying assets	152
Successions and successive transfers: the rules in ss 108 and 109	152
Liquidations and reconstructions	155
The meaning of 'quoted'	156
Recognised stock exchange	156
Other trading exchanges	157
Which stock exchanges are actually recognised?	157
The winding-up exclusion	160
Chapter 7: Relevant business property: exceptions to the relief: excepted assets	**162**
Quick guide	162
Excepted assets (*s 112*)	162
Use versus ownership	166
Personal use	166
Valuation principles applied where there are excepted assets	166
Valuation where assets used in part for business purposes	167
Special rules for corporate structures?	168
The meaning of 'business'	169
The leading case: *McCall v HMRC*	170
HMRC commentary	170
Hybrid company	171
Case law commentary	172
Barclays Bank Trust Co Ltd v IRC	172
Salaried Persons Postal Loans Ltd v HMRC	174
The meaning of 'group' for BPR	174
Interaction of excepted assets and debt	175
Chapter 8: Relevant business property: aspects of valuation and the treatment of liabilities	**177**
Quick guide	177
Basic principles	177
Quoted shares and securities	177
Alternative investment market (AIM)	178
Unquoted shares	178
Caton's Administrators v Couch	179
Hawkings-Byass v Sassen and related appeals	180
Denekamp v Pearce	180

CVC/Opportunity Equity Partners Limited and Another v Demarco Almeida 181
Phoenix Office Supplies Limited v Larvin 181
So which basis applies? 182
Commentary 183
A basis for discounting? 184
Minority interests in property: what discount? 184
General rules applied in ss 110 and 111 186
Executors of Eighth Marquess of Hertford, deceased v CIR 187
The treatment of liabilities: the new rules 189
Liabilities of sole traders 195
Guarantees by business owners 196
The related property rules 196
Valuation of shares and loan notes 200

Chapter 9: Relevant business property: exceptions to the relief: 'wholly or mainly holding investments' 201
Quick guide 201
Introduction 201
The *Martin* case 203
The *Burkinyoung* case 203
Principles emerging from the comments of the Special Commissioner 204
The level of activity 204
Zetland 205
Subsequent decisions on s 105(3) 206
Hall v CIR 206
The *Pearce* case 207
Furness: a victory for the taxpayer 207
Weston v IRC 208
Stedman's Executors v IRC; CIR v George and Another 209
Review of the law in the light of George 211
Holiday lettings 212
Holiday homes: Pawson, Green and Ross 212
Livery business: *Vigne v HMRC* 214
Mixed use estate: *Farmer v IRC* 215
The *Earl of Balfour's* case 216
Moneylending: the *Phillips* case 217
Building and developing: *Piercy* 217
Carrying on a business 'for gain' 217
Property management as a business: *Clark* 218
Property management: *Best v HMRC* 219
Overview 220

Chapter 10: Clawback 222
Quick guide 222
Introduction 222

Clawback of BPR: the basic rule 222
Computational problems relating to clawback 222
Additional tax: changes to the rules on computation introduced in 1992 225
Partial clawback: the operation of s 113A(5) 225
Interaction of reliefs: clawback and 'pre-death loss relief' 226
Disregard of BPR: the rule in s 113A(7A) 227
Additional conditions 227
Definitions employed in s 113A: 'the original property' and 'the transferee' 228
Relief from clawback 228
Problems of identification: the rule in s 113A(6) 228
Replacement property for the purposes of clawback (BPR) 229
Further conditions: s 113B 230
Clawback of APR: the basic rule 231
Partial clawback: the operation of s 124A(5) 233
Disregard of APR: the rule in s 124A(7A) 233
Definitions employed in s 124A: 'the original property' and 'the transferee' 233
The effect of clawback 235
An exception to clawback 236
Relief from clawback 236
Problems of identification: the rule in s 124A(6) 236
Replacement property: the conditions 237
Clawback flowchart 239
Replacement property: how the rules work 241
Gifts with reservation and the application of the clawback rules 243
Countering clawback in relation to BPR and APR 244
Replacement of agricultural property by business property 244
Surrender of farm tenancies and FBTs 244
Further hazard: discretionary trusts 245
Further difficulty: value-shifting 248
Comparison of potentially exempt transfers (PET) with immediately chargeable transfers (CLT) 249
Clawback: a loophole? 250

Chapter 11: Woodlands **252**
Quick guide 252
Commercial occupation of woodlands: overview of tax treatment 252
Is there a trade? 253
Case commentary on timber: whether income or capital 254
The former woodland relief 255
Relief on death 256
Prior relief as agricultural property 259
Prior claim to BPR 259
Property qualifying for relief 260

The charge to tax on disposal 261
The basis of charge 262
The rate of tax 262
Net value: deduction of expenses 262
Credit for tax charged 263
50% reduction under s 127 265
Relief for lifetime gifts 266
The treatment of liabilities: the new rules 266

Chapter 12: Overlap of APR and BPR **268**
Quick guide 268
Historical perspective 268
Priority/election 269
Territoriality 269
Amount 269
Replacement assets: switching 270
Where APR is available but BPR is not 271
The situations where BPR is more favourable than APR 272
The farm: applicability of BPR 272
Sheltering proceeds of sale of a business through agricultural property 272
The entrepreneur who 'turns his hand' to farming 273
Use a corporate 'wrapper'? 274
Family limited partnerships 275
Treatment of FLPs for law and tax 275
The general partner 276
Regulation and protection 276
Share farming or contract farming 277
Granting agricultural tenancies 279
Security of tenure 279
BPR on an agricultural property company 280
Farmers diversifying into non-agricultural enterprises 280
Holiday lettings 282
Is there a need for two reliefs? 283

Chapter 13: Tax shelters **284**
Quick guide 284
Introduction 284
The General Anti-Abuse Rule (GAAR) 285
Farms 286
As landlord 286
Share-farming and contract farming 286
Tenant buying out landlord 286
Businesses 288
Listed unquoted trading companies 288
Caveat: secondary market listings 289

Fully private companies 289
Conversion of debt into equity 290
Conversion of cash into equity 291
Buildings or machinery 291
Nursing home businesses 291
Lloyd's interests 291
Investment properties 292

Chapter 14: General planning points **293**
Quick guide 293
Unfortunate gifts 293
The taxation of relevant property trusts of relieved property 296
Gifts with reservation 297
Gifts with reservation: de minimis 297
Gifts with reservation: treatment where BPR is available 298
Further substitution rules affecting gifts with reservation 299
Deductibility of debts 304
Planning since *FA 2013* 310
Restriction on offset of debt 311
Buy and sell 312
A way out, perhaps? 313
Options in relation to shares 313
Retirement 314
Consider alternatives to an immediate sale, especially where ill-health may be an issue? 314
What lessons can learnt from Swain Mason? 315
Inactive partner still a partner 316
Leaving money in 317
Money box 318
What is possible? 318
Various structures 319
Wholly or mainly 319
Dual/multiple company structure: the hybrid company 320
Use of relevant property trusts to secure 100% relief 320
Associated operations: Reynaud v IRC 321
Associated operations: the Rysaffe case: now or never? 322
Transfer between settlements 325
The maiden aunt 326
CGT planning: exploiting ss 191 and 179 326
Stonor (executors of Dickinson deceased) v IRC 326
Death bed planning 327

Chapter 15: Will planning **329**
Quick guide 329
The nil rate band 329
Infinite nil rate band 329

Double nil rate band 330
Partly exempt transfers: interaction with APR and BPR 330
Non-exempt 'gross' legacy 330
Non-exempt 'free of tax' legacy 331
A mixture of legacies, with exempt and non-exempt shares of residue 332
Difficulties arising under s 41(b) 336
The impact of additional charity relief 337
Will planning for an elderly couple 340
Drafting points – the use of a cap 344
Drafting points: defining the legacy 345
Attribution rules 345
The effect of s 39A 347
Avoiding the difficulties of s 39A 348
Routing through non-exempt beneficiaries 351
Gifts of businesses 352
The meaning of 'spouse' 353

Chapter 16: APR and BPR: the new regime 354
Quick guide 354
What new regime? 354
Excluded property trusts 354
Reform 355
Use of 100% reliefs while they last 356
Locking into the 100% rate even on failed PETs 357
Loss of CGT free uplift on death 358
The last resort: life assurance 360

Chapter 17: Heritage property: what qualifies, undertakings and access 361
Quick guide 361
Overview 362
Finding the law 363
Heritage property: qualification 363
The quality of heritage property: rules after 30 July 1998 364
Acceptance in lieu 365
Land and buildings: England 367
Land and buildings: Scotland 368
Associated objects 368
Preliminary approach 369
Public access and publicity 369
New undertakings 370
Chattels 370
Land and associated chattels 370
Amenity land 371
Publicity of the terms of undertakings 371

Public access under the 'new' rules 372
What amounts to reasonable access 372
Access: the chattel, not the owner 374
Means of access 376
Damage to chattels 376
Other access details: charges, limited numbers of visitors and guided tours 377
Venue for provision of access 377
Publicity 378
The register of conditionally exempt works of art 378
What the owner need not disclose to the public 379
Disclosure: pre-eminent chattels and pre-eminent collections 379
Disclosure: outstanding buildings and amenity land 379
Disclosure: land 379
Footnote: foreign-owned works of art 380
Cultural Gifts scheme 380

Chapter 18: Heritage property: application for exemption and variation of undertakings 381
Quick guide 381
The old law 381
Requirements under the new law: the timing of the claim 382
Heritage property in discretionary trusts 382
Protective or provisional claims 383
The claim itself 384
No intention to sell 386
Triggering a chargeable transfer on which to hang a claim 386
Claims processing 387
Supervision of owners' compliance with undertakings 387
Variation of undertakings 388
Managing change: the transition from the pre FA 1988 regime to the current regime 389
Museum quality properties 390
Sales of heritage property 390
Review and variation of undertakings 391
Undertakings of A and B 391
Previous viewing 394
Appeal 398
Preserving earlier exemption: replacement undertakings 398
Does the system work? 398

Chapter 19: Heritage property: loss of the exemption 400
Quick guide 400
The general rule 400
Breach of undertaking under s 32 400
Death or disposal 402

The composite heritage estate: 'associated properties' under s 32A 403
Selling (only) the family silver 406
'Relevant disposal' 407
Part disposals: effect limited to the item sold 407
Loss or destruction of heritage property 408
The charge to tax 410
Relevant person 411
The rate of IHT applied: lifetime transfers 412
No allowances 412
Transfers on death 412
Increasing rates of IHT 414
Set-off and credit 415
The 'clock' of chargeable transfers 418
The old rules in relation to heritage property 419
Transferable nil rate band 419
Clawback of heritage relief whilst surviving spouse still alive 419
Clawback of heritage relief after the second death 421
Sting in the tail: ATED liability on top of recapture charge 422
Compliance 422

Chapter 20: Heritage property: the regime for relevant property trusts 423
Quick guide 423
Treatment of relevant property trusts: rules applicable after 8 March 1982 423
Clawback charge on relevant property trusts 424
Calculating the tax charge 425
The 'relevant person' and special rules 425
Exemption from the 10-year periodic charge 427
Two exemption regimes 427
Calculating the rate of IHT on new heritage property in a relevant property trust 428
Anti-avoidance: prevention of the purchase of heritage property to keep down periodic charges 430

Chapter 21: Maintenance funds 431
Quick guide 431
Outline of the scheme 432
Summary of the relief 432
Conditions for the relief 433
The first condition: the trust instrument 434
Qualifications to the first condition 436
Maintenance of heritage property 436
Public access requirements 438
Expenses and other items 438
The second condition: the trustees 438

The third condition: the content of the maintenance fund 439
Further condition: anti-avoidance 'charity' rules 439
Procedural points 440
Supervision of the fund 445
Accounts 445
Entry of property to the fund 446
Contributions to maintenance funds from discretionary trusts: anti-avoidance rules 447
Property entering a maintenance fund from an interest in possession trust 449
Anti-avoidance rules 450
Property leaving the maintenance fund 451
Reverter to settlor 452
Use for heritage purposes 452
Transfer to another maintenance fund 452
Other acceptable uses 453
Events giving rise to clawback of relief 453
The rate of tax: general cases 454
Death of life tenant 457
Special property reliefs 460
Calculation of the clawback charge where property was previously comprised in a relevant property trust 460
The impact on maintenance funds of other taxes 461
Income tax 461
Income tax treatment of maintenance funds 461
Stamp duty and stamp duty land tax 462
Capital gains tax 463

Chapter 22: Compliance **465**
Quick guide 465
Chargeable lifetime transfers (CLT) 465
Duty of the personal representative 467
HMRC guidance 468
Property details 468
The use of estimates 470
Robertson v IRC 470
Robertson v IRC (No 2) 471
Valuation: compliance issues 472
Cairns v HMRC 472
Personal applications for grants of representation 473
Penalties: the current regime 474
Overview of the penalty provisions 475
Levels of penalty 476
Mitigation of 30% penalties 478
Mitigation of 70% penalties 478
Mitigation of 100% penalties 478

The criminal offence of tax fraud 478
Forms and procedures 479
Agricultural relief 479
Business relief 480
Woodland relief 480
Heritage relief 480
Disclosure of Tax Avoidance Schemes (DOTAS) 480

Chapter 23: Payment of IHT **483**
Quick guide 483
The general rules 483
'Free of tax' legacies 484
Payment of tax on failed PETs 484
Due date for payment 485
Payment of tax by instalments 486
'Section 228 shares' 486
Interest on instalments 487
Termination of the right to pay tax by instalments 487
The HMRC charge 488
Purchase of own shares by an unquoted trading company 489
Public purchase of heritage property by private treaty sales: outline 490
'Douceur' 491
Stamp duty and stamp duty land tax 493
Value added tax 493
Interest and the valuation date option 494
'Supermarket' heritage offer 494
Offers 'in situ' 495
Gifts to the nation: the Cultural Gifts Scheme 496

Appendix 1: Statutory Instruments **498**

Appendix 2: Extra-statutory Concessions **499**

Appendix 3: HMRC Statements **500**

Appendix 4: HMRC and other press releases **504**

Appendix 5: Heritage property: the old rules **510**

Index *517*

Table of Statutes

[References are to paragraph numbers and appendices]

Administration of Estates Act 1925
s 34(3) 23.1
46 .... 6.11
Sch 1 .... 23.1
para 8 .... 23.1
Agricultural Holdings Act 1948 .. 3.18
s 1(1), (2) 3.18
Agricultural Holdings Act 1986 .. 3.4, 3.15, 3.16, 3.28
s 1 .... 12.18
Sch 3 .... 3.26
Agricultural Tenancies Act 1995. 3.15; 4.21; 13.5; App 3
s 4 .... 12.18
38(1) 2.2

Companies Act 1985
s 459 .... 5.15; 8.9
736 .... 8.14
Corporation Tax Act 2010
s 26 .... 7.7
1000 .... 23.10
Pt 23 Ch 3 (ss 1029–1063) 23.10
s 1033 .... 23.10
(3) 23.10
1044 .... 23.10
1045(1), (5) 23.10

Disability Discrimination Act 1995 .... 18.15

Finance (1909–10) Act 1910
s 61(5) 11.4
Finance Act 1930
s 40 .... App 5
Finance Act 1960
s 64 .... App 5
Finance Act 1969
s 39 .... App 5
Finance Act 1975 .... 1.2
s 20(2) 4.15
31(2) 18.14
32, 34 .... App 5
Sch 8 .... 3.9
para 1(2A) 3.10
(3) 3.14
3 .... 3.7
(1) 3.9
(3) 3.8
(5), (6) 3.9
4 .... 3.12
5 .... 3.12, 3.28
(2) 3.12
(3) 3.10
14(2) 8.21
Finance Act 1976
s 73 .... 1.2
74(6) 3.12
Sch 10 .... 1.2
Finance Act 1980
s 98 .... 21.31
Finance Act 1981
s 96 .... 1.2; 12.1
Sch 4 .... 1.2
Sch 14 .... 3.12, 3.28; 12.1
para 2(ii)(b) 3.28
(iv) 3.28
Finance Act 1982
s 49 .... 21.3
Finance Act 1984
Sch 14
para 1(1) 6.2
Finance Act 1986 .... 10.1
s 101(3) 17.3
102(1) 8.23
102B(4) 5.6; 14.3
106 .... 15.12
(4), (6), (8) 5.8
Sch 15
para 11(5)(b) 14.3

Finance Act 1986 – *contd*
Sch 19
para 8 ... 17.3
40(1) ... 18.10
46 ... 11.4
Sch 20
para 1A ... 14.5
2(4) ... 14.6
6(1)(a), (b) ... App 4
8 ... 14.5, 14.6
(1) ... 14.5
(c) ... 14.6
(1A) ... 14.6
(2) ... 14.5, 14.6
(3) ... 14.5, 14.6
Finance Act 1987 ... 21.17
s 58 ... 10.19
Finance Act 1988 ... 11.1
Sch 6
para 2(1) ... 11.1
3(5)(b) ... 11.1
Finance Act 1995
s 154(2), (3) ... 2.25
155 ... 3.16; App 3
Finance Act 1996 ... 5.27
s 184 ... 6.12
(4) ... 10.30
(6) ... 16.5
(b)(ii) ... 16.5
Sch 19
para 21 ... 6.12
Finance Act 1998 ... 1.3; 17.4; 18.1, 18.11, 18.12, 18.14, 18.15; 20.1
Sch 25
para 10 ... 18.14
Finance Act 2000
s 144 ... 13.1; 22.17
Finance Act 2003
s 68 ... 23.13
69 ... 21.31; 23.13
Sch 8 ... 23.13
Finance Act 2006 ... 4.4; 5.42; 10.9, 10.16; 12.12; 13.6; 14.21; 20.0
s 49 ... 3.5
Finance Act 2007 ... 22.11
Sch 24 ... 22.1, 22.11, 22.12, 22.13
para 4(1)(c) ... 22.13
Finance Act 2007 – *contd*
Sch 24 – *contd*
para 10(3), (4) ... 22.15
11 ... 22.13
12(4) ... 22.13
Finance Act 2008 ... 15.10; 21.32; 22.13
s 122, 123 ... 22.12
Sch 36 ... 22.1
Sch 37 ... 22.1
Sch 38 ... 22.1
Sch 39 ... 22.1
Sch 40 ... 22.1, 22.11, 22.12
Finance Act 2009
s 96 ... 22.1
105 ... 23.4
122 ... 1.5, 2.29; 3.7
Sch 48 ... 22.1
Finance Act 2010 ... 9.13; 21.15
Finance Act 2012 ... 23.0; App 5
s 49 ... 17.1; 23.18
209 ... 15.8
Sch 14 ... 17.1; App 5
para 1(3) ... 23.19
2(2) ... 23.18
33(5) ... App 5
Sch 33 ... 15.8
Sch 38 ... 22.1
para 3(1) ... 22.11
Finance Act 2013 ... 1.4; 5.19; 8.20, 8.21; 11.16; 14.7, 14.8
Pt 3 (ss 94–174) ... 19.22
(2)–(4) ... 19.22
176 ... 5.45; 8.16; 11.16
206 ... 13.2
207 ... 13.2
(1) ... 13.2
208 ... 13.2
(a) ... 13.2
209–215 ... 13.2
Sch 36 ... 5.45; 7.14; 8.16; 11.16
Sch 43 ... 13.2
Finance Act 2014 ... 22.23
Finance Act 2015 ... 14.22; 18.3; 22.23
Finance Act 2019 ... 18.3
Finance (No 2) Act 1992 ... 16.5
s 73 ... 16.5

Finance (No 2) Act 1992 – *contd*
Sch 14 .... 10.4; 16.5
para 2, 8 .... 6.18
9(1) 10.4; 16.5
Finance (No 2) Act 2015 .... 14.2; 20.6
s 62 .... 1.7
Finance (No 2) Act 2017 .... 16.2
Financial Services Act 1986 .... 5.8

Housing Act 1988 .... 2.6
Human Rights Act 1998
s 3 .... 15.17

Income and Corporation Tax Act 1988 .... 11.8
s 13ZA 7.7
16 Sch B 11.2
18 Sch D Case I 11.2
53(4) 11.8
219(2) .... App 4
415 .... 6.19
690 .... 21.30
691 .... 21.30, 21.32
692–694 .... 21.30
701(5)–(7) 6.11
Sch 18 .... 14.13
para 5 .... 14.13
Income Tax Act 2007
s 67(2) 2.30
Pt 9 Ch 10 (ss 507–517) 21.30
508 .... 21.30
(3) 21.30
509, 511, 512, 513 .... 21.30
515 .... 21.30
1005 .... 6.4, 6.19; 13.6
Income Tax (Earnings and Pensions) Act 2003 .... 16.1
Income Tax Law 1954
s 8(o) 5.3
Income Tax (Trading and Other Income) Act 2005
s 323(2), (3) 12.21
325, 326 .... 12.21
Pt 5 Ch 5 (ss 619–648) 5.31
Inheritance Tax Act 1984 .... 2.4, 2.25; 3.0; 4.0; 5.21; 6.10; 7.13; 15.4; 18.10; 21.1
s 3 .... 14.22
(4) 6.14

Inheritance Tax Act 1984 – *contd*
s 3A 18.3
4 .... 14.2
(1) 2.17; 6.14
5 .... 5.18
(1) .... 5.39; App 3
(b) 17.28
(1B) 21.15
(3) 14.10
(5) 8.22
6 .... 16.2
(2) 14.15
7 .... 21.25, 21.26
(1) 14.1
(4) 10.6
8A 15.10; 19.19
(2), (4) 19.20
8B 15.10; 19.19
8C 15.10; 19.19
(2), (3) 19.20
(4), (5) 19.21
8D–8M 15.10
10 .... 5.24; 21.15, 21.24
12(3) 7.1
13A(3B) 6.18
16 .... 12.17; 21.24
(2) 4.27
18 .... 15.17
19 .... 1.6
20 .... 1.6
(1) 1.6
21 .... 1.6
23(2), (3) 21.11
24 .... 4.27
(4) 21.11
24A, 25, 26 .... 4.27
26A 21.23
27 .... 21.1, 21.2, 21.15, 21.23
(1)(a) 21.12
(2) 21.11
30 .... 17.2; 18.10; 19.22; App 4
(3BA) 18.2, 18.5
31 .... 17.2, 17.3; 18.8, 18.9; 19.3, 19.22; 20.1
(1) 17.3, 17.5
(a) 17.8
(b) 17.6

Inheritance Tax Act 1984 – *contd*
s 31(1)(c) 17.8
(d) 20.1
(e) 17.8
(1A) 17.3
(2) 17.11
(b) 18.6
(3) 17.11, 17.14
(4A) 17.13
(4C) 17.13
(a), (b) 17.13
(4D) 17.13
(4FA) 17.16
(4FB) 17.14
32 .... 17.2; 18.10; 19.2, 19.22; 20.1; 23.9; App 5
(1) 19.1
(d) 19.3
(3)(a), (b) 19.3
(4) 19.3; 23.14
(5)(a) 19.3
(b) 18.10
(5AA) 18.10
32A .... 19.4, 19.5, 19.7, 19.21; 20.1; 23.9; App 5
(1) 19.4
(5) 19.6; 23.14
(a), (b) 19.6
(7) 19.6; 23.14
(8), (8A) 19.6
(9) 19.7
(10) 19.5
33 .... 17.2; 19.10
(1) 20.1
(b)(i), (ii) 20.4
(3)–(7) 20.1
(2) 19.14
(2A) 19.12
(5) 19.10
(7) .... 20.1; App 5
34 .... 17.2
(1)–(3) 19.17
35 .... 17.2
(2) 20.1
35A 17.2, 17.10; 18.0, 18.10, 18.15, 18.16; 20.1
(2) 18.10

Inheritance Tax Act 1984 – *contd*
Pt II Ch III (ss 36–42) 15.12
s 36 .... 15.3
37 .... 15.3
(1) 15.6
38 .... 15.3
(3) 15.5, 15.6
(5) 15.6
39 .... 15.3
39A 5.32; 15.12, 15.13, 15.14
(2)–(4) 15.13
40 .... 15.3
41 .... 15.3
(b) 15.7
42 .... 15.3
43 .... 14.23, 14.24
(1) 14.24
(2) 14.23
(b) 14.24
49 .... 8.18
(1) 3.2, 3.20; 4.4; 5.44
50(5) 4.5
52(1) 5.45
53 .... 21.15
56(3)(a) 21.15
57 .... 21.15
(1)(a), (b) 21.17
(5) 21.2
57A 21.1, 21.15, 21.17
(1) 21.17
(1A) 21.15
(3) 21.17
(4) 21.18
(6) 21.18
Pt III Ch III (ss 58–85) 4.4; 6.3
s 58 .... 21.1
62 .... 14.2, 14.22
(1) 5.39
(2)–(4) 14.2
62C 14.2
64 .... 14.23; 17.28; 18.3; 20.8
65 .... 21.1, 21.15, 21.16
67(s) 22.1
68(5)(a) 14.2
76 .... 21.23
77 .... 21.1
78 .... 20.1

Inheritance Tax Act 1984 – *contd*
s 78(1) 20.1
(3) 20.4
(4)(a), (b) 20.3
(5) 20.4
79 ..... 4.4; 6.3; 18.3; 20.5
(3) 20.7; 23.9
(4) 20.6
(5A) 20.7
(9) 20.8
81 ..... 14.24
91 ..... 3.9; 6.9, 6.10, 6.11
(1) 6.10, 6.11
(2) 6.11
94 ..... 5.24; 10.29
98 ..... App 4
(1) ..... App 4
(a), (b) ..... App 4
102(1)–(4) 14.3
102A(3), (5) 14.2
102B(4) 14.2
103 ..... 5.14; 12.3; App 3, App 4
(1) 6.3
(b) 6.3
(2) 7.13
(3) 2.30; 7.13; 9.16; 12.10; 14.16
(4) 2.24
104 ..... 5.2; 6.1, 6.2; 12.2; App 3, App 4
(2) 6.3
105 ..... 5.2, 14.28; App 3, App 4
(1) 1.2; 4.1; 5.27; 6.15; 10.30; 14.6
(a) ..... 5.3, 5.15, 5.33, 5.38, 5.44, 5.45; 12.8; 14.16, 14.17; App 4
(b) 5.25; 10.3, 10.30; 14.6; 16.5
(bb) 5.27; 10.3, 10.5, 10.30; 13.6, 13.8, 13.9; 16.4, 16.5
(cc) 5.28, 5.29; 9.23; 14.6; 16.4
(d) 5.26, 5.33, 5.38, 5.40, 5.42, 5.43, 5.44; 7.1; 10.8, 10.11; 12.4; 13.5; 14.7

Inheritance Tax Act 1984 – *contd*
s 105(1)(e) 5.40, 5.44, 5.45
(1ZA) 6.18; 13.6
(2) 2.14; 5.29
(3) 5.6, 5.8, 5.21; 6.20; 7.4, 7.6, 7.10; 8.14; 9.1, 9.2, 9.3, 9.4, 9.5, 9.7, 9.9, 9.10, 9.11, 9.16, 9.17, 9.18, 9.20, 9.22; 10.30; 12.11, 12.19; 13.6, 13.9; 14.18, 14.19
(4) 5.8; 8.14; 9.23
(b) 5.21
(5) 6.22
(6) 5.33
(7) 5.8
106 ..... 6.5 ,6.6, 6.7, 6.8, 6.9, 6.16; 7.2; 10.2, 10.12; 14.6, 14.28; App 3, App 4
107 ..... 6.7, 6.9 ,6.16; 9.17; App 3, App 4
(1) 6.8
(a) 6.7
(2) 6.7, 6.8, 6.16
(4) 6.7, 6.8; 13.10
108 ..... 4.10; 5.20; 6.16; 15.14, 15.16; App 3, App 4
(b) 6.9, 6.16
109 ..... 5.20; 6.16; 7.1; 15.16; App 3, App 4
(3) 6.16
109A 6.16
110 ..... 5.14, 5.19, 5.44; 7.14; 8.14, 8.15, 8.20, 8.21; 13.14; 14.7; App 3, App 4
(b) 5.18, 5.38; 8.15, 8.16, 8.21
(c) 7.10; 14.20
111 ..... 7.1, 7.5, 7.13; 8.14; App 3, App 4
112 ..... 2.4; 5.10, 5.17; 6.1, 6.2; 7.1, 7.2, 7.4, 7.5, 7.6, 7.13; 8.15; 9.16; 13.14; 14.18; 15.9; App 3, App 4

Inheritance Tax Act 1984 – *contd*
s 112(1) 7.4
(2) 5.43; 7.1, 7.10; 8.15; 14.20
(b) 7.2, 7.11
(3) 5.33, 5.43
(b) 5.43
(4) 7.5
(5) 7.1
(6) 7.3, 7.6
113 .... 6.13, 6.14, 6.15, 6.17, 6.22; App 3, App 4
(a) 6.3
(2), (3) 10.2
(3A) 10.31
(7A) 10.16
113A 6.12, 6.17; 10.2, 10.4, 10.5, 10.7, 10.8, 10.9, 10.11, 10.12; 16.5, 16.6
(1) 10.7, 10.30
(2) 10.8, 10.30
(3) 8.15; 10.8
(3A) 10.3, 10.30; 16.5
(a), (b) 10.3, 10.30
(3B) 10.7
(5) 10.5
(6) 6.17; 10.10, 10.12
(7A) 10.7, 10.15
(8) 10.9
113B 10.4, 10.11, 10.12; 12.5
(1)(b) 10.11
(2)(a) 10.12
(b) .... 10.11; App 4
(3) 10.12
(a) 10.12
(c) .... 12.5; App 3
(4) 10.12
(5) 10.12
(a), (b) 10.12
(6) 10.12
(7) 10.12
(8) 10.16
114 .... 8.16, 8.21; 12.2; App 3, App 4
(1) .... 1.9; 10.17; 12.2, 12.5, 12.19; App 3

Inheritance Tax Act 1984 – *contd*
s 114(2) 12.2
115 .... 2.4, 2.17, 2.18
(1) 4.4
(2) 2.1, 2.4, 2.6, 2.16, 2.17, 2.24; 3.25; 4.1; 13.3
(b) 2.5
(3) 3.1; 4.23
(4) 2.1, 2.24; 3.7
(5) 12.3
116 .... 2.28; 4.12; 10.16; 14.5
(2) 3.2, 3.28
(a) 3.4; 4.24, 4.28
(b) 3.12, 3.13, 3.28
(c) 3.15
(3)(a), (b) 3.13
(4), (5) 3.12, 3.28
(5A)–(5D) 3.15
(6) 3.15; 4.2
(7) 1.7
(8) 2.29
117 .... 2.2, 2.20, 2.24; 3.3, 3.24, 3.25; 4.3, 4.8, 4.10, 4.11
(a) 3.20; 4.3, 4.4, 4.6, 4.10; 13.5
(b) 4.3, 4.8, 4.10; 13.3
118 .... 4.12
(1) 4.6
(2), (3) 4.3, 4.8
(4) 4.8
119 .... 4.3, 4.11; 11.8
(1) 3.25; 4.9, 4.10
(2) 4.3; 5.24
120 .... 3.3; 4.3, 4.10, 4.11
(1)(a), (b) 4.10
(2) 3.13
121 .... 4.11
(1) 4.11
(b) 4.11
(2) 4.11
(3) 3.10; 4.11; 12.2
122 .... 2.28; 3.25
(1) 10.13, 10.16
(b) 14.6
(2) 2.28

Inheritance Tax Act 1984 – *contd*
s 122(3) 2.28
123 .... 2.28; 14.6
(1)(a) 3.25
(i) 4.12
(b) 2.28; 4.12
(3) 4.12
124(1), (2) 4.1
(3) 10.13
124A 10.13, 10.14, 10.16, 10.26
(1) .... 10.25; 12.5; App 3
(2) 10.27
(3) 10.13, 10.14, 10.19, 10.20
(a) 10.16
(4) 10.13
(5) 10.14
(6) 10.19, 10.22
(7A) 10.15
(8) 10.16, 10.18
124B 10.13, 10.20, 10.22, 10.26; 12.5
(1) 10.22
(b) 10.20, 10.26
(2), (3) 10.22
(4) 10.20, 10.22
(5) 10.22
(7) 10.22
124C 2.25, 2.27; 3.25
125 .... 11.5, 11.8, 11.9, 11.13, 11.14
(1) 11.8
(a) 11.5
(2)(a) 11.5
126 .... 11.5, 11.7, 11.11, 11.13; 23.3
(1) 11.9
127 .... 11.0, 11.5, 11.9, 11.11, 11.14
(2) 11.7, 11.10
128 .... 11.5, 11.9, 11.11
129 .... 11.5, 11.13; 23.5
130 .... 11.5, 11.12
141(1)(a) 6.1
(3) 6.1
(b) 6.1
142 .... 21.18
(3) 15.14
144 .... 5.32; 15.8

Inheritance Tax Act 1984 – *contd*
s 147(10) 19.21
160 .... 4.14
161 .... 4.21, 4.22, 4.27; 5.26; 8.0, 8.13, 8.23; 21.18; App 3
(1) 4.22
(2) 8.23
(b) 21.18
(3) 4.21, 4.22; 8.23
(4) 4.21
162 .... 7.14; 8.21; 14.7
(1) 8.22
(4) 8.20; 14.8
162A 8.19; 11.18
162B 8.16, 8.17, 8.18, 8.19; 11.18; 14.7
(1)(b) 8.16, 8.19
(i), (ii) 8.16
(2) 8.16
(5) 11.16
(c) 11.18
(i)–(iii) 11.16
(7) 8.17; 11.17
(8) 8.18
162C 8.19; 11.18; 14.7, 14.8
(2) 14.7
164(4) 4.17
165 .... 6.2
(1) 6.2
(3) 6.2
168 .... 8.4
169(1) 3.26
171 .... 4.21
(1), (2) 4.21
176 .... 2.28; 4.27; 5.29
177 .... 4.15
(1) 4.15
(2) 4.15
(i), (ii) 4.15
178 .... 4.25
(1) 4.25
179 .... 4.25, 4.26; 14.26
180(1) 4.25
190 .... 4.19
(1) 4.26; 14.27
191 .... 4.26; 14.26, 14.27
(1), (2) 4.26
(3) 4.26

Inheritance Tax Act 1984 – *contd*
s 191(3)(b) 4.26
192–198 .... 4.26
199(1) 23.3
204 .... 23.3
207(3) 20.7
208 .... 23.3
211 .... 23.1
(1) 23.1
213 .... 23.8
216 .... 22.12
(1)(a) 22.1
(3) 22.4, 22.6
(3A), (3B) 22.4, 22.5
217 .... 22.12
225 .... 15.9
226(1), 93A) 23.4
227 .... 23.5
(1C) 23.5
(4) 23.8
228 .... 23.5, 23.6
(3A) 23.5
(4) 23.6
229 .... 11.13
230 .... 17.5; 21.23
(1) 23.11
(5) 17.5
234(2) 23.7
(3)(b), (c) 23.7
(4) 23.7
237 .... 23.8
(1) 23.8, 23.9
(3), (3A), (3C) 23.9
238(1) 23.8
245(2), (7) 22.11
247(3), (4) 22.11
248 .... 22.11
256 .... 22.12
267A 5.12, 5.21; 6.9
(d) 5.24
268 .... 14.22, 14.23
(1) 14.23
(2) 3.16
(3) 14.23
269 .... 5.26
(1), (2) 5.26
(3) 3.5; 5.26
(4) 5.30
272 .... 22.4

Inheritance Tax Act 1984 – *contd*
Sch 1
para 5 .... 23.9
Sch 1A 15.4
Sch 2 .... 19.21
para 6 .... 21.26
Sch 3 .... 21.4, 21.23; 23.11
para 5 .... 23.9
Sch 4 .... 19.8; 21.1, 21.3, 21.5, 21.30
para 1 .... 18.3; 21.24, 21.32
(2) 18.3
(3) 21.3
2 .... 21.3
(1)(a)(ii) 21.12
(b) 21.9
(2) 21.9
3 .... 21.3, 21.24
(1)(a) 21.4, 21.24
(i) 21.4, 21.21
(ii) 21.21
(b) 21.4
(2), (4) 21.4
(5A) 21.4
4 .... 21.3, 21.5
5 .... 21.13, 21.24
7 .... 21.14
8 .... 21.19; 23.9
(2) 21.24
(a) 21.24
(4) 21.25
9 .... 21.19
(2) 21.28
(4) 21.25
10 .... 21.19
(1), (2), (4), (5), (8) 21.20
11 .... 21.25
(2), (4) 21.28
12 .... 21.25
13 .... 21.25, 21.26
(2)–(4) 21.25
14 .... 21.25, 21.26
(1) 21.25
(2), (3) 21.25, 21.26
(4), (5) 21.26
(6) 21.26
(a) 21.26
(7) 21.26
(8) 21.25
(9) 21.26

Inheritance Tax Act 1984 – *contd*
Sch 4 – *contd*
para 15 .... 21.25
15A 21.4, 21.18, 21.25
16 .... 21.16
(1) 21.2, 21.15
(2) 21.16
Sch 5 .... 23.14; App 5
para 1(ii) .... App 5
2(2)(b) .... App 5
(3), (4) .... App 5
3(ii), (iii) .... App 5
4(b) .... App 5
6 .... App 5
Sch 6
para 1 .... 23.2
4 .... App 5
(3) .... App 5
Interpretation Act 1978 .... 2.29

Landlord and Tenant Act 1954 .... 9.4
Law of Property Act 1925
s 188(1) 8.13
Law of Property (Miscellaneous Provisions) Act 1989
s 2(5) 6.14
Limited Liability Partnerships Act 2000 .... 5.21, 5.22
s 11 .... 6.9
Limited Partnerships Act 1907 .... 5.21

National Heritage Act 1980
s 11 .... 23.13

Partnership Act 1890
s 4(2) 4.3; 5.24
26, 32 .... 4.28
Perpetuities and Accumulations Act 2009 .... 5.41
Planning (Listed Buildings and Conservation Areas) Act 1990 .... 17.6

Rent (Agriculture) Act 1976 .... 2.6; 3.1

Settled Land Act 1925
s 64 .... 21.16

Taxation of Chargeable Gains Act 1992
s 23 .... 19.8
77, 78 .... 21.32
126 .... 6.7, 6.8, 6.17; 10.10, 10.12, 10.19; 13.10
(1) 6.17
127–136 .... 6.7, 6.8, 6.17; 10.10, 10.12, 10.19; 13.10
165 .... 6.2; 12.13; 13.11; 14.21; 16.6
169I(8)(c) 13.5
223 .... 3.21
257 .... 16.4
258(4) 20.5
(8) .... App 5
260 .... 12.20; 14.21; 16.6; 21.2
(2)(b)(iii) 21.32
272 .... App 3
(3) 8.2
(a), (b) 8.2
274 .... 5.39; 14.26; 22.7
282 .... 6.2
(2) 6.2

Value Added Tax Act 1994
Sch 9
Group 11 .... 23.14
item 1–4 .... 23.14
Variation of Trusts Act 1958 .... 21.17

IRELAND
Income Tax Act 1967
s 4 .... 5.7

Table of Statutory Instruments

[References are to paragraph numbers and appendices]

Finance Act 2009, Schedule 51 (Time Limits for Assessments, Claims etc) (Appointed Days and Transitional Provisions) Order 2010, SI 2010/867
  art 2 .... 2.29
Inheritance Tax Avoidance Schemes (Prescribed Description of Arrangements) Regulations 2011, SI 2011/170 .... 22.23
Inheritance Tax (Delivery of Accounts) (Excepted Estates) Regulations 2004, SI 2004/2543 .... 22.1
Inheritance Tax (Delivery of Accounts) (Excepted Settlements) Regulations 2008, SI 2008/606 .... 22.1
Inheritance Tax (Delivery of Accounts) (Excepted Transfers and Excepted Terminations) Regulations 2008, SI 2008/605 .... 22.1
Inheritance Tax (Double Charges Relief) Regulations 1987, SI 1987/1130 .... 8.23
  reg 5 .... 8.23
Inheritance Tax (Market Makers) Regulations 1992, SI 1992/3181 .... 5.8; App 1
Special Commissioners (Jurisdiction and Procedure) Regulations 1994, SI 1994/1811
  reg 21 .... 22.7
Units of Measurement Regulations 1995, SI 1995/1804 .... 3.7

Table of Cases

[References are to paragraph numbers and appendices]

A

ACC Green v HMRC [2015] UKFTT 236 (TC).. 9.13, 12.21
Alexander v IRC [1991] STC 112, (1991) 23 HLR 236, [1991] 2 EGLR 179 ... 4.0, 4.17
American Leaf Blending Co Sdn Bhd v Director General of Inland Revenue [1979] AC 676, [1978] 3 WLR 985, [1978] 3 All ER 1185................ 5.7; 7.6, 7.7
Arbuckle Smith & Co Ltd v Greenock Corpn [1960] AC 813, [1960] 2 WLR 435, [1960] 1 All ER 568 .... 3.25
Arkwright (William's Personal Representative) v IRC [2004] EWCA 1720 (Ch), [2005] 1 WLR 1411, [2004] STC 1323............................. 4.21, 4.22; 8.23
Arnander (executors of McKenna, dec'd) v R & C Comrs [2007] RVR 208, [2006] STC (SCD) 800, [2007] WTLR 51 .... 2.10, 2.14, 2.18, 2.20, 2.21; 3.17, 3.22, 3.23
Arnold (Sir Malcolm), Re; Day v Harris (unreported, 2010) 22.2, 22.3
Assessor for Lothian Region v Rolawn 1989 SC 329, 1990 SLT 433............... 3.17
Atkinson (Inspector of Taxes) v Dancer [1988] STC 758, 61 TC 598, [1988] BTC 364... 5.13; 7.1
Attwood (Inspector of Taxes) v Anduff Car Wash Ltd [1997] STC 1167, 69 TC 575, [1997] BTC 454.. 5.37
Ayrshire Pullman Motor Services & Ritchie v IRC (1929) 14 TC 754, 8 AC 531 .... 13.2

B

Baird's Executor v IRC 1991 SLT (Lands Tribunal) 9 .... 4.0, 4.15
Barclays Bank Trust Co Ltd v IRC [1998] STC (SCD) 125............................. 7.0, 7.6, 7.11; 14.18
Barrett v Morgan [2000] 2 AC 264, [2000] 2 WLR 284, [2000] 1 All ER 481.. 4.24
Barrs v IRC [1957] 1 WLR 529, [1957] 1 All ER 832, 1957 SC (HL) 1 .... 11.3
Beckman v IRC [2000] STC (SCD) 59, [2003] WTLR 773, [2000] STI 162.. 14.0, 14.17
Bedford (Duke of) (dec'd), Re; Russell v Bedford [1960] 1 WLR 1331, [1960] 3 All ER 756, (1960) 39 ATC 340.. 19.10
Best v R & C Comrs [2014] UKFTT 77 (TC), [2014] WTLR 409........ 9.1, 9.14, 9.22
Bestway (Holdings) Ltd v Luff (Inspector of Taxes) [1998] STC 357, 70 TC 512, [1998] BTC 69.. 5.37
Bradley (Inspector of Taxes) v London Electricity plc (No 2) [1996] STC 1054, [1996] EG 147 (CS), (1996) 93 (37) LSG 27.. 5.37
Braithwaite v R & C Comrs [2008] STC (SCD) 707, [2008] STI 1233........ 5.11, 5.36
Brander v R & C Comrs; Earl of Balfour (dec'd), Re [2009] UKFTT 101 (TC), [2009] SFTD 374, [2009] WTLR 1117 .... 1.6; 9.17; 12.9
Brown's Executors v IRC [1996] STC (SCD) 277 .... 14.18
Brutus v Cozens [1973] AC 854, [1972] 3 WLR 521, [1972] 2 All ER 1297 .... 11.8
Buller (dec'd) R & C Comrs *see* Best v R & C Comrs [2014] UKFTT 77 (TC), [2014] WTLR 409

Burden v United Kingdom (Application 13378/05) [2008] STC 1305, [2008] 2 FLR 787, [2008] All ER (D) 391 (Apr) 15.10
Burkinyoung (Executor of Bukinyoung) v IRC [1995] STC (SCD) 29 .... 9.3, 9.4, 9.5, 9.8
Bushell v Faith [1970] AC 1099, [1970] 2 WLR 272, [1970] 1 All ER 53 .... 5.26

C

C & E Comrs v Lord Fisher [1981] 2 All ER 147, [1981] STC 238 .... 3.1; 5.6, 5.7; 7.7; 9.4; 10.17
CVC / Opportunity Equity Partners Ltd v Demarco Almeida [2002] UKPC 16, [2002] BCC 684, [2002] 2 BCLC 108 .... 8.8, 8.9, 8.12
Cairns v R & C Comrs [2009] UKFTT 67 (TC), [2009] STC (SCD) 479, [2009] WTLR 793 .... 22.9
Carson (Inspector of Taxes) v Cheyney's Executor [1959] AC 412, [1958] 3 WLR 740, [1958] 3 All ER 573 .... 5.38
Caton's Administrators v Couch [1995] STC (SCD) 34.................. 8.5, 8.7, 8.8, 8.11
Charkham (dec'd) v IRC [2000] RVR 7 .... 8.13
Christie v Davis [1945] 1 All ER 370 .... 11.2
Chu Lip Kong v Director-General of Inland Revenue [1981] STC 653 .... 11.3
Clark v R & C Comrs [2005] STC (SCD) 823, [2005] WTLR 1465, [2005] STI 1758 .... 5.5; 9.20, 9.21
Collins v Fraser [1969] 1 WLR 823, [1969] 3 All ER 524, 46 TC 143 .... 11.2
Comr of Income Tax v Hanover Agencies [1967] 1 AC 681, [1967] 2 WLR 565, [1967] 1 All ER 954 .... 5.3, 5.7
Cook (executor of Watkins dec'd) v IRC [2002] STC (SCD) 318, [2002] WTLR 1003, [2002] STI 936 .... 4.5
Courthorpe, Re (1928) 7 ATC 538 .... 8.10, 8.11, 8.12

D

DWC Piercy's Executors v R & C Comrs [2008] STC (SCD) 858, [2008] WTLR 1075, [2008] STI 1647 .... 5.5; 7.10; 9.19
Daffodil (Administrator of Daffodil, dec'd) v IRC [2002] STC (SCD) 224, [2002] WTLR 795, [2002] STI 786 .... 5.38; 22.4
Denekamp v Pearce (Inspector of Taxes) [1998] STC 1120, 71 TC 213, [1998] BTC 413 .... 8.7, 8.12
Dixon v IRC [2002] STC (SCD) 53, [2002] WTLR 175, [2002] STI 43........ 2.6, 2.14, 2.30; 3.17; 11.8
Dugan-Chapman (Mary) (executors of) v R & C Comrs (2008) SpC 666....... 6.5, 6.7, 6.17; 13.1, 13.10
Duke of Buccleuch v IRC [1967] 1 AC 506, [1967] 2 WLR 207, [1967] 1 All ER 129 .... 4.0, 4.14, 4.18
Duke of Westminster v IRC [1936] AC 1, 19 TC 490 .... 13.2
Duomatic Ltd, Re [1969] 2 Ch 365, [1969] 2 WLR 114, [1969] 1 All ER 161 .... 6.7
Dyson Holdings Ltd v Fox [1976] QB 503, [1975] 3 WLR 744, [1975] 3 All ER 1030 .... 15.17

E

Earl of Ellesmere v IRC [1918] 2 KB 735 .... 4.0, 4.18
Earl of Normanton v Giles [1980] 1 WLR 28, [1980] 1 All ER 106, (1980) 39 P & CR 478 .... 3.1

Executors of Atkinson (dec'd) v R & C Comrs [2011] UKFTT 108 (TC).... 3.20, 3.22, 3.24; 4.3

F

Family Golf Centres Ltd v Thorn (Inspector of Taxes) [1998] STC (SCD) 106... 5.37
Farmer (Farmer's executors) v IRC [1999] STC (SCD) 321 .... 7.1; 9.1, 9.16, 9.23; 12.9, 12.20, 12.22
Farmer (Valuation Officer) v Buxted Poultry Ltd [1993] AC 369, [1993] 2 WLR 34, [1993] 1 All ER 117 .... 1.6; 2.5; 7.1
Faulks v Faulks [1992] 1 EGLR 9, [1992] 15 EG 15 .... 2.26
Featherstonehaugh (formerly Finch) v IRC [1985] Ch 1, [1984] 3 WLR 212, [1984] STC 261... 5.33, 5.44, 5.45; 8.14, 8.15, 8.21; App 4
Finch v IRC *see* Featherstonehaugh (formerly Finch) v IRC [1985] Ch 1, [1984] 3 WLR 212, [1984] STC 261
Fisher's Executors v IRC [1926] AC 395 .... 13.2
Fitzpatrick v Sterling Housing Association Ltd [2001] 1 AC 27, [1999] 3 WLR 1113, [1999] 4 All ER 705 .... 15.17
Fry (Inspector of Taxes) v Salisbury House Estate Ltd [1930] AC 432, 15 TC 266.. 5.3, 5.7; 9.5
Furness v IRC (1999) SpC 202... 9.9; 14.19

G

Gladstone v Bowers [1960] 2 QB 384, [1960] 3 WLR 575, [1960] 3 All ER 353 .... 3.4; 4.13
Gray v IRC; Lady Fox's Executors v IRC [1994] STC 360, [1994] 38 EG 156, [1994] RVR 129... 4.0, 4.16, 4.18, 4.22; 8.8
Griffiths (Inspector of Taxes) v Jackson [1983] STC 184, 56 TC 583, [1983] BTC 68 .... 13.12
Grimwood-Taylor v IRC [2000] WTLR 321 .... 7.13; 9.20
Gunter (Valuation Officer) v Newtown Oyster Fishery (1977) 244 EG 140 .... 2.4, 2.30

H

Hall (executors of Hall, dec'd) IRC [1997] STC (SCD) 126................. 9.7, 9.8; 12.10
Hanson v R & C Comrs [2013] UKUT 224 (TCC), [2013] STC 2394, [2013] BTC 1900 .... 1.2; 2.8, 2.16, 2.18, 2.19, 2.20, 2.21, 2.22; 3.17
Hardcastle (executors of Vernede dec'd) v IRC [2000] STC (SCD) 532, [2001] WTLR 91, [2000] STI 1520.. 4.23; 5.18; 15.16
Harrold (executors of Harrold dec'd) v IRC [1996] STC (SCD) 195, [1996] EG 70 (CS).. 2.15, 2.27; 3.24, 3.25
Harrison Broadley v Smith [1964] 1 WLR 456, [1964] 1 All ER 867, [2008] BTC 7085 .... 4.28
Hawkings-Byass v Sassen (Inspector of Taxes) [1996] STC (SCD) 319 .... 8.6, 8.8, 8.11
Hemens (Valuation Officer) v Whitsbury Farm & Stud Ltd [1988] AC 601, [1988] 2 WLR 72, [1988] 1 All ER 72 .... 2.24
Hertford v CIR *see* Seymour (Ninth Marquess of Hertford) v IRC
Higginson's Executors v IRC [2002] STC (SCD) 483, [2002] WTLR 1413, [2002] STI 1400... 2.12, 2.19; 12.10

Holland (executor of Holland dec'd) v IRC [2003] STC (SCD) 43, [2003] WTLR 207, [2003] STI 62 .... 15.10, 15.19
Holmes v McMullen (Re Ratcliffe dec'd) [1999] STC 262, [1999] BTC 8017, (1999) 96 (11) LSG 70 .... 15.7, 15.14
Hopwood v CN Spencer Ltd (1964) 42 TC 169, 43 ATC 334, [1964] TR 361 .. 11.3
Howarth's Executors v IRC [1997] STC (SCD) 162 .... 1.8; 23.8, 23.9
Hutchings (Timothy Clayton) v HMRC [2015] UKFTT 9 (TC), [2015] WTLR 1359, [2015] STI 1260 .... 22.13

I

IRC v Barclay Curle & Co Ltd [1969] 1 WLR 675, [1969] 1 All ER 732, [1969] 1 Lloyd's Rep 169 .... 5.37
IRC v Crossman [1937] AC 26, [1936] 1 All ER 762 .... 4.14, 4.15; 8.1
IRC v Forsyth Grant 1943 SC 528, 1944 SLT 89, (1943) 25 TC 369 .... 2.21; 4.13
IRC v George; Stedman's Executors v IRC [2002] STC (SCD) 358, [2002] WTLR 1357, [2002] STI 1155 .... 9.11, 9.12, 9.14, 9.15, 9.22, 9.23; 14.20
IRC v John M Whiteford & Son 1962 SC 229, 1962 SLT 269, 40 TC 379 .... 4.23
IRC v Korner [1969] 1 WLR 554, [1969] 1 All ER 679, 1969 SC (HL) 13....... 4.23
IRC v Lloyds Private Banking [1998] STC 559, [1999] 1 FLR 147, [1998] 2 FCR 41 .... 15.9
IRC v Stannard [1984] 1 WLR 1039, [1984] 2 All ER 105, [1984] STC 245.... 23.8
IRC v Williamson Bros 1950 SC 391, 1950 SLT 326, 42 R & IT 358.............. 11.2

J

J Sainsbury plc v O'Connor (Inspector of Taxes) [1990] STC 516 .... 14.13
Jaggers (t/a Shide Trees) v Ellis (Inspector of Taxes) [1997] STC 1417, 71 TC 164, [1997] BTC 571 .... 2.3; 11.8
Jewell v McGowan [2002] EWCA Civ 145, [2002] 3 EGLR 87, (2002) 99 (11) LSG 39 .... 3.18
Jones v IRC [1997] STC 358, [1997] BTC 8003, [1997] EG 21....................... 4.26
Jowett (Inspector of Taxes) v O'Neill & Brennan Construction Ltd [1998] STC 482, 70 TC 566, [1998] BTC 133 .... 7.7f

K

Kauri Timber Co Ltd v Taxes Comr [1913] AC 771 .... 11.3
Khan v Miah [1998] 1 WLR 477, [1998] CLC 210 .... 12.4
Korner v IRC [1969] 1 WLR 554, [1969] 1 All ER 679, (1969) 45 TC 287 ...... 2.14

L

Land Management Ltd v Fox (Inspector of Taxes) [2002] STC (SCD) 152, [2002] STI 403 .... 5.4
Lindsay v IRC (1953) 34 TC 289 .... 4.23
Lloyds TSB Private Banking plc (Personal Representative of Rosemary Antrobus (dec'd)) v IRC [2002] STC (SCD) 468, [2002] WTLR 1435, [2002] STI 1398 .... 2.8, 2.13, 2.14, 2.17, 2.19
Lloyds TSB Private Banking plc (PR of Rosemary Antrobus Dec'd) v Twiddy [2006] 1 EGLR 157, [2006] RVR 138, [2005] WTLR 1535 .... 4.23
Lockhart v Harker; Re Benham' Will Trusts [1995] STC 210..................... 15.7, 15.14
Lynall v IRC [1972] AC 680, [1971] 3 WLR 759, [1971] 3 All ER 914............ 4.14

Lyon's Personal Representatives v R & C Comrs [2007] STC (SCD) 675, [2007] WTLR 1257, [2007] STI 1816 .... 14.4

M

McArthur's Executors v R & C Comrs [2008] SSCD 1100 .... 5.25; 8.24
McCall (personal representatives of McLean, dec'd) v R & C Comrs [2009] NICA 12, [2009] STC 990, [2009] NI 245 .... 1.6, 1.9; 5.6; 7.8; 9.14, 9.15, 9.22; 12.4, 12.16
McGregor (Inspector of Taxes) v Adcock [1977] 1 WLR 864, [1977] 3 All ER 65, [1977] STC 206 .... 5.13
Mallender v IRC [2000] STC (SCD) 574, [2000] STI 1656 .... 7.13; 9.18
Mallender (executors of Drury-Lowe, dec'd) v IRC [2001] STC 514, [2001] BTC 8013, [2001] WTLR 459 .... 5.19; 8.15
Mannion v Johnston [1988] STC 758 *see* Atkinson (Inspector of Taxes) v Dancer [1988] STC 758, 61 TC 598, [1988] BTC 364
Manzur (Dr KMA) v R & C Comrs [2010] UKFTT 580 (TC), [2011] STI 1219 .... 5.4; 7.7
Martin (executors of Moore, dec'd) v IRC [1995] STC (SCD) 5 .... 9.2, 9.3, 9.4, 9.5, 9.8; 21.18
Mehjoo v Harben Barber (a firm) [2013] EWHC 1500 (QB), [2013] BTC 509: 13.1; 14.15
Midland Bank Trust Co Ltd v Hett Stubbs & Kemp [1979] Ch 384, [1978] 3 WLR 167, [1978] 3 All ER 571 .... 14.15
Mitchell v IRC 1943 SC 541, 1944 SLT 95 .... 4.13
Munby v Furlong (Inspector of Taxes) [1977] Ch 359, [1977] 3 WLR 270, [1977] 2 All ER 953 .... 5.36
Murray v IRC (1951) 44 R & IT 270, 32 TC 238, (1951) 30 ATC 53 .... 11.3

N

Nelson Dance Family Settlement Trustees v R & C Comrs [2009] EWHC 71 (Ch), [2009] STC 802, 79 TC 605 .... 1.9; 5.11, 5.13, 5.14, 5.16, 5.38

O

Oakley (Personal Representatives of Jossaume dec'd) v IRC [2005] SpC 460.. 5.16
Owers, Re (No 1) [1941] Ch 17, [1940] 4 All ER 225 .... 23.2

P

Pattersons Will Trusts, Re; Lawson v Payn [1963] 1 WLR 623, [1963] 1 All ER 114, (1962) 41 ATC 412 .... 23.2
Pawson (dec'd) v R & C Comrs [2012] UKFTT 51 (TC), [2012] WTLR 665 rev'sd [2013] UKUT 50 (TCC), [2013] STC 976, [2013] BTC 1605 ..... 1.10; 5.6; 9.14, 9.22; 12.21
Pearson v IRC [1981] AC 753, [1980] 2 WLR 872, [1980] 2 All ER 479 .... 5.41
Pepper (Inspector of Taxes) v Hart [1993] AC 593, [1992] 3 WLR 1032, [1993] 1 All ER 42 .... 9.5
Phillips' Executors v R & C Comrs [2006] STC (SCD) 639, [2006] WTLR 1281, [2006] STI 2094 .... 5.5; 9.18
Phoenix Office Supplies Ltd v Larvin [2002] EWCA Civ 1740, [2003] BCC 11, [2003] 1 BCLC 76 .... 8.9

Powell (personal representatives of Pearce, dec'd) v IRC [1997] STC (SCD) 181 .... 9.8; 13.12
Price (executor of Price, dec'd) v R & C Comrs [2010] UKFTT 474 (TC), [2011] SFTD 52, [2011] WTLR 161 .... 4.14, 4.21, 4.22
Prosser v IRC [2001] RVR 170, [2002] WTLR 259 .... 4.20

R

R v Allen [1999] STC 846 *see* R v Dimsey (Dermot Jeremy) [2000] QB 744, [2000] 3 WLR 273, [1999] STC 846
R v Dimsey (Dermot Jeremy); R v Allen (Brian Roger) [2000] QB 744, [2000] 3 WLR 273, [1999] STC 846 .... 16.2
R v IRC, ex p Bishopp [1999] STC 531, (1999) 11 Admin LR 575, 72 TC 322 5.23
R & C Comrs v Atkinson [2011] UKUT B26 (TCC) 3.24, 3.25
R & C Comrs v Golding [2011] UKFTT 351 (TC), [2011] WTLR 1183 .... 2.9, 2.24; 3.22
R & C Comrs v Hanson *see* Hanson v R & C Comrs [2013] UKUT 224 (TCC), [2013] STC 2394, [2013] BTC 1900
R & C Comrs v Pawson (dec'd) *see* Pawson (dec'd) v R & C Comrs [2012] UKFTT 51 (TC), [2012] WTLR 665 rev'sd [2013] UKUT 50 (TCC), [2013] STC 976, [2013] BTC 1605
RSPCA v Sharp [2010] EWHC 268 (Ch), [2010] STC 975, [2010] WTLR 855 revs'd [2010] EWCA Civ 1474, [2011] 1 WLR 980, [2011] PTSR 942.... 14.15; 15.8, 15.11
Rahman v Malik; Re Gate of India (Tynemouth) Ltd [2008] EWHC 959 (Ch), [2008] 2 BCLC 403 .... 5.15
Raikes v Lygon [1988] 1 WLR 281, [1988] 1 All ER 884, (1988) 85 (3) LSG 34 .... 21.16
Rashid v Garcia (Status Inspector) [2003] STC (SCD) 36, [2003] STI 35 .... 7.7
Reynaud v IRC [1999] STC (SCD) 185 .... 14.2, 14.22
Robertson v IRC (No 1) [2002] STC (SCD) 182, [2002] WTLR 885, [2002] STI 766 .... 22.6
Robertson v IRC (No 2) [2002] STC (SCD) 242, [2002] WTLR 907, [2002] STI 899 .... 22.7
Ross (Marjorie) (Dec'd) v HMRC [2017] UKFTT 507 .... 9.14; 12.21
Rosser v IRC [2003] STC (SCD) 311, [2003] WTLR 1057, [2003] STI 1152.. 2.17, 2.18, 2.19; 3.21
Rysaffe Trustee Co (Channel Islands) Ltd v IRC [2003] EWCA Civ 356, [2003] STC 536, [2003] BTC 8021 .... 14.2, 14.23

S

St Barbe Green v IRC [2005] EWHC 14 (Ch), [2005] 1 WLR 1772, [2005] STC 288 .... 5.44; 14.10
Salaried Persons Postal Loans Ltd v R & C Comrs [2006] EWHC 763 (Ch), [2006] STC 1315, [2006] BTC 423 .... 7.12
Scott (No 2), Re [1916] 2 Ch 268 .... 19.10
Seymour (Ninth Marquess of Hertford) (executor of Eighth Marquess of Hertford, dec'd) v IRC [2004] STC (SCD) 177, [2005] WTLR 85, [2004] STI 2546 .... 8.11, 8.15
Silber (Lerner's Personal Representative) v R & C Comrs [2012] UKFTT 700 (TC), [2013] WTLR 113, [2013] STI 326 .... 5.25; 22.4

Starke (Executors of Brown, dec'd) v IRC [1994] 1 WLR 888, [1994] STC 295, [1994] STI 161 [1995] 1 WLR 1439, [1996] 1 All ER 622, [1995] STC 689 *aff'd* [1996] 1 WLR 1439, [1996] All ER 622, [1995] STC 689 .... 2.2, 2.11, 2.14, 2.16, 2.17; 4.23
Stedman's Executors v IRC *see* IRC v George [2002] STC (SCD) 358, [2002] WTLR 1357, [2002] STI 1155
Stoner (executors of Dickinson, dec'd) v IRC [2001] SpC 288, [2001] SWTI 1501 .... 4.26; 14.27
Sudley (Lord) v A-G [1897] AC 11 .... 3.9
Sutherland (dec'd), Re *see* Winter v IRC [1960] Ch 611, [1960] 3 WLR 491, [1960] 3 All ER 270
Swain Mason v Mills & Reeve [2011] EWHC 410 (Ch), [2011] STC 1177, [2011] WTLR 1589 revs'd in part [2012] EWCA Civ 498, [2012] STC 1760, [2012] 4 Costs LO 511 .... 10.24; 12.9; 14.0, 14.14, 14.15; 16.7

T

Tapp v Ryder (Inspector of Taxes) [2008] RVR 340, [2011] WTLR 363 ..... 4.14, 4.22
Thompson (Thompson's Executor) v IRC SpC 429 .... 22.4
Tyser v A-G [1938] Ch 426, [1938] 1 All ER 657 .... 19.9

V

Vigne (Maureen) (dec'd) v HMRC [2017] UKFTT 632 (TC) 2.24; 9.14
Vinton v Fladgate Fielder (a firm) [2010] EWHC 904 (Ch), [2010] STC 1868, [2010] PNLR 26 .... 5.32; 6.5; 13.1

W

WT Ramsay Ltd v IRC [1982] AC 300, [1981] 2 WLR 449, [1981] 1 All ER 865 .... 14.22
Walding v IRC [1996] STC 13 .... 5.26, 5.33
Walker's Executors v IRC [2001] STC (SCD) 86, [2001] WTLR 773, [2001] STI 734 .... 5.26
Weston (executor of Weston, dec'd) v IRC [2000] STC 1064, [2000] BTC 8041, [2001] WTLR 1217 .... 9.10
Wheatley's Executors v IRC [1998] STI 559 .... 2.6, 2.24, 2.30
Wight v IRC (1982) 264 EG 935, [1984] RVR 163 .... 8.13
Willett v IRC (1982) 264 EG 257, [1984] RVR 163 .... 4.20
Williams (Personal Representatives of Williams, dec'd) v R & C Comrs [2005] SpC 500 .... 2.3, 2.4, 2.5; 3.17
Woodhall v IRC [2000] STC (SCD) 558, [2001] WTLR 475, [2000] STI 1653 4.5

Y

Yarmouth v France (1887) 19 QBD 647 .... 5.34
Yaxley v Gotts [2000] Ch 162, [1999] 3 WLR 1217, [2000] 1 All ER 711 .... 6.14

Z

Zetland Settlement, Trustees of v R & C Comrs [2013] UKFTT 284 (TC), [2013] WTLR 1065, [2013] STI 2424 .... 9.6

Table of Examples

[*References are to paragraph numbers and appendices*]

Alan – clawback and reserved benefit (*Example 14.4*)	14.6
Alan Bennett – liability and BPR offset (*Example 8.4*)	8.17
Algernon – credit relief (*Example 11.2*)	11.13
Alice and John – simple grossing up (*Example 6.3*)	6.3
Andrew – donee death before clawback event (*Example 10.9*)	10.13
Arlington estate – market value premium (*Example 3.1*)	3.1
Arthur Nexis – combined ownership period (*Example 6.9*)	6.9
Asset switch to earn relief (*Example 14.11*)	14.28
Ben – tenant to owner (*Example 13.1*)	13.5
Barry Bolt – lifetime termination of lifetime interest (*Example 5.9*)	5.45
Boris – protecting the farmer son (*Example 13.2*)	13.5
Boris Morris – mixed relievable property (*Example 8.3*)	8.16
Brendan – debt allocation (1) (*Example 14.5*)	14.7
Camila Davies – indirect debt (*Example 8.1*)	8.16
Cedric – APR rate change (*Example 10.13*)	10.17
Cedric – control shareholding (*Example 16.1*)	16.4
Church Farm – replacement property (*Example 10.15*)	10.22
Clancy brothers (1) – accumulated cash profits (*Example 6.1*)	6.1
Clancy brothers (2) – accumulated cash profits tax thrown away (*Example 6.2*)	6.2
David – GWR and 'deemed' ownership (*Example 11.1*)	11.5
David Bannister – limited relief (*Example 4.8*)	4.11
Denis – assets changing character (*Example 14.3*)	14.6
Diana – substantial charitable gift (*Example 15.4*)	15.8
Donald and Evelyn – organic growth (*Example 6.5*)	6.5
Donald Drake – treatment of liabilities (*Example 7.5*)	7.14
Ellen McGregor – movements in capital account (*Example 5.2*)	5.15
Enthusiasts together (*Example 19.1*)	19.2
Farley family (1) – organic development of a business (*Example 7.1*)	7.1
Farley family (2) – examination of the business (*Example 7.2*)	7.4
Fergus Shannon – constructed occupation (*Example 4.7*)	4.10
Fiona – cumulative sales (*Example 19.5*)	19.15
Fred – partial reinvestment (*Example 10.6*)	10.11
Fred – woodlands and BPR (*Example 11.3*)	11.14
George – sale of lifetime gift (*Example 19.4*)	19.14
Gerald – debt allocation (2) (*Example 14.6*)	14.7

HMRC illustration updated (1) – Appendix 11 (*Example 23.1*) 23.12
HMRC illustration updated (2) – Appendix 12 (*Example 23.2*) 23.12
Harry – successive deaths (*Example 18.1*) 18.7
Hector – hypothetical example (*Example 21.5*) 21.32
Henry – credit against later charge (*Example 19.7*) 19.16
Henry Brown – fudging the issue (*Example 1.1*) 1.6
Henry and Ian – purchased interest (*Example 6.4*) 6.5
Hetty – imaginative use of the rules (*Example 10.17*) 10.30
Hubert – adoption of the higher tax rate (*Example 21.3*) 21.25
Hugh – lifetime gifting and death (*Example 14.1*) 14.1
Hugh Bannon – period of occupation (*Example 4.5*) 4.10

Interaction of clawback and transferable nil rate band (*Example 19.9*) 19.2
Investment into trading company (*Example 14.10*) 14.28
Isaac Walton – APR analysis (*Example 2.1*) 2.4

James Benjamin Holdings Ltd – mixed development (*Example 7.4*) 7.10
James Cook – occupation test (*Example 4.1*) 4.3
James Grant – restricted relief (1) (*Example 6.9*) 6.16
Jeff and Jill – BPR maximised (*Example 15.5*) 15.9
Jennifer – partial clawback (*Example 10.10*) 10.14
Jennifer Price – restricted relief (2) (*Example 6.10*) 6.16
Jessica – gift of NRB legacy (*Example 15.8*) 15.13
Jessica – separation values (*Example 10.14*) 10.18
Joanna – gift bearing its own tax (*Example 15.1*) 15.4
John and Helen – acquisition of established business (*Example 5.8*) 5.43
John Bull – occupation of successive properties (*Example 4.2*) 4.3
John – clawback after flotation (*Example 10.5*) 10.7
John Jones – FA 2013 debt repositioning (*Example 8.2*) 8.16
Jonathan – BPR asset no longer held by trust (*Example 10.16*) 10.27
Josiah – a living collection (*Example 19.2*) 19.7
Judith – preservation of APR (*Example 3.2*) 3.13
Julienne (1) – BPR position pre FA 1986 (*Example 15.6*) 15.12
Julienne (2) – BPR position post FA 1986 (*Example 15.7*) 15.13

Kenneth – BPR legacy (*Example 15.9*) 15.14
Kylie – single grossing (*Example 15.2*) 15.5

Lucy – tax credit (*Example 19.6*) 19.16

Mary – costly health problems (*Example 10.8*) 10.13
Matthew – roll-over (*Example 14.8*) 14.8
Mary Brown – transferable nil rate band (*Example 19.8*) 19.20
Maud – dilution of BPR (*Example 15.10*) 15.14
Myers Discretionary Trust – transfer between trusts (*Example 21.1*) 21.16

Nellie – charges on selling silver (*Example 20.1*) 20.4
Nigel – serial investments (*Example 14.9*) 14.8
Nigel Derby – guarantee in place (pre-FA 2013) (*Example 5.3*) 5.19
Nigel Derby – guarantee secured on farmland (pre-FA 2013) (*Example 5.4*) ... 5.19

Official example (1) (*Example 17.2*) 17.17
Official example (2) (*Example 17.3*) 17.17
Official example (3) (*Example 17.4*) 17.17
Oscar – takeover intervening (*Example 10.1*) 10.2
Overlooked masterpiece (*Example 19.3*) 19.9
Owen – gift with reservation (*Example 14.2*) 14.6

Paul – APR exchanged for BPR (*Example 10.12*) 10.17
Percy and Maureen – mixed trade (*Example 9.1*) 9.1
Peregrine Took – change of use (*Example 3.3*) 3.17

Quentin – debt allocation (3) (*Example 14.7*) 14.7

Rhine Gmbh – share classes (*Example 5.5*) 5.26
Richard Andrews – multiple farming enterprise (*Example 2.2*) 2.19
Roger – farming partnership (*Example 2.3*) 2.22
Roger (1) – Bateman collector (*Example 17.1*) 17.17
Roger (2) – Restricted access (*Example 17.5*) 17.21
Roger Lapin – late incorporation (*Example 6.7*) 6.8

St John's Retreat – ransom strip (*Example 8.5*) 8.23
Sally – failure to advise fully? (*Example 16.2*) 16.6
Samantha – occupation for agricultural purposes (*Example 2.4*) 2.24
Sarah (1) – partial clawback (*Example 10.3*) 10.5
Sarah (2) – partial clawback trust (*Example 10.4*) 10.5
Sebastian – withdrawal of fund (*Example 21.4*) 21.26
Sir Frederick Blaise (1) – assignment of royalties (*Example 5.6*) 5.38
Sir Frederick Blaise (2) – holding of royalties (*Example 5.7*) 5.38
Spencer family – occupation by widow (*Example 4.6*) 4.10
Stanley – control shareholding (*Example 10.2*) 10.3
Susan – exchange of debt for shares (*Example 13.3*) 13.9
Susan Furniss – sale of estate land (*Example 4.9*) 4.26
Susanna Green – double BPR (*Example 6.6*) 6.7
Swallow family – scaling back of farm activities (*Example 5.1*) 5.6
Swift purchase of replacement assets (*Example 10.7*) 10.13

Tanya – double grossing (*Example 15.3*) 15.6
Terence – transferee ownership (*Example 10.11*) 10.16
The Vittles family – family limited partnership (*Example 12.2*) 12.12
Thomson Fabrication Ltd – limited relief (*Example 21.2*) 21.18
Trevor Black – replacement farm (*Example 4.4*) 4.8

Waterfall Ltd – Spanish villa (*Example 7.3*) 7.6
Wayne Bright – farming by trustees (*Example 4.3*) 4.4
Wilfred – Market value premium (*Example 12.1*) 12.4
Woodland inheritance (*Example 15.11*) 15.15

Chapter 1

Introduction

THE PLACE OF BPR/APR IN INHERITANCE TAX (IHT)

1.1 Agricultural property relief (APR) and business property relief (BPR) are the most significant inheritance tax (IHT) reliefs (as opposed to exemptions) affording the ability to shelter up to 100% of real value from exposure to IHT. It is therefore to be expected that both reliefs carry significant weight in any tax planning routine and as a consequence command keen attention from tax advisors and HMRC.

The owner of agricultural/business property may have acquired it by purchase, self-generation or by inheritance but now wishes to pass it on intact and without diminution of value or loss of relief.

As part of a standard IHT mitigation strategy, the tax advisor will often suggest gifting assets early (thus during lifetime), generously and without reservation or strings attached. However, both APR and BPR have the potential to offer the donor something much more attractive: the ability to retain both control over and the income of the assets until death without incurring a tax penalty. Given the apparent generosity of the twin reliefs, **Chapter 16** questions how long that situation may last and considers what the prudent family involved in business or agriculture should do to protect itself in the event of change.

A significant attraction of holding assets that qualify for APR or BPR is the 'two-year rule' under which, even late in life, action may be taken to mitigate the burden of IHT on an estate. This is considered in **Chapter 3** at **3.3** (APR) and **Chapter 6** at **6.5** (BPR) and illustrated in numerous examples.

All references are to IHTA 1984 unless otherwise specified.

HISTORY OF BPR/APR

1.2 In disputed cases, reliance has been placed on the history of and background to both reliefs as an aid to understanding the intended meaning of obscure or complex legislation. However, that approach does not always succeed and indeed failed, for example in *Hanson v HMRC* [2013] UKUT 0224 (TCC), where the Upper Tribunal did not find it helpful. The concept of the 'working farmer', discussed in **Chapter 3**, was first introduced in *FA 1975,* but the definition was so wide as to include farmers who had retired and full-time students. Even so there is a trend in recent cases to import some of the spirit of 'working farmer' when considering relief on farmhouses.

In the past APR did not extend to let farms, which were generally worth much less than land with vacant possession. Farmers therefore opted to enter into partnership with one or more of their children and let the land to that partnership business. This secured the generous 'double discount': the land itself was devalued; and access was gained to the then valuable APR rate of 50%.

This common practice was halted by *FA 1981, s 96* and *Sch 4* which for the first time enabled agricultural landlords to gain access to relief. The 'working farmer' limits were removed but with a sweetener provided in the form of 'transitional relief' – that transitional period is very long indeed since it continues to this very day.

The history of BPR is far more recent. There was no relief under *FA 1975* except that tax could be paid by interest-free instalments. *FA 1976, s 73* and *Sch 10* introduced limited relief up to 30% for transfers made after 6 April 1976, with multiple reliefs being introduced in October 1977. From 10 March 1992 onwards, the reliefs were increased further.

From 6 April 1996, matters were simplified. *IHTA 1984, s 105(1)* was amended to allow 100% relief on all shareholdings of unquoted shares of any size transferred on or after that date. That change also applied to any charge to tax on an earlier transfer which arose as a result of an event occurring on or after that date.

THE NEED FOR HERITAGE PROPERTY RELIEF

1.3 Unlike APR and BPR, which reliefs by their very nature are familiar in any IHT planning strategy, heritage property relief is not often encountered in practice and for that reason is far from well known. There is no question that the rules are complex; in many cases the application for relief may be governed by strict deadlines and there is an ongoing 'observation' by HMRC. Nevertheless the aim of the relief is very simple, namely to provide favoured treatment to the otherwise chargeable transfer of funds set aside to provide for maintenance of certain specific types of property (realty, land and chattels) important to the history, heritage and culture of the UK, but which remain in private not public ownership.

The process and grant of relief, known as conditional exemption, is rigorously controlled and covers six broad asset categories as set out in **Chapter 17**. Conditional exemption is dependent on continual fulfilment of certain conditions (undertakings) as to the maintenance and preservation of the heritage property but the provision of reasonable access to the public remains a critical component as discussed in **Chapter 18**. The precise meaning of 'reasonable access' was considerably tightened under *FA 1998* with the aim of countering blatant abuse through disregard or poor observance under the 'by appointment only' system of public access – all owners of conditional exemption heritage property must now provide some measure of open public access that is both reasonable and practicable.

The favoured tax treatment is also extended to trusts ('maintenance funds') established to meet the cost of looking after designated conditional exemption heritage property. The rules are quite complex and are discussed in detail at **Chapter 21**.

Until *FA 1998*, the principal guidance for heritage property was to be found in HMRC booklet IR67 'Capital Taxation and the National Heritage'. Following consultation between HMRC and key professional and government organisations, a revised, updated and much expanded guidance was published also under the title of 'Capital Taxation and the National Heritage' in September 2011. This remains the main control document.

OUTLINE OF THE SCOPE OF THIS BOOK

1.4 This book follows the statute direction by considering APR first and then BPR – this mirrors the priority afforded to APR within the legislation. **Chapter 2** first considers the kind of property in respect of which the relief may (or may not as is the case) apply whilst **Chapter 3** examines both the value in respect of which the relief may be given and of the rate of the relief. **Chapter 4** identifies and examines some of the more complex areas of concern.

This same structured approach is then adopted in relation to BPR, with **Chapter 5** considering the kind of property that now qualifies for the relief and **Chapter 6** setting out the most important conditions for eligibility for relief whilst identifying the factors that may preclude that relief. This theme is continued in **Chapters 7, 8** and **9** which cover the treatment of liabilities (and impact of the debt repositioning rules contained in *FA 2013*, that operate to reduce the relief) and two critical areas of concern, namely excepted assets and 'wholly or mainly holding investments', which may prompt partial or entire loss of the relief. **Chapter 10** provides a detailed commentary of the so oft misunderstood application of clawback relief.

Chapter 11 considers woodlands relief which whilst closely related to APR and BPR is very much the poor relation taking second place to those reliefs.

In **Chapters 12** to **16**, the book adopts a broader overview by looking at what may be done to minimise the IHT burden both in relation to the reliefs and generally.

Chapters 17 to **20** examine the access to and complexities of the reliefs conditionally available in respect of heritage property, and **Chapter 21** considers the (rather rarefied) world of maintenance funds. The weight of the legislation is such that these chapters do not make for easy reading but they will provide good guidance and a useful reference point.

Chapter 22 is, from a practical point of view, where one should start in dealing with a client problem. It outlines the compliance requirements and considers key points of dispute with HMRC. Finally **Chapter 23** looks at facilities for payment of tax or, in the case of heritage property, the use of this property itself to satisfy the liability to tax.

TERRITORIALITY OF APR, BPR AND HERITAGE RELIEF

1.5 This is considered at **2.29** in relation to APR, and at **5.3** in relation to BPR. In broad terms, APR and woodlands relief were, until 2009, restricted to property in the UK, the Channel Islands or the Isle of Man but the position was relaxed by *FA 2009, s 122* and the relief is now extended to land within the European Economic Area (EEA). By contrast BPR was never so restricted and always operated on a worldwide basis. There are, however, strict territorial aspects to relief on heritage property where such assets must be connected with and located within the UK.

INTERACTION OF APR AND BPR WITH OTHER RELIEFS

1.6 BPR applies to the value transferred by a transfer of value which is attributable to business property. Importantly, the relief is applied before other reliefs or exemptions; thus a gift of property valued at £500 qualifying for 50% BPR would leave a value of £250 which might thus be exempt as a 'small gift' under *IHTA 1984, s 20*. (Note, incidentally, that the value transferred is not grossed up for this purpose.) However, care should be taken to avoid the use of *s 20* where it is not appropriate, since a transfer of £251 is not exempt at all under *s 20*. Likewise, HMRC may refuse relief under *s 20* where the gift is made into settlement, since *s 20(1)* makes it clear that the transfer must be 'by outright gifts'.

At a more realistic level, the same principle applies in relation to the combination of the use of APR or BPR and relief under *s 21* (normal expenditure out of income) and *s 19* (annual exemption). Occasionally, in order to test the value of an asset from the point of view of a larger proposed gift, a transferor would make a gift whose value net of BPR deliberately exceeded the available exemption by a small margin. If the gift is a potentially exempt transfer (PET), this would not trigger debate with HMRC, since the possibility that the transfer might gain full exemption after seven years would in the short-term negate any immediate tax impact from HMRC viewpoint. The same holds true in the case of a chargeable transfer which falls within the nil rate band of the transferor – HMRC would not be drawn into the debate. Valuation may be in point for the purposes of capital gains tax (CGT) in such a circumstance, but that is a matter to be regulated by self-assessment and the CG34 procedure for post-transaction valuation checks. If the issue is the availability of BPR, it may be possible to get a non-statutory ruling, as is explained later, but again only in relation to chargeable lifetime transfers, never PETs.

The availability of APR or of BPR to personal representatives will have a direct bearing on any IHT payable on the estate; here the availability of the nil rate band (to include the transferable nil rate band(s)) will be affected but not the residence nil rate band which relief is calculated ignoring the availability of APR/BPR. Likewise, the availability of APR/BPR will affect the calculation of the 36% IHT rate where property has been left to charity. Sorting out APR or BPR may delay the administration; thus proper preparation of the claims to relief may save valuable time.

The interaction of APR, BPR and woodlands relief is complicated and this is considered in detail in **Chapter 11**.

There is some interaction between APR and BPR, on the one hand, and heritage relief, on the other. Here, the relationship is not so much between legal rules as a practical consideration.

Example 1.1: Henry Brown – fudging the issue

Henry Brown, a widower, died in December 2017 leaving an estate which included a working farm and some let lands together with a substantial Grade II listed house with good quality contents in which Henry lived until his death. At some distance from the main house, there is another property that is occupied by Henry's son Tom, who also acts as the farm manager.

The executors might initially claim BPR on the whole estate, perhaps claiming the kind of relief that was allowed in the case of *Farmer* (examined in **Chapter 9**) and encouraged by elements of the decisions in *McCall* and *Brander* (see below and in **Chapter 5**). A claim might also be made that the main house qualifies for APR as the 'farmhouse' on the ground that the deceased was, actively involved in the farming business.

Applying the tests that are examined in **Chapter 2**, it is quite likely that the claim to APR on the principal house might fail because it is really the other house (where the son/farm manager lives) that is the 'farmhouse'. It might, therefore, be appropriate, if the principal house is sufficiently grand, to seek relief on it as heritage property, particularly if the son is happy living in his present circumstances and does not want to move up to the main house. As will be seen in **Chapter 18**, a provisional or protective application for heritage property relief may be made even before the issue of APR or of BPR has been concluded.

HOW APR AND BPR APPLY COMPUTATIONALLY

1.7 The starting position is always the full, open market value because relief may not be allowed, or may only apply in part (as in the case of APR which is limited to agricultural value). The value transferred by a transfer of value is then reduced by the appropriate percentage of the relief: see, for example, *IHTA 1984, s 116(7)*. In lifetime this applies equally to outright gifts to individuals and those occasions on which tax is chargeable under the relevant property trusts regime. As stated above, APR and BPR are given before deduction of the annual exemption, before taking account of any other exemption and, where the transferor (in the case of relevant property trusts) pays the tax, before grossing up.

Where the percentage of the relief has in the past been increased, for example from 30% to 50% and from 50% to 100%, those increases apply both to:

- transfers of value made after the dates on which the increases became effective; and
- charges, or further charges, to IHT that arise on a death that occurs after the effective date of the increase in respect of any transfer made before 10 March 1992.

The treatment of liabilities used to differ between agricultural property and business property. If liabilities are charged on agricultural property, they are deducted from the value of that property but if the value is only partly agricultural, it becomes necessary to apportion the liabilities between the part of the value that qualifies for APR and the part that does not. However, the situation is slightly different where the liabilities are not charged on the agricultural property and this is discussed in greater detail in **Chapter 8**. Special rules apply to disregard both APR and BPR in a clawback situation: this is more fully explained in **Chapter 10**.

Particular complications apply to relief in respect of qualifying property in the ownership of a relevant property trust. Note, in particular, that:

- a reference to 'a transfer of value' in relation to such a trust can include an occasion of charge on the settled property;

- a reference to 'the value transferred' is treated as being a reference to the amount on which IHT is actually chargeable; and
- a reference to 'the transferor' in such a situation can include a reference to the trustee.

As a result of this, BPR may be available by treating the trustees as the owners of that part of the trust fund which might qualify for the relief.

However, the way that the charging rules operate for relevant property trusts can create the effect that whilst APR/BPR may reduce the *amount* on which tax is charged it may not reduce the *rate* at which that tax is charged. On other occasions, depending on the circumstances, the effect may be to reduce the rate of tax but not the amount on which that tax is charged. A detailed exposition of the charges to IHT on relevant property trusts is outside the scope of this book: the reader should refer to *Inheritance Tax 2017/18* (Bloomsbury Professional) or *Trusts & Estates 2017/18* (Bloomsbury Professional).

Where there is a potential exit charge then APR or BPR, if in point, will affect the value subject to charge but not the exit charge rate itself. On the other hand, the periodic charge (ten-year anniversary charge) made post 17 November 2015 is concerned with a hypothetical transfer made up of:

- the sum of the amount in respect of which the charge is now made (the relevant property);
- the historic value of property comprised in any related relevant property settlement immediately after that settlement commenced; and
- the added value of pilot trusts taken into account under the 'same day addition' rule introduced in *F(No 2)A 2015, s 62*.

Accordingly, where the periodic charge is reduced because BPR or APR applies, that will also serve to reduce the exit charge during the next ten-year period, whether or not that exit charge relates to property which itself qualifies for BPR or APR.

INTERACTION WITH IHT INSTALMENT RELIEF

1.8 Under the estate duty regime, there was some relief for 'industrial hereditaments' by way of a reduction in the rate of tax but as previously commented except for the option of paying tax by interest-free instalments there was no relief for business assets until 6 April 1976. There are inherent dangers in paying tax by instalments, as the lawyer executor found to his acute cost in *Howarth's Executors v IRC* [1997] STC (SCD) 162, an account of which is included at **23.9**.

THE DISTINCTION BETWEEN BPR AND APR

1.9 The same property may, and often does, qualify both as agricultural property and as relevant business property but there are differences of application and calculation.

APR is subject to the notable, and in practice oft overlooked, restriction to agricultural value only. There may be circumstances, such as the existence of valuable mineral rights on a farm worked in hand, where BPR may be available to mop up the premium value not covered by APR (but not in the case of the farmhouse). The premium for the development value of land encouraged the unsuccessful claim to BPR in *McCall and Another (Personal representatives of McLean Deceased) v HMRC* [2009] NICA 12.

Another key distinction between APR and BPR, as examined in the book, is the 'use' or 'occupation' test. In general terms, APR may be available in respect of land which is occupied for agriculture irrespective of whether this is by the landowner or his tenant, whereas BPR is closely restricted to assets actually used for the purpose of the business of the owner alone.

Under *IHTA 1984, s 114(1)* APR is automatically applied in preference to BPR; thus the latter is not available on any value which has been reduced by the former. See **Chapter 12** for examples of overlap of the reliefs and of situations where BPR is available in the absence of APR.

The greatest difference between the two reliefs is that APR is given on specific items of property, whereas BPR is given on the business itself. In the past it was generally accepted – and driven by a series of cases, as noted in **Chapter 5** – that a 'mere asset' of the business did not itself qualify for BPR. However, this approach was successfully challenged in the landmark case of *Nelson Dance v HMRC* [2009] EWHC 71 (Ch). This analysis also has knock-on effects in relation to the disposal of specific assets (but not the business itself) for the purpose of clawback: see **Chapter 10**.

IS THERE A NEED FOR BOTH APR AND BPR?

1.10 A strong argument could be made for simplification of the law by extending the scope of BPR and repealing APR in its entirety. Farming as a viable and sustainable industry has experienced serious difficulties in recent years and rural employment remains as scarce as ever. This has led to increasing diversification in the use of the land and redundant buildings for the leisure and commercial letting trades in order to improve the financial position. Julie Butler in her book *Tax Planning for Farm and Land Diversification*, 4th Edition (Bloomsbury Professional), has done excellent work in exploring the possibilities here. Holiday cottages do not qualify for APR and, since *HMRC v Pawson* [2013] UKUT 050, supported by later case decisions decided in favour of HMRC, they are now effectively in all but the very rare circumstance, excluded from BPR.

Chapter 2

Agricultural property: qualifying property

QUICK GUIDE

- This chapter defines the property that will qualify for APR with a bias towards land rather than to buildings.
- A major and contentious area of interest is the qualification of the farmhouse: see the discussion at **2.8** onwards.
- Issues arise as to what actually constitutes farming and whether any such farming is being done. The other major problem area concerns farming companies where APR is available only to controlling shareholders.
- The scope of APR is both narrow and restrictive; thus some specific assets do not qualify: see **2.30**.

2.1 At first sight *IHTA 1984, s 115(2)* seems straightforward. 'Agricultural property' means agricultural land or pasture but specifically includes:

- woodland *provided* occupied with agricultural land or pasture and that occupation is ancillary to, ie does not dominate, the occupation of the agricultural land or pasture;
- any building used in connection with the intensive rearing of livestock or fish *provided* that building is occupied with agricultural land or pasture and the occupation is ancillary to that of the agricultural land or pasture;
- such cottages together with the land occupied with them as are of a character appropriate to the property;
- such farm buildings together with the land occupied with them as are of a character appropriate to the property; and
- such farm houses together with the land occupied with them as are of a character appropriate to the property.

The scope of *s 115(2)* is extended by *s 115(4)* in relation to horses, and further extends to short rotation coppice (see **2.25**) and land in conservation schemes (see **2.27**).

AGRICULTURAL LAND OR PASTURE

2.2 'Agriculture' is defined in the *Agricultural Tenancies Act 1995, s 38(1)* as including:

> '...horticulture, fruit growing, seed growing, dairy farming and livestock breeding and keeping, the use of land as grazing land, meadow land, osier land, market gardens and nursery grounds, and the use of land for woodlands where that use is ancillary to the farming of land for other agricultural purposes...'

It is irrelevant that the property is of a type which might qualify for agricultural land; rather the test is that it must be occupied for the purposes of agriculture or deemed to be so occupied under a 'set aside' or similar agreement (*IHTA 1984, s 117).*

In *Starke and Another (Executors of Brown Deceased) v IRC* [1995] STC 689 (see **2.16)** it was stated that a building cannot be 'agricultural land or pasture', rather the latter means bare land, fields or land which is used either for the grazing of animals or for the cultivation of crops.

WOODLAND

2.3 Woodland relief is discussed in **Chapter 11** but woodland to the extent that it does not qualify for APR may also attract BPR (see **Chapter 5**). For an account of an attempt to treat a plantation as woodland, see *Jaggers (t/a Shide Trees) v Ellis* [1997] STC 1417, discussed in greater detail in **Chapter 11**. For the occupation of woodland adjoining pasture, see comments *obiter* in *Williams v HMRC* [2005] SpC 500, discussed at **2.4** below.

BUILDINGS USED FOR INTENSIVE REARING OF LIVESTOCK OR FISH

2.4 HMRC always carefully examine the merits of any APR claim but pay particular attention to those where the farming business is conducted on a small level (ie occupies less than 30 acres) which might not be a 'viable' business entity. These cases present a problem for the practitioner, especially where the owner is no longer actively farming perhaps due to old age or incapacity.

This issue was tested before the Special Commissioner in *Williams v HMRC* [2005] SpC 500. Mrs Williams had until her death owned 7.41 acres comprising:

- approximately 0.5 acres on which the residence stood;
- a paddock of 0.33 acres;

- grazing land of just under four acres; and, most important to the case,
- 2.99 acres on which stood three broiler houses.

BPR was not in point as Mrs Williams no longer owned the business that was run from the broiler houses, having let the premises some years earlier. However, since it was agreed that the buildings were used for the intensive rearing of livestock (*s 115(2)*), the personal representatives made a claim for APR. It was held that the buildings dominated the land they stood on and as a result access to APR was denied.

The case discussion merits a detailed review as it examines the meaning in *s 115(2)* of 'occupation is ancillary to that of the agricultural land and pasture', both as to the relevant land and the nature of the occupation. It also contains several propositions which, though not binding on others, are useful in considering borderline situations:

- 'Occupation', as shown by earlier case law, exists where four elements are present, namely:
  - physical possession;
  - control of use;
  - power to exclude others; and
  - some form of right to enjoyment of the land.
- The tenant occupied the broiler houses.
- The other land was not relevant to the 'ancillary' test, because the same person must be in occupation of the land and the buildings to satisfy the test.
- The *nexus* for this part of *s 115* is occupation rather than ownership.
- The occupation of the broiler houses was not ancillary to the occupation of the land on which they stood.
- 'Ancillary occupation' requires that the use of buildings be ancillary to the use of the land.
- Occupation of woodland, for example, could be ancillary to occupation of land where there was a common purpose, as where the woodland provided shelter, for instance, for grazing cattle.
- The buildings need not be used only for purposes ancillary to the use of the land (so, in the case of woodland, it might also produce timber or support wildlife).
- 'Intensive' buildings within *IHTA 1984, s 115(2)* are to be considered as one with the land on which they stand and do not on their own qualify as agricultural land or pasture.

- Ancillary control does not require any test of the area of the land.
- The right to exclude others did not decide the issue here.
- Minor rights of enjoyment did not decide the issue here.
- The 'character appropriate' test is to be applied to farm buildings, namely:
  - size, content and layout by reference to the farmland;
  - size proportionate to farming use;
  - overall impression;
  - the 'educated rural layman' test; and
  - the history of the building.

Applying these principles, the Commissioner reviewed both the legislation and *Hansard*, but gained little help from either. The broiler houses would be ancillary only if occupied as an 'add-on' to larger agricultural enterprises on the other land, here meaning only the 2.99 acres. They failed that test.

On the other hand fish farms, or stew ponds may be treated as farm buildings in certain circumstances.

Example 2.1: Isaac Walton – APR analysis

Isaac Walton is the owner of the Compleat Estate, through which runs a natural stream giving a very high-quality water supply. Isaac excavates near the stream, extracting gravel, and turns the excavations into a series of stew ponds:

In pond A, he rears rainbow trout.

In pond B he breeds brown trout.

In pond C, he breeds mirror carp.

In pond D, he offers some 'easy' fly fishing for rainbow trout.

Isaac operates a small shop from which in addition to selling fishing permits, he also sells fresh and smoked rainbow trout to passing trade. He also sells brown trout and carp to specialist angling clubs.

An analysis of the eligibility of the estate for APR would conclude that the entirety of the stream and pond complex is occupied with the agricultural land that constitutes Compleat Estate and in that respect, subject to the comments below, the stew ponds may be treated as akin to farm buildings. However, strictly, a fish farm which rears fish only for sport is not an agricultural pursuit.

Isaac could put forward a case that the part of the business which rears rainbow trout and carp is rearing livestock for food because they are sold as such at the farm gate. It could also be argued that fishing for rainbow trout in an overstocked pool, even without bait, is so predictable as to be, not sport but merely a delayed means of eating the fish in question. The fish breeding tanks or ponds would qualify for APR where the fish are reared for food, but pool D and the stream that feeds the whole farm would not (see IHTM24033 and in particular IHTM24039), though *de minimis* unless the fishing rights in respect of the stream are valuable, the value, if any, of pool D and the stream itself may be 'lost' in the valuation of Compleat Estate as a whole.

The distinctions are to an extent academic as Isaac has run the whole operation on business-like lines. Despite his famous name, he is not an angler himself and no part of the operation is excluded as an excepted asset under *s 112*. In so far as APR is not available, Isaac would secure BPR at 100%.

Whilst the traditional type of trout farm could be said to have a dependence on the land and therefore fall within the meaning of 'husbandry' not every kind of fish farm may be considered to be 'farming'. HMRC have indicated (see BIM55105) that in the context of other taxes, fish farming in natural lakes or rivers will be farming, but fish farming where the stock are kept entirely separate from the land, as in artificial tanks or in the sea, and fed entirely on purchased feed, will not. For BPR the difference is immaterial since, in either case, the operation may be a business regardless of its linkage to land but the brief summary in BIM55105 does not sit easily with *s 115(2)* and indeed makes no reference to it – all the cases there referred to pre-date *IHTA 1984*.

For APR, which is an asset-based relief, the linkage with land is explicit in the wording of *s 115(2)*, as interpreted in the *Williams* case (see **2.4**). The exclusion of sea-borne activities may be consistent with the exclusion from relief of oyster bed (see *Gunter v Newtown Oyster Fishery* (1977) 244 EG 140). However, the exclusion of conventional 'onshore' fish farming from relief does seem hard to justify. A 'building' used for the intensive rearing of fish is specifically within *s 115*, so, in an appropriate case, the argument might turn on the meaning of 'building' which the Concise Oxford Dictionary defines as 'permanent fixed thing built for occupation (house, school, factory, stable, etc)'. Although a building commonly has a roof, that may not be essential and a tank constructed specifically for the rearing of fish, no doubt with inbuilt protection from predators, ought to fall within the definition – this also accords with the HMRC view set out in IHTM24033. Notwithstanding, there is no specific case which tests this point but in any event, in the light of the decision in *Williams*, the installations relating to fish would still have to be part of a larger overall establishment.

Ancillary occupation

2.5 The same issue of occupation arises in relation to woodland and until *Williams* there was, little direct authority on the interpretation of *s 115(2)(b)*. *Williams* did take note of the House of Lords' decision in *Farmer (Valuation Officer) v Buxted Poultry Ltd* [1993] AC 369, where the meaning of the expression 'occupied together with' (a phrase from rating law) was considered.

To succeed in the dispute, Buxted Poultry Ltd had to show that a poultry processing factory was 'occupied together with' buildings used for the keeping or breeding of poultry. As is the custom for the industry, the buildings were scattered over a very wide area. The House of Lords held that the factory was not an agricultural building and that, for one building to be 'occupied together with' another building (or land), the following tests must be satisfied:

- the buildings must be in the same ownership at the same time and jointly managed or controlled; and
- the buildings must be occupied together so as to form a single agricultural unit in a real sense. Evidence that the buildings are close together may well show that they are so occupied. If they are separated, that may indicate that they are not.

Woodland often adjoins farmland and this may provide a benefit to the farmer by offering a degree of privacy or may form a habitat for birds not reared as livestock (ie pheasants). That benefit does not cause APR to be denied because the APR code has no real parallel to the 'excepted assets' rule for BPR. It is only necessary for the farmer to show that the woodlands are occupied with the land that they shelter, that such occupation is ancillary to that of the farmland, and that such occupation is, in fact, not for the purpose of an annual shoot but is for the purpose of agriculture. Again, this approach was confirmed in *Williams*. For the practitioner the issue will be in gathering the available evidence and recording the precise use of each element of a landed estate might and perhaps should form part of the practitioner's 'IHT health check' for the clients.

FARM COTTAGES

2.6 The provision in *s 115(2)* relating to farm cottages is extended by ESC F16 (see also IHTM24034). That Concession provides that, on a transfer of value comprising agricultural property which includes a cottage occupied by a retired farm employee or by the widow or widower of such an employee, the condition as to occupation for agricultural purposes is regarded as satisfied with respect to the cottage if either:

- the occupier is a statutorily protected tenant (for example, under the *Housing Act 1988*, the *Rent (Agriculture) Act 1976* or similar legislation in Northern Ireland or Scotland); or
- the occupation is under a lease granted to the farm employee for his or her life and the life of any surviving spouse as part of the contract of employment of the employee. That contract of employment must be by the landlord for agricultural purposes.

The following principles emerge relating to farm cottages:

- there is no relief if the cottage is not 'of a character appropriate';
- 100% relief is available if the tenancy began on or after 1 September 1995 and is an unprotected service tenancy or an assured shorthold tenancy; this is however subject to certain qualifications and is examined in more detail at **3.18** below;
- APR will be limited to 50% of the tenancy in an assured agricultural occupancy, unless it can be shown that suitable alternative accommodation is available, or that the tenant is in breach of the tenancy such that vacant possession could be obtained at the time of transfer, albeit after court delays; see again at **3.18** below; and
- 'retired' means 'no longer working full time': see IHTM24034. Just moving to another job is not good enough.

Attempts are made from time to time to test the boundaries of the definition of 'agricultural property', even in relation to quite small parcels of land. One example is *Wheatley's Executors v CIR*, considered at **2.24** below.

A second example was *Dixon v IRC* SpC 297, [2002] SWTI 43, where the Special Commissioner, adjudicating a claim to APR on a share in a small residential property, applied the test of 'the educated rural layman'. The property comprised a cottage, garden and orchard: total area 0.6 acres. It adjoined a nursery garden and orchard where vegetables, fruit and flowers were grown commercially. The damson trees in the orchard were picked by the neighbour and sold with his own and the proceeds, some £70 per annum, were not returned as income on the taxpayer's tax returns. Geese grazed under the damson trees; a goat was kept for milk; chickens were kept both for their eggs and for the table; a neighbouring farmer had been allowed to graze six ewes and lambs but again the consideration for that grazing did not appear on the taxpayer's tax return. Rather, in return for the grazing, the farmer's wife did some household chores for the taxpayer.

By 2001 the orchard was overgrown and the damson trees were not pick-able commercially (but there was potential if the orchard could be cleared). On the death of a part-owner of the property, HMRC determined that the property was not 'agricultural property' within *s 115(2)* and the Special Commissioner agreed.

Fruit growing and grazing could be agriculture, but it was a question of fact and of degree to be determined in the light of the purposes of the *Act*, which was designed to give relief for land and pastures used for agriculture but not for a private residence. On the facts, the activities were consistent with the use of the property as a residence and garden, not for agriculture. The issue was not whether the cottage was appropriate to agricultural land, but rather whether the orchard and garden were appropriate to the cottage. There was no history of agricultural production – the property consisted primarily of a cottage used for domestic and residential purposes, with a garden and orchard that were of a character appropriate to such a cottage. The appeal was dismissed.

FARMHOUSES

2.7 In practice few difficulties arise when claiming APR on a house which is modest in size and which has for generations been owned and occupied by farming families together with adjoining land. Many claims are properly evidenced and succeed but the problems, as always, arise at the margins of what is reasonable. One problem here is that the claim to APR is all or nothing although, in practice, where it can be argued that part of the house is the farm office, a claim to BPR may give some scope for 'fudging the issue', since such a claim is for a proportion of the open market value of the house, not of the agricultural value. Even so, a substantial amount of value may remain subject to tax and even where an APR claim succeeds, the taxpayer's victory is not total: relief is always restricted to the agricultural value, as seen at **3.1** and following.

Tests of a farmhouse: historical background to the current law

2.8 Advocates in farmhouse cases often refer back to the old law but agriculture is changing and that approach may no longer be justified. The Upper Tribunal, in *HMRC v Hanson* [2013] UKUT 0224, said '... We do not, however, find this exercise of legislative archaeology of assistance and doubt, in any case, that this is an appropriate area in which we should even admit the earlier definitions as an aid to construction ...'

This area of the law is hotly contested, both in the courts and in the political arena. The Institute for Public Policy Research, in its tax commentary 'Fair Dues', noted:

> '... The Agricultural Relief also provides an exemption for farmhouses, which anecdotal evidence suggests provides a cover for a large number of country cottages [or], "fake farmhouses" to be passed on tax free. At the least, it may be desirable to tighten the definition of agricultural property to close this loophole ...'

It is worth noting the observations of the Lands Tribunal in *Lloyds TSB Private Banking plc (PR of Rosemary Antrobus Deceased) v Peter Twiddy (Inland Revenue Capital Taxes)* [2002] STI 1398, discussed in detail below and better known as '*Antrobus Number 2*'. The Lands Tribunal were not considering the availability of APR and the comments were *obiter*, but in order to arrive at a value for APR purposes, they did consider what house should qualify for the relief in the first place. They concluded that:

> 'A farmhouse is the chief dwelling house attached to a farm, the house in which the farmer of the land lived. There is, we think, no dispute about the definition when it is expressed in this way. The question is: who is the farmer of the land for the purposes of the definition of s 115(2)? In our view it is the person who lives in the farmhouse in order to farm the land comprised in the farm and who farms the land on a day to day basis … We do not think that the house occupied with the farm is a farmhouse simply because the person living there is in overall control of the agricultural business conducted on the land; and in particular we think that the lifestyle farmer, the person whose bid for the land is treated by the appellant as establishing the agricultural value of the land, is not the farmer for the purposes of the provision.'

Interestingly, the fact that those comments were *obiter* does not seem to dissuade HMRC from relying on them: see IHTM24036.

The historical context of the law and of the subject dwelling

2.9 In 1975, the concept of the 'working farmer' (discussed in greater detail in **Chapter 3**) was far more important than it is now. The farmhouse was seen to be of a character appropriate to that 'working farmer' test. The classification of rural residences included mansion houses, farmhouses and farm cottages but of these, only the latter two qualify for relief and the test of the house was not necessarily an economic one.

There are many smallholdings in the UK of between 25 and 50 acres which are centred around a good-sized detached house. It might be thought impossible to carry on a viable farming business on such a small scale, yet many continue to do so. Notwithstanding, under the old rules this kind of farming operation would enable the farmhouse to qualify for relief where the owners continued to try to eke out a living from the land and this was highlighted in *HMRC v Golding* at **3.22** below.

Consider, however, that some change of economic circumstance caused such a family to give up farming and to sell the house plus its land, both situated in a prime location, to the family of, say, a solicitor. Assume that the acreage remained the same, and that the solicitor arranged a team of contractors to work the land. It might appear that the land was still put to the same use, however, it is much less certain that APR would be available to the solicitor than it was to the original owners. The building would be a farmhouse by virtue of historical association, but the failure of the solicitor's family to rely upon the land wholly for their livelihood would very much weaken the case for relief. Furthermore, the prime location would likely have given the house, standing as it does in open land, a premium value well above agricultural value. That premium would not in any event qualify for APR: see **3.1** relating to agricultural value, and the commentary on valuation at **4.23**.

Big farmhouses

2.10 Is there an upper size limit of a farmhouse, above which, no matter how large the estate, no relief will be given in respect of the principal residence on the estate? The parameters are not clearly defined within the legislation but in theory a grand house could qualify for APR as a 'humble' farmhouse. However, it will not help the claim if the true centre of the farming operations is an estate office situated in or near the residence of the farm manager, and the owner of the estate goes to that office to supervise farming and estate matters. It was on this very point that *McKenna* floundered; see **2.19**.

Tests applied in earlier cases

2.11 Traditionally, there were three tests as to whether a house was 'of a character appropriate to the property' namely: historical association, geographical comparables and financial viability. For a time it seemed, certainly from an HMRC perspective, that the last of these tests was the most important prompting the question of whether:

- the income of the enterprise actually was, at the time of the relevant transfer, sufficient to support the farmhouse or farmhouses for which relief was claimed and to provide a livelihood for their occupants; or
- the activities of the enterprise could be so arranged as to produce such a result.

In Autumn 1994, an unpublished decision of the Special Commissioners rejected the traditional approach stating '...whether or not one can make a living from a given acreage of land cannot be determinative of the question whether a building is a farmhouse...'. Subject to what follows, this view is now reflected in the treatment of claims for APR on farmhouses with HMRC leaning towards a functional test (see IHTM24036).

The starting point was laid down in *Starke v IRC* [1994] STC 295, where it was held that:

> 'cottages, farm buildings and farmhouses ... will constitute "agricultural property" if used in connection with agricultural land or pasture provided

> that they are of a character appropriate to such agricultural land or pasture (that is, are proportionate in size and nature to the requirements of the farming activities conducted on the agricultural land or pasture in question).'

Higginson: an extreme case

2.12 The facts in Executors of *Higginson v CIR* [2002] STI 1400 were that Mr Higginson's home, Ballyward Lodge, was set in a landed estate of 134 acres comprising:

- 63 acres of agricultural land;
- three acres of garden; and
- 68 acres of woodland and wetland.

Sales particulars for the house described it as having 'enormous potential as a charming family home or as a small active farm'. The deceased had bought the property in 1954 and farmed it until 1985 after which it was let, with the woodland and wetland made into areas of nature conservation. The heir to the estate realised that the farming operation was not economically viable, and sold it for £1.15 million.

The Commissioners considered that the intention of APR was to allow families to continue farming by removing the threat of having to sell assets used in that business to settle the death duties thereon. There was no precondition that the deceased should have been farming but it was a very significant factor; land and house together must be an agricultural unit and within that unit, the land must be the dominant element; farm buildings and farmhouses must be ancillary to the land. However, that was not the case here but notwithstanding, the single most significant fact was the price for which the family were able to sell the property. A figure of approximately £1 million for the property as a whole was beyond the means of many who might otherwise be interested in farming and it would represent 'an appalling investment in terms of yield' from a farming perspective. APR was refused on the house.

This case is helpful in illustrating another point, addressed in **Chapter 3** at **3.1** and **Chapter 4** at **4.23**, concerning the restriction of APR to the agricultural value of the relevant asset. The comments are clearly *obiter*, but the Commissioner did say:

> 'It seems to me that the notional restrictive covenant would have much less of a depreciatory effect in a case where the property has a value greater than ordinary, not because of development potential but rather because of what I might call "vanity value" on account of its site, style or the like.'

In other words (subject to the discussion below) there is no automatic 30% discount of the open market value of a farmhouse to arrive at the agricultural value. In the light of *Higginson*, the practitioner must now ask, in relation to a dwelling for which APR is sought, whether a purchase of it and its land would, at today's values, be 'an appalling investment in terms of yield'.

Antrobus No 1: lifestyle farming

2.13 *Lloyds TSB as personal representative of Rosemary Antrobus Deceased v CIR* [2002] STC (SCD) 468 concerned Cookhill Priory, with 126 acres of freehold land and 6.54 acres of tenanted land. The deceased's father had purchased the property in 1907: it had six bedrooms and dated to the mid-16th century – the original nunnery had been founded in 1260 and in 1785, the chapel of the nuns was rebuilt. An extension had been added in about 1765 and in 1910 this Georgian extension to the property was again extended. The father had farmed the land until 1942 – he died in 1943 and his widow died in 1959. Miss Antrobus had a tenancy of 104 acres of other land and a licence over 20 acres. She acquired the property from relatives in 1959 and it was then listed Grade II*, not because it was particularly attractive but because of its history.

Miss Antrobus was clearly a farmer, working with both arable and stock – the farm accounts showed losses from time to time although she had made profits in 1993 and 1994; but on balance the holding was not an economic one. On her death, it was agreed that the land and buildings were agricultural property, but APR was disputed in respect of two let houses and Cookhill Priory itself. An expert witness described the property as a working farm and produced much evidence of comparable buildings in that part of England. HMRC acknowledged that this was not 'a rich man's considerable residence' but argued that it did have a dual purpose – it had become the dwelling of a prosperous person, was surrounded by parkland and was a family home of some distinction.

It was acknowledged by both parties that Cookhill Priory was a farmhouse and the expert advice was accepted that Cookhill Priory was not a gentleman's residence but had been a working asset integral to the farm for at least 100 years. It was relevant that the visual impact of the Georgian front was somewhat impaired by large corrugated iron farm buildings not too far away, and that the property was in a poor if not dilapidated state of repair. Even if, at one time, this had been 'a family home of some distinction', it had become, by the date of death, both in appearance and in use the farmhouse of a working farm and that was sufficient to decide the matter in favour of the taxpayer.

The Antrobus checklist

2.14 The Commissioner in *Antrobus* (Dr Brice) very helpfully summarised the relevant principles for deciding the test set in *s 105(2)*. Whilst decisions of the Special Commissioner are not binding on the Upper Tribunal (nor on any higher court) the principles in *Antrobus*, (as further amplified in *McKenna*) now form the basis of a checklist for any enquiry namely:

(1) *Is the house appropriate, by reference to its size, content and layout, to the farm buildings and the particular area of farmland being farmed?*

This test is derived from *Korner and Others v CIR* (1969) 45 TC 287, an income tax case which whilst not considered to be directly relevant (as HMRC had conceded that the house in question was a farmhouse), did consider comparables. It required the dwelling to be judged by 'ordinary ideas of what is appropriate in size, content and layout, taken in conjunction with the farm buildings and the particular area of farm land being farmed, and not part of a rich man's considerable residence'.

(2) *Is the house proportionate in size and nature to the requirements of the farming activities conducted on the agricultural land or pasture in question?*

This is the test from *Starke and Others (Executors of Brown Deceased) v IRC* [1996] All ER 622. It was noted in *Antrobus* that most farmers were now making less profit than previously.

(3) *Is the subject property 'a family home of some distinction'?*

It may not be possible to describe a farmhouse which satisfies the 'character appropriate' test nevertheless it can be recognised when seen – this is the test in *Dixon* [2002] SWTI 43 where the Commissioner was guided by the expert evidence that Cookhill Priory was a farmhouse of a character appropriate. This is the 'elephant' test referred to below.

(4) *Would an educated rural layman regard the property as a house with land or as a farm?*

This is 'the man on the Clapham omnibus' test. The Commissioner heard the evidence, looked at photographs and considered that the whole use and visual presentation of Cookhill Priory supported the view that it was a farmhouse. Practitioners should always consider the merits of visiting the house in respect of which a claim to APR is made.

(5) *How long has the house in question been associated with the agricultural property? Was there a history of agricultural production?*

These are also tests from the *Dixon* case. Can the taxpayer point to 95 years of agricultural production within the same family, as could the family of Miss Antrobus? If not, what period can be shown, bearing in mind the relaxed time scales that may apply in rural communities?

The 'elephant' test

2.15 These are not the only criteria (that had earlier been described by the late Mr Twiddy on behalf of HMRC when describing his approach to the issue) as HMRC would also consider whether there had, for example, been a large reduction in the area farmed from the house. Having applied the test of the rule equivalent of the 'man on the Clapham omnibus' it was necessary to '… stand back and take a balanced view in the round, and that may be described as the elephant test – difficult to describe, but you know one when you see one …'.

It follows that each case will turn on its own facts and for example, *Harrold (executors of Harrold deceased) v CIR* [1996] STC (SCD) 195 identified a six-bedroomed farmhouse as appropriate for the substantial acreage of the farm with which it was 'enjoyed' albeit that case also turned on another issue, as is explained at **3.12**.

Review of s 115(2) in Starke

2.16 One difficulty for the practitioner was HMRC reliance on the decision in *Starke v IRC* [1994] STC 295. However, that decision was always subject to limitations but now of much less importance since the decision of the Upper Tribunal in *Hanson*, reviewed at **2.18** below. However, *Starke* was the first occasion on which it was noted that *s 115(2)* may conveniently be divided into three parts or 'limbs' (separately numbered in the extract below):

The text of *s 115(2)* with the subdivisions as adopted in *Starke*, reads:

> 'In this Chapter "agricultural property" means:
>
> [limb 1] agricultural land or pasture; and
>
> [limb 2] includes woodland and any building used in connection with the intensive rearing of livestock or fish if the woodland or building is occupied with agricultural land or pasture and the occupation is ancillary to that of the agricultural land or pasture; and
>
> [limb 3] also includes such cottages, farm buildings and farmhouses, together with the land occupied with them, as are of a character appropriate to the property.'

The difficulties in *s 115(2)* pivot on the following:

- Does 'land' also include any buildings on it?
- Does the requirement that woodland etc be occupied 'with agricultural land or pasture' mean with any such land or pasture, or must that land or pasture belong to the owner of the woodland or building?
- Must cottages, farm buildings and/or farmhouses with their land be of a character appropriate to the agricultural land in limb 1?
- Is it sufficient if cottages etc are generally of an agricultural type, or must they be appropriate to other agricultural land that is in the same occupation?

The Court of Appeal decided the case on the fairly narrow ground that the property in question did not fall within limb 1:

> 'Thus the question, whether the property with which this appeal is concerned is excluded from part 3 because there is no other property in the same ownership to which its character may be appropriate, does not arise for decision. Counsel for the Crown indicated that the official view is that there must be some nexus between the property alleged to fall within part 3 and other agricultural land or pasture and that such nexus must be derived from common ownership as the structure of Inheritance Tax legislation deals with the diminution in value of the estate of the transferor. The alternative view might be that the nexus, which must surely be required, may be provided by common occupation without common ownership thereby recognizing the reality of the agricultural unit of which, as in this case, the buildings evidently formed part …'

Rosser v IRC

2.17 The issue of *nexus* came before the Special Commissioner in *Rosser v IRC* SpC 368, [2003] STI 1152 but the decision is limited when properly analysed. The case is useful because it considers whether a barn, rather than a house, might be of a character appropriate to the property, and applied the checklist approach that was adopted in *Antrobus*.

The deceased and her husband set up a farming partnership in the 1930s, later buying a farm comprising 41 acres of agricultural land, a house and a barn. In 1989, they gave 39 acres to their daughter, retaining on a two acre site the house, which was their home, and a barn. The daughter and her husband farmed the 39 acres and, from 1990 onwards, also farmed the two acre site that had been retained by the deceased and her husband. In 1996, the farming partnership was dissolved and the only agricultural activities carried on at the retained property were the provision of refreshments for those working on the farm and the storage of farm tools and pesticides. The barn was still used for agricultural purposes in connection with the adjoining two acres of land and served the needs of the 39 acres that had been given to the daughter (Mrs Rosser). Following the death of the deceased and her husband, HMRC issued a notice of determination that the house and barn did not fall within the definition of 'agricultural property' in *s 115(2)*, not being 'of a character appropriate' to the agricultural land and pasture that remained in the estate of the deceased.

The first issue was whether the house and barn were agricultural land within limb 1 of *s 115(2)*. The Special Commissioner considered that he was bound by the specific decision of the Court of Appeal in *Starke* to the contrary but he did review the issue that the court had not considered in *Starke*, namely whether the farmhouse and/or the barn should be considered by reference to the two acres of agricultural land that were retained at the time of death, or to the original

41 acres. He acknowledged that there was no decided authority on the nature of the *nexus* between the buildings and the property and heard argument from Mr Peter Twiddy for HMRC that the *nexus* must be derived from common ownership, not common occupation as supported by the following analysis:

- IHT is charged on the value transferred by a chargeable transfer, which is itself a transfer of value made by an individual where not exempt.
- That transfer is defined as a disposition as a result of which the value of the estate of the transferor is reduced.
- *Section 4(1)* charges tax as if there were a transfer of the value equal to the value of the estate immediately before death.
- The common connection is the word 'estate' and, on that basis, the only *nexus* acceptable for the interpretation of *s 115* is common ownership.

The Special Commissioner agreed, stating that the '… alternative view that the farm buildings are in the estate that the property to which they refer is not is untenable as this would seriously undermine the structure for Inheritance Tax and create considerable uncertainty about when tax is chargeable and the amount of the value transferred …'.

The second issue, concerned whether the house was in use as a farmhouse at the date of death and it was concluded that it was not. The activities carried on there were incidental to the prime function of the house as a retirement home and were of themselves not enough to characterise the property as a farmhouse. As the property failed the primary test it was not necessary to consider whether it was of a character appropriate.

HMRC v Hanson

2.18 It is no overstatement to remark that *HMRC v Hanson* [2013] UKUT 0224 was a landmark case – it did not follow the decision in *Rosser* at either the First-tier Tribunal, nor on appeal to the Upper Tribunal. It concerned several issues, some of which did not concern APR and which went against the taxpayer on the evidence and those points are touched upon in the review of compliance issues in **Chapter 22**.

The material facts were that a family farm of some 800 acres was gradually partitioned among family members. In 1947 the deceased and his wife moved into the house in question, then owned by the father of the deceased who, ten years later, settled it on the deceased for life with remainder to the son of the deceased. The family land was partitioned in several stages, in the course of which the son acquired 126 acres, which he farmed. In 1978 the deceased moved out of the house and the son took up occupation – the son was still farming his 126 acres when his father died in 2002. There was insufficient evidence to support a claim that there had been a gift to the son in 1978 and the issue on appeal was the availability of APR on the house in which the son lived but where the supporting land was not owned by the estate owner in respect of whose death the claim to APR arose.

The First-tier Tribunal considered *Rosser* but decided not to follow it, having found that the wording of *s 115* was circular and to that extent ambiguous. It was held that there was no need for the farmhouse and land to be in the same ownership: common occupation was enough to support a claim to APR and that approach was confirmed on appeal by HMRC to the Upper Tribunal.

The very detailed judgment considered the meaning of 'ownership', including owning interests in land and determined that allowing claims to APR on the basis of common ownership would not, as had been held in *Rosser*, undermine the structure of IHT. It had been the approach of HMRC that, even where a building was, as here, acknowledged to be a farmhouse, and had therefore passed tests as to its use, it must also be in the same ownership as its supporting land to qualify for relief. The taxpayer had, in his skeleton argument, questioned the need for this extra hurdle but HMRC could provide no policy reason for insisting upon it. Reviewing the possibilities, the Upper Tribunal thought that a test of common occupation might sometimes actually be easier to administer than one of common ownership. The Tribunal also added comments on the meaning of 'farmhouse' which are noted in **2.19** below following the review of *McKenna*.

The Upper Tribunal agreed with the main point made, 'that it is appropriate to look at the situation on the ground in order to establish the reality of the farming unit'. The common occupation between house and land provided the *nexus* between limb 1 and limb 3 by which the 'character appropriate' test is to be assessed.

The leading case: McKenna

2.19 *Arnander (Executors of McKenna Deceased) v HMRC* [2006] STC (SCD) 800 provides a valuable and very thorough examination of the issues outlined above and a full reading of the case is strongly recommended.

Rosteague House, Cornwall, overlooked the sea. It was in poor condition but was nevertheless listed Grade II* and 'at the very top end of the size of a farmhouse in Cornwall', even in the context of many of the listed farmhouses in the district. It stood in 187 acres which included 52 acres of coastal slope (from which at one time seaweed was collected as fertiliser) and over a mile of sea frontage. Its origins went back perhaps as far as the 13th century and the present house back in part to 1597 according to contemporary maps. It was a very substantial house with extensive and impressive gardens – there were also farm buildings.

The farming history could be traced back to 1365, though recently it had been conducted by agents. The gardener also briefly engaged for a time in pig breeding for the owner but from 1984 onwards, there were contract farming arrangements. The contractor would have grown maize, but Mr McKenna disliked that and prevented it. The farm buildings were little used as such, but did house garden implements. The woodland area was extended with the help of a planting grant. There were good farm records etc which took the owner perhaps an hour a day to prepare but, from 1995 onwards, illness prevented him from inspecting the land. In 1997, he put all the property then remaining into joint names with his wife who was then in poor health. Mr McKenna died in January 2003 and his wife in June 2003. The estate was eventually sold, but not as a farm, for £3.05 million and later evidence showed that urgent repairs would cost the purchaser nearly £200,000.

It was determined by the commissioner (Dr Brice):

- It was not a farmhouse. She quoted with approval the decision in *Rosser v IRC*. She approached the views of the Lands Tribunal in *Antrobus No 2* 'with some caution' as *obiter*, but held that the principle that 'the farmer of the land is the person who farms it on a day-to-day basis rather than the person who is in overall control of the agricultural business conducted on the land is a helpful principle'. After a review of all the cases, she derived the following principles:
  - A farmhouse is a dwelling for the farmer from which the farm is managed.
  - The farmer is the person who farms on a day-to-day basis, not the manager.
  - The purpose, not the status, of occupation is what matters.
  - If the premises are extravagantly large, then even though occupied for the purpose of agriculture they may have become something much more grand.
  - Each case turns on its own facts, judged by ordinary ideas of what is appropriate in size, content and layout, taken in context with the buildings and the land.

  Based on this, Dr Brice found that the actual farming was not done from the house, which was too grand for the much-reduced farming operations carried on.

- It was not of a 'character appropriate' anyway. After referring to the *Antrobus* principles and to *Higginson*, Dr Brice reviewed the elements as follows:
  - There were many fine farmhouses in Cornwall, some quite large and with very pleasant gardens.

- Rosteague House was large and, despite its condition, had 'an interior of grace and charm'.
- The farm buildings were not really used as such.
- There was not enough land.
- A layman would think of it as a large country house; it was in effect marketed as such.
- On sale, 65% of the price realised was for the house, part for tenanted properties and only 12% for the agricultural land. It would not attract commercial farmers needing to make a living from the land.

- It was not occupied for the purposes of agriculture. Neither Mr McKenna nor Lady McKenna had been able to farm actively during the relevant two-year period.
- Most of the farm buildings were not occupied for the purposes of agriculture. One was a dung stead, latterly use in connection with livery, but livery of horses is not an agricultural purpose. On the facts, however, some were used for agriculture.

This decision makes a claim for a 'lifestyle' farmer in respect of a charming country house that much harder to pursue.

Example 2.2: Richard Andrews – multiple farming enterprise

Richard Andrews is a successful and wealthy man. His home, 'Hawthorn Manor' is a Grade II listed building comprising: ten bedrooms (all en suite); five spacious reception rooms; a well-equipped state of the art office from which Richard runs his business operations; more than adequate ancillary kitchen, utility and boot rooms; coach houses that have been adapted tastefully to accommodate the large range of cars. The manor is beautifully located, set in 75 acres of land, mainly woodland with some parkland immediately adjoining the house.

Richard has three farming enterprises:

- Home Farm lies only three miles from the Manor. Richard manages the farm himself which consists of 400 acres of bare land and some farm buildings – he has long since sold off the Victorian pile that was the farmhouse as being surplus to his requirements.
- Low Farm is 30 miles away in the next county. It is 'carrot' land: there is irrigation but there are no buildings. Richard uses contractors to farm this land, saving himself the need to have either employees or farm buildings for the effective management of the land.

- Top Farm is a 450-acre holding lying 100 miles from the Manor. In the early years before the devastation of BSE (bovine spongiform encephalopathy), Richard ran a very successful stock enterprise. He employed a farm manager who lived in the substantial Dales farmhouse and farm workers who lived in the various farm cottages. Post BSE, Richard took the decision to terminate the risks of maintaining his own stock and used part of the government BSE compensation to pay off the various farm workers. The farm manager received the house in which he was living as part of a compensation package. Most of the former farm cottages are now used for holiday lets managed by a local holiday letting agency.

In this context, the Manor may well be the operational centre of Richard's business enterprise but is it really the farmhouse in relation to any of the more distant acreage? Richard cannot, in the physical sense, actively farm the other land from it on a day-to-day basis. The Manor is a substantial property, exceptionally well maintained and comfortable but it neither looks nor smells nor has the feel of a farmhouse. However, a part of the property is clearly devoted to the administration of a farming business – Richard is the farmer and he does not live anywhere else. None of the farms now have a building with any better claim to be regarded as the farmhouse.

The question that has not yet properly been examined by recorded cases is the size of property that is appropriate in relation to a large and successful farming operation. It may seem unfair to deny Richard APR on the Manor simply because he has so ordered his farming arrangements that he does not, each morning, step into his gumboots by the kitchen door and walk across a 30-foot courtyard to tend his stock. However, it may only just satisfy the functional test in *Hanson* described at **2.20** below.

The courts have, however, accepted evidence that farming can be viable only with the benefit of certain subsidies. It is perhaps no exaggeration to argue that farming success lies as much in the successful management of paper as of crops and stock. Farming has moved a long way from the traditional image of a rotund middle-aged farmer of cheerful disposition with mud on his boots and a straw in his mouth who constantly lives close to the land. It is perfectly acceptable and reasonable for farmers to leave home and drive to work just like the majority of the working population.

The meaning of 'farmhouse' in Hanson

2.20 Aside from the issue of nexus, the judgment in Hanson adds useful a gloss on what is a 'farmhouse'. The word is not defined in the legislation and the description in *McKenna* in **2.19** was confirmed. In the past the focus had been on the house, close/adjacent to an identifiable farm, where the farmer, who was responsible for the operation of that farm, lived. There had to be some functional connection between the house and the farm and it was held that '… If that connection ceases to exist, we consider that the house ceases to be a farmhouse …'.

The Tribunal then considered possibilities, very similar to those imagined in **2.9** above, where a house loses its status as a farmhouse and becomes a mere family home: '… even if it remained a "farmhouse", the conditions of *s 117* would not be satisfied. However we prefer the conclusion that the reason the house does not attract relief is because it ceases to be a farmhouse …'.

The fine house with 'a bit of land'

2.21 *McKenna,* as confirmed by *Hanson,* demonstrates that substantial houses will struggle to qualify for APR, certainly if enjoyed with relatively small acreage with or without farm buildings, or where the land is occupied under a grazing licence. Even where the transferor can meet the 'character appropriate' tests in relation to the mansion house, he faces a further hurdle in relation to the land. If he has the right to obtain vacant possession within 24 months, he may qualify for APR at 100% but if he grants a farm business tenancy (FBT) then, in relation to the farmhouse in which he still lives, he is no longer a farmer but a mere landlord. He is not occupying that property for the purpose of agriculture and loses APR on the house. His alternative is to grant a right of herbage to a grazier but unless the owner of the house personally undertakes work on the land it seems that, however much pasture he owns, he will get no relief on the house.

The landowner must take care – he must ensure that any contract must direct and recognise, that the responsibility for any manuring, seeding and fertilising the land are retained by him and that either he or someone living in his house, does the actual work. Following *IRC v Forsyth-Grant* (1943) 25 TC 369, the owner will be considered to be occupying the land for the purposes of husbandry and will be accepted as such both for the purposes of income tax and IHT, qualifying for 100% APR on the agricultural value of both farm and farmhouse. The income will be taxed as trading income and the taxpayer will preserve any entitlement he already has to roll-over relief for CGT as well as in so far as other tests are satisfied, entrepreneurs' relief.

Not all owners of mansion houses are expert in the arts of fertilising or of hedging or ditching and it would be fatal to the APR claim to employ the person who is also party to the agreement as grazier. The following is a basic checklist:

(a) the owner must cultivate for, sow and establish the grass crop;

(b) the owner must harrow and roll the grass as necessary;

(c) the owner must fertilise the grass crop in the Spring and through the season as necessary;

(d) the owner must cut or spray all weeds to prevent seeding;

(e) the owner must do any mowing that may be required for whatever reason or purpose on the ground to be grazed;

(f) the grazier must covenant 'not to mow or cut the grass'; and

(g) the owner must do any hedging, fencing and ditching needed and any other work of a proprietorial nature.

The Country Land and Business Association (CLA) prepared a form of deed to grant a 'right of herbage' that was approved by HMRC as long ago as March 1997 which covers all of these points.

Land and farmhouse not in identical ownership

2.22 This is no longer such a problem since *Hanson* as where house and land are not in the same ownership there may be some 'wriggle room', as the next example illustrates.

Example 2.3: Roger – farming partnership

Roger and his wife Stephanie own Bridge Farm, comprising 300 acres, the farmhouse in which they live and various farm buildings. They have taken their two daughters into the farming partnership and transferred the bulk of the land (80%) into the names of the four partners. Post transfer, Roger and Stephanie remain the joint owners of the farmhouse and 20% of the land, some 60 acres.

APR on the farmhouse is likely not lost by reason of the transfer. However, if the combined ownership by Roger and Stephanie (and of the survivor of them) were to fall below 10% of the total farmland (less than 30 acres), APR on the farmhouse would be seriously at risk. In the light of *Hanson*, it will be important to show the extent to which Roger and Stephanie still run the farm: the moment that they retire, allowing their daughters control and management, the deciding factor will be to show where the daughters live.

Farmers who fall ill

2.23 Even though a dwelling might in other respects qualify as a farmhouse, it will not qualify for APR unless occupied for the purposes of agriculture. Given the current HMRC view that the farmhouse must be the centre of farming operations, there will be a difficulty where perhaps the farmer is no longer well enough to look after the land. The various possibilities are considered in detail at **3.23** and **3.24**.

HORSES

2.24 It had been held in *Hemens v Whitsbury Farm and Stud Ltd* [1988] AC 601 that the breeding and rearing of horses was not, under the general law, 'agriculture' but for the purposes of IHT, that rule is specifically amended by *s 115(4)*. That section allows the breeding and rearing of horses on a stud farm, and the grazing of horses in connection with such breeding and rearing, to be regarded as agriculture. A stud farm can, in theory at least, qualify within *s 115(4)* even if there is no stallion permanently present at the farm. Furthermore, any buildings used in connection with those activities can be treated as farm buildings.

With any other horses, it is important to establish the agricultural connection (such as the controversial horse meat trade), because mere livery is not agriculture (though it may be a business – see the recent successful case of *Vigne v HMRC* (2017 UKFTT 632 TC) discussed in **Chapter 9**). The HMRC guidance to their Inspectors (see IHTM24068) is to carry out the following checks before allowing relief for a stud farm:

- age of taxpayer at the time of purchase of the farm and at the time of transfer, eg death;
- period of ownership of the stud;
- number of horses and breeding record;
- advertising and publicity;
- accounts of the business, showing purchases and sales; and
- where the horses pass into the food chain, evidence from the horse passport scheme.

Naturally, HMRC are looking to see evidence of a viable business. That is appropriate for BPR, where *s 103(4)* excludes hobby enterprises, but there is no statutory parallel in the APR code which does not demand the business be run for gain, as supported in the case of *Golding v HMRC* [2011] UKFTT351 (see **Chapter 3**).

In *Wheatley's Executors v CIR* SpC 149, [1998] STC (SCD) 60 a meadow used for grazing horses failed to qualify for APR. Although the meadow had been owned by the deceased throughout seven years prior to his death, subject during all that time to a grazing agreement and as such constituted 'pasture' within *IHTA 1984, s 115(2)*, it was not 'occupied for the purposes of agriculture'. On the evidence, the horses which grazed the meadow were not connected with agriculture but were used by their owner for leisure pursuits and thus were not 'livestock'. The grazing by horses would only fall within the provisions of *s 117* if the horses were connected with agriculture, which was not the case in this appeal.

That decision has been criticised for concentrating on the nature of the animals that graze rather than on the main purpose of occupation of the land. Where land is used for grazing, the issue is whether the landowner is growing a crop of grass – if the landowner is treated as being in occupation of the land, and if his profits are assessed as trading income, it really makes no difference what eats the grass.

Example 2.4: Samantha – occupation for agricultural purpose

Samantha's farm includes pasture and several willow trees. She regularly harvests the willow as feed for the giraffes at a nearby zoo and supplies hay for the bison. The land is occupied for agriculture because Samantha is producing a crop of feed but it would be different if those same animals visited Samantha's farm to graze.

There is an issue here for landowners who wish to ensure that their land should qualify for APR by reason of the underlying usage by the tenant. The FBT may well contain a covenant by the tenant to occupy only for agricultural purposes but the landlord's access to relief will be based on what actually happens not what the terms of the agreement dictate should happen. Failure to adhere to the requirements of the agreement may give right of action against the tenant but such a claim might be substantial and that the tenant might not have the resources to meet it. Where the land is held in trust, the trustees should put in place some safeguards against the risk that, at the time of a chargeable transfer, the land might not be occupied for the purposes of agriculture. If land agents are employed, their periodic inspections should be enough; otherwise the trustees ought really to visit the land themselves from time to time.

SHORT ROTATION COPPICING

2.25 By *FA 1995, s 154(3)*, 'short rotation coppice' means a perennial crop of tree species planted at high density, the stems of which are harvested above ground level at intervals of less than ten years. *Section 154(2)* allows land on which short rotation coppice is cultivated to be regarded as agricultural land, and buildings which are used in connection with the cultivation of short rotation coppice to be regarded as farm buildings. Examples of this activity are trees harvested regularly where the branches are used as a renewable fuel, for certain types of power station.

Section 154(2) is expressed to be 'for the purposes of the *Inheritance Tax Act 1984*' and its application is more general than that of the relief for horses which is expressed to be 'for the purposes of this chapter', and of the 'set aside' relief of *s 124C* which, again, is expressed to be 'for the purpose of this chapter', ie the APR Code.

MILK QUOTA

2.26 Milk quota is now of little practical or commercial importance in the context of IHT as it ended on 31 March 2015. However, for historic valuation dates its value might still be significant and in those cases needs to be included as an asset of an estate or one of the assets of a dairy farming company, partnership or sole business.

The aim of the milk quota system, introduced on 2 April 1984, was to control the supply of milk within the EU and it limited the specific quantity of dairy product which a farmer could produce without creating liability to pay a levy for over-production. The quota was not an interest in land but rather was a licence that could be sold with freehold land, and with certain tenancies, and which might be leased. Its ownership could change involuntarily on the termination of a tenancy and on certain other grants, assignments and transfers of interests in land. Its value was commonly treated as if part of the land (see below).

In *Faulks v Faulks* [1992] 1 EGLR 9, it was held that milk quota should not be included among the assets of a partnership. It was available to a deceased's partner in his capacity as a tenant of land, not as a partner *per se*. Quota was not, as such, an asset; it was the elimination of a liability. *Faulks* was a case on partnership law rather than on revenue law. For IHT, it thus becomes important to identify the precise interest of the transferor. The value of milk quota could be regarded as included in the value of land transferred. That helped owners of tenanted land who accordingly qualified for APR on the value of the milk quota, assuming the land itself qualified for APR.

LAND WITHIN CONSERVATION AND SIMILAR SCHEMES

2.27 Farmland and related buildings which have been dedicated to wildlife habitats on or after 26 November 1996 qualify for 100% APR where the normal ownership tests have been satisfied. Land used in such a scheme must be managed in accordance with the scheme. The effect of *s 124C* is, notwithstanding precisely the fact that land is taken out of agricultural use, to satisfy the occupation test; thus where land is farmed in an organic way, the occupation test is satisfied without the need for *s 124C*. The extension of the relief under *s 124C* to buildings used in connection with management of the land as a habitat scheme is helpful, in that those buildings are to be regarded as farm buildings. Given the test of use for the purposes of agriculture as examined in *Harrold (Executors of Harrold Deceased) v CIR* [1996] STC (SCD) 195, which is considered in **Chapter 3**, it may be that the test under *s 124C* is easier to satisfy than the test that was appropriate in *Harrold*.

The habitat scheme in England closed in January 2000 when it became subsumed into the existing Countryside Stewardship Scheme (CSS). This does not ban agricultural activity and nor does it include any specific option for taking land out of agricultural production. However, the CSS is not limited to farmers or agricultural land, so that someone accepted into the scheme would not automatically qualify for agricultural relief. In Scotland, the habitat scheme closed on 31 December 1996 and was subsumed into the Countryside Premium Scheme.

There are many other agricultural/environmental schemes that can apply to agricultural land. The Farm Woodland Premium Scheme (FWPS) exists to promote the establishment of woodland by offering annual payments to compensate for the agricultural income foregone. The Scheme is open to farmers who, either personally or through a manager, run an agricultural business that includes the land to be converted to woodland. Agriculture relief is not available as the purpose of the scheme is to take the land out of agriculture.

FARMING COMPANIES

2.28 *Section 122* allows APR on the value of shares in, or securities of, a company only if:

- the agricultural property forms part of the assets of the company, and part of the value of the shares or securities can be attributed to the agricultural value of that agricultural property; and
- the shares or securities gave the transferor control of the company immediately before the transfer.

The definition of 'control' for *s 122* provides (see *s 122(2)*) that there is no control if:

- the shares or securities would not have been sufficient, without other property, to give the transferor control of the company immediately before the transfer; and
- the value of the shares or securities is taken to be less than the value previously determined under the related property provisions of *s 176*,

ie where there is a sale within three years after the death of the transferor (see **Chapter 4** and, in the context of BPR, at **5.31**).

It is rare for control to be achieved by owning securities other than shares.

To calculate the value on which APR is allowable, it is necessary to deduct any debts secured on that property and to spread unsecured debts (or debts which are secured only by a floating charge) across all the assets of the company, whether agricultural or not. That calculation reveals the part of the value of the shares or securities that is attributable to the farm and qualifies for APR. Where the rest of the company is engaged in trading rather than investment, the exercise is of little practical value, since BPR will usually be available on that part of the value which does not qualify for APR.

Section 116 is modified by *ss 122(3)* and *123* requiring confirmation that the interest of the company in the land carries the right to vacant possession (or to obtain it within 24 months). If the company could obtain that possession, 100% relief is available on the shares of the company (or the agricultural proportion of them). Likewise, that relief is available if the company can satisfy the 'working farmer' test.

To satisfy *s 123* (minimum period of occupation or ownership) the company must show:

- that it has occupied the farm for the purpose of agriculture throughout the last two years before the date of transfer; or
- that it has owned the farm throughout the last seven years before the date of transfer, and that either the company or someone else has occupied the farm for that purpose throughout that period.

Even that is not enough to satisfy the substituted test as to ownership and occupation as by *s 123(1)(b)*, the transferor must have owned the shares or securities throughout the same period. As noted in **Chapters 3** and **4**, a company is treated as having occupied a farm which it owns at any time when that farm was occupied by a person who subsequently controls the company. There are provisions as to replacement of property in relation to farming companies in *s 123(3)*, which are considered in the examination of replacement property in **Chapter 4**.

TERRITORIALITY

2.29 With effect from 22 April 2009 and by virtue of *FA 2009, s 122*, APR is available on the transfer of agricultural property in:

- the UK, the Channel Islands and the Isle of Man; or

- any EEA State (as defined by the *Interpretation Act 1978*).

There is thus a distinction between APR and BPR since a business may qualify for BPR if it is situated anywhere in the world (see **Chapter 5**). There are currently 31 EEA territories, namely Austria, Belgium, Bulgaria, Croatia, Republic of Cyprus, Czech Republic, Denmark, Estonia, Finland, France, Germany, Greece, Hungary, Iceland, Ireland, Italy, Latvia, Liechtenstein, Lithuania, Luxembourg, Malta, Netherlands, Norway, Poland, Portugal, Romania, Slovakia, Slovenia, Spain, Sweden and the UK. *Finance Act 2009, s 122* introduced provisions requiring equivalence, in local law, to certain UK rules: see *s 116(8)*.

If tax was paid under the old rules which is no longer payable, a claim for repayment had to be lodged by the later of the normal repayment deadline and 21 April 2010.

The statutory deadline for repayment claims is four years from the payment of the tax, or of the last instalment. See the 'appointed day' provisions of *SI 2010/867, para 2* and commentary in IHTM24257.

WHAT DOES NOT QUALIFY FOR APR?

2.30 The restrictions on APR will be apparent from the discussion in this chapter of its limitations, and the identification of further problems in **Chapters 3** and **4**.

In an important distinction from BPR, it is not necessary for the carrying on of agriculture to be profitable and carried on for gain as *s 115* does not contain a restriction parallel to that for BPR in *s 103(3)*. The restriction for income tax relief in respect of farming and market gardening losses contained in *ITA 2007, s 67(2)* does not rob a farming enterprise of APR.

Chiefly, APR will be lost in what might otherwise have seemed to be 'likely situations' where, when viewed as a whole, it cannot be said that there is an agricultural enterprise being carried on. This was the case in *Dixon v CIR* and in *Wheatley's Executors v CIR*, considered earlier in this chapter. If the profits of the farming enterprise are low and a claim is made to APR, it may be necessary to examine just why the transferor is (or was) in business. It may very well appear, by rigorous examination, that the enterprise taken as a whole is really an investment and this point is considered again in the context of BPR.

The distinction between what does and what does not qualify for APR is examined at various points in this and the next two chapters. In brief, particular attention should be given to the following areas:

- woodland not occupied with agricultural land;
- buildings used for rearing birds or fish for sport;
- farm cottages neither used nor occupied by farm workers nor former farm workers nor their dependants;
- mansion houses occupied with small let acreages;
- farmhouses, the ownership of which is separated from that of the land, where the land is not held by a company controlled by the owner of the house;
- farmhouses that lack the purposive connection noted in this chapter;
- paddocks for ponies used for riding;
- 'unusual' property, for example oyster beds (see *Gunter v Newtown Oyster Fishery* (1977) 244 EG 140); and
- farms which, for whatever reason other than set-aside, are not occupied for agriculture.

Chapter 3

Agricultural property: agricultural value, the rates of relief, occupation and other aspects

QUICK GUIDE

- Relief is either at 50% or 100%, pivoting on whether land is let on a 'new' tenancy, ie one beginning on or after 1 September 1995, or under older tenancies.
- The relief is restricted to and applied against the agricultural value of the asset which is often lower than its market value.
- The agricultural property must be occupied for agricultural purposes and various rules can extend the nature of that occupation, particularly in the context of trusts: see **3.5**.
- The old 'working farmer' relief continues to remain in point despite its classification as a 'transitional relief' see **3.6**. CGT has, through the introduction of entrepreneurs' relief, revived the concepts that were relevant to retirement relief: perhaps there is a place for the view that agricultural relief is 'really' for those few people who make their sole living from working the land.

All references are to *IHTA 1984* unless otherwise specified.

AGRICULTURAL VALUE

3.1 APR is limited by *s 115(3)* to 'the value which would be the value of the property if the property were subject to a perpetual covenant prohibiting its use otherwise than as agricultural property', but if that property is located elsewhere within the EEA the limitation is modified to what that value would be if subject to an equivalent provision. Where the market value of agricultural property exceeds the agricultural value, ie farm land with shale gas, that premium may, subject to qualification, be franked by BPR. However, it must be shown that the relevant assets are actually used for a business: an unused farm shed will not qualify for either APR or for BPR (see IHTM24151).

Sporting rights may have considerable influence on the value of agricultural land. Shooting rights were considered in *Earl of Normanton v Giles* [1980] 1 All ER 106 which concerned the *Rent (Agriculture) Act 1976* in connection with a gamekeeper's tenancy and established the important principle that the use of land for sport is not agriculture. The House of Lords considered that pheasants reared to be shot for sport were not kept for the production of food – the eating of the pheasants was incidental to the shooting. A shoot may, however, be managed as a business, paving the way for access to BPR. That was the issue in *Customs and Excise Comrs v Lord Fisher* [1981] STC 238, a case on whether Lord Fisher had conducted a sporting shoot with such efficiency from a financial point of view that it was a business in respect of which VAT should be charged. The landowner will not secure APR on the value of sporting rights but may succeed in claiming BPR if the exploitation of those rights amounts to a business.

Trees and underwood are included within 'agricultural value' and crops growing on the land at the date of transfer are legally part of that land until severed. IHTM23180 directs the Valuation Office to include pasture, cultivations and unexhausted manure but exclude the value of growing crops from the valuation report. If the taxpayer has included the value of growing crops without a separate figure, enquiries will be made as to the separate value.

The definition of that value which applied for working farmer relief was similar to that now applied in *s 115(3)*. For working farmer relief, the value would include trees, underwood and growing crops, even though they might be subject to a contract for sale. The important point for working farmer relief was not the sale but whether the crops had been severed from the land.

Example 3.1: Arlington estate – market value premium

The Arlington Estate originally comprised a five bedroomed house, with barns, stockyard and sheds adjoining, three cottages and 450 acres. However, taxes and school fees have taken their toll and the farming activity has been severely curtailed with the best 100 acres long since sold off. The Estate now comprises:

- house and garden: 0.75 acre;
- farm buildings still used as such: 1 acre;
- redundant farm buildings: 1.25 acres;
- meadows and paddocks: 5 acres;
- pasture subject to grazing licence and (trout) stream: 25 acres;

- arable land subject to a contract farming agreement: 275 acres;
- woodland parcels adjoining pasture: 5 acres;
- woodland subject to a sporting lease that will soon expire: 35 acres;
- cottages let to commuters: 1 acre; and
- roadways: 1 acre.

Much of the value lies in the situation of the land – there are good rail links to London, the house has strong amenity appeal and there is ample opportunity for fishing and shooting. The agricultural value would be severely restricted and if the house qualifies for APR at all (see IHTM24165) agricultural value could expect to be no more than 60% of open market value. APR may be available on the five acres of woodland adjoining the pasture (see **Chapter 2**) but not the 35 acres of woodland subject to a sporting lease. Neither the cottages nor farm buildings are occupied for the purpose of agriculture, so APR is not in point.

The valuation of farmhouses for the purpose of APR can be problematic (see **4.22** below).

THE RATES OF RELIEF

100% APR

3.2 *Section 116(2)* provides for APR at 100% if the basic conditions of ownership and/or occupation are satisfied. The life tenant of a qualifying interest in possession trust (*s 49(1)*) is regarded as the beneficial owner of the settled property (see **3.5** below).

The ownership or occupation test

3.3 *Section 117* states that no relief is available unless:

- the property was occupied by the transferor for the purposes of agriculture for two years ending with the date of transfer; or
- the property was owned by the transferor throughout the period of seven years ending with the date of the transfer *and* was, throughout that period, occupied for the purposes of agriculture, whether by the transferor or by some other person.

The meaning of 'occupation' for this purpose is considered at **3.17**, in the case of partnerships at **4.3** and under mere licence at **4.29**. Ownership may also be deemed under the succession provisions of *s 120* (see **4.11**).

The test determinative of the rate of relief

3.4 100% relief is available where *any* of the following conditions are met:

(*a*) (i) the transferor had, immediately before the transfer, the right to vacant possession of the property; or

(ii) the right to obtain it within the next 12 months; or

(iii) the right to obtain it within the next 24 months; or

(iv) the property is valued at an amount broadly equivalent to vacant possession value, notwithstanding the terms of the tenancy;

(*b*) the transferor has been entitled to his interest in the property since before 10 March 1981 and the 'working farmer' conditions are satisfied (see below); or

(*c*) the property, not falling within (*a*) or (*b*) above, is let on a tenancy beginning on or after 1 September 1995.

A sole owner has vacant possession but joint owners are also treated as having vacant possession if their interest and the similar interest of all the co-owners carry that right: see IHTM24141. It has been successfully argued that vacant possession can be given even if there is a tenancy in place where land is occupied by:

- a partnership;
- a company controlled by the taxpayer; or
- a third party.

However, this is not a foregone conclusion and in such cases the precise position and evidence will be crucial to the outcome.

The conditions (*a*)(iii) and (iv) are not statutory but are contained in ESC F17 which extended the condition in *s 116(2)(a)* in two circumstances where:

- the transferor's interest carried a right to vacant possession within 24 months of the date of the transfer; or
- the interest of the transferor in the property is valued at an amount broadly equivalent to the vacant possession value of the property, notwithstanding the terms of the tenancy.

Situation (*a*)(iii) will commonly arise where there is a tenancy which lasts for more than one year but less than two years, as was the case in *Gladstone v Bowers* [1960] 3 All ER 353. Alternatively, this may arise where the landlord, whilst having the right to serve notice on the tenant to vacate the property within 12 months, cannot actually obtain vacant possession for a further 12 months after the termination of the agreement, owing to the provisions of the *Agricultural Holdings Act 1986*.

The (*a*)(iv) situation will commonly apply where land is let to a company which is controlled by the transferor – HMRC refer to this as a 'packaged' valuation (see SVM108210). The vacant possession value must be considered in the completion of the estate return IHT400.

The increased rates of relief are generally applicable from 10 March 1992 et seq. However, if a chargeable lifetime transfer was made before 10 March 1992, the higher rate of relief applies only for the purpose of calculating the 'top up' tax charged by reference to the death of the transferor – there is no change to the value for cumulation purposes which is calculated at the time of the transfer by reference to the rate of relief which applied at that time. The reference to the rate of relief which is applicable at a particular time is examined in **Chapter 16**.

FARMING AND TRUSTS

3.5 The life tenant with a qualifying interest (as defined in *FA 2006, s 49*) in settled property is treated as the owner of the settled property. The termination of his interest, in whole or in part therein, is thus a transfer by him of the settled property. If the transfer is an occasion of charge, the claim to APR in respect of, say, shares in the family farming company, is dependent on whether the life tenant had control of the company (*IHTA 1984, s 269(3)*) – in that respect voting powers on trust shares or securities are treated as being given to the life tenant except where the trust is discretionary.

The tests as to ownership or occupation, mentioned at **3.3**, therefore apply to the situation of that life tenant and thus 100% relief is available where, in his capacity as beneficiary, that life tenant can obtain vacant possession or where he qualifies under the 'working farmer' relief provisions (see **3.7**). Following *FA 2006*, the life tenant may have a non-qualifying interest such that the trust in which that interest subsists will be a relevant property trust for IHT purposes. Under a relevant property trust, it will be the trustees, and not the life tenant, who are deemed to own the underlying assets.

For the purpose of satisfying the test as to control of shares in the family farming company, the holder of an interest in possession may take account not only of the shares in which he has an interest in possession but also his personal shares and any related property. The tests as to ownership and occupation of company land are then applied.

Must the holder of an interest in possession himself be the farmer?

3.6 A life tenant who farms personally is obviously in occupation of the land but he may also be the occupier where he is not personally involved, for example, where the trustees employ a farm manager and merely pay the life tenant any profit made. Certainly that would be consistent with the situation of the outright owner of land – he is still in occupation even where he employs a manager.

Many permutations will be encountered in practice but consider the straightforward case where the land has been held by the trust subject to the interest in possession (whether a qualifying interest or not) of the life tenant for more than seven years. APR will be available either because:

- the life tenant has farmed the land; or
- the trustees have owned the farm for the requisite period and it has been occupied for the purposes of farming during the whole of that time.

Equally, where the ownership period exceeds seven years, it ought not to matter who is employed to do the farming and that could be the life tenant himself.

Where the period of ownership is less than seven years but more than two years and the trustees employ the life tenant (who holds a qualifying interest) as the farmer, he will likely be treated as being in occupation qualifying under the two year rule. This will equally be so where the life tenant is merely allowed to farm the land on an informal basis.

Where the trustees have owned the land for less than seven years but employ a farm manager rather than the life tenant, the life tenant does not automatically qualify for 100% relief because he has not occupied it for long enough. If the manager is employed by the trustees, such that the trustees do the farming, it is they who occupy the land and the land will be valued on the basis of vacant possession with relief allowed at 100%.

Consider that the trustees grant a lease to the life tenant (who holds a qualifying life interest) who then either farms the land himself or employs a manager. The life tenant occupies the property and therefore qualifies for 100% relief after two years. This is because, on the one hand, he is deemed to be the owner of the freehold of the land and, on the other hand, he is personally entitled to the lease. The interests together comprise a value equal to that of vacant possession.

Trustees of relevant property trusts (discretionary or non-qualifying interest in possession trusts) must apply slightly different rules. If they themselves farm the land or appoint a manager, they will qualify after two years for 100% relief, because they have the right to vacant possession. If they let the land the relief is available only after seven years. Even if they have owned the land since before the 'working farmer' rules first applied, they cannot claim the 'double discount', because they do not have vacant possession of the land and could never have qualified for working farmer relief.

WORKING FARMER RELIEF ('DOUBLE DISCOUNT CASES')

3.7 This subject is of limited interest and busy practitioners should move straight to **3.15**.

For land let before 10 March 1981, 'working farmers' gained two advantages:

- the value of the land was less because it was let;
- APR was available on agricultural property.

It should be noted that stud farms do not come within working farmer relief because *s 115(4)*, which extended APR to stud farms, was not in force prior to 10 March 1981.

Although the relief is described as 'transitional', it is still relevant albeit infrequently met in practice. The conditions for working farmer relief are that:

(*a*) the transferor has owned his interest in the land since before 10 March 1981; and

(*b*) the transferor would on that date have been entitled to the old relief; and

(*c*) at no time between that date and the transfer has there been a right to vacant possession; and

(*d*) the transferor did not fail to obtain a right to vacant possession by any act or deliberate omission.

Relief is limited by reference to the greater of £250,000 and 1,000 acres (1,000 acres equals 4,046,856.4224 square metres of land: see the *Units of Measurement Regulations 1995 (SI 1995/1804))*. The further detailed conditions for the relief (see *FA 1975, Sch 8, para 3)* are:

(*a*) the transferor was wholly or mainly engaged in the UK (the later extension to APR in *FA 2009, s 122* does not apply) in one of the capacities set out below in not less than five of the seven years ending on 5 April immediately before the transfer; *and*

(*b*) (i) the agricultural property was, at the time of the transfer, occupied by the transferor for the purposes of agriculture; *and*

(ii) was so occupied by him throughout the two years immediately preceding the transfer; *or*

(iii) replaced other agricultural property and was so occupied by him for a period which, when added to any period during which he so occupied the replacement property, made up at least two years in the five years immediately before the transfer.

The favoured capacities required for the relief are:

- farming as a trade, alone or in partnership;
- being employed in farming as a trade by someone else;
- being a director of a company carrying on farming in the UK (which includes, for this purpose, the Isle of Man and the Channel Islands) as its main activity; or
- in full-time education.

'Wholly or mainly engaged'

3.8 The main (but not exclusive) test of 'wholly or mainly engaged' is defined in *FA 1975, Sch 8, para 3(3)*. The test is satisfied where at least three-quarters of the relevant income of the transferor was derived directly from his engagement in agriculture in the UK. The 'relevant income' is the aggregate of income in any five of the last seven years of assessment immediately preceding the transfer, including unearned income but excluding income from a pension, superannuation, other allowance, deferred pay or compensation for loss of office.

Where a transferor failed the '75% test', he might still show that he was a working farmer. In certain situations (see below), a farming widow or widower enjoyed the benefit of 'transferred years' to help satisfy the seven-year test.

The occupation test for working farmers

3.9 There is further help in satisfying the occupation test mentioned above. Where the agricultural property had been occupied by the transferor for the purposes of agriculture by a member of the transferor's family throughout the 'gap period', linking the transferor's own occupation to the date of the transfer, the conditions in *FA 1975, Sch 8, para 3(1)* are treated as satisfied if they would have been satisfied had the transfer occurred at that time.

Difficulties could arise where the transferor occupied agricultural property at the time of transfer but had not occupied that property throughout the two-year period immediately before the transfer. If that was replacement property, and the earlier agricultural property had had a lower value than the replacement property, taking the value at the time that the replacement property was first occupied, the relief was reduced to the fraction of the agricultural value of the replacement property represented by the agricultural value of the earlier held land (see *FA 1975, Sch 8, para 3(5)*).

Working farmer relief also contained provision, in *Sch 8 para 3(6)*, for the situation where the transferor inherited agricultural property, providing that the occupation by the transferor of the agricultural property was deemed to have begun on the death of the person from whom he inherited it. Where:

- that person was actually the spouse of the transferor; and
- at the time of the death of the transferor's spouse the working farmer conditions were satisfied,

it should be noted that the above conditions are treated as being satisfied in respect of the inheritor who is now the transferor.

The date on which a person was deemed to have become entitled to a farm for the purposes of working farmer relief was not the same as it is now. This is because the legislation, which now forms *s 91* in relation to extending the period of ownership (see **6.11**), did not apply to cases falling within *FA 1975, Sch 8*.

Instead, following the rule in *Sudley v A-G* [1897] AC 11, a residuary legatee becomes 'entitled' to a specific asset which formed part of residue only when the estate had been administered. Through delay in administration followed by the early death of the residuary beneficiary, relief under the working farmer relief rules may well have been lost through no fault of the relevant transferor.

Other difficulties and restrictions relating to working farmer relief

3.10 A farmer could not increase the amount of working farmer relief by 'trading up' shortly before the chargeable event, such as death. A restriction applied – the part of the value of the new farm which qualified for relief was the proportion which the value of the previous farm bore to the value of the new farm when acquired.

The relief was, by contrast, quite generous in the situation where the working farmer retired since relief was not denied merely because he no longer qualified as the occupier of the farm. As long as he had previously occupied the farm for the purposes of agriculture, relief was still available if the farm was now occupied by a member of his family from the time that he last occupied it up to the time of transfer. This was so, even if members of the transferor's family in succession occupied the farm. Strikingly, if the farmer found that retirement did not suit him and again took occupation, the relief which he had previously enjoyed was lost until he had completed two years of reoccupation.

A further relief was available where, for example, the recipient of agricultural property himself died within two years. Relief was available on the recipient's death, provided the conditions as to 'working farmer' and 'occupation' were satisfied and by a deeming provision, those conditions were satisfied where:

- there had been a transfer of value not more than two years before the later transfer, and one of those transfers was on death;
- the value transferred related to the same farm;
- the working farmer and occupation conditions were satisfied on the first transfer;
- the farm was being farmed at the time of the later transfer, either by the recipient of the land or by the personal representatives of the first transferor; and
- the farm became the property of the recipient (or of his spouse) as a result of the earlier transfer.

Slightly different rules from those just set out applied before 29 July 1976. From and after that date, a restriction applied, relating, in particular, to the situation where the recipient of the land had in part purchased it from the earlier transferor. *FA 1975, Sch 8, para 1(2A)* worked in much the same way as is now provided by *IHTA 1984, s 121(3)* in relation to successive transfers so as to restrict the relief.

Transfers made by the same person on the same day were treated as one transfer for working farmer relief. Where those transfers, taken together, exceeded the limits, the excess was spread across the farms according to agricultural values or areas. That difficulty could, of course, be avoided by making the transfers on different days, with the result that the first transfer would have priority over the second or later transfers: see *FA 1975, Sch 8, para 5(3)*.

Working farmer relief and companies

3.11 Occupation by a company which is controlled by the transferor is treated as occupation by the transferor himself. Occupation of property by a Scottish partnership is treated as occupation by the partners.

Limitation of working farmer relief by value or by acreage

3.12 The conditions for relief under *FA 1975, Sch 8, para 4* were:

- the farm must form part of the assets of the company, and part of the value of the shares or debentures of the company must be attributable to the agricultural value of the farm;
- the transferor must have control of the company through its shares or debentures immediately before the transfer;
- the main activity of the company must be farming and must have been farming for the last two years;
- the farm must be in the UK, Channel Islands or Isle of Man (but relief is not extended to the EEA as for APR); and
- the farm concerned must be occupied at the time of the transfer for farming and must have been so occupied throughout the last two years before the transfer or must have replaced another farm giving the company, over the five years before the transfer, two periods of occupation together making up at least two years.

The limit described below relating to value or acreage applied to a company so as to reduce the relief proportionately in respect of the excess over the 1,000-acre limit.

Where the shares or debentures of the farming company were related property, relief might be lost where, as a result of a sale within three years after a death, the value was reduced. If, notwithstanding the relevant sale, the holding of shares or debentures gave the deceased control of the company without relying on the related property provisions, relief was preserved.

As with agricultural property generally, working farmer relief was affected by the existence of a binding contract for sale but a form of relief applied where the death of the vendor occurred between contract and completion. Similarly, if the purchaser died between contract and completion, relief might be available (if other conditions were satisfied) on the amount which the purchaser, at the date of his death, had paid, ie normally the deposit. The rules were different in Scotland: where death occurred between missives of sale and the delivery of a disposition, the rule used to be that relief was available to the buyer rather than the seller; that changed for the short period from 1978 to the expiry of applicability of the rules in March 1991, during which time the seller might secure working farmer relief even if he had received the purchase money.

The limitation to £250,000 or 1,000 acres for working farmer relief is not absolute. By *FA 1975, Sch 8, para 5(2)* (inserted by *FA 1976, s 74(6)*), the area of any rough grazing land was to be counted as one-sixth of its actual area, at least in relation to chargeable transfers made after 5 April 1976.

Where the restriction bites, the percentage of APR available is:

- 100% in respect of the unrestricted portion; and
- 50% in respect of the remainder (see *s 116(4)*).

Section 116(5) provides that *FA 1975, Sch 8, para 5* must be interpreted as if the references to relief given in *FA 1975, Sch 8* in respect of earlier chargeable transfers included references to relief given by virtue of *IHTA 1984, s 116(2)(b)* or *(4)* or by *FA 1981, Sch 14* as far as the earlier chargeable transfers were concerned. This means that the farmer entitled to double discount, who would have benefited from the old relief, will get 100% on the unrestricted portion of the transfer and 50% on the rest. He may not claim BPR instead of the 50% APR.

Working farmer relief: successions

3.13 *IHTA 1984, s 120(2)* extends working farmer relief such that where a person became entitled to an interest on the death of a spouse on or after 10 March 1981:

- the survivor is deemed, for the purposes of *s 116(2)(b)*, to have been beneficially entitled to the agricultural property for any period in respect of which the spouse had been entitled to it;
- if (and only if) the condition in *s 116(3)(a)* (that the earlier holder would have qualified for the relief) is satisfied, then that condition is satisfied in relation to the surviving spouse; and
- the condition in *s 116(3)(b)* (no recovery of vacant possession) is treated as satisfied only if there was no opportunity of recovering vacant possession available either to the original holder or to the spouse.

In practice this means that, if the original holder (the deceased spouse) could have claimed working farmer relief, then so can the survivor, provided that neither of them could have recovered vacant possession of the land after 9 March 1981. This relief is available only where one of the spouses has died.

Importantly it does not apply to lifetime gifts between spouses: there is, thus, scope for negligence through advising accidentally on a transfer between spouses which causes the relief to be lost.

Example 3.2: Judith – preservation of APR

Jim White had been a full-time working farmer and on his death in 1977, his daughter Judith inherited Grange Farm comprising 600 acres of hill land in Wales. Prior to marriage to her cousin Henry, Judith farmed the land herself but from Michaelmas 1979 onwards, the land was let and occupied by the tenant for the purposes of agriculture.

In June 1983, Judith, who was terminally ill, and Henry consulted their solicitor, Jack, for IHT planning advice. Following Jack's recommendation to equalise their wealth, Judith made a lifetime transfer to Henry of the tenanted farm and under the terms of her will left a nil rate band legacy of the non-qualifying APR assets to her niece with the balance to Henry. Judith died less than six months later – no IHT was payable on her estate due to the combination of the nil rate band and spousal exemption.

However, as a result of acting on Jack's advice, the working farmer extension provisions of *s 120(2)* no longer applied to Henry and he was concerned that, on his own death, the APR rate would be restricted to 50%. In 2000, he consulted his accountant, Cerys, who, with some care, negotiated a variation of the lease with the tenant which fell to be treated as a new lease created after 31 August 1995. As a consequence Henry, having owned the land in his own right for 17 years, throughout which it had been occupied for the purpose of agriculture is entitled to APR at 100%.

Restrictions on working farmer relief

3.14 The relief was applied by way of reducing the value of the 'part eligible for relief'. That part might be less than the agricultural value if not all of the property had been transferred. Such a situation could arise where the agricultural property was sold at an undervalue or where the transfer was partly exempt and the result was then a complicated multi-stage calculation. First, the value transferred had to be established without working farmer relief and as if no tax were chargeable on it. That was the value after exemptions and before grossing up: the 'unreduced value' in *FA 1975, Sch 8, para 1(3)*. Relief was available on such part of that value as was attributable to the agricultural value. It was that figure that qualified for 50%.

The deduction of exemptions before applying 50% relief to the agricultural value was less favourable than the current APR and BPR rules where the reliefs apply before deduction of exemptions. This has made working farmer relief less valuable than the new reliefs in certain circumstances.

LAND LET ON NEW TENANCIES

3.15 There is a neat and useful 'potted history' of the law of farming tenancies in IHTM24211.

'New' tenancy for this purpose means a farm business tenancy (FBT) granted on or after 1 September 1995 (that does not grant security of tenure); thus the straightforward case of a transfer of land subject to a new FBT will give little difficulty. More interesting is the situation where there is an existing tenancy under the *Agricultural Holdings Act 1986* which gives no right to vacant possession within either 12 months or 24 months. To secure 100% APR, that tenancy should come to an end – there must be a clear surrender of the old tenancy and a grant of the new.

Interestingly, a tenancy which begins on or after 1 September 1995, only because the statutory succession rules apply, will not for that reason be excluded from full relief – for the purposes of IHT, the commencement of the new tenancy is taken to be the date of death of the tenant who is succeeded (*s 116(5A)*). *Section 116(2)(c)* applies to all agricultural tenancies throughout the UK which began after 31 August 1995 and it is thus not limited to tenancies under the *Agricultural Tenancies Act 1995*.

Section 116(5B) covers the situation where the tenant, or the last surviving tenant, of the property died on or after 1 September 1995 and, as a result:

- a person obtained the tenancy of that property under statutory rules; or
- a person is granted a tenancy in circumstances where he is already entitled statutorily to obtain that tenancy, except that the tenancy to which he is entitled under statute takes effect at a later date; or
- a person is granted the tenancy who is, or who has become, the only (or only remaining) applicant or the only (or only remaining) person who is eligible to apply for such a tenancy under a particular enactment.

In any case covered by these circumstances, the tenancy is treated as beginning on the death of the tenant, except, see *s 116(5C)*, where the property is situated in Scotland. *Section 116(5B)* applies where the property is outside Scotland and, although the point is not expressed within the legislation, *s 116(5A)* deals with a similar situation where the property may be situated in Scotland and provides that:

- on the death on or after 1 September 1995 of the tenant (or last surviving tenant) of any property;
- where that tenancy becomes vested in a person as a result of beneficial entitlement under the will of the deceased tenant or under other testamentary disposition; and
- the tenancy becomes binding on the landlord and the inheritor respectively as landlord and tenant,

s 116(2)(c) is treated as having effect as if that inherited tenancy had begun on the date of the death of the tenant.

A special deeming provision, *s 116(5D)*, applies where the transferor dies on or after 1 September 1995 and, before the death:

- a tenant has given notice of an intention to retire in favour of a new tenant; and
- that retirement in favour of a new tenant takes place after the death but not more than 30 months after the giving of the notice.

In these circumstances, the tenancy granted or assigned to the new tenant is treated as if it began immediately before the deemed transfer on death. By way of interpretation, 'the new tenant' in *s 116(5D)* means the person or persons who are identified, in the notice of intention to retire in favour of a new tenant, as the person or persons who it is desired should become the tenant of the property to which the notice relates. Alternatively, 'the new tenant' may mean the survivor or survivors of the persons identified in the notice, either alone or with others. The notice must be in writing.

Section 116(6) provides that, for the APR code, the interest of joint tenants is taken to carry the right to vacant possession of the land if, taken together, all the joint owners have vacant possession or the right to obtain it within 12 months. The section has wording to accommodate Scottish equivalents.

The example set out in IHTM24241shows that it is possible for a tenancy not to be a FBT and yet for the land still to qualify for APR at 100%.

'RENEWING' THE TENANCY TO SECURE 100% RELIEF

3.16 There is a suggestion that it is possible to secure 100% APR without losing the protection of the *Agricultural Holdings Act 1986* if a small piece of land can be added to the holding or if a small extension can be made to the term of the contract. It is not easy to extend the term of an annual tenancy, but it might be relatively easy to add a small area of land. The addition of a new party to the tenancy has been regarded merely as a variation but for technical reasons, it would appear that the addition of land or the extension of a term must cause a surrender and re-grant.

Rules introduced by *FA 1995, s 155* increased the rate of relief for transfers of tenanted farmland from 50% to 100% where the tenancy commenced on or after 1 September 1995 'in relation to transfers of value made *and other events occurring*, on or after 1 September 1995'. The reference covers the situation where the landowner had made a PET of tenanted land before 1 September 1995 and died thereafter in circumstances where the previous tenancy has been replaced by a new tenancy commencing after 31 August 1995. In these circumstances, APR which would previously have been available only at 50% will be available at 100%. This makes the practical effect of the failure of the PET a non-event, subject of course to problems of clawback and reservation of benefit.

This situation can perhaps be exploited where the practitioner can identify PETs of this nature made within the last seven years, ie those cases where the donor and donee are still on risk. The donee should take steps to bring the old tenancy to an end and to grant a new one. Not only will this mean the donee is in the position of having 100% relief himself in respect of any transfer made by him, but it will also avoid a charge to IHT if, on the death within seven years of the gift of the donor, the PET should fail.

The interaction of IHT and CGT is very relevant here as a simple surrender of lease is a disposal by the tenant, and can give rise to a chargeable gain. That is recognised by ESC D39, which provides that the extension/grant of a lease will not be regarded as a disposal or part disposal of the old lease where the transaction meets the following conditions:

- it is on arm's-length terms, even if between connected persons;
- it is not part of a larger transaction;
- the lessee does not receive a capital sum;
- the same area of land is let; and
- the lease terms are substantially the same, except for duration and rent.

To this list should be added two provisos when dealing with Extra Statutory Concessions:

- that they are of general application only, so might not relieve a particular case; and
- that the transaction is not entered into for tax avoidance.

It would be usual practice for the tenant to ask for a tax indemnity before agreeing to the variation. Whilst this may be the price of 'doing the deal' such tax indemnity should not be lightly accepted without at the very least qualifying the potential CGT exposure.

Where a tenancy is granted for full consideration, *s 268(2)* provides that a gift of the freehold reversion more than three years later shall not be taken to be associated with the earlier grant (that section also provides, though this will now be of academic interest in most cases, that no operation effected after 27 March 1974 shall be taken to be associated with an operation effected before that date). It follows that care should be taken not to make a gift within the three year period, since the transfer of value will be treated as taking place at the date of the later of the associated operations and only where, for example, the rate of relief had become more valuable in the interim would the taxpayer score an advantage.

For IHT purposes, the parties should keep in mind the strong hint in IHTM24243: '*...So long as there has been a clear surrender of the old tenancy (or the* *necessary notice and counter notice has taken place to formally determine the old tenancy) there should not be a problem…'* So the paperwork must support the reality of the actions taken.

OCCUPATION FOR THE PURPOSES OF AGRICULTURE

Agricultural purposes

3.17 The cases of *Dixon*, *Williams* and *McKenna* for analysis of the law and the 'farmhouse' cases discussed in **Chapter 2** illustrate the principles now applied. There is a strong link (see *Hanson* **Chapter 2**) between these issues:

- Is it a farmhouse?
- Is it of a 'character appropriate'?
- Is it occupied for the purposes of agriculture?

The meaning of 'agricultural purposes' was first considered in connection with rating cases. A useful summary of earlier principles in Scottish law is *Assessor for Lothian Region v Rolawn Ltd* 1990 SLT 433, 10 STNR 158. The company grew and sold turf and the Assessor argued that this was not an agricultural purpose but the local valuation appeal committee disagreed. The land valuation appeal court approved the decision of the appeal committee.

The case draws out the following pointers:

- Is the land being cultivated? That may be an end to the matter unless that cultivation is for the purposes of horticulture or forestry.
- Is the crop an unusual one? The decision whether the land is occupied for agriculture may depend on the general impression gained from the evidence as a whole.
- If the land is not being cultivated, is it used to support livestock? If so:
  - Does the livestock produce or contribute to producing the means of human sustenance?
  - Are the animals of the kind found on an ordinary farm?
- Agriculture has its ordinary meaning, which is to be decided by the tribunal of first instance rather than the court, especially where the members of the tribunal are drawn from an agricultural community and understand agricultural questions.
- How is the land treated for rating purposes? An IHT appeal would lie to the Special Commissioner who might not be an expert in agriculture. The treatment of land for rating purposes would be relevant.

It may be necessary to look at the use of the land from more than one standpoint. Land of poor quality may yield no more than basic income for much of the year but may be capable of producing extra revenue from occasional use.

Example 3.3: Peregrine Took – change of use

Peregrine Took was for many years the owner of a small sporting estate in Scotland. The land supported small-scale sheep farming and, more importantly to Peregrine, it had the benefit of excellent shooting and one of the finest runs of sea trout on the West Coast of Scotland. At the time of his death in December 2009, it was argued successfully that the occasional use of the land for the non-agricultural pursuit of shooting did not rob the land of its essential quality as grazing for the sheep. However, the premium market value attributed to the sporting rights so greatly exceeded the agricultural value that APR was severely restricted.

Peregrine's son, Sam, had abandoned his agricultural studies at the local college in favour of his love of music. He brought new wealth to the estate which he inherited on his father's death, establishing an annual three week summer festival which coincided with the start of the annual grouse shooting season on 12 August. On Sam's premature death in December 2017, it was again argued that APR should be available in respect of the estate.

Quite apart from the cessation of agricultural activity during the three week duration of the 'rock fest', evidence indicated the land was taken out of use for agricultural purposes for a longer period so that it could be prepared for the annual influx of visitors. Considerable alterations had been made to the estate to provide the necessary facilities for the visitors and indeed the entire character of the estate had changed to such an extent that the income from the grazing of sheep was insignificant compared with the 'rock fest' income. The use of the land for grazing had been interrupted so much that the land was not occupied for the purposes of agriculture. Similarly, whereas farm cottages had been occupied by retired farm workers or their surviving spouses, annual noise pollution prompted them to seek council accommodation in the nearby town, with the result that relief was not available in respect of any of the cottages, which were now occupied by recording engineers for the period of the festival and by holiday makers at other times.

The land was no longer being used for the purposes of agriculture so as to qualify for APR, though there could be grounds to mount an argument that part of it might (just) qualify for BPR.

Diversification

3.18 For the purposes of *s 1(1)* and *(2)* of the *Agricultural Holdings Act 1948*, an agricultural holding is 'land used for agriculture'. The tenancy of such land is an agricultural one if the land is let for use as agricultural land, but subject to such exceptions as do not substantially affect the character of the tenancy. The question of occupation for 'agricultural purposes' was examined, though not in a fiscal context, by the Court of Appeal in *Jewell v McGowan* [2002] EWCA Civ 145. The tenant of land in Gloucestershire was bound by his agreement to use it 'for agricultural purposes only'. Most of the tenant's income was derived from the 'open farm' open to the public and not from the operation of farming activities and the landlords objected. The tenant initially complied but later sought a declaration from the court that he could continue the 'open farm' without being in breach of the agreement.

At first instance, the court thought that 'open farm' was agricultural, partly on the ground that it was essential for the financial viability of the farm but the Court of Appeal disagreed. 'Agricultural purposes only' in the agreement was not intended to refer exclusively to the definition in the *1948 Act*. The wording of the particular agreement, to use the land 'for agricultural purposes only', prevented the tenant from enjoying any exception to that agricultural use, even where such exception did not affect the character of the tenancy. The 'open farm' was a separate commercial enterprise and its purpose was to educate the public and to promote both organic farming and organic farm produce. Whilst that might direct the agricultural activity, it did not comply with the terms of the agreement – the' open farm' activities were not agricultural in nature and did not become agricultural merely because they enabled farming to be carried out on the holding.

That case turns on its own particular facts and, at a time when it seems that the continuation of farming in many parts of England and Wales is dependent on diversification, it is to be hoped that the less restrictive interpretation of what constitutes occupation for the purposes of agriculture will prevail when tax is in issue. HMRC are certainly alive to the issues: see their Rural Diversification Project (more fully described, with commentary, on the website of CLA) which urges landowners to comply with their obligations as to income tax, VAT, the intricacies of incorporation, PAYE and the correct treatment of overheads.

The Basic Farm Payment (BFP)

3.19 There is a trend, noted at **2.8**, for HMRC to examine closely the nature of farming activities when deciding whether or not to allow APR on the farmhouse. The range of subsidies that used to apply has been replaced by the Basic Farm Payment (BFP). The BFP is payable not only to 'working farmers' but also to those who just happen to own land that was once farmed, leading to a natural tendency to view the BFP more in the nature of investment income than trading income.

BIM55130 sets out some principles for the general tax treatment of BFP as does IHTM24254 for APR purposes but there are also implications for income tax and CGT. The most serious effect may be the loss of APR where the farmer decides not to farm the land and sells all/nearly all of his live/dead stock and keeping only such equipment as is necessary to comply with the conditions for the BFP, which will require him to keep the land in good condition but not actually to farm it. BIM55130 argues that in that situation, there is:

- no agricultural production; so
- no occupation for the purposes of husbandry; so
- no farming trade.

HMRC recognise that a new trade could arise of managing the land on a commercial basis with a view to profit, but that would not be farming – the receipt of a farming subsidy does not make someone a farmer. The requirements for BFP, to keep the land in good agricultural and environmental condition involve:

- soil management;
- maintenance of habitat and landscape; and
- protection of hedgerows and watercourses.

Merely doing that would not, enable the landowner to claim APR on his residence (see IHTM24254). As far as the land itself is concerned, HMRC consider that 'there must be an intention or expectation that the land will be back in production at some time in the future'. That is a fairly vague test.

Who is the occupier?

3.20 The landowner who lets it to a tenant farmer under a full FBT is not in occupation. A farm manager does not occupy the farm rather it is the employer, who benefits from profits or suffers losses, who is the occupier. The life tenant who holds a qualifying interest in possession in settled property does not occupy it where the farm is operated exclusively by the trustees on behalf of the trust and this might also occur where the trustees form a company for the purpose or employ a manager – it is the trust that takes the profits or suffers any losses. The end result is that when the life tenant dies, he has not been in occupation under *s 117(a)*.

On the other hand, the person who actually farms the land for his own benefit is indeed in occupation and so too if he farms under an agricultural tenancy or an FBT. The farmer is treated as occupying not only the house in which he lives but also the houses or cottages in which his farm manager and employees live (see later). Similarly, where the farming is carried on in partnership, each partner is treated as occupying the property and this rule can be very useful in certain circumstances but did not salvage the case of *Atkinson* on appeal (see **3.24**).

Where the land is settled and the farming is carried on not by the trustees as was described above but by the holder of a qualifying interest in possession, that life tenant will be in occupation by virtue of *s 49(1)* – his farming profits, if he makes any, are for consistency to be treated as his earned income and not as trust income. However, where the trustees join in the farming operation, perhaps as limited partners, the two year occupation test cannot be satisfied by reference to them as *s 49(1)* does not apply to the trustees; rather they will need to satisfy the seven-year ownership rule.

Occupation in partnership situations

3.21 A common scenario is the family partnership of mother, father and son/daughter, where the parent partners take son/daughter into partnership not only as to the profits or losses from year to year but also ownership of the farmland itself. Where the parent partners retain sole ownership of the farmhouse and give the son/daughter an interest in the farmland only, the complications examined in *Rosser v IRC* arise (see **2.17**). Where, as an alternative, the son/daughter acquires a one-third interest in the residence but does not live there, no principal private residence (*TCGA 1992, s 223)* CGT shelter will be available to the son/daughter, albeit to the parent partners. The family will probably accept that restriction as in most cases it is unlikely that the farmhouse will be sold.

Does the son/daughter occupy the farmhouse for APR purposes?

- Although the son/daughter no longer lives there, does he/she visit his parents' house every day? Is there a farm office at the house?
- What is the pattern of working: does the father still work the land, or is his involvement more akin to that of a landlord?
- Matters of family relationships could be relevant: does the daughter-in-law resent the time that her husband spends at the farmhouse?

If the case came before the First-tier Tribunal, evidence that parents and son/daughter did not get on and spent virtually no time together could have a significant effect on a claim for APR in respect of the son's share in the house.

Most farming families put the farm before the family, and put the farming members of the family before the non-farming members. Most farming families will therefore be happy to confirm, in the situation described above, that the son/daughter does indeed occupy the farmhouse on a daily basis because it is from there that he/she and the parent partners run the farm. It may be difficult to show that all the partners take an active role in managing the farming enterprise and great care should be taken to avoid the suggestion that in fact neither parent partner has any active role (see *Rosser v IRC*).

Occupation of the farmhouse for agriculture

3.22 Proving the occupation of the farmhouse for the purposes of agriculture can give rise to problems. Certainly where the occupier lets the land that he previously farmed, APR on the farmhouse where he continues to live is lost – the farmland then becomes a mere investment enjoyed by a person who just happens to live in what was previously recognised to be a farmhouse; they are not occupying in the character of a farmer because they are no longer farming. IHTM24013 also directs further investigation where the deceased is described (on form IHT400) as retired and tax practitioners have long since recognised the weaknesses of the 'retirement' situation.

Assuming that the hurdle of occupation of the farmland can be overcome, the secondary question is whether the taxpayer was occupying the farmhouse itself. Occupation is a matter of fact and if a person farms land for his own profit, he is in occupation not his employees and contractors. This is true whether the owner of the business is in physical occupation or not thus it is possible for the owner to occupy the entire farm, including the tied cottages of employees but this will not help a claim to APR since for farmhouses, the test is by reference to the person who actually does the work and not the person supervising it (see *McKenna*).

Allowing other people to use the land for part of the year on seasonal grazing or mowing licences can be compatible with retaining occupation for the purposes of agriculture, although much will depend on the terms of the licence. There is a distinction between physical occupation and a legal right to occupy, but it is not reasonable, in the context of APR, to limit the meaning to physical occupation.

In *Executors of D Golding Deceased v Revenue & Customs Commissioners* [2011] UKFTT 352 (TC), HMRC had accepted that the house in question was a farmhouse but had denied relief on grounds it was not of 'character appropriate' – the argument rested to a great extent on the profitability of the farm.

The deceased had for many years until his death at age 81, farmed a smallholding of just 16.29 acres – the diverse farming trade comprised some 600 chickens, seven to ten cattle, harvesting of fruit from trees and growing vegetables. The farm produced milk and crops of wheat and barley were produced for sale. It was accepted that old age and infirmity led to scaling back of activities confined to selling of eggs, fruit and vegetables which produced a net profit of approximately £1,500 pa on average over the last five years – equal to less than 25% of the deceased's sustainable income. Even so, in the early years the farm had been sufficient to sustain a family albeit not to any great state of luxury. The farmhouse was small in size, without electricity in parts and in a very poor state of repair. In a refreshingly simple and clear analysis of the facts as presented, the tribunal in finding for the taxpayer stated 'we do not accept that the lack of substantial profit is detrimental to the decision that the farmhouse was of character appropriate …'.

Rather belatedly and in response to the outcome of the *Golding* case HMRC updated the guidance to their IHT manual (see IHTM24036 et seq and HMRC Guidance reported at STEP UK *News Digest* 13 May 2013) which considers the APR application to the farmhouse under the character appropriate test.

The farmer who falls ill

3.23 This is a difficult area with little case history other than *Atkinson* (see **3.24**). It is not uncommon for the farmer to suffer ill-health, effectively give up living at the farmhouse and move to a nursing home/hospital and in such circumstance access to relief will be dependent on how the farming is carried on. In the absence of help from family members it may be possible to show that there is a contracting agreement in place but all too often the arrangements will be informal. The key to the analysis is that 'occupation' does not strictly mean 'residence', although the requirements set out in *McKenna* virtually lead to that conclusion.

It may be possible to argue that a neighbouring farmer, under an unwritten but settled relationship, looks after the land and for that purpose has access to the farmhouse and uses part of the farmhouse for managing the land. On this basis, an argument that the business of the farmer includes part use of the farmhouse may form the basis for a modest BPR claim where the full claim for APR is strenuously disputed by HMRC. This may not actually be as poor a substitute as it sounds: the claim to APR would only ever have been in respect of the agricultural value of the farmhouse, whereas a BPR claim in respect of the 'business part' of the house will be on the appropriate proportion of the full value of the house. HMRC are loathe to accept BPR claims in such circumstances but even so it may form the basis of a compromise that will allow the executors to wind up the estate. In these cases, much will turn on the specific evidence:

- What is the extent of the farming operations?
- When did the farmer last direct farming operations from the farmhouse?

- Has anyone else occupied the farmhouse as a mere residence since the farmer went into the nursing home/hospital?
- How often is attendance at the property required for the running of the business?
- Is this a business which, following the illness, is progressively run down?

Atkinson: what constitutes 'occupation for the purposes of agriculture?'

3.24 In *Executors of Atkinson Deceased v HMRC* [2011] UTUK FTC/61/2010, the Upper Tribunal had to consider the issue outlined above. The deceased had been in partnership with his son and daughter-in-law from 1980 and had granted a tenancy of the farm to that partnership. The partnership was increased in 1994 when the grandson joined, but reduced the following year when the son died. In 1996, the deceased, his daughter-in-law and grandson agreed that they farmed in partnership and that the earlier tenancy was a partnership asset – the tenancy was assigned to the deceased, his daughter-in-law and grandson. A bungalow had been built in 1966 and was occupied by the deceased but he became ill, moving first to hospital and thence to a care home where he later died. The bungalow remained furnished and his personal things were left there.

The tribunal held that the bungalow was 'occupied for the purpose of agriculture' within *IHTA 1984, s 117*. It, like the whole farm, was occupied by the partnership. For 22 years, the bungalow had been the residence of the deceased. For the last four years of his life, it was not, but he remained a partner, taking an interest in the business of the farm, visiting the bungalow occasionally and keeping his possessions there. The tribunal noted that the facts were broadly the reverse of those in *Harrold v IRC* [1996] STC (SCD) 195 (see **3.25**) where a farmhouse had not been brought into use and was not therefore occupied for agriculture. Here, the bungalow had been so occupied and nothing was done to disturb that.

On HMRC appeal (*HMRC v Atkinson and Another* [2011] UKUT B26 (TCC)), it was held that the occupation was not by the farmer but by the partnership and that partnership had ceased to occupy the bungalow when Mr Atkinson moved into the care home. It was necessary to identify a strong connection between the use of the bungalow and the agricultural activities being carried out on the farm – the mere housing of personal possessions and occasional attendance by the partners to oversee the property could not be said to amount to occupation in the real world.

Occupation of dwellings for agricultural purposes

3.25 Difficulty may arise where a farming company owns a farmhouse and other residential property occupied by members of the family. By *s 119(1)*, occupation by a company controlled by a transferor is regarded as occupation by the transferor himself but under *s 122*, APR is allowed on shares in a company only if the transferor controlled the company. Under *s 123(1)(a)*:

(1) the property must have been occupied either by the company for agricultural purposes throughout the previous two years; or

(2) the property must have been owned by the company throughout the seven-year period and occupied throughout (by the company or another) for agricultural purposes; and

(3) the shares must also have been owned for the corresponding period.

The test in each case is whether the occupation is accepted as being for agricultural purposes. Clearly, a contract of employment requiring occupation of a particular dwelling is helpful in negotiations with HMRC to support a claim that occupation, perhaps by a member of the family, has indeed been for agricultural purposes. If the occupier does not have a controlling holding, relief will be under BPR rather than APR (see **Chapter 6**).

The issue of occupation was considered in *Harrold (executors of Harrold deceased) v IRC* [1996] STC (SCD) 195, SpC 71, upon which reliance was place in *Atkinson*. A Norfolk farmer and his son bought a farm which included a six-bedroomed farmhouse much neglected and in need of substantial renovation. This was destined to be the home of the son when fit for occupation. Following initial running repairs, a major renovation was carried out but it was not until 64 months after the purchase that the son took up residence. Some two years prior to taking up residence, the father assigned his share of the farm to the son, dying some 18 months later. APR was claimed on the whole farm but this was disputed by HMRC on the grounds that such part of the value transferred by the deed of gift as was attributable to the farmhouse:

- was not agricultural property within the meaning of *s 115(2)*; or alternatively;
- having regard to *s 117(a)*, was not agricultural property occupied by the assignor for the purposes of agriculture throughout the period of two years ending on the transfer.

The Special Commissioner held that the farmhouse did indeed fall within the meaning of 'agricultural property' as it passed all the tests but the stumbling block was *s 117* which required the property to be occupied for the purposes of agriculture. An intention to occupy at a future date was not enough – the owner was not in occupation until he moved in. The judgment relied in part on the test as to future intention in *Arbuckle Smith & Co Ltd v Greenock Corpn* [1960] AC 813, prior to the enactment of *s 124C*, that land lying fallow within a conservation scheme might be regarded as not occupied for the purposes of agriculture (see **2.27**).

On this evidence, the Commissioner could not find that either the father or the son occupied the farmhouse for the purposes of agriculture before the date on which the son took occupation.

Occupation by a partnership or company is considered at **4.9** below.

Farm cottages: the rate of relief

3.26 100% relief will be available (provided the two-year occupation/ seven-year ownership conditions have been satisfied) if:

- the occupation began on or after 1 September 1995; or
- the tenancy began before 10 March 1981 and the 'working farmer' tests are satisfied.

Relief may therefore be available in the case of an occupier with no protected service tenancy, perhaps a partner or an occupier with the benefit of an assured agricultural tenancy, such as a farmworker employed on the farm or an occupier who has an assured shorthold tenancy. HMRC have confirmed that a person with the benefit of an assured shorthold tenancy will normally be regarded as occupying the cottage for agricultural purposes.

The difficulty arises in relation to cottages first occupied after 9 March 1981 post access to the working farmer relief and before 1 September 1995 when the current rules began to operate. In these cases, the nature of the occupation determined the rate of the relief. Where there is an unprotected service tenancy or an assured shorthold tenancy, 100% relief is available but in the case of an assured agricultural occupation (whether by a farmworker, a retired farmworker or by the surviving spouse of a farmworker) the rate is 50%. Thus the farmer-landowner can be in a worse situation than, for example, the owner of a forestry business whose estate includes a woodman's cottage where that cottage will attract 100% BPR.

There is one exception (see IHTM24034) which applies where there is evidence that suitable alternative accommodation was available for the tenant at the date of transfer. Where the tenant was in default of the terms of his occupation, then 100% relief may likewise be available, because the landlord can recover possession but where the tenant was not in default the rate of relief will depend on whether possession may be obtained by another route, as for example under the *Agricultural Holdings Act 1986, Sch 3*.

The 50% relief applies where the landlord is unable to obtain vacant possession within 24 months. Thus where, for example, a cottage had been part of a farm but was let temporarily to an outsider before being returned to farm use and was transferred within less than two years after such resumption of farm use, APR would not be available.

Farm cottages must be valued on the basis that there is a planning tie restricting occupation to agricultural use only (see *s 169(1)*). The law in this area is not entirely satisfactory but the values involved may, in the context of whole landed estates, be marginal, such that litigation, even before the First-tier Tribunal, is unlikely. See IHTM24165 for a discussion of the issues that the District Valuer is asked to address and the extent to which they are to apply the 'character appropriate' test to farm cottages, with a hint as to leniency.

Share farming and contract farming

3.27 Ever fewer farmers now keep and maintain the equipment needed for the most basic farming operations. In the past, the earlier generation were content with lower-powered machinery and would wait for the weather to work in their favour whilst in current times the benefit of more modern tractors/ state of the art equipment means work can be carried out in all weathers. The by-product is that it is common to find that the farm buildings stand idle because all the heavy gear is owned by the contractor. Under a share farming arrangement, the landowner provides the land, buildings and fixed equipment whilst the working farmer supplies the labour and working machinery and both pay a share of the costs – profits are divided between the landowner and the working farmer according to an agreed formula.

In July 1991, HMRC revised their earlier practice and stated that whilst they continue to regard the working farmer as the farmer, the landowner would not be regarded as in occupation of the land for the purpose of husbandry and thus would not qualify for the various reliefs from capital taxation. This was not well received and following debate HMRC agreed that the landowner may establish trading status – he must show that he concerns himself with the details of farming policy and goes onto the land for a 'material purpose', eg to inspect crops or stock and to decide the farming policy in the light of that inspection.

50% APR

3.28 *Section 116(2)* allows relief at 50% in any case where the conditions for 100% relief are not met. This mainly means land let under an *Agricultural Holdings Act* tenancy.

One specific situation where 50% relief will apply is in relation to working farmer relief where the £250,000/1,000 acres test cannot be met (see *s 116(4)*). In these circumstances, relief is at 100% for the part of the value which qualifies for the old relief and 50% for the remainder. There is a complication at the point of overlap between APR and BPR arising from *s 116(5)* (noted at **1.8**) and discussed more fully in **Chapter 12**. That subsection requires *FA 1975, Sch 8, para 5* to be construed as if references to working farmer relief included references to relief given under *IHTA 1984, s 116(2)(b)* or *(4)* or under *FA 1981, Sch 14* by virtue of *FA 1981, Sch 14, para 2(ii)(b)* or *(iv)*. The effect is that a transferor who would have been entitled to working farmer relief secures the new 100% relief up to the limit of the qualifying value or acreage, and 50% on the balance. There can be no claim to BPR as an alternative to the 50% APR.

Chapter 4

Agricultural property: areas of particular concern

QUICK GUIDE

- APR is lost if there is a binding contract to sell the asset at the time of the transfer.
- For trusts and partnerships, there are complicated rules as to occupation of land so as to qualify for APR (see the example at **4.4**).
- There are two circumstances in which the two-year ownership rule is varied:
  - the replacement of qualifying property by other qualifying property (see **4.6** et seq);
  - the replacement of one qualifying owner by another under the rules relating to successive transfers (see **4.11** et seq).
- A spouse acquiring APR property from a spouse on their death automatically acquires the period of ownership of that the deceased spouse. This relief cannot be applied to spousal lifetime gifts.
- Complicated rules affect replacement property in order to limit APR to the value of property that has been held for the two-year period.
- The valuation of individual components of agricultural estates is complex. There are established precedents which derive from the key decided cases of *Buccleuch*, *Baird*, *Ellesmere*, *Lady Fox* and *Alexander*. If a value is properly agreed, it will not normally be reopened, even where the land is later sold for a higher figure. The rules as to hope value are considered at **4.21**.
- The impact of APR on sales of quoted shares/securities within 12 months of death, and of land within four years of death, is examined in detail at **4.64** and **4.27**. A problem applies with the interaction of the related property rules: see **4.28**.

All references are to *IHTA 1984* unless otherwise specified.

NO CONTRACT FOR SALE

4.1 *IHTA 1984, s 124(1)* states that APR is forfeit if, at the time of the transfer, the transferor has entered into a binding contract to sell the property. The principal exception is where:

- the sale is to a company; and
- the consideration is wholly or mainly shares in or securities of the company; and
- those shares or securities give the transferor control.

Similarly, APR is forfeit under *s 124(2)* (which allows APR in respect of farming companies) if, at the time of the transfer of the shares or securities, the transferor has entered into a binding contract of sale but with an exception where that sale is for reconstruction or amalgamation of the farming company.

A sale for less than full consideration which itself gives rise to a claim for tax may qualify for APR: (see IHTM24040).

The APR and BPR treatment of sales is different – APR applies essentially to particular assets from or by which farming is carried on, whilst BPR tends to attach to the business entity rather than the underlying assets. Accordingly, APR may still be given in respect of the shares of a farming company which is in liquidation where the farm itself is not being sold and that contrasts sharply with the BPR position where by virtue of *s 105(5)* relief is blocked. However, consider that APR will be denied where the sale of the asset giving rise to the relief takes place before the transfer, whereas for BPR the exchange of one business asset for another does not on its own jeopardise the relief: it is only necessary to show that the business continues and that the new asset is used or, if cash, that it is required or will be required for the purposes of the business. In the case of farmland, the replacement provisions may help. Where a growing crop is sold, the crop itself no longer qualifies for APR, but BPR will usually be available on the proceeds.

Section 115(2) defines property as 'agricultural land' etc, rather than a mere interest in such land and other asset categories ie farm cottages. A school of thought, not yet tested, could interpret agricultural property as meaning the underlying asset itself. Running this argument in a partnership scenario, consider that the vendor does not agree to sell the land itself but merely to sell an interest in it – this would appear to avoid recognition of a binding contract that would otherwise exclude relief under *s 116*. This argument, if it worked, could only apply to partnerships and not to the direct ownership of assets such as the land itself or a controlling shareholding in a farming company. The argument could not be used to secure BPR because that relief relates not to the underlying asset but to the interest in the business itself.

This argument might now fail because of the comments *obiter* in *Hanson* on the meaning of 'ownership':

> 'Whether something less than a legal or equitable interest of that sort amounts to ownership is not something we need to decide although we would suggest that anything which the law recognises as an interest in land and which would give rise to an inheritance tax charge on death as an asset of the deceased's estate ought, in principle, to be capable of attracting relief …'

JOINT OWNERSHIP

4.2 *IHTA 1984, s 116(6)* provides that the interest of one of two or more owners of land, whether they hold as joint tenants or as tenants in common, carries the right to vacant possession if the interests of all of them carry that right. This also holds good for Scottish estates, where there are joint owners or owners in common.

PARTNERSHIPS

4.3 Occupation of land by an English partnership is treated as occupation by the partners, however inactive that partner may be, and that very point was keenly illustrated in *Atkinson's Executors v HMRC* [2010] UKFTT 108 (see **3.24**). This rule can prove helpful when reliance is made on *IHTA 1984, s 117(b)*.

Example 4.1: James Cook – occupation test

James Cook bought Bridge Farm many years ago and in 2015, he took his wife Mary into the farming partnership but the freehold of the farm remained in his name. James was not in good health and in 2016 he retired from the partnership, at the same time that their son Charles became a partner. In 2017, James transferred the freehold of the farm as to 75% to his wife Mary and 25% to his son Charles.

Section 117(a) would not be satisfied in relation to the transfer to Charles (being irrelevant to the other transfer) because, at the time of transfer, James was no longer occupying the land for the purposes of agriculture, but *s 117(b)* would be satisfied by virtue of the occupation by Mary and Charles together.

A Scottish partnership is a legal person in its own right but notwithstanding, *s 119(2)* provides that occupation by any such partnership is to be treated as occupation of the land by the partners irrespective of *s 4(2)* of the *Partnership Act 1890.*

There is a link, though not explicitly expressed, between *s 119* (occupation by a company or partnership) and *s 120* (successions) which allows successive ownership to 'bank' periods of occupation.

Example 4.2: John Bull – occupation of successive properties

In 2009, John Bull, an accountant by profession, bought 'Manor Farm' which adjoined the 'Manor House' in which he lived – the farm was subject to a long-term agricultural tenancy. There was no link in occupation between the 'Manor House' and the farm at John's death in 2012 as whilst the farm was occupied for the purposes of agriculture, it was not so by reference to him. At the time of his death he had not owned the land for seven years so as to comply with *s 117(b).*

Under the terms of John's will, the house and the farm was left on trust to his widow Sarah for life (qualifying interest in possession). The will contained a power to 'trade down' and, at Sarah's request, the trustees sold both the house and farm using the proceeds to purchase 'Low Farm'. This comprised a smaller house but a larger acreage with that acreage again being let to a local farmer but producing a greater income.

At the time of Sarah's death in December 2017 the conditions of *s 117(b)* were not satisfied – whilst she was treated as the owner of the land by virtue of her interest in possession, Sarah had not owned the property for seven years at the time of her death. The replacement provisions of *s 118(2)* were also not satisfied because, in her capacity as life tenant, it could not be shown that Sarah had owned the land for seven years. However, the situation is saved by the operation of *s 120,* as Sarah is deemed to have owned 'Manor Farm' (but not the 'Manor House' itself) from the date of its purchase by her husband John within *s 117.* As a result, the qualification under *s 118(2)* is met.

However, it should be noted that the proceeds of the 'Manor House' as well as 'Manor Farm' have been applied in the purchase of Low Farm. The 'Manor House' was never occupied for the purposes of agriculture – it was simply the residence of a retired accountant and then his wife as life tenant. On that basis, *s 118(3)* operates so as to restrict APR to the proportion of the value of 'Low Farm' which relates to the proceeds of 'Manor Farm'.

Limited liability partnerships (LLPs) give rise to difficult issues (see **5.23**) but the basic principle which must be grasped is that LLPs are, like other partnerships, treated as opaque for IHT.

TRUSTS

4.4 It was noted in **Chapter 3** that the tests as to occupation give rise to problems in relation to trusts.

Example 4.3: Wayne Bright – farming by trustees

Wayne Bright is 21 at the time of the death of his father Stephen in July 2017. Under his father's will Wayne has an immediate post-death interest (qualifying interest) in the 500-acre family farm and the trustees are the family lawyer, Jack and Stephen's accountant, Cerys. Wayne is no scholar nor even, yet, a competent farmer and Jack is concerned that the farming assets should not be exposed to risk whilst Wayne is learning his trade. He therefore suggests that, for the next few years at least, the farming operations should be conducted or supervised by Cerys and himself, employing a manager for that purpose.

Cerys observes that since Wayne has a qualifying interest in possession, the arrangement which Jack proposes will mean that, for the time being at least, the condition in *s 117(a)* will not be satisfied, though she recognises that there is every prospect that Wayne is likely to live for many years yet. It is therefore agreed, subject to suitable indemnities, that the farming operations be conducted as Jack proposes. The trustees, in the exercise of the power given them by the will, take out term assurance on Wayne's life for an amount which they consider will compensate the fund for a charge to IHT made in the event of Wayne's death within the next, say, ten years.

Where property is held in a settlement in which the life tenant has a qualifying interest in possession, the land can qualify for relief by reference to that life tenant since they are treated as beneficially entitled to the property in which the interest subsists (*s 49(1)*).

Post *FA 2006*, the creation of any *inter vivos* life interest trust (special category trusts excepted) will be a relevant property trust whereby the trustees of such trusts are treated as the legal owners of the land and not the life tenant who holds the non-qualifying interest. By retaining trust ownership of that APR qualifying land/property, the trustees will avoid a positive ten-year charge on the agricultural value thereon and distributions of such assets will escape a positive exit charges once the trustees have held the land for seven years (or two years if they do the actual farming).

More specifically, *IHTA 1984, s 115(1)* provides that references to a transfer of value include references to an occasion on which tax is chargeable under the relevant property code of *Chapter III* of *Part III* of the *Act* excluding the rare situation where *s 79* (exemption from the ten-year charge where there has been conditional exemption for heritage property) applies. The value transferred by a transfer of value includes the amount on which tax is chargeable where that tax is met by the transferor, which term includes the trustees of the settlement. Where the whole or part of the value transferred by a transfer of value is attributable to the agricultural value of agricultural property, *s 116(1)* provides that the value (or part of the value) so transferred is reduced by 'the appropriate percentage', subject to the general provisions of the APR code.

Occupation by trust beneficiaries: Woodhall v IRC and Watkins v IRC

4.5 The case of *Woodhall v IRC* [2000] STC (SCD) 558 considered the identity of the person who holds the qualifying interest in possession when the trust deed permits a class of beneficiaries to occupy.

The will of the deceased provided a trust for sale of residue with power to postpone, but there was to be no sale of the family home for as long as any of the children of the testator wished to live there. There were three children – Annie, Alan and Eric. The will provided that, until such sale, the trustees (two of whom were children of the deceased) should permit any of the children to occupy the house subject to payment of ten shillings per week rent and all outgoings, and to keep the house in good repair. Following a sale, the three children would share the proceeds equally.

At the time the of the testator's death in1957, Eric had already left home and Annie followed a year later on her marriage – only Alan took up occupation. Annie died in 1971 but Alan lived on in the house until his death in 1997. Eric, as Alan's personal representative, argued that for as long as the house was settled property, Alan was not entitled to an interest in possession in it; rather the trustees under the terms of the will had the discretion to give an interest in possession to any one of the three children. What was more, even if Alan did have an interest in possession, it was in respect of only half the house, within *s 50(5)*, because Eric in his personal capacity was entitled to the other half. HMRC disagreed, on the grounds that Alan was entitled, at the time of his death, to an interest in the whole property as settled property.

The Special Commissioner considered that the trustees had no discretion as to which of the three children should occupy the house; rather the purpose and effect of the will was to give each of the three children a determinable life interest. The trustees did not have a dispositive power which allowed them to decide that any one child should occupy the house alone but had only an administrative power to allow any child to occupy it and if more than one had wanted to live there, they could have done so. Following Annie's death, Alan held a determinable life interest in one half only. The appeal, therefore, was partly successful.

The outcome should be contrasted with *Brian Maxwell Cook and Elaine Margaret Dawn (Executors of Joyce Elizabeth Watkins Deceased) v IRC* [2002] STC (SCD) 318 in which the deceased left the former matrimonial home, then divided into two separate flats of which one was let, to her husband to occupy for as long as he wished. After his wife's death, the widower stayed with his daughter but he fell ill and later died not having taken up occupation in the property.

The issue was whether the widower had an interest in possession in all of the property or only in part of the house. The Special Commissioner could find nothing from the facts to show that the widower never again intended to live at the flat nor was there was any disclaimer of the life interest, by conduct or in writing. It was held that the widower did acquire an interest in possession in the former home, but did not take such an interest in the flat that was let.

REPLACEMENT PROPERTY

4.6 The occupation condition of *IHTA 1984, s 117(a)* is extended under *s 118(1)* where, on the date of transfer, the agricultural property occupied by the transferor had replaced qualifying agricultural property.

Test of replacement property as to occupation

4.7 The occupation condition is treated as satisfied if the agricultural property transferred, the property which it replaced and any agricultural property directly or indirectly replaced by that other property were occupied by the transferor for the purpose of agriculture for periods which, taken together, comprise at least two years falling within the five years which end with the date of transfer.

Test for replacement property as to ownership

4.8 *Section 118(2)* is concerned with satisfaction of the ownership test under *s 117(b)*. Where the agricultural property owned by the transferor at the date of the transfer replaces other agricultural property, that ownership test is satisfied if:

- that agricultural property;
- the other property which it replaced; and
- any agricultural property directly or indirectly replaced by that other property

were, for the periods which together comprised at least seven years falling within the ten years counting back from the date of transfer, both owned by the transferor and occupied (by him or another) for the purposes of agriculture.

The provision operates to preserve the status quo and thus the relief in relation to new property, cannot exceed the relief that would have been available on the old property (*s 118(3)*); but any changes which result from the formation, alteration or dissolution of a partnership are to be disregarded (*s 118(4)*).

The relief is computed by reference to the agricultural values at the date of the death (or of the transfer) of respectively the old farm (in respect of which the *s 117* test was satisfied) and the replacement farm with APR is limited to the lower of:

- the agricultural value of the old farm, if it were still owned by the transferor; and
- the agricultural value of the new farm.

From a practical viewpoint, it may be difficult to value the old farm and indeed the land may have been broken up and/or put to different use. HMRC may well be influenced by evidence (or the lack of it) that the transferor sold the old farm and bought the new one in order to exploit APR. However, the point may be academic as BPR may be available in respect of excess non APR relievable value.

The replacement property rules cannot be used to circumvent the base two year/seven year rules of *s 117*.

Example 4.4: Trevor Black – replacement farm

Trevor Black's farm adjoins a new town development, the extension of which results in the compulsory sale of the whole farm to the local council. The farm had an agricultural value of £1 million but the sale price, which reflected the development premium, was £6 million. After gifts to the non-farming members of his family, Trevor applied £3.5 million in buying a new farm some distance from the town but the agricultural value is £3 million. His death occurs within two years of the purchase of the new farm, at which time the new farming business is worth £4 million.

The natural assumption made is the market value of the original farming business run by Trevor was equal to its sale price of £6 million. Following the gifts and the repurchase, the value of the remaining farming business is £4 million of which £1million qualifies for APR and £3 million for BPR as replacement of the business assets.

CONTROLLED COMPANIES

4.9 Under *IHTA 1984, s 119(1)*, occupation by a company which is controlled by the transferor is treated, for the purpose of the rules relating to minimum period of ownership or occupation or replacement, as occupation by the transferor.

SUCCESSIVE TRANSFERS AND INHERITED OWNERSHIP

Successions

4.10 Aspects of *IHTA 1984, s 120*, which deals with successions and provides a means of extending the minimum period of occupation or ownership under *s 117*, have already been considered in relation to occupation by partnerships.

In other situations where the transferor of property became entitled to it on the death of another person, *s 120(1)(a)* provides that his ownership starts on the date of that earlier death albeit a residuary beneficiary cannot claim any particular asset until the administration is complete. By the same provision, the transferor of land is deemed to have occupied land from the date of death if he later occupies it. The relief in *s 120(1)(a)* is valuable but the requirements of the section must be strictly complied with.

In addition to that relieving provision which can apply where the transferor has inherited property from any person, *s 120(1)(b)* extends the relief in the case of transfers between husband and wife. By *s 120(1)(b)*, if the earlier owner of the property was the spouse of the inheritor/transferor, the transferor is deemed to have occupied the land for the purposes of agriculture for any period for which the spouse so occupied it. The inheritor/transferor is also deemed to have owned the land for any period for which the spouse owned it. There is a mirror provision in *s 108* which applies to qualifying business property.

Example 4.5: Hugh Bannon – period of occupation

Hugh Bannon occupied Top Farm for many years, living and working on the farm right up to the date of his death in October 2015. He bequeathed it to his brother Alex, a solicitor, who lived 75 miles away and Alex made immediate arrangements for the farm to be managed in his absence. Alex then took steps to sell his high street practice with the intention of retiring to Top Farm.

In fact the sale progressed slowly and it was 18 months before Alex could take up residence at Top Farm. Sadly, he died in a farming accident two months later in June 2017.

The conditions of *s 120(1)(a)* are satisfied. The occupation of the farm by Alec was for agricultural purposes and, though transient, it was his intention to live at the farm for the rest of his days, as indeed he did. Further, the farm was not occupied for any non-agricultural purpose at any time between the earlier death (Hugh) and the later transfer on death (Alex).

The position would have been otherwise had Alex immediately wound up the farming trade with a view, for example, to a sale of the land for some non-agricultural purpose. In that situation, he would not have satisfied the requirements of *s 120(1)(a)*.

Example 4.6: The Spencer family – occupation by widow

George Spencer had farmed in partnership with his sons Robert and Tom since April 2009 but continued to own the farm house and land personally. George died in November 2015 and under the terms of his will left the farm house and land on trust for his widow Katherine. Robert and Tom carried on farming but in April 2016 asked Katherine to join them in the partnership, which she did. At the date of Katherine's death in June 2017, the farm qualified for APR. By virtue of *s 120(1)(a)*, she had occupied the farm since George's death in 2015 and by *s 120(1)(b)*, she was deemed to have occupied the farm ever since George bought it. In her own right, she has occupied it as a partner for only 15 months, but that does not matter in view of the alternative qualifying periods.

Example 4.7: Fergus Shannon – constructed occupation

Fergus Shannon is a stockbroker based in Manchester. He inherited his father's farm in July 2010 and immediately let it to one of his father's neighbours who farmed it through a private company which shortly afterwards suffered financial decline. In March 2017 Fergus bought a controlling interest in the farming company but having no real interest in farming, a month later he transferred the farm to a discretionary trust accepting that this would be a chargeable transfer for IHT purposes. Although the farming profitability was poor at the time of the transfer the company was still operating the land.

Fergus cannot secure APR on the transfer to the trust by placing reliance on *s 117(a)*, not having set foot on the farm for many years, nor on *s 117(b)*, since his period of ownership from July 2010 is short of the seven years required by that section. However, by the deeming provision of *s 119(1)*, the occupation by the farming company which he controlled at the time of the transfer is regarded as being his own occupation. Examination of *s 120(1)(a)* shows that Fergus is deemed to have owned the farm from the date of his father's death in July 2010 and, although there was a gap during which he did not actually occupy it, ie July 2010 to March 2017 when he purchased the controlling interest, he is deemed to have occupied the farm from the date of his father's death. The farm has actually been occupied throughout, however unprofitably, by the company for agricultural purposes. It therefore seems that, in fact, APR will be allowable in respect of the transfer into the trust (for a similar example see IHTM24124).

Successive transfers

4.11 The rules in *s 121* as to successive transfers relate to a slightly different situation from those in *s 120*. By *s 121(1)*, relief is available where the conditions of *s 117* as to the minimum period of occupation or ownership would not otherwise be satisfied provided *all* of the following four conditions are complied with:

- The property concerned qualified for APR (or would have done had APR been available at the time of the transfer). This applies where the condition is satisfied as to the whole or part of the value transferred.
- The whole (or part) of the property qualifying for APR under the earlier transfer became, as a result of that transfer, the property of a person (or spouse) who transfers that property on a later transfer and is, at the time of that transfer, occupying the land for the purposes of agriculture. That occupation may either be in person or through the personal representatives of the earlier transferor.
- The property, or part of any replacement property, would qualify for APR on the later transfer, but for the 'minimum period' of *s 117*.
- One of the transfers was made on death.

Although the legislation does not specifically say so, it is understood that a life tenant who holds a qualifying interest in farmland can come within the second of these conditions by virtue of the deeming provisions: the property is treated as that of the life tenant for the purposes of satisfying *s 121(1)(b)*.

The position is not so clear where farmland is occupied by a company controlled by the second of the two transferors since this scenario is not specifically addressed within *s 121*. Plainly speaking it would be awkward if *s 121*, which is apparently intended to extend *s 117*, did not benefit from the kind of extension to the rule which has been examined already in relation to *s 119*. This could result in a situation where the treatment, for example, between English and Scottish partnerships might be different: a partner in an English partnership might satisfy the condition of *s 121(1)(b)*, whereas the partner in a Scottish partnership might not.

Section 121 'should' – at least in theory – also apply to replacement property, although there is no decided case on the point. However, *s 121(2)* provides that where property qualifies for relief under *sub-s (1)*, that relief is not to exceed what it would have been if the replacements had not been made. If there is replacement property, how is the occupation requirement in *s 121(1)(b)* satisfied? It seems that the occupation test should apply by reference to the replacement property. *Section 121(3)* restricts the relief where, under the first of the transfers to be considered, APR was available on only part of the value. There will be a proportionate restriction on the relief on the second transfer.

Example 4.8: David Bannister – limited relief

David Bannister had occupied Loam Pit Farm for three years when, in January 1992, he resolved to transfer it to his son Tom. He did not feel that he could afford to make an outright gift, so he sold it to Tom for £400,000 at a time when it was worth £600,000, thereby making a gift of £200,000. At that time, the APR rate was 50% thus resulting in relief of £100,000.

Tom died in July 1993 when the farm was worth £650,000. On his death, the basic condition in *s 117* could not be met but relief was available under *s 121(1)*, although restricted under *s 121(3)*, of £108,333 (£650,000 × £100,000/£600,000).

Replacement property

4.12 If shares or securities owned at the date of transfer replace other property qualifying for APR, ie a farm itself, or shares or securities of a farming company, then *s 123(3)* comes into play. It operates to ensure that the condition in *s 123(1)(b)* as to ownership which qualifies the transfer for relief under *s 116* is satisfied if the replacement property and the property which it replaces (directly or indirectly) were owned by the transferor for periods which, taken together, comprised:

- at least two years within the five years ending with the date of transfer, where the situation falls within *s 123(1)(a)(i)* and the agricultural property was occupied by the company for the purposes of agriculture throughout the two years ending with the date of transfer; or
- at least seven years within the least ten years up to the date of transfer, where the agricultural property was owned by the company throughout the period of seven years ending with the date of transfer and was throughout that period occupied either by the company or by someone else for the purposes of agriculture.

This relieving provision is itself subject to the restriction contained in *s 118*, which prevents relief from being extended by virtue of replacement: replacement does not increase the relief that would have been available had there been no replacement.

PROBLEMS ARISING OUT OF SUCCESSIVE *GLADSTONE V BOWERS* AND GRAZING AGREEMENTS

4.13 A mere casual 'letting' of land does not deny the qualification of the owner of the land to be regarded as the occupier as demonstrated in *IRC v Forsyth-Grant* (1943) 25 TC 369 and *Mitchell v IRC* (1943) 25 TC 380. However, a problem may arise where there is a significant interval between one casual letting and another prompting the question is the land 'occupied for the purposes of agriculture' in that interval? The answer will be in the positive where it can be shown that the casual letting was for the part of the year when the land could effectively be used for agricultural purposes, and the gap occurred during a time when the land could not well be so used – the existence of a gap will not on its own be grounds for showing that the land has not been occupied for the purposes of agriculture.

A problem arises where consecutive grazing agreements or consecutive *Gladstone v Bowers* agreements have been entered into over a period of time. If from an examination of the conduct of the parties over that time period, it was evident and understood that the agreement would be renewed, it may well be argued against the taxpayer that effectively the land is let. That in itself will not create too much of a problem provided it can be shown that the letting began on or after 1 September 1995 thereby providing access to 100% relief. Rather the difficulty will arise where the situation has existed since before that date, as it will become arguable that the relief should be restricted to 50%. This could lead to problems where the land in question falls within the circumstances of 'the fine house with a bit of land' and the existence of a long-term letting demonstrates that the occupier of the mansion does not occupy the land (see **Chapter 2**).

VALUATION OF AGRICULTURAL PROPERTY

General considerations

4.14 A full discussion of IHT valuation principles is outside the scope of this book and reference should be made to *Inheritance Tax 2017/18* (Bloomsbury Professional) but as a general rule the value on death is the price which the property might reasonably fetch if sold on the open market at that time. The price is not to be reduced on the ground that, if the whole of the property to be valued were placed on the market on the date of valuation, the price would be naturally depressed (*s 160*). An 'adversarial' valuation only provokes enquiry and a 'probate' valuation reflecting an unsupported market discount, will provoke not merely enquiry but the risk of imposition of penalties (see **Chapter 22** in relation to the valuation of chattels etc).

The general rule is varied both in relation to certain circumstances and in relation to particular forms of property. Among the circumstances which will concern the practitioner in connection with APR and BPR are the rules relating to transfers within seven years before death which require valuation based on the loss to the estate principle and those relating to the valuation of land sold at a lower price within four years after death.

Among the other situations where the general rule does not apply, the practitioner must also consider:

- any restrictions on freedom to dispose of the asset;
- the valuation of related property;
- the situation where a lease is treated as a settlement;
- the treatment of debts;
- farm cottages; and
- unquoted shares and securities.

The 'open market' is, to some extent, entirely hypothetical; rather the courts envisage a market with certain ideal attributes. Some of these are best replicated in an auction, although sales by private treaty or by tender may be the best available guides where there has been a sale of similar property at similar date. For example, the sale of small parcels of agricultural land by tender can throw up unpredictable and potentially extreme sale values, where farming neighbours compete with each other to secure land that will not come up for sale again within 50 years. The court assumes that there has been adequate publicity or advertisement before the sale and that the market envisaged is one in which the property is offered for sale to the world at large, so that all potential purchasers have an equal opportunity to make an offer as a result of it being widely known what is being offered for sale (consider the observations in the House of Lords in *Lynall v IRC* [1972] AC 680).

Certain restrictions, notably pre-emption clauses in the articles of association of a company, also have an impact in relation to IHT valuation. In *IRC v Crossman* [1937] AC 26, the House of Lords considered what assumptions are to be made in relation to the hypothetical sale:

- What is valued is the interest in the hands of the transferor at the relevant time.
- Restrictions that would prevent or hamper a sale are not taken to do so in considering what a purchaser would have paid for the asset.

- It is assumed that the purchaser would be subject to those restrictions after he had made his purchase.
- The price to be paid by the purchaser would take into account the fact that those restrictions would then apply to him.

A clear summary was given by Lord Guest in the later case of *Duke of Buccleuch v IRC* [1967] 1 AC 506:

> 'It is not necessary to assume an actual sale; a hypothetical market must be assumed for all the items of property at the time of death. The impossibility of putting the property on the market at the time of death or of actually realising the open market price is irrelevant. In other words, you do not have to assume that the property had actually to be sold; the assumption is that it is sold at the moment of death.'

Importantly, it need not be assumed that the sale has occurred; rather the question is, 'if it were sold today on the open market, what would the asset fetch?'.

When considering the valuation of a control holding in respect of a company, it is appropriate to consider the best price that a purchaser might get on liquidation of the company immediately after purchase. Such a buyer would take account of liabilities to tax within the company that would crystallise on liquidation and would allow for that in the price that he offered. A nominal value should never be used for unquoted shares when completing an HMRC IHT400/100 account. A reasoned value should be shown, noting if appropriate that it is an estimate and the basis of that estimate. As and when more reliable figures are available, they should be notified to HMRC, particularly where the adjustment increases the tax by £1,000 or more (see **Chapter 22**).

The case of *Tapp v Ryder* [2008] TMA/284/2008 and reaffirmed in *Price v HMRC* (see **4.23**) confirms that the expenses of sale may not be deducted in arriving at the IHT value on death. In *Tapp*, a house had been occupied by the deceased, Mr Atkinson, from 1974 until December 2005 when, aged 83, he went into hospital. The house was filthy and smelt; most of the night storage heaters did not function; the toilet did not work properly; and the wiring was defective. Mr Atkinson's niece thought that Social Services might not let her uncle move back there; further, not only was the property in its present dilapidated condition unlikely to sell for anything over £150,000 but it would also be difficult to sell. She therefore arranged essential repairs, the cost of £15,000 being funded partly by Mr Atkinson and partly by the niece and her husband. Mr Atkinson died in April 2006 and the niece, losing enthusiasm for the project, had the house valued in June 2006 with a view to sale. Even though much of the finishing was still to be done, it was now worth between £210,000 and £220,000 and the niece and her husband carried on renovating the property such that by October, the house was ready for sale. It was sold in November 2006 for £267,000.

For IHT purposes, the niece assumed straight line growth from £150,000 to the sale, and valued the house at April 2006 at £195,000 and also asked the Tribunal to take into account the costs of sale. This value was challenged by HMRC and based on comparable properties, they put the date of death value at between £230,000 and £250,000.

The tribunal considered that work was less advanced in April 2006 than the Valuation Office may have thought. The valuer saw the property only long after the work had been done but even so, the value reached by the Valuation Office had been heavily discounted. The Tribunal fixed the value at £230,000, allowing no discount for the costs of sale, relying both on *s 160* and on the principles set out in the 'Red Book'.

Tenancies: Baird's Executors v IRC

4.15 It is rare for an agricultural tenancy to be assignable. Nevertheless tenancies established before 1 September 1995 are regarded as having a value, even though the tenant cannot sell because of the security of tenure provisions which apply. Importantly, the fact that the tenancy cannot be assigned does not prevent it from having a taxable value. Applying the principle in *Crossman*, the tenancy is valued on the assumption that it is capable of being sold notwithstanding the restriction. The assumed new tenant will be subject to the same restriction.

This very point was considered in *Baird's Executors v IRC* 1991 SLT (Lands Tribunal) 9, where George Baird had relinquished his 1921 tenancy of an agricultural holding in favour of his daughter-in-law and grandson. The transfer of it took place in 1977 with the consent of the landlord and Mr Baird died in 1985. HMRC determined that the transfer of the lease was a chargeable transfer for capital transfer tax (CTT) under *FA 1975, s 20(2)*.

The Lands Tribunal in Scotland determined that the value of the interest assigned is the price which the interest might reasonably be expected to fetch if sold on the open market at the date of the transfer. The tribunal disregarded the fact that the tenancy could not be assigned, but assumed that the assignees would be subject to the same restrictions. The District Valuer had taken note of four factors:

(1) the price that a landlord might pay to a tenant to recover vacant possession;

(2) the price that a prospective tenant might accept from a prospective landlord in a sale and leaseback transaction;

(3) the price paid to a tenant where there was compulsory acquisition of part of an agricultural holding; and

(4) the rents commanded by equivalent holdings when let in the open market on terms similar to those of the lease in question.

Having come to a figure of 25% of the vacant possession value, the District Valuer did indicate that the value could be as high as 50% of vacant possession value, but for the uncertainty in the hypothesis he was applying. That approach was largely accepted by the tribunal.

As a result, the right of a tenant in an agricultural lease is an asset which can have a value but not necessarily 25% of the vacant possession value of the land itself. However, under *IHTA 1984, s 177*, which applies to Scottish agricultural leases such as that considered in *Baird*, it is provided (see *s 177(1)*) that, where any part of the value of a person's estate immediately before his death is attributable to the interest of a tenant in the unexpired portion of a lease for a fixed term of agricultural property in Scotland, any prospect that that lease might be renewed by 'tacit relocation' is to be left out of account in determining the value of the lease.

Further, where any part of the value of a person's estate immediately before his death is attributable to the interest of a tenant of agricultural property in Scotland which is held by virtue of tacit relocation and which is, on the death, acquired by a new tenant, in general the estate is valued without including the value of that tenancy. That is provided by *s 177(2)*, except that neither *sub-s (i)* nor *(ii)* is to apply unless the deceased had been tenant of the property in question for at least two years continuously preceding the death or had become a tenant by succession. The part of the value to be left out of account does not include compensation to a tenant for improvements. It would seem that the upshot of this is that there is no value transferred where there is no new tenant by succession.

Section 177 applies on the death of individuals. If the tenancy is held by a partnership which continues or by a company, *s 177* will not apply. In those circumstances, it will become necessary to value the lease.

The situation becomes more complicated where the same person has an interest both in the tenancy and in the freehold, as will be seen in the next section.

Gray (surviving executor of Lady Fox) v IRC

4.16 This was a decision by the Lands Tribunal which was subsequently considered by the Court of Appeal and reported at [1994] STC 360 – it still remains good law. Lady Fox owned a substantial agricultural estate farmed by a partnership in which she was a partner and that partnership had an agricultural tenancy. Lady Fox was entitled to 92.5% of the partnership profits. On her death on 27 March 1981, HMRC assessed the then capital transfer tax (CTT) on the basis that both the land and the share in the partnership could be valued together, with the vacant possession value being discounted in a number of ways. That produced a much higher figure than merely taking the investment value (if the partnership share was disregarded). Although the Lands Tribunal disagreed with HMRC on this point, the Court of Appeal reversed the tribunal on the question of principle. The tribunal had already considered what should be the method of valuation if it was permitted to lot the two together, the method which was approved by the Court of Appeal.

The vacant possession value of the land was agreed at £6.125 million and from that figure, the Cambridge District Valuer deducted his estimate of what the partners (other than Lady Fox) would accept to sell their share of the tenancy. He reckoned that figure was £100,000, based on valuing the tenancy as a whole at nearly 25% of the vacant possession value. He took 45% of the difference between vacant possession value and tenanted value (which had been agreed at £2.751 million based on comparables that were available to him. Having arrived at a value, he discounted it by 7.5% of the vacant possession value to compensate for the risk and delay in obtaining vacant possession of the whole after a purchase of the freehold. This gave him a figure of £5.565 million for the two interests taken together, namely the freehold and the partnership share.

In considering the evidence, the tribunal felt that there was no guarantee that the minority partners would accept £100,000 nor that they would come to the table quickly. It seemed to them that 7.5% discount was not enough, but the valuer for Lady Fox gave no evidence on that particular point, so the tribunal had no better evidence than that of the District Valuer.

Each valuation is likely to turn on its own facts and on another occasion, there might be a strong argument for a greater discount than that applied in the *Lady Fox* case. Further, where the agricultural tenancy is held by trustees or by executors in their own right and not merely because they are trustees or executors, and yet they hold the freehold in that fiduciary capacity, there is no obligation to dispose of the tenancy with the freehold so as to get a better price for the trust. In these circumstances, only a tenanted valuation would apply.

The effect of ESC F17 (Relief for Agricultural Property) was noted earlier, in considering the rates at which APR is allowed. On a transfer of tenanted agricultural land, 100% relief will be allowed where the interest of the transferor in the property is, notwithstanding the terms of the tenancy, valued at an amount broadly equivalent to the vacant possession value of the property. That concession takes some of the sting out of the *Lady Fox* case.

Alexander v IRC

4.17 In *Alexander v IRC* [1991] STC 112, the deceased owned the long lease of a flat in the Barbican which had been bought under the 'right to buy' legislation. The price, net of discount, was £24,600 but under the terms of the lease, the discount was repayable on a sliding scale if the property should be disposed of within five years of the purchase. The deceased died in January 1984, one year after the purchase. Had the flat been sold at the date of death, all of the discount would have been repayable. The Lands Tribunal ignored the discount and valued the flat at £63,000.

The Court of Appeal held that the charge in respect of the discount should be taken into account. It was an encumbrance on the property within what is now *IHTA 1984, s 164(4)*. The court referred the matter back to the Lands Tribunal to fix the amount that a purchaser would pay, even taking into account the discount to be repaid. As a result, the deduction from the full value of £63,000 was less than the full discount of £24,600 claimed by the taxpayer.

Lotting for sale: 'natural units'

4.18 It is an established principle of valuation that the seller would be considered to have marketed the property in the way that would produce the best price which where appropriate would involve dividing it into lots or assembling disparate items to sell them together. This approach may be particularly relevant (see **4.25**) in relation to farmhouses.

In *Earl of Ellesmere v IRC* [1918] 2 KB 735, the estate was sold as a single entity after wide advertisement. The purchaser kept part of the land but sold of the estate remainder making a profit (after allowing for the retained portion) of 19% against his purchase price. HMRC noting this premium assessed the estate value for estate duty purposes at more than the original sale price achieved by the executors. The court partially agreed with this approach but reduced the HMRC figure to one which represented a smaller 11% premium.

In *Duke of Buccleuch v IRC* [1967] 1 AC 506, the trustees did not dispute the valuation basis put forward by HMRC which divided an estate into 532 separate lots and yielded a figure for the entirety of £868,129. The trustees did, however, argue that, after taking out the readily saleable lots, amounting to about 9% of the total value, the remainder should be valued at the price that would be paid by an individual purchaser who would expect a profit (of 20% it was argued) for taking on the whole estate.

The House of Lords accepted that:

> 'We must take the estate as it was when the deceased died; often the price which a piece of property would fetch would be considerably enhanced by small expense in minor repair or cleaning which would make the property more attractive to the eye of the buyer. Admittedly that cannot be supposed to have been done: and I can see no more justification for requiring supposition that natural units have been divided. This subsection applies to all kinds of property. A library was instanced by Wynn LJ. Generally there would be little difficulty, delay or expense in getting someone knowledgeable to pick out valuable books for separate valuation, and I would therefore regard such books as natural units. Suppose, however, that the deceased had bought a miscellaneous and mixed lot of surplus stores intending to sort out and arrange them in saleable lots. That might involve a great deal of work, time and expense, and I see no justification for requiring the supposition that that had been done and then valuing the saleable lots that would have emerged.'

The general principle that emerges is not that there must be lotting into 'natural units', but that the property should be considered to be marketed in such a way as to produce the largest price, provided that that did not involve undue expenditure of time or effort. This principle is particularly relevant to the farmhouse and garden and practitioners should resist any attempt by the District Valuer to regard these assets as a separate valuation unit isolated from the farm as a whole, even though IHTM09715 urges the Inspector to apply whatever lotting will yield the highest taxable value. The correct lotting must be a matter of fact in each case. A situation could arise where a unit which is uneconomic on its own is particularly valuable to two neighbouring owners, such that it may be natural to lot the farmhouse and garden separately from the land but this must not be assumed to be a natural lotting. There is more specific consideration of the valuation of farmhouses at **4.25**.

Returning to the case of *Gray (surviving executor of Lady Fox) v IRC* [1994] STC 360, the Lands Tribunal considered that the land and the partnership interest taken together must be a 'natural unit', as that expression was used in *Buccleuch*, if they are to be valued as if sold together. The tribunal did not think that the land and the partnership interest did together constitute a natural unit, but the court indicated that the hypothetical vendor must be regarded as having done whatever was reasonably necessary to get the best price, to include taking separate properties and selling them together, as had been done in the current case with the freehold and the partnership share. Certain technical objections had been raised by the tribunal, in that it had exclusive jurisdiction over questions as to the valuation of land but not of other kinds of property. The court held that the tribunal could consider the value of assets which were not land as part of the means by which it arrived at the value of the land itself.

Other aspects of valuation

4.19 The established valuation process for multiple estate components is for the Valuation Office to negotiate with the taxpayer, either directly or, more rarely, through HMRC. Where no agreement can be reached, the Board may issue a notice of determination which the taxpayer can appeal, as appropriate, to the First-tier Tribunal. The tribunal, as has just been noted, may determine the value of the land, taking into account not only the value of assets other than land, where relevant, but also charges relating to contingent liabilities. Although there is no exact parallel to the CG34 procedure for CGT, the District Valuer can sometimes informally agree the value of land before delivery of the IHT account.

It is not routine HMRC practice to reopen a valuation once it has been agreed, provided the basis of that agreement is founded on full and accurate information supplied by the accountable persons, and where HMRC have either confirmed that agreement and made an assessment of tax on the basis of that value, or issued a letter of clearance. This is so, even though the property may later be sold for a higher price. On the other hand, if the property is later sold for less than the agreed valuation, the taxpayer may have two lines of argument upon which to trigger a revision of the valuation. The first might be that he could show the original valuation was based on inaccurate information and the second, considered at **4.28**, are the provisions of *IHTA 1984, s 190* which may apply where the sale takes place at a loss over probate valuation within four years of death.

Let land is usually valued as a multiple of the net annual value, calculated by deducting from the gross rent the annual outgoings such as repairs, insurance, ground rent etc but not the expense of management. If the property is vacant, it is usual to compare similar properties to arrive at a value but this is not easy where there are no comparables or where the situation of the property is in some respects unique. Where there is no formal tenancy, it is necessary to establish whether there is a reasonable prospect that vacant possession will be obtained in the not too distant future. It has already been noted that this is an aspect of valuation where land is let to a company which is controlled by the transferor.

Planning permission

4.20 The existence or prospect of obtaining planning permission cannot be ignored but must be taken into account in arriving at the valuation.

Prosser v IRC [2001] RVR 170 illustrates the principles to be applied where there is hope value (prospect) but no formal planning consent (existence). The deceased owned a house with a garden big enough for an extra building plot but there had been no application for planning permission by the date of death, although the executors did later get planning permission on the land. The District Valuer proposed a date of death figure of 80% of the value that the plot would have commanded with the benefit of such consent. The tribunal considered that whilst there was at least a 50% chance of obtaining planning permission in the absence of such permission a speculator would offer no more than 25% the development value – this reduced the probate value of the affected land to £12,500. The tribunal observed:

> 'if a prospective purchaser was seeking a plot on which to construct a house for his own occupation, he would probably prefer to await the outcome of a planning application and, if appropriate, pay the full value rather than pay half of that value and run the risk of losing his entire investment. Similarly, a speculator would in my view not be interested in a purchase unless the potential profit resulting from a successful planning application were significantly greater than the potential loss if permission were not forthcoming.'

Farm cottages are valued ignoring any additional market premium that might apply if they were not occupied for agricultural purposes and this will be particularly relevant where a tied cottage is occupied by an agricultural worker.

The agricultural value of agricultural property is the price it would fetch if it were sold subject to a perpetual covenant confining its use to agriculture (or equivalent restriction where not situated in the UK). A freehold interest which is subject to an 'old' agricultural tenancy is clearly worth much less than vacant land, even though the tenant might well be a special purchaser. The discount to vacant possession value varies widely across the country. The profitability of farming must be a factor.

Willett v IRC [1984] RVR 163, 264 EG 257, concerned the May 1977 value of a 48 acre arable farm in Staffordshire where the issue was the value of the freehold reversion. The taxpayer's figure was £28,000 and that of the District Valuer £45,000 but the tribunal fixed it at £39,000, slightly above the midway point between the two parties. The 60-ish year old tenant had refused to surrender his tenancy at the time of the May 1977 valuation but did so barely 18 months later and the land eventually sold for over £126,000. As at the valuation date, it was considered unlikely that the family of the tenant could claim succession, so it fell to the tribunal to consider how soon the tenancy would fall in. The tenant might be likely to retire in eight years, and might live nearly four years after that.

The vacant possession value was fixed by the tribunal at £66,000. It then calculated the value of the freehold reversion by adding together a capitalisation of the existing rent, a figure for an increase in rent for the remainder of the term until possession might be acquired and the value for the reversion discounted because possession could not be obtained immediately to produce a total of a little under £36,000. The tribunal considered that there had always been a possibility that the tenancy might fall in before the natural date of retirement or death of the tenant (as was indeed the case) and that the value of the reversion should be enhanced to allow for that by 10%. The resulting figure of £39,000 was 59% of the vacant possession value, as fixed by the tribunal.

Joint property: *Arkwright v IRC*

4.21 It is an established principle that the half share of a freehold property jointly owned by husband and wife/civil partners must be equal to 50% of the value of the entirety but that principle was robustly challenged in two cases: *Arkwright v Another (PRs of Williams deceased) v IRC* [2004] STI Issue 3 and *Price (Executor of Price, Deceased) v HMRC* [2010] UKFTT 474.

In *Arkwright v IRC*, Ashland Farm had been owned by Mr and Mrs Williams as tenants in common in equal shares. At the date of the death of Mr Williams, aged 83, Mrs Williams was aged 79 and in good health. Her husband had left her a life interest in his half share of Ashland Farm but some months after the death, Mrs Williams and her daughters entered into a deed of variation that gave that half share to the daughters. The value of that half share immediately before Mr Williams' death was the price that his interest might reasonably be expected to fetch if sold in the open market at that time. The parties agreed that, immediately before death, the house was worth £550,000 and HMRC considered that the value of Mr Williams' share was £275,000. *IHTA 1984, s 171* provides that changes in the value of the estate of the taxpayer that occur by reason of the death and which fall within *s 171(2)* are to be taken into account as if they occurred before the death. *Section 171(2)* provides that a change falls within *s 171(1)* if it is an addition to the property comprised in the estate, or is an increase or decrease in the value of that property, unless the decrease arises from an alteration in the rights of shares within *s 98(1)*. However, *s 171(2)* specifically excludes from the operation of *s 171(1)* the termination on the death of any interest or the passing of any interest by survivorship.

As a separate argument, HMRC applied the related property provisions of *s 161*, the general effect of which is to deny the personal representatives any discount in respect of a share in joint property where the other share is owned by the spouse. The executors took a different view. Immediately before the death of Mr Williams, a purchaser of his share of the property would have taken account of the fact that Mrs Williams owned the other half of the property and was in good health and had a right to occupy it. A purchaser would have recognised that he could not sell the property without the consent of Mrs Williams, and that would devalue the interest of Mr Williams. HMRC regarded these arguments as being essentially arguments of value that should be heard before the Lands Tribunal, rather than fall to be determined by statutory interpretation.

Dr Brice, as Special Commissioner, accepted the argument of the taxpayers that a purchaser would not pay full value for the share of Mr Williams but would discount it. Under the normal rules of valuation, the values of the shares owned respectively by Mr and Mrs Williams were not identical. Dr Brice then reviewed *s 171*. The proviso to *s 171(2)* had the effect that *s 171(1)* did not apply to the termination on death of any interest or the passing of any interest on survivorship but, in the present case, the interest of Mr Williams did not terminate on his death. It passed to his daughters. It passed to them not by survivorship but by the will, as varied. The value of his interest in the house would have changed by reason of his death, without regard to his actual death, because it would then have been a certainty and not merely a possibility that he would die before his wife. There was a decrease in the value of his interest, and his interest did fall within *s 171(2)*, so *s 171(1)* did not apply. As a result, the decrease in the value of his share that occurred by reason of his death should be taken into account in valuing his interest.

Next, Dr Brice considered the related property provisions of *s 161*. The scheme of *s 161* is, first, to fix the value of the entirety; and then to value each share in the property to establish a ratio. That ratio is then applied to the aggregate and, if the value of the share of the deceased is greater than the value of his separate share, the higher figure is taken. *Section 161(4)* provides a proportionate rule in relation to 'shares of any class', and applies similarly to stock debentures and units of any other description of property. HMRC argued that the shares of Mr and Mrs Williams were such units and that it followed that the ratio for the purposes of *s 161(3)* was one half. The personal representatives disagreed. Undivided shares in land were not 'units of property'. Dr Brice agreed with them, with the result that the value of the share of Mr Williams was not necessarily one half.

Dr Brice then considered the application by HMRC in respect of jurisdiction. Was this an appeal as to the value of land that should have gone to the Lands Tribunal? Where an appeal had been brought to the Special Commissioners and included fixing the value of land, the correct procedure (see *s 222(4)–(4B)*) was not to strike out the appeal, as HMRC would have liked, but for the Special Commissioners to refer the matter to the Lands Tribunal. This arose from an amendment to *s 222* by *FA 1993* to correct a problem that had occurred in the past, in which an appeal must be made either to the Commissioners or to the Lands Tribunal. Under the new rules, it was no longer necessary, if the appellant picked the wrong tribunal, to start again with the other one. The Special Commissioners could accept the appeal, referring valuation questions to the Lands Tribunal, but all issues other than questions as to the value of land were properly to be determined by the Special Commissioners. HMRC appealed to the High Court, arguing that the Commissioner had exceeded her jurisdiction and that the case should be referred to the Lands Tribunal. Gloster J in part agreed with HMRC. The Special Commissioner was entitled to conclude that the value of the deceased's interest in the property was not inevitably a mathematical one half because, in the circumstances, *s 161(4)* did not apply. Where the Special Commissioner had gone too far was in determining that, as a matter of fact, the value of the interest was less than 50%. That was a matter for the Lands Tribunal.

Price (Executor of Price, deceased) v HMRC

4.22 In the second case of *Price (Executor of Price, Deceased) v HMRC* [2010] UKFTT 474 the facts were far more straightforward. Mrs Price died in 1999, survived by her husband and four children, leaving her interest in the London family home, owned as tenants in common in equal shares with her husband, to her children. The whole house was worth £1.5 million but subject to a mortgage of £364,164. The will was later varied to leave the share in the house to the children, subject to the mortgage and to a charge of £260,000 in favour of residue, which passed to the husband. It was agreed that the deceased's half share in the house valued on its own would be worth £637,500 but there was no evidence that either the deceased or her husband contemplated a sale of the house.

Essentially, the case turned on the meaning in *s 161(1)* of the words 'the value of that and any related property'. The taxpayer argued that the two shares in the house should be aggregated, with each being valued in isolation leading to a 15% discount, but HMRC said that one should value the whole as a single entity. HMRC first argued that this was a matter for the Lands Chamber of the First-tier Tribunal the successor to, the Lands Tribunal referred to above, rather than the Tax Chamber, but that conflicted slightly with the statement of case. To interpret *s 161(1)*, one should, he said:

- fix 'the value of any property comprised in a person's estate'; then
- fix 'the value of the aggregate of that and any related property'; then
- fix the 'appropriate portion' of the latter value, using *s 161(3)*, ie the proportion which the value of the property in the estate bears to the sum of that value and the value of related property, but each being valued separately (in this summary, the 'last' value); and
- compare the values, taking the last value if higher than the first.

The shares must, HMRC argued, be aggregated before valuation, not after. As to the costs of sale, they referred to *Tapp v Ryder* [2008] RVR 340, where the same advocate had persuaded the Lands Tribunal to disregard such items (see **4.15**). Deduction of one half of the mortgage debt should be a separate step: the taxpayer's approach involved double counting.

The first part of the tribunal's decision concerned jurisdiction in the light of HMRC's review procedure and the reorganisation of tribunals, and concluded that it did have power to decide the case.

Without feeling bound to accept the decision in *Arkwright*, and taking note of the comments of Hoffmann LJ in the Court of Appeal in *IRC v Gray (Lady Fox)*, the tribunal decided that it was right to value two assets held in the same estate on the basis that they would achieve a greater price if offered together, and applied that reasoning to *s 161*. Where *s 161* applies, the two items of property are to be valued as if offered for sale together and at the same time. There should be no deduction for expenses of sale. Debts are to be deducted separately. The detailed values, if not agreed, should be referred to the Lands Chamber.

THE VALUATION OF FARMHOUSES IN THE LIGHT OF *ANTROBUS NO 2*

4.23 Valuation of the farmhouse can be problematic. In the same manner as the principle established in *Hardcastle (executors of Vernede deceased) v IRC* [2000] STC (SCD) 532, which applies a separate commercial basis and a 'relief' basis of valuation of the interest in a business for BPR purposes, so too the valuation of a farmhouse for general probate and IHT purposes may well be substantially different from its valuation for the purposes of APR. It is quite common for the District Valuer to argue that the agricultural value of a farmhouse represents no more than 70% of the value of the house on the open market.

The issues set out at **4.18**, relating to 'natural units' of valuation are again relevant here. Consider the scenario of a farmhouse, which although occupied as such with a small acreage, does not in truth form part of a particularly viable agricultural unit but happens to adjoin a number of large and extremely viable farming units. In such a case, it might well be argued that almost any of the adjoining owners would be keen to add to their acreage but would not want the farmhouse which, like so many, is thus surplus to requirements. Such a situation might lead to lotting the house with, say, a pony paddock or other non-farming element, and to offer the land separately.

There is good argument to support the idea that the value for APR should be as high as the open market value, on the grounds that in certain parts of the UK there is a ready market for a house with a sufficient land to allow the owner both privacy and protection of attractive viewpoints. For many years, the law did not provide any administratively convenient rule of thumb and different percentages applied for income tax and for VAT. However, the Lands Tribunal did carefully consider the matter of agricultural value of a farmhouse in *Lloyds TSB Private Banking plc (Personal representative of Rosemary Antrobus Deceased) v Peter Twiddy (HMRC Capital Taxes)* [2006] 1 EGLR 157, better known as *Antrobus No 2*. This was the sequel to the case discussed at **2.13**, where it was recognised by the Special Commissioner that the 'grand' house in question, Cookhill Priory, was a farmhouse. This is an important case which remains influential in current times and for that reason it is worth setting out the facts in some detail.

This reference to the Lands Tribunal was on the issue of value, though the tribunal did go on to consider wider issues. The open market value of the farmhouse and gardens was £608,475 and HMRC argued that of this amount, agricultural relief should apply to only £425,932.50, ie 70%. The tribunal set the agricultural value at this figure, but further decided that, if 'lifestyle' purchasers were included in the notional market, the right figure would be £517,203 (85%) of the open market value.

The taxpayer argued that the market for a farmhouse or farm that qualified for APR should include buyers who might be described as 'lifestyle' farmers with a career in other pursuits. If 'lifestyle' farmers were excluded, the value of a house might be a great deal less because the other buyers would be farmers who were more focussed on the overall return on capital. There was evidence in this particular case of the people who might have been interested in buying the property – one offered £935,000, intending to refurbish the house and sell on for a profit, another offered £1 million and there was also a former chartered surveyor who was a specialist farmer, who wished to purchase the estate as a whole. The latter gave evidence that he would still have offered for the farmhouse provided that he was free both to farm the land and to carry on his business elsewhere; but that he would not have been interested if the property had been subject to a strict agricultural occupancy condition.

There was much expert evidence – Savills argued that there was no other non-agricultural use of the property that would be in conflict with a perpetual covenant prohibiting its use otherwise than as agricultural property, so there should be no discount from market value. A covenant for agricultural use would have no adverse effect on the value, because the lifestyle buyer wanted privacy such as would be available where a farmhouse was surrounded by land. Such a buyer would actually welcome a restriction on use. However, it was clear from Savill's evidence that 'lifestyle' farmers bid up the market, and that property which would eventually benefit from agricultural relief 'must top the lifestyle farm market, given the potential for 40% tax saving'.

HMRC Valuation Office argued for a discount of 30% on the basis that not only could the agricultural covenant not be lifted but that the effect of the law was to require 'the hypothetical purchaser to continue to occupy the house as a farmhouse, together with sufficient agricultural land or pasture to make the house of a character appropriate to the land'. He concluded that, for the purposes of discount, it was necessary to look at 'a working farmer, mainly engaged in agricultural operations on the land' but a covenant in compliance with *s 115(3)* would have the effect of excluding lifestyle farmers from the market: 'Restricting demand to full-time working farmers would result in the price realised for the residence being significantly less than it would otherwise have been.'

The Valuation Office looked at the profit that could be earned from farming, the land that a non-farming purchaser would have to acquire and the cost of that acquisition concluding that the return available from a joint farming venture would be extremely modest in relation to the investment involved. It produced evidence of discounts between 25% and 40%, for properties subject to planning tie citing specific examples of discounts at 32.4% and 23.6%. The argument ran that the agricultural value should exclude the overbid for the farmhouse, with a small area of land adjoining it, made by non-farming residential purchasers where the amount of that excess would be dependent on location, setting, views etc.

The taxpayer's argument considered that a lifestyle farmer would satisfy the requirements of *s 115(3)* as there was no need for a farmhouse to be occupied by someone who was on the premises during the working day nor was there a requirement for the occupier to be the person who was managing the farm on a day-to-day basis. Farming could be carried on in different ways, including short-term grazing lets, share farming, contract farming and in effect the farmer could do the paper work but leave others to do the 'hands-on' work.

The HMRC argument considered that the dwelling in question must be occupied by the person who was managing the farm on a day-to-day basis. Drawing an example from earlier legislation, HMRC highlighted the distinction between a working farmer, who lived in a farmhouse, and a landed proprietor, who lived in a mansion house – although the occupier of the mansion house might still technically carry on a farming business that did not mean that his dwelling house was a farmhouse.

The tribunal, in leaning towards the evidence of the Valuation Office and the HMRC arguments, held that:

> 'a farmhouse is the chief dwelling house attached to a farm, the house in which the farmer of the land lived. There is, we think, no dispute about the definition when it is expressed in this way. The question is: who is the farmer of the land for the purpose of the definition of section 115(2)? In our view it is the person who lives in the farmhouse in order to farm the land comprised in the farm and who farms the land on a day to day basis … We do not think that a house occupied with a farm is a farmhouse simply because the person living there is in overall control of the agricultural business conducted on the land; and in particular we think that the lifestyle farmer, the person whose bid for the land is treated by the appellant as establishing the agricultural value of the land, is not the farmer for the purposes of the provisions.'

The tribunal concluded that, 'the concept of the farmhouse is as the dwelling of a working farmer who requires a suitable house to support his working life'. The tribunal carefully considered the old legislation for 'working farmers' and whilst the taxpayer argument that the working farmer conditions had long since been removed when the law was changed, the tribunal did not feel that that in itself altered the meaning of 'farmhouse' or what amounted to 'agricultural property'. The tribunal also took the time to review the meaning of 'cottages' and buildings on a farm used only to a minor extent for agricultural purposes as well as the limitations commonly applied in planning restrictions, namely to restrict the occupation of the dwelling to a person solely or mainly working, or last working, in the locality in agriculture or in forestry, or the widow or widower or dependants of such a person. Taking these factors together, the tribunal concluded that the HMRC approach was right, even conservative.

The tribunal also reviewed the cases of *Lindsay v CIR*, *CIR v John M Whiteford & Son*, *Starke v IRC*, and *IRC v Korner* coming to the conclusion that the purpose of the legislation 'is to accord relief to the operational components of agricultural units'. On that basis, the agricultural value of a farmhouse fell to be determined on the assumption that the perpetual covenant would have prevented its use in any other way and would therefore exclude the 'lifestyle' purchaser.

The tribunal therefore accepted the argument of the District Valuer and fixed the agricultural value of Cookhill Priory at £425,932. Having done that, the tribunal then considered what the value might be if their interpretation of the legal position was incorrect, and if demand from lifestyle farmers might be taken into account when calculating the agricultural value. Based on the evidence, the tribunal considered that the lifestyle purchaser would still offer less for a property subject to a restrictive covenant than one that was not so subject. On that basis, they concluded that, if lifestyle purchasers were in the market, the agricultural value would be £517,203.

PLANNING EXERCISES RELATING TO TENANCY VALUATION

4.24 Prior to enactment of the *Agricultural Tenancies Act 1995* a landowner and a farmer might make arrangements which were designed to ensure that the farmer had no security of tenure. One scheme involved the grant of a tenancy to an intermediary, who later granted a sub-tenancy to the farmer and the intermediary, on receiving notice to quit from the landowner, would fail to serve a counter-notice, thus losing his security of tenure. Under the common law, the sub-tenancy would fall with the tenancy.

This scheme was reviewed by the House of Lords in *Barrett v Morgan* [2000] 2 AC 264. Lord Millett there also considered the situation where the intermediary surrendered the tenancy to the landlord. He considered that there would be no automatic termination of the sub-tenancy, because surrender of the end tenancy is not the unilateral act of the landlord but occurs by agreement between landlord and tenant thus the intermediary could not by this means prejudice rights that had been created in favour of the farmer sub-tenant.

However, do these arrangements amount to collusion between the freeholder and the intermediary? This was the question considered in the Court of Appeal in *Barrett v Morgan* which concluded that any agreement by the intermediary not to serve a counter-notice is void and as a result, in pre-1995 Act cases, there must always be a risk that the intermediary will fail to serve a counter-notice, following service on him of a notice to quit by the freeholder.

For as long as that situation applies, there must be a risk that the basis of valuation for APR is the lower 50% rate on the ground that the freeholder is not entitled to vacant possession within 12 months, as required by *s 116(2)(a)*. Upgrade to 100% relief in such circumstances would only be available by reliance on ESC F17 whereby (subject to the general point that a Concession cannot be relied upon where tax avoidance is in issue) F17 extends the period, within which the landowner must obtain vacant possession, from 12 months to 24 months.

Against this background, if it is essential for planning purposes to ensure that APR is available at 100%, it is not safe to rely upon an old 'intermediary'-style structure. Rather it is preferable to bring the existing sub-tenancy to an end and create a new FBT direct to the farmer thus bypassing the intermediary. In the light of the decision in *Barrett v Morgan*, that the sub-tenant does not have security, it may now be easier than hitherto to persuade the farmer to co-operate. He must now accept that he no longer has the security that he may have thought he had. He should, therefore, be more ready to accept a new farm business tenancy, giving the landowner APR at 100%.

SALES OF QUOTED SHARES WITHIN 12 MONTHS OF DEATH

4.25 Under *IHTA 1984, s 179* a claim may be made by 'the appropriate person' for relief so as to reduce the value of the estate for IHT by the amount of the loss suffered on selling quoted shares (but excluding those which attract APR or BPR) within 12 months of death. Where the loss arises because the securities are sold, it is the date of the contract which determines the availability of the relief and not date of receipt of the net proceeds. The rule relates to 'qualifying investments', defined by *s 178* as quoted shares or securities or holdings in unit trusts which, as is explained in the chapters on BPR, includes shares listed on the Alternative Investment Market (AIM) and ISDX.

The difficulty of a claim lies in the fact that, if there is more than one sale of a qualifying investment, all the sales must be aggregated for the purposes of the relief: the taxpayer can 'cherry pick' his losses only by not selling, within the relevant period, those securities which stand at a gain. He may, however, 'cherry pick' in a different way. He may, as executor, transfer those securities which stand at a gain to the beneficiaries *in specie*, such that the gains are realised by the beneficiaries personally. He can then let the losses arise within the estate, where they may form the basis of a claim for relief.

The relief relates only to sales effected by the 'appropriate person', who is the person liable for the tax. Where there is more than one person who falls within the definition in *s 178(1)*, the relief is available to the person who actually pays the tax. The relief is also restricted where new shares or securities are purchased by the appropriate person during a period which starts with the date of death and ends not within the 12-month period but within two months after the last sale in respect of which a claim is made: see *s 180(1)*. That is unlikely to happen where APR is claimed on shares in a farming company. There are special rules relating to capital receipts, payment of calls and changes or exchanges relating to holdings.

LAND SOLD WITHIN FOUR YEARS AFTER DEATH

4.26 Whilst the rule in *IHTA 1984, s 179* will seldom apply to property qualifying for APR or BPR, in contrast the mirror relief in *s 191* may often be relevant. Where an interest in land forming part of the deceased's estate is sold by 'the appropriate person' within the period of four years immediately following the date of death and that person makes a claim (stating the capacity in which he makes it), the value of the interest in land, and of any other interest in land comprised in the estate which is also sold within that period by the person making the claim, is the sale value – sales made at a gain in the fourth year are ignored for this purpose. That general rule in *s 191(1)* is qualified by *de minimis* provisions in *s 191(2)*. There is no relief where the fall in value is the lower of less than £1,000 or less than 5% of the probate value. Similarly, the relief does not apply (see *s 191(3)*) on the sale by a personal representative or a trustee to:

- a beneficiary; or
- the spouse, child or remoter descendant of the beneficiary; or
- trustees of a settlement under which a beneficiary (or spouse, child or remoter descendant) has an interest in possession in the property sold; or
- a sale under or by virtue of which either the vendor or any of the persons mentioned in the categories above obtains a right to acquire the interest in the land or any other interest in the same land.

These exclusions are absolute and there is no saving provision, even where the land in question is sold at market value. It will be noted that the restrictions apply neither to a sale to a parent of the beneficiary nor to trustees of a discretionary settlement. The 'appropriate person' defined in *s 190(1)* means, as in relation to the relief for shares, the person liable for the tax or, where there is more than one such person, the person who actually pays it.

An interesting attempt to exploit the relief was examined in *Jones (Balls' Administrators) v IRC* [1997] STC 358. In that case, the deceased died on 26 June 1988 owning a farm. Letters of administration were granted to his son in December 1988 and in June 1991, the son contracted to sell the property for £300,150. Completion was to take place a month later, but did not in fact take place and in August 1991, the son forfeited the deposit but died weeks later. In December 1991, letters of administration were granted and a contract for sale of the farm was exchanged in January 1992 and completed in February 1992.

In May 1995, the probate value of the farm was agreed in the sum of £447,000, and the executors made a claim under *s 191* on the ground that there had been a sale in 1991. The claim was rejected on grounds that for the purposes of the relief, a sale did not mean the exchange of contracts which did not proceed to completion. That would be inconsistent with the wording of *s 191(3)(b)*, which refers to the reduction of the value of an interest 'for any other reason'. The whole tenor of the code in *ss 191–198* pointed to a completed transaction. The draftsman clearly understood the distinction between a completed sale and a mere contract. The relief was refused.

Example 4.9: Susan Furniss – sale of estate land

Susan Furniss died on 20 December 2015 leaving an estate which included three parcels of land:

- Low Cottage was valued for probate at £75,000 and sold in April 2016 for £74,001 realising a loss of £999.
- Top Pightle, valued for probate at £110,000, sold for £10,000 more than that in June 2016.
- Broadacres was valued for probate at £230,000 but proved difficult to sell, realising only £218,000 in September 2017, a loss of £12,000.

The loss of £999 on Low Cottage is ignored under the *de minimis* provisions. The profit of £10,000 on Top Pightle is set against the loss of £12,000 on Broadacres with the result that a claim may be made to reduce the estate overall by £2,000 to achieve an IHT saving of £800 (£2,000 @ 40%).

Had the net proceeds on the sale of Low Cottage been for £1 less at £74,000 then the revised £1,000 loss (£74,000 – £75,000) could have been claimed in addition to the £2,000 net loss to achieve a potential max. IHT saving of £1,200 (£3,000 @ 40%).

Is it possible to exploit *s 191* so as to achieve a beneficial result for the purposes of CGT? One learned commentator certainly used to think so, but the law has since been revised by the case of *Stoner (executors of Dickinson deceased) v IRC* [2001] SpC 288, [2001] SWTI 1501.

SALES OF RELATED PROPERTY WITHIN THREE YEARS AFTER DEATH

4.27 *IHTA 1984, s 176* is a useful and potentially valuable relief which applies where, within three years after the death of the transferor, there is a qualifying sale of property which:

(*a*) (i) was comprised in his estate immediately before his death; and

(ii) was valued in accordance with the related property provisions of *s 161*; or

(*b*) (i) was valued in conjunction with property which was also part of the estate; but

(ii) has not at any time since the death been vested in the vendors.

Section 161 itself provides that, where the value of any property comprised in the person's estate would be less than the appropriate portion of the value of that property, if taken together with related property, the value of the portion of the property in the estate is to be regarded as the appropriate portion of the value of the entirety.

This related property valuation rule applies (*s 16(2)*) so as to regard property as being related to the property in a person's estate where:

- it is part of the estate of the transferor's spouse; or
- it is, or has within the preceding five years been, the property of a charity or held on trust for charitable purposes only or has been the property of one of the bodies mentioned in *s 24, 24A, 25* or *26* (political parties, housing associations, national purpose gifts, public benefit gifts) and that property became the property of the charity or other public body on a transfer made by the transferor or his spouse after 15 April 1976 and was exempt to the extent that the value transferred was attributable to the property.

If a claim is made under *s 176*, the property may be valued, as at the date of death, without reference to the related property. The conditions for this treatment are:

- the sale must be by the persons in whom the property is vested immediately after the death or by the personal representatives;
- the sale must be at arm's length, the price freely negotiated and unconnected with the sale of any of the related property;

- no seller is to be connected with any purchaser, nor is any person having an interest in the sale proceeds to be connected with any purchaser nor, in either case, is there to be any connection with any person who has an interest in the purchase; and
- neither the vendors nor anyone having an interest in the sale proceeds is, in connection with the sale, to secure the property 'by the back door', by acquiring an interest in the property sold or an interest created out of that property.

A special rule applies to shares or securities of a close company. Relief under *s 176* will not apply where, between the death and the sale, the value is reduced by more than 5% and that reduction arises from an alteration in the share capital of the company, or its loan capital, or any rights attaching to either.

Finally, the *s 176* rules may also apply where the property was not part of the estate of the deceased, but was valued at his death in conjunction with other property which he did not devolve on the vendors. This might, for example, be property comprised in a settlement where the deceased was tenant for life.

VACANT POSSESSION OF LAND HELD OUTSIDE A PARTNERSHIP

4.28 Quite often, the owner of agricultural land will be a member of a partnership which farms the land. The situation may arise where the partnership occupies the land on a mere licence, without any partnership agreement or any document setting out the terms of occupation referred to as a partnership at will. The partnership can usually be brought to an end on the next accounting date following the giving of notice – *Partnership Act 1890, s 32* relates to dissolution of a partnership by notice (see *Lindley and Banks on Partnership* (17th edn) at 24–19). Effectively, the dissolution occurs on the dates:

- specified in the notice; and
- on which the notice is communicated to all partners.

The basic rule, in *Partnership Act 1890, s 26* is that, subject to any contrary agreement between the parties (and often there will be none), retirement is achieved simply by notice.

Relying on *Harrison-Broadley v Smith* [1964] 1 All ER 867, HMRC argue that, until the partnership can be determined, the landowner does not have vacant possession. Under *IHTA 1984, s 116(2)(a)*, 100% APR might be denied in such a circumstance. The period of 12 months has been extended by ESC F17, as already noted, but that Concession relates to 'tenanted agricultural land' and not to land which is subject to a mere licence.

Clearly, where the practitioner is aware of the difficulty, it is relatively easy to put in place a partnership agreement which gives the landowner the right to vacant possession within 12 months. In practical terms, the difficulty may be in persuading families operating in an area not noted for its formality that they should have a partnership agreement at all. A full-blown partnership deed may not be needed: it may be enough to record an agreement as to what will happen in the event of the death of the owner of the land occupied by the partnership.

Chapter 5

Business property: property qualifying for the relief

QUICK GUIDE

- There are six categories of potential BPR property (see **5.2**) of which those met most commonly in practice are a sole trader, a partner or director, and unquoted shares in trading companies.
- Only those businesses carried on with the prospect of gain will have the potential to qualify for BPR. The exceptions are considered in later chapters.
- It was long thought that relief attached to an interest in a business rather than a 'mere asset'. The CGT retirement relief definition was adopted (see **5.10**), but that view was successfully challenged.
- Special and detailed rules apply to Lloyd's Names and 'Namecos' (see **5.19** et seq).
- With limited liability partnerships, there is scope to 'look through' to the owner.
- Land and buildings, machinery and plant that are owned outside a business but used in it attract relief at 50% (see **5.34** et seq).

INTRODUCTION: THE CONCEPTS

5.1 The scope of BPR has been developed over time and but it is now relatively stable with little real change made in recent years. However, the impact of the relief, operating to shelter up to 100% of value from a maximum IHT charge of 40%, makes it vulnerable to review and amendment. It is therefore critical to closely keep abreast of any developments in this area.

RELEVANT BUSINESS PROPERTY: THE SIX CATEGORIES

5.2 *Section 105* provides that 'relevant business property' comprises the following:

(*a*) property consisting of a business or an interest in a business;

(*b*) securities of a company which are unquoted and which, either by themselves or together with other such securities owned by the transferor and any unquoted shares owned by the transferor, gave the transferor control of the company immediately before the transfer;

(*c*) any unquoted shares in a company;

(*d*) shares in, or securities of, a company which are quoted and which, either by themselves or together with other such shares or securities owned by the transferor, gave the transferor control of the company immediately before the transfer;

(*e*) any land or building, machinery or plant which, immediately before the transfer, was used wholly or mainly for the purposes of a business carried on by a company of which the transferor then had control or by a partnership of which he then was a partner; and

(*f*) any land or building, machinery or plant which immediately before the transfer was used wholly or mainly for the purposes of a business carried on by the transferor and was settled property in which he was then beneficially entitled to a qualifying interest in possession.

By virtue of *s 104*, only property which falls into the above three categories qualifies for relief at 100% with the remainder securing relief at the lower 50% rate.

SECTION 105(1)(A): BUSINESS OR AN INTEREST IN A BUSINESS

5.3 The term 'business' is not defined; it thus takes its ordinary wide meaning and may include the exercise of a profession or vocation. BPR differs from its APR sibling in two major respects: first the business under consideration need not be confined to the UK or EEA and second, it must be carried on for gain.

'Business' has a wider meaning than 'trade' – in the Privy Council case of *Comr of Income Tax v Hanover Agencies Ltd* [1967] 1 AC 681 it was held that 'business', must be given its ordinary meaning unless the context otherwise required. Since the company was carrying on a business of negotiating leases and collecting rents from their properties for the purpose of *s 8(o)* of the *Income Tax Law 1954*, the profits on which they were assessed were profits arising from that 'business'. The court distinguished the case of *Fry v Salisbury House Estate Ltd* [1930] AC 432, which had held that annual income from the ownership of land could be assessed only under Schedule A, and that the option for HMRC to assess under whatever schedule they might prefer did not exist. The schedules were mutually exclusive.

Activities constituting a business

5.4 The corporation tax case of *Land Management Ltd* [2002] SpC 306 provides a useful insight into the activities considered to constitute a 'business'. It was held that in considering the activities of the company in the round, a business included:

- The management and letting of freehold property.
- The activities of a company whose income was mainly derived from the making of investments.
- The making of a loan in order to make profits which made the best use of its assets.

In the income tax case of *Dr KMA Manzur v HMRC* [2010] TC 00830, a retired surgeon failed to persuade the First-tier Tribunal that his losses on share transactions should be set against other income. On the facts it was held that the transactions were more in the nature of management of an investment portfolio.

Non-investment businesses

5.5 In *Phillips' Executors v HMRC* [2006] SpC 555, a company lent funds to other family companies. HMRC had rejected a claim to BPR on a majority shareholding on the basis that the company was making or holding investments (see **Chapter 9**). The company did not invest in loans but, rather, made the loans which were not investments for their own sake but the provision of finance for the other companies. It was held that the company was the banking arm for in-house transactions and few would regard the activities of a moneylender as investment; thus BPR was allowed.

DWC Piercy's Executors v HMRC [2008] SpC 687 concerned a property development company that owned land on which it had built workshops for letting. The executors of a major shareholder claimed BPR, but HMRC denied it on the ground that the company received substantial investment income and its business was therefore mainly making or holding investments, an issue examined in **Chapter 9**. The executors claimed that the company was still trading; it still held undeveloped land that it wished to develop for housing, but had to wait for uncertainty to be resolved concerning proposals for a new railway line. The Commissioner found that the land was still held as trading stock, not as an investment, and allowed the appeal.

Also refer to **Chapter 9** for a review of *Clark and Another v HMRC* [2005] SpC 502 which held that management of properties was a business.

Business, but of the wrong kind for relief: McCall v HMRC

5.6 The case of *McCall and Keenan v Revenue and Customs Commissioners* [2009] NICA 12 involved a parcel of land which had an agricultural value, at the date of death, of £165,000 but a market value of £5,800,000. It was therefore important for the executors to show that at the date of death there was a business qualifying for BPR to soak up the development value rather than a mere agricultural interest with its lower APR coverage.

The evidence showed (just) that the son-in-law of the deceased had spent time looking after the land, which lay on the edge of a town. The work involved walking the land, repairing the fencing, cleaning drinking troughs, clearing the drainage systems of mud and leaves, cutting and spraying weeds. For these purposes, the son-in-law had a tractor and reaper and a knapsack sprayer. He probably worked no more than 100 hours per year but some work was contracted out. The grazier fertilised the fields.

At first instance the Commissioner reviewed the authorities, including several mentioned above and in particular the *indicia* noted by Gibson J in *Customs and Excise Commissioners v Lord Fisher* [1981] STC 238. He concluded that the son-in-law's activity of tending the land:

> '... was, just, enough to constitute a business ... but was it a trade? The letting of the land was earnestly pursued, the work of tending the land was modest but serious, the letting and tending were pursued with some continuity, the income was not insubstantial, the letting was conducted in a regular manner although the use of [the son-in-law's] time was something which is not a feature of an ordinary business, and the letting of land for profit is a common business. To my mind the Lord Fisher indicia point towards a business ...'

However, the taxpayers' case fell, both at first instance and on appeal, because it was held that the business consisted wholly or mainly of holding investments, within *IHTA 1984, s 105(3)*. After a detailed consideration of the nature of farming arrangements under the law of Northern Ireland which may not be of general application the Commissioner held that:

> 'The test to be applied is that of an intelligent businessman, not a land lawyer. Such a person would be concerned with the use to which the asset was put. The deceased was making the land available, not to make a living on it but from it; the management activities related to letting the land; it was unlike "hotel accommodation for cattle" as suggested by Mr Massey; nor was it like a "pick your own" fruit farm after months of weeding, fertilising spraying and pruning, customers are licensed to enter to take the produce and pay by the pound for what they take away; in the business of letting the fields there was less preparatory work, the fields were let for the accommodation of the cattle as well as for the grazing, and the rent was paid by the acre rather than by the ton of grass eaten ...'

The Court of Appeal in Northern Ireland agreed – the Commissioner had properly understood the law. The test was that of the intelligent businessman (which phrase also surfaced in the recent *Pawson* case discussed in **Chapter 9**), who would look at the use to which the land was put – a landowner who derives income from land or a building will be treated as having a business of holding an investment even though he may carry out incidental work to obtain the income. Here, the activities of the son-in-law were in the nature of maintenance work to achieve a successful letting. The use of the land by graziers was exclusive, so the owner could not use the land for any purpose that interfered with the grazing. The appeal was dismissed.

Example 5.1: The Swallow family – scaling back of farm activities

Willow Farm has belonged to the Swallow family for three generations.

From 1926 to 1950, grandfather ran cattle and sheep on the pasture, some of which, by the stream, was quite boggy, so grazed for only part of the year. His sons followed, replacing the heavy horses with second-hand tractors and farming without paid help.

The elder son, Ted, became disillusioned and in 1970 the business was split, and some of the better land and two cottages were sold off. The younger son, Phillip, burdened with debt, carried on with (unpaid) help from his wife, Laura and daughter, Ann.

From 1995 onwards, the greater part of the land was put into a contracting agreement, leaving the farmer with some of the pasture on which to fatten cattle; but even that became too much and the cattle were sold and a grazing agreement put in place. Only the basic payment scheme (BPS) now provides funds to service the outstanding mortgage, because the contractor in effect takes all the income from the land. The daughter now lives with her family in the village nearby. Following the death of Laura, Phillip continued to live on in the farmhouse which is both lonely and neglected. The farmhouse is no longer the centre of farming operations, even though Phillip still walks the land each day and supervises the animals for the grazier. He continues to maintain the hedges and fences but only in so far as the weather and his health allow.

The APR position could be improved in a number of ways:

- Regularise the way in which the farm accounts are presented. Ensure that any positive aspects are emphasised and income properly characterised, ie not shown as 'rent'. Check what records are kept of the work the farmer still does. Does the daughter, Ann visit regularly? Does she do the books?

- Would it be feasible for Ann and her family to move in with Phillip? That, coupled with a gift of a share in the house covered by the rule in *FA 1986, s 102B(4),* would compensate her family for having to live with granddad, and the proceeds of sale of her present house might finance renovation of the farmhouse, with provision of a good farm office as well as privacy for all. If she has teenage children, the extra space will be some compensation for living further out. If there is no local employment, even working on the farm is better than nothing.
- If the family can increase their involvement in farming operations, they may be able to save the business by increasing turnover and direct profit. They will also strengthen the APR claim on the house itself, where the tax may be significant. The big problem is likely to be investment in new machinery, the need for which was the spur for the contracting agreement in the first place, and learning the skills to use it. If the adviser can make a good case to a lender that the changes may save significant tax, finance may be available.

Corporate v non-corporate

5.7 There is an important distinction between the activities of individuals and partners of a partnership, when compared to those of a company. The mere receipt of rents by an individual is not regarded as carrying on a business, but for a company the rule is different as demonstrated in the Malaysian case of *American Leaf Blending Co SDN BHD v Director General of Inland Revenue* [1979] AC 676. After abandoning a tobacco business which had made losses, the company let a warehouse which was part of its business premises followed three years later by the remainder of the premises which had become empty. The company was assessed to income tax in respect of the rents received but claimed set-off for the earlier losses.

The Privy Council allowed the relief on the wording of *s 4* of the *Income Tax Act 1967*, applying *Hanover Agencies Ltd* and distinguishing *Salisbury House Estate Ltd*. Where a company had been incorporated for the purposes of making profits, any gain from a use to which it put its assets *prima facie* amounted to the carrying on of a business. Although the fact that the letting of its premises was included in the objects of the company, that was not conclusive in deciding that the company was carrying on a business, since the only conclusion of fact which any reasonable Commissioners could reach from the evidence was that the company was carrying on a business of letting its premises for rent. Notably, it was observed that:

> 'the carrying on of "business", no doubt, usually calls for some activity on the part of whoever carries it on, though, depending on the nature of the business, the activity may be intermittent with long intervals of quiescence

in between. In the instant case, however, there was evidence before the Special Commissioners of activity in and about the letting of its premises by the company during each of the five years that had elapsed since it closed down its former tobacco business.'

The VAT case of *C & E Comrs v Lord Fisher* [1981] STC 238, sets out helpful guidance by considering six key criteria which paint a picture of a 'business':

(1) Is there 'a serious undertaking earnestly pursued' or 'a serious occupation, not necessarily confined to commercial or profit-making undertakings'?

(2) Is there 'an occupation or function actively pursued with reasonable or recognisable continuity'?

(3) Has the activity 'a certain measure of substance as measured by the quarterly or annual value … of supplies made'?

(4) Was the activity 'conducted in a regular manner and on sound and recognised business principles'?

(5) Is the activity 'predominantly concerned with the making of … supplies to customers with consideration'?

(6) Are those supplies 'of a kind which, subject to differences in detail, are commonly made by those who seek to profit by them'?

These criteria have been relied on regularly in debated BPR and APR cases.

The restrictive effect of s 105(3)

5.8 Access to BPR is denied in totality if the business falls within the exclusion contained in *IHTA 1984, s 105(3)* which applies 'if the business or, as the case may be, the business carried on by the company consists wholly or mainly of one or more of the following, that is to say, dealing in securities, stocks or shares, land or holdings or making or holding investments …'. The subject is considered in greater detail in **Chapter 9**.

Section 105(4) contains a narrow exception to the exclusion for a business or an interest in a business or shares or securities of a company where the business is:

- wholly that of a 'market maker' or of a discount house, and in either case is carried on in the UK; or
- that of a holding company of one or more companies whose business does qualify.

'Market maker' means, by virtue of *s 105(7)*, a person who holds himself out at all normal times in compliance with the rules of the Stock Exchange as willing to buy and sell securities, stocks or shares at a price specified by him and is recognised as doing so by the Council of the Stock Exchange. By *FA 1986, s 106(4), (6)* and *(8)* HMRC may make regulations to modify that definition. That power has been used (see the *Inheritance Tax (Market Makers) Regulations 1992, SI 1992/3181*) in relation to transfers of value and other events occurring after 22 March 1992. The effect of those regulations was to include, within the reference to the Stock Exchange, 'LIFFE (Administration and Management)', which had become a recognised investment exchange within the meaning of the *Financial Services Act 1986*.

Woodlands

5.9 Woodlands managed as a business qualify in the same way as other property for BPR. The relief, as is explained in **Chapter 11**, applies to the land and other assets used in the business, trees and underwood.

Sole proprietor's business

5.10 This category is the simplest form of property qualifying for the relief with eligibility determined by the ability of the taxpayer to demonstrate that there is actually a business, ie that the activities carried on have sufficient commercial purpose. However, there is a distinction between the sole trader and the incorporated business.

The sole trader balance sheet may include significant assets that are not directly employed in the generation of profit and the value of such assets may fall within the excluded asset rule of *s 112* – see **Chapter 7**. In some cases assets employed in the business are simply not reflected in the balance sheet or indeed for less sophisticated businesses, a formal balance sheet may simply not exist at all.

It should be noted that if the successor spouse inherits the assets of and carries on the profession of their deceased spouse, the successor spouse is deemed to have started a new 'business'. The successor will not be able to count the period of ownership of the deceased spouse towards his or her own period of ownership.

Authors, artists, sculptors and musicians

5.11 In the past BPR was often denied on gifts made during the author's lifetime on grounds that the unique nature of the business of an author made it incapable of transfer and indeed that business ceased on their death. The real value lay in copyrights producing royalties, and that all that could therefore be given away was the mere copyright the holding of which was an investment. Therefore, the 'mere asset' test applied to deny relief, because it was impossible for the author to transfer the business as a whole. BPR was intended to minimise the risk that the payment of capital taxation might disrupt or liquidate a healthy business which could otherwise have continued after the death of the owner and on interpretation such a policy consideration would not and could not apply to authors or indeed other creative artists.

HMRC's attitude is now a little more relaxed and the facts may support the BPR claim where, prior to the death, business accounts show that the royalties are actively managed also supported by VAT registration. Even though in their later years an artist may no longer be as creative there may be a substantial body of past work which the artist could promote and sell. An elderly composer who no longer produces new material may be active in attending concerts of their works, encouraging the young to perform their music and vigorously promoting a back catalogue – all of this shows an intent to stimulate a profit stream and keep the business alive.

BPR may be allowed in respect of 'personal copyrights' of a deceased or of a majority shareholding in a company which owns the copyrights. However, this will not help the gift of a copyright during the lifetime of the author, which will probably not qualify for relief, unless the decision in *Nelson Dance v HMRC* [2009] EWHC 71 (Ch) (discussed at **5.14**) can be extended. For an illustration of the difficulty of assigning royalties, see the income tax case of *Braithwaite v HMRC* [2008] SpC 674, where an attempt to transfer 50% of the income from royalties failed.

Partnership interest

5.12 It is important to recognise that the interest of a partner in a business is not an interest in the underlying assets but in the proceeds of the sale and division of the assets of the partnership. An interest in a business includes the interest of a partner but only provided that interest is not merely by way of a loan which in itself is a liability of a business (see **Chapter 8**). The interest may also be in any joint venture, including foreign forms of association not recognised in UK law.

In recent years many businesses have 'converted' to the LLP structure and the LLP may itself own shares in one or more unquoted companies. The use of an LLP has advantages for income tax and CGT because the structure is transparent, but for IHT *s 267A* treats the LLP as opaque, akin to a general partnership (see **5.21**).

The 'mere asset' exclusion: review of the early cases

5.13 In the past the legislation appeared to draw a distinction between an interest in a business and the underlying assets but that view has been firmly repositioned following the *Nelson Dance* decision (see **5.14**).

Sale of a small part

In *McGregor v Adcock* [1977] STC 206, there was a disposal of five acres from a holding of 35 acres on which the taxpayer had been carrying on mixed farming. The taxpayer continued to farm the rest of the land, but claimed that the sale of the five acres, on which he had obtained planning permission, was a disposal of part of his business entitling him to retirement relief. It was held that nothing suggested that the sale had seriously altered the scale of the business or interfered with the farming which could be described as amounting to a disposal of part of the business. There was simply the sale of an asset, and not part of the business.

Was there a major change in the business?

In *Mannion v Johnston* [1988] STC 758, the Commissioners found on the facts that, after the sale, there had been a major change in the business of the taxpayer and, on that basis, there had been a disposal of part of the farming business. This view was replicated in *Atkinson v Dancer* [1988] STC 758, where various parts of the mixed farming business depended on each other and following the sale of part of the land, it was not possible to continue with the remainder. The principle that emerged in both cases was that a change in the activities of a business, or in the assets of the business occasioned by anything other than sale, was irrelevant. It was correct to consider the business activities before and after the sale. If the position after that sale was wholly different as a result of changes caused by the sale, then the sale might be regarded as of a business or of part of a business. Two separate disposals should not be taken together, as if a single transaction, unless both formed part of the same contract.

Nelson Dance

5.14 The main challenge to the 'mere asset' rule is in *Trustees of the Nelson Dance Family Settlement v HMRC* [2009] EWHC 71 (Ch), which is worthy of detailed review.

Mr Dance, a farmer, transferred land to a settlement. APR was not in issue; here rather the taxpayer wanted BPR because the land had development value. HMRC denied relief on the basis that the subject matter of the transfer was not an interest in his farming business but a 'mere asset'. The taxpayer successfully argued that it was necessary to look not at the actual asset transferred, but at its transfer of value based on the 'loss to the estate' principle. The estate of Mr Dance was reduced by the transfer: what was reduced was the value of his farming business; he had been in farming for many years; so the transfer of value fell to be reduced by BPR.

At first instance the Special Commissioner could see a distinction between those cases, such as spousal relief where the recipient is important, and the present situation where the basis of relief is determined by the transfer, not the underlying asset, commenting:

> 'All these form part of an overall scheme. Everything turns on the loss in value to the donor's estate, rather than what is given or how the loss to the estate arises, except where the identity of the recipient is crucial to a particular exemption …'

As expected, HMRC appealed to the High Court contending that the reference in the legislation to the value of business property implied that the transfer must be of business property, not merely business assets. Sales J dismissed the appeal.

It was sufficient that a possible and proper characterisation of the value transferred was that such value was attributable to the value of a business. The source of that value did not require an exclusive type of characterisation and thus it did not matter that the attribution could be to the value of the land transferred, as long as it could be said that the attribution could also be to the value of a business. That interpretation was simple and certain and sat well with *s 110* and with the 'loss to donor 'principle in *s 103*.

Disposal of part of a partnership share

5.15 *Section 105(1)(a)* grants relief in respect of 'a business or interest in a business' but that does not mean the *whole* of a business or the *whole* of an interest in a business. BPR may be allowed where a sole proprietor of a business takes another person into partnership and transfers to that person part of the capital, ie a share both of the assets and liabilities as identified in the balance sheet of the business, as the capital of that sole trader. By extension of that principle, one partner among many may (doubtless only with the consent of the others) divide his existing interest, as identified by his capital account in the balance sheet of the partnership, between himself and another person and claim BPR on that transfer.

What may be less certain is whether a mere book entry is enough.

Example 5.2: Ellen McGregor – movements in capital account

Ellen McGregor farmed in partnership with her sons Fred and George. At Michaelmas 2014 the capital account showed:

Ellen	£30,000
Fred	£20,500
George	£18,000
	£68,500

By Michaelmas 2015, the balances were:

Ellen	£34,000
Fred	£22,000
George	£17,000
	£73,000

During 2016 the farming merely broke even, but a road-widening scheme resulted in compensation of £100,000 paid to Ellen in respect of farmland held outside the partnership. She paid the compensation into the farm account and instructed her accountant to transfer £80,000 of the capital to Fred and George equally. He showed the balances at Michaelmas 2016 as:

Ellen	54,000
Fred	62,000
George	57,000
	173,000

On Ellen's death in 2017, her executors wished to claim BPR on the £80,000 gift. The claim failed as to £60,000, because Ellen's capital from the two year period prior to transfer had been increased by £20,000. As to the £20,000 itself, the claim succeeded. It was helped by the fact that there was a clear instruction to make a gift, which was promptly acted on. Ellen had in fact signed off the 2016 accounts. It was not merely a case of writing up the books after her death with the benefit of hindsight.

For a case that crossed the boundaries between partnership, quasi-partnership and a company, see *Rahman v Malik and Others* [2008] EWHC 959 (Ch), concerning interests in a restaurant. The son of one of the founders of the business, having been thrown out by his uncle, four years later petitioned under *Companies Act 1985, s 459* alleging that his exclusion, the non-payment of dividends and under-declaring of profits all amounted to unfair prejudice. The court agreed, notwithstanding the nephew's delay, and ordered the uncle to buy the nephew's shares.

One business or several? What is the interest?

5.16 A person with multiple businesses may keep one and gift away the other and this is particularly so in farming. It will be a question of fact whether the gift is truly of a business in its own right or is merely a gift of a bundle of assets. If the latter, BPR may still be in point if it can be brought within the terms of *Nelson Dance*.

The identification of the exact nature of an interest arose in *Oakley and Another (PRs of Jossaume Deceased) v IRC* [2005] SpC 460 which demonstrates the crossover between the basic principles of land law and the law of trusts in the tax analysis. The issue was whether a building yard was to be treated as part of the estate of the widow on the basis that she had an interest in possession. It was held that the trusts of a will (as substituted by deed of variation on the earlier death of the husband) came to an end on the death of the widow. The substituted will curiously gave an interest in possession to the family company, and not to the widow, even though on the earlier death the husband's estate had been taxed as if there had been a spouse-exempt transfer to the widow. The case turned on whether the direction in the substituted will that the company need pay no rent was dispositive or merely administrative. It was dispositive: this may in future be relied upon by HMRC, in circumstances such as non-payment of rent for occupation of a residence.

Lloyd's underwriters: the assets that qualify for BPR

5.17

Premium income deposit fund; special reserve fund

The premium income deposit and the special reserve funds of a Lloyd's underwriter qualify for BPR. Commonly, assets are deposited with Lloyd's as security for any obligations in respect of underwriting business. The assets so deposited will qualify for BPR provided that they are not excessive in value compared with the risks underwritten with such excess regarded as an excepted asset under *IHTA 1984, s 112* (see **Chapter 7**). Here a distinction is made between sums in respect of which accounts are closed and sums deposited to support open years.

Bank guarantees

A shareholder in an underwriting company at Lloyd's may provide collateral to support the activities of that company, the 'Nameco'. HMRC considered that BPR is not available on the collateral, but that the valuation of the collateral assets for IHT purposes must take account of the negative value to the estate of the fact that those assets are provided as security for the underwriting.

The formula is complicated but in simple terms it is necessary to: first, value the open years and the possibility that the collateral assets will be drawn upon; then second, discount the value of the assets to take account of the period for which they will be subject to the collateral guarantee to arrive at the value for BPR purposes.

Assets securing both Nameco and personal underwriting

The arrangements are different where assets secure not only collateral security for the Nameco but also for the personal underwriting by the shareholder. These assets are known as 'inter-available funds at Lloyd's'.

Deaths on or after 1 January 2005

BPR is allowed on the value represented by the inter-available funds as needed for the underwriting of the individual Name but only in so far as they are proportionate to the unlimited liability underwriting and not on assets put up as third-party funds. The IHT valuation of those assets must take account of the negative value of the arrangement for the estate of funds at Lloyd's.

Lloyd's Name losses

5.18 Clarification of the treatment of losses, in relation to BPR claims on behalf of the estates of deceased Names, emerged in *Hardcastle (Executors of Vernede deceased) v IRC* [2000] STC (SCD) 532, although HMRC consider the case to be one decided on its own particular facts (see below).

At the date of his death, Mr Vernede belonged to 11 Lloyd's syndicates. His estate protection plan policy, with an excess of £251,900, indemnified his estate against losses arising in respect of underwriting results not notified before death. The executors completed the HMRC account so as to show that the net value of the interest in the business was £265,507. That was the amount of the funds held at Lloyd's. That was also the sum in respect of which the executors claimed BPR.

As a separate item, the executors showed in the HMRC account that there were underwriting losses for years prior to death which amounted to £301,301. These were certain deferred losses and the excess under the estate protection plan policy. No demand had been received for payment of the deferred losses at the date of death. The executors claimed to offset the losses, not against the value of the interest in the business, but against the estate as a whole. HMRC thought that to be the wrong treatment. The executors appealed.

By the time of the hearing before the Special Commissioner, the dispute concerned only the amounts owed by the deceased in respect of years that were open at the date of his death. By virtue of the insurance policy, the amount in issue was therefore £251,900. The question was whether the money owed on these open years was, within *s 110(b)*, 'a liability incurred for the purposes of the business ...'. HMRC had argued that the deceased's trading contracts, which might have given rise either to profits or to losses, were choses in action. They were assets because they were the means by which the business was carried on. It followed, their argument went, that, if the contracts were profitable, those contracts would increase for BPR purposes the net value of the interest of the deceased in the business. It should also follow that, where a trading contract went the other way, it gave rise to a liability incurred in the course of carrying on the business, and it was therefore incurred for the purposes of the business.

The Special Commissioner disagreed. Two tests were to be applied here: first, one must calculate the valuation of a business for general IHT purposes; and, second, one must calculate the value of the business for the purposes of BPR. The first basis of the valuation (described below as the 'commercial' basis) is to be determined by reference to *s 5*, which defines the meaning of 'estate' for IHT. The commercial basis includes not only assets and liabilities but trading profits or trading losses.

The second basis of valuation, is more restrictive and concerned only with the assets used in the business and the liabilities incurred for the purposes of the business. HMRC had agreed with the executors of Mr Vernede that trading losses were not liabilities incurred as such for the purposes of the business. The question was rather whether the trading contracts that gave rise to those losses were assets used in the business.

For the purposes of the relief basis of valuation, some help could be drawn from the distinction, in income tax, between expenditure which brings into existence an asset for the enduring benefit of the trade and ordinary commercial contracts. In that context, insurance contracts in which a result has not been notified are ordinary commercial contracts made by the deceased in the course of carrying on his underwriting business. They are contracts for the disposal of his 'product'. The product dealt in is the assumption of a risk in return for a premium. Seen in that light, a contract which produced a loss was not a liability incurred for the purposes of a business. On that basis, the Special Commissioner found for the executors.

The point that seemed to have been established in *Hardcastle* was that ordinary liabilities incurred in the day-to-day running of the business would not reduce the value of the business for BPR.

Assets supporting the underwriting business

5.19 Generally, the assets deposited will be stocks and shares but occasionally, underwriters deposit letters of credit or bank guarantees instead, because the underwriters may wish to use their personal capital in another way or do not wish to realise certain assets such as gold. BPR will apply to assets taken by way of collateral security or bank guarantees or letters of credit but subject to the requirement that the asset in question is restricted in use and cannot be used for personal benefit.

Where an individual Lloyd's Name provides a guarantee as part of his funds, 100% BPR is allowed up to the amount of the guarantee that is used or required as funds at Lloyd's.

The value of an underlying asset is reduced by the amount of the guarantee for the purpose of giving any other reliefs or exemptions.

HMRC offer two examples both pre the *FA 2013* rules on repositioned debt.

Example 5.3: Nigel Derby – guarantee in place (pre-FA 2013)

Nigel Derby is an individual Member and Lloyd's Name; he is also a farmer. The farm, which qualifies for APR at 100%, is worth £1 million and Nigel has other assets of £1.25 million on which a bank guarantee of £400,000 is secured.

The valuation of the Lloyd's interest, which qualifies for 100% BPR, is ignored for the purposes of this example, as are any other reliefs but the calculation of APR and BPR on death is as follows:

	£
Farmland	1,000,000
Other assets	1,250,000
	2,250,000
Less BPR (£400,000 on bank guarantee)	(400,000)
APR on farmland	(1,000,000)
Net estate after reliefs	850,000

There is no reduction in the value of the farmland, because the bank guarantee is not secured on it.

Example 5.4: Nigel Derby – guarantee secured on farmland (pre-FA 2013)

The facts are as in the above example but this time the guarantee is secured on the farmland. The IHT calculation, ignoring valuation of open years and other reliefs, is as follows:

	£
Farmland	1,000,000
Other assets	1,250,000
Total	2,250,000
Less BPR (on bank guarantee)	(400,000)
APR (£1 million – £400,000)	(600,000)
Net estate after reliefs	1,250,000

This issue was examined by the Special Commissioner in *Mallender (Executors of Drury-Lowe deceased) v IRC* [2001] STC 514.

The deceased had a personal reserve of quoted shares and there were also bank guarantees of £100,000 which were based on an indemnity given by the deceased to the bank against all liabilities. That indemnity was in turn secured by a charge over freehold property which the deceased owned. The property itself, 'the Denby land', which had been worth £1.2 million in May 1991, was worth £1 million in July 1993 when the deceased died. It was subject to a coal mining lease and the rent was £170,000 per annum but British Coal was the only realistic operator and the lease itself was a depreciating asset. The executors claimed BPR on the value of the freehold but HMRC determined that, for the purposes of *s110*, the value of the property itself should be excluded as its purpose was simply to provide the guarantee.

The Special Commissioner took a realistic approach that no bank guarantee would have been forthcoming without security. The minimum security would have been £142,858 but, in fact, security was provided of over £1 million. That was enough to satisfy the Special Commissioner that the land itself was one of the assets used in the business and that the net value of the freehold should be included for the purposes of relief.

HMRC succeeded on subsequent appeal. The Denby land was not used in the underwriting business and in calculating net value, the liabilities must be set against assets used in the business. It did not matter that the guarantee was supported by a charge on the land.

Succession

5.20 Is there succession to the business of a Name? The succession provisions in *ss 108* and *109* (see **Chapter 6**) do not in general apply.

The right to a share in the profit of the syndicate is an asset of the business qualifying for BPR. The business of corporate Lloyd's Members may likewise qualify for the relief. The taxation of Lloyd's Names is a specialised subject. It is dealt with at HMRC through Shares Valuation Division at Fitzroy House, PO Box 46, Castle Meadow Road, Nottingham, NG2 1BD.

Limited partnerships and limited liability partnerships (LLPs)

5.21 Under the *Limited Partnerships Act 1907* a partner may contribute a defined amount of capital to the partnership and his liability may be limited to that amount. The partnership must be registered with the Registrar of Companies showing the details of the amount of the capital of the limited partners. Such a partnership must contain at least one general partner whose liability is not so limited.

As to LLPs, *s 267A* was inserted into *IHTA 1984* by the *Limited Liability Partnerships Act 2000* (*LLPA 2000*) with effect from 6 April 2001 and provides that, for the purposes of *IHTA 1984* and of any other enactment relating to IHT:

- the property of the limited liability partnership is treated as the property of its members and that property used by such a partnership is treated as being used by its members as partners; and
- the business carried on by the partnership is treated as carried on in partnership by its members; and
- incorporation of the partnership, or any change in its membership, or its dissolution, is treated as the formation, alteration or dissolution of a partnership; and
- any transfer of value made by or to a limited liability partnership is treated as being made by or to the members of that partnership and not by the partnership itself as such.

Whilst a full consideration of *LLPA 2000* is outside the scope of this book, it should be noted that LLPs will be governed by certain provisions that are imported from company law rather than from partnership law: the participators will be members rather than partners.

HMRC guidance on LLPs states the following in relation to BPR (IHTM25094):

> '… an interest in a LLP is deemed to be an interest in each and every asset of the partnership, while an interest in a traditional partnership is a "chose in action", valued by reference to the net underlying assets of the business. This may require you to consider issues of situs of property. In cases of doubt refer to Technical Group (TG) for advice.
>
> However, in considering if an LLP is an investment business … you should look at the nature of the business underpinned by those assets, rather than the nature of the assets themselves, to see whether IHTA84/S105(3) is in point.

There has been an increase in the use of LLPs in commercial structures, and sometimes there can be a different outcome for Business Relief purposes than that available from a conventional corporate structure. In the case of an LLP simply taking the place of a holding company, s.267A has the effect of preventing the LLP from benefitting from *s.105(4)(b)*. In cases where the LLP itself also carries on a qualifying business, the business may be regarded as a hybrid, and if the shares in the subsidiary companies are used in the business (rather than being held as investments), then it is possible that the interest in the LLP may qualify for relief if it does not fail the "wholly or mainly" test (IHTM25264). The question of whether an asset is used in the business or held as an investment will be highly fact specific …'

LLP accounts and formalities

5.22 LLPs must prepare accounts in a style similar to that of companies – they must be true and fair, published and filed but the requirements are different according to the size of the partnership. There is provision in *LLPA 2000* for audit. The limited partner is not allowed to take any part in the management of the firm's business or to bind the firm. He may give advice to the other partners. Importantly for tax purposes, a limited partner is treated in the same way as a general partner, except that, for the purposes of income tax, there is a restriction on the amount of loss for which he may claim relief.

The interest of a member in an LLP may be capital or a loan. The capital may attract BPR but the loan will not, any more than a loan to an unlimited partnership would, unless for some reason it is 'an interest' in the business, but that will be rare. That accords with the SORP *Accounting by Limited Liability Partnerships*.

Jersey LLPs

5.23 A Jersey LLP may be a partnership or a company (see *R v IRC ex p Bishop* [1999] STC 531). However, HMRC seem inclined to treat them as 'opaque' (see *Tax Bulletin* Issue 83 (June 2006)), which is more in the nature of a body corporate than a partnership but this view has yet to be tested.

Transition to LLP

5.24 Tax Bulletin issue 50, December 2000, clarified issues somewhat in relation to the incorporation of a partnership as an LLP.

IHTA 1984, s 94, which relates to the charge on participators where a transfer is made by a close company, will not apply because transfers of value are treated (see *s 267A(d)*) as being made not by the partnership but by its members. Thus the reliefs available to the 'old' partners are available to members of the LLP. This will be relevant where a disposition occurs which is not intended to confer any bounty such that *s 10* will apply.

The position as far as Scottish partnerships are concerned is slightly clearer. A Scottish partnership is a separate legal person distinct from the individual partners. But for the saving rule in *s 119(2)*, a difficulty could therefore arise for a member of a Scottish partnership, whether limited or otherwise, as to occupation of land, since it would be the partnership that occupied the land rather than the individual. That would flow from *Partnership Act 1890, s 4(2)*. As a result of *s 119(2)*, this difficulty is removed, and occupation of any property by a Scottish partnership is treated as occupation by the partners.

SECTION 105(1)(B): UNQUOTED SECURITIES GIVING CONTROL

5.25 The subject matter discussed here relates to unquoted securities ie loan stock, which either by themselves or combined with other securities/ unquoted shares, gave the transferor control of the company immediately before the transfer (it is not necessary, except where clawback of relief is in point, for either the transferor or the transferee to have control post transfer). The reference to control applies only to securities under this heading and not to unquoted shares themselves.

BPR will only apply to this asset class if they have votes attached to them and form part of a controlling holding. It follows that, where a transferor already has enough ordinary shares to have control, the securities do not attract BPR because they do not give him control. It is the shares that give control in their own right.

For an interesting twist on the issue of company securities see *Silber (Lerner's PR) v HMRC* [2012] TC 02369, reviewed at **22.4** below. Also note the valuation case of *McArthur's Executors v HMRC* [2008] SpC 700 in which the taxpayer held shares in three family investment companies and had lent money to two of them. The loans carried rights to acquire further shares in the form of convertible loan stock. The Special Commissioner rejected the taxpayers' argument that those rights had little value.

Example 5.5: Rhine Gmbh – share classes

Rhine Gmbh is a middle-sized German company. The shares are held by the founding family and there are in issue one million 'A' shares, carrying ten votes each, and five million 'B' shares carrying one vote each. The 'A' shares and 'B' shares in other respects carry the same rights, particularly as to dividends but whilst the A shares are unlisted the B shares are listed on the Frankfurt Stock Exchange which is a recognised exchange for IHT purposes.

The shares are held as follows:

- Helmut, aged 78, holds 700,000 'A' shares and the same number of 'B' shares, giving him 7.7 million votes (7 million (A) + 700,000 (B)). Helmut is domiciled and resident in Germany so BPR is of no concern to him.
- The remainder of the 'A' shares are held by Helmut's daughter, Elke, who also holds 2 million 'B' shares – taken together, her shares give her 5 million votes (3 million (A) + 2 million(B)). Elke, is married to an Englishman and has lived in the UK for 30 years such that she has UK deemed domiciled status for IHT purposes.
- The remaining 2.3 million 'B' shares (5 million – (700,000 + 2 million)) are held by employees of the company and by trusts for the benefit of employees.

Elke appreciates that her 'A' shares qualify for BPR because they are unquoted. However, the 'B' shares are regarded as quoted being listed on a recognised exchange, so Elke is at risk of a charge to IHT on their value.

Elke consults her solicitor, Jack, who is keen to persuade Helmut to hand over the bulk of his shareholding to Elke, on the general principle that it must be a good idea to pass value down the generations, but Helmut is not in agreement. Fortunately, Jack refers the detail of the matter to Cerys, who does the sums. Cerys is able to advise Elke that she needs only 2,500,001 more votes to have control (more than 75%). If Helmut is willing to give Elke 250,000 'A' shares and one token 'B' share, Elke will have control and Helmut will still retain, for the purposes of dividends, the bulk of his original holding. The saving will arise because Elke's 'B' shares will now, when taken with her 'A' shares, qualify for BPR at 50% under *s 105(1)(b)*.

The meaning of 'control'

5.26 The definition of 'control' is set out in *IHTA 1984, s 269*. The basic test is that a person has the control of powers of voting on all questions affecting the company as a whole which, if exercised, would yield a majority of the votes capable of being exercised on those questions. By *s 269(2)*, the related property rules of *s 161* (described more fully in **Chapter 8**) operate so that securities are deemed to give a person control of a company if they would be sufficient to give control if taken together with related property. Where shares or securities are comprised in a settlement in which there is an interest in possession, the life tenant is deemed to have the powers of voting which the shares or securities give to the trustees of the settlement. This does not apply (see *s 269(3)*) to a discretionary trust.

A limited rule applies where shares or securities have powers of voting limited to the question whether the company should be wound up, or a question which primarily affects shares of securities of that particular class. In those circumstances, *s 269(1)* is treated as meaning a reference not to all questions affecting the company as a whole but to all questions other than any in relation to which those powers are capable of being exercised. In other words, the fact that a company has one or more classes of shares with restricted voting rights will not prevent a person from having control.

Certain questions of fact may arise here. For example, two unconnected persons may each hold half the shares of a company, with the chairman of the meeting having the casting vote. It is quite possible that one of those shareholders may act as chairman but consider that person suffers an illness leading to death, during which, naturally, the other shareholder carries on the business and acts as chairman of future meetings. At the time of the death, it cannot be said that the first shareholder has been exercising control of the company through his position as chairman. He does not have control at the time of his death.

This was substantially the issue in *Walker's Executors v IRC* [2001] STC (SCD) 86. The deceased owned land used for the purpose of a company in which she held 50% of the shares. She was chairman of the board of directors and, by virtue of the articles of association, enjoyed a casting vote at general meetings. To secure BPR, her executors needed to show (under *s 105(1)(d))* at the date of her death, the deceased had control of the company. The executors argued that the casting vote gave control, because the deceased had control of powers of voting on all questions affecting the company as a whole which, if exercised, would yield a majority of the votes capable of being exercised on them. HMRC had argued that, when exercising her casting vote as chairman, the deceased owed a fiduciary duty in her capacity as an officer of the company. Had there been any conflict between her personal interest and that of the company, she could not have used her casting vote.

The Special Commissioners found for the taxpayer. The mere fact that the deceased might be trustee of the casting vote did not mean that she did not have control of the powers of voting and importantly, *s 269(3)* makes no reference to fiduciary duties. The deceased, therefore, did control the company and was entitled to 50% BPR on the value of the land it used.

Occasionally, shareholders will have special voting rights, in particular enough extra votes to outvote all the other shareholders on the question whether that person be removed as a director. Such rights were considered in *Bushell v Faith* [1970] AC 1099. The existence of such additional rights effectively dilutes the voting control of a shareholder who, on all other issues, has a majority holding of voting shares. This is because that majority shareholder will not have a majority on all questions affecting the company as a whole.

Walding v IRC [1996] STC 13 determined that the capacity of a shareholder to exercise votes does not affect control. The deceased held 45% of the voting share capital of the company, and her four-year-old grandson held 24%. On her death, her executors argued that the deceased had control of the company because the grandson could not exercise votes attached to his shares. The executors failed.

Where shares are held in a relevant property trust (ie a trust subject to the IHT regime of ten-year and exit charges) the related property rules cannot apply and unless the trustees hold more than 50% of the company, they will not qualify for BPR. Consider that two majority shareholders jointly settle substantial holdings on relevant property trusts with identical trustees. Although the trustees are the same, there are actually two separate settlements and to exercise control, the trustees need to control more than 50% of the votes of the company but they hold two separate holdings, neither of which gives control.

SECTION 105(1)(BB): UNQUOTED SHARES

5.27 The expressions 'quoted' and 'unquoted' remain in the wording of *s 105(1)* even though, in the changes introduced by *FA 1996*, the expression 'listed' was generally used. Shares listed on Alternative Investment Market (AIM) are unquoted for this purpose.

The legislation does not limit relief to ordinary shares as such but rather refers to 'any unquoted shares in a company'. Whilst loan stock is not 'shares', preference shares and/or redeemable shares are not specifically excluded from BPR. For a commentary on the meaning of 'quoted' and of 'recognised Stock Exchange', see **Chapter 6**.

Unquoted shares will not qualify for BPR if the company is an investment business, if the company is in liquidation or if there is at the time of the chargeable event, a binding contract for sale – see the flowcharts in IHTM25022, 25023 and 25024.

SECTION 105(1)(CC): CONTROLLING HOLDINGS IN A QUOTED COMPANY

5.28 Shares which are listed or quoted on a recognised Stock Exchange and which gave the transferor control of the company immediately before the transfer qualify for the lower 50% BPR rate.

SECTION 105(2): RESTRICTION ON RELIEF

5.29 The meaning of 'control' takes in all of the aspects mentioned in relation to unquoted securities set out above, with one additional consideration under *s 105(2)*. Shares in or securities of a company do not fall within *s 105(1)(cc)* if:

- those shares or securities would not have been enough, without other property, to give the transferor control of a company immediately before the transfer; and
- their value is taken to be less than the value previously determined by reliance on *s 176*.

Section 176 is a specific rule concerning related property where, within three years after the death of the transferor, there is a qualifying sale of any property comprised in the estate of that person immediately before death ('the property concerned') valued in accordance with *s 161* (related property rules), or valued in conjunction with property which was also comprised in the estate but which has not at any time since the death been vested in the vendors of the property. Accordingly, if the shares were part of a control holding at death and qualified for BPR at 50%, that relief is lost if those shares are sold and revalued as part of a minority holding.

Preference shares: effect on control

5.30 The effect on the definition of control of the existence of preference shares is contained in *s 269(4)* (see **5.26** and **5.27**). In brief, where a company has a class of shares with limited voting rights, the definition of control, which requires powers of voting on 'all questions affecting the company as a whole', is regarded as a reference to all questions except those contained in the specific class rights of preference shares.

Clearances for BPR

5.31 A limited but valuable form of help by way of clearance is available in situations of genuine difficulty and importantly where there is IHT at stake. The procedure is described by HMRC on their website (www.gov.uk/guidance/non-statutory-clearance-service-guidance) and requires an application, preferably by e-mail, with supporting information. The following information must be supplied:

- particulars of the applicant;
- IHT reference, if any;
- details of the business, including UTR or company registration number;

- details of interest in the business held;
- contact details;
- what the application is about;
- why the transaction arises;
- description of the proposed transaction;
- what HMRC are being asked to clear;
- likely date of the transaction;
- contingencies involved;
- how important tax is to the transaction;
- the reason for doing the deal this way;
- other related clearances;
- the commercial importance of the issue to the business;
- two years' business accounts;
- the legislation in point;
- the doubt as to the application of the legislation, with reference to published guidance and case law;
- legal advice obtained that the taxpayer is willing to disclose;
- details of previous HMRC advice.

The broad description of the service suggests that the process has limited application and is aimed predominantly at those cases of genuine difficulty – in essence it is no substitute for a thorough study of the law and of guidance as to its application. However, it is the experience of the author that applications, properly made and fully supported with appropriate documentation, are dealt with promptly and can give much valued certainty. Some practitioners may simply find the completion of the application for clearance a useful tool for considering the matter in more detail and that, having thought the issues through, it may not then be necessary to proceed with clearance.

On 13 June 2017 HMRC updated their general online guidance and also the BPR checklist Annex C). The section 'When HMRC will not provide advice under this service' has been amended to state that HMRC will confirm the reason why advice requested has not been provided and listed examples include where:

- all necessary information has not been provided;
- HMRC do not consider there are genuine points of uncertainty;

- the advice requested amounts to tax planning advice or approval of tax planning products or arrangements;
- the application is about the treatment of transactions that HMRC considers are for the purpose of avoiding tax;
- HMRC is already checking the tax position for the period in question;
- any return for the period in question has become final;
- there is a statutory clearance procedure;
- the advice sought relates to the application of the settlements legislation in *ITTOIA 2005, Pt 5, Ch 5* or the tax consequences of executing non-charitable trust deeds or settlements.

Annex C (BPR checklist) has been updated to make clear that the clearance process may not be used for gifts (PETs) to individuals, ten-year anniversary charges, conditional dispositions of property under a will, deeds of variation, the tax consequences of executing a trust deed, the application of *ITTOIA 2005, Pt 5, Ch 5*, valuations, or general confirmations of the status of businesses for business property relief (see www.gov.uk/government/uploads/system/uploads/attachment_data/file/377648/annex-c.pdf).

Situations not covered by the clearance procedure

5.32 HMRC will not be drawn on the availability of BPR if no IHT is at stake and that makes it difficult for executors to know with certainty what assets to appropriate to the trusts of the nil rate band, where so directed by the terms of the will. If too much value has been transferred to the trust, because BPR is on later testing held not to be available, HMRC will treat that as a transfer by the residuary beneficiary on grounds that the executors had power only to transfer such value as would be within the nil rate band, so any excess was not a transfer by them and must have been by the surviving spouse.

To prompt HMRC debate the following might be considered:

- a will could be drawn (or varied by deed of variation) so as to include a discretionary trust;
- a provision of the will or variation specifically gives the target asset to a discretionary trust (the specific gift takes it outside *s 39A* – see **Chapter 15**);
- include a gift of the nil rate band to the discretionary trust, being defined as such sum as would not attract IHT, but for this purpose ignoring the prior gift of the target property.

If BPR is found to be available, the asset can be appointed out to chargeable beneficiaries, such as children. If BPR is refused, and if a decision can be reached within two years of death, the trustee still have the option of appointing the asset to the spouse, in reliance on *s 144*, so as to avoid the immediate payment of IHT.

The difficulty is, of course, in getting everything done quickly enough that a decision is reached well within 24 months from the death. The procedure could easily be frustrated by hesitation within the family or pressure of work on the adviser (as was seen in slightly different circumstances in *Vinton v Fladgate Fielder* [2010] EWHC 904 and its preceding litigation at **6.5**).

SECTION 105(1)(D): LAND AND BUILDINGS ETC USED IN THE BUSINESS

5.33 Relief is available (see *s 105(1)(d)*) where first, the asset concerned is used wholly or mainly for the purposes of a business and second, that business is carried on by:

- a company of which the transferor had control at the time of transfer; or
- a partnership of which the transferor was at the time of transfer a partner.

A further condition is added by *s 105(6)*. Land, a building, machinery or plant owned by the transferor, and used wholly or mainly for the purposes of the business carried on in compliance with *s 105(1)(d)*, will not qualify for the relief unless:

- the business; or
- the transferor's interest in the business; or
- the shares or securities of the company carrying on the business

are relevant business property in relation to the transfer immediately before the transfer takes place. In practice, this means that the owner of, say, a valuable printing press used in a business in which he is neither partner nor, if incorporated, a shareholder cannot obtain relief on his deathbed by buying a controlling interest in the company which operates the press or by joining the partnership which operates it. The minimum period of ownership in relation to that company or partnership would not be satisfied.

Where the relevant asset is used by a company, it is important for the transferor to have control, and this was the point at issue in *Walding v IRC* [1996] STC 13 (see **5.26**). The estate of Mrs Walding included the shares in the family trading company and the factory units used by that company. No relief was available in respect of the factory units under *s 105(1)(d)* because, as noted, it was held that the deceased did not control the company.

Notwithstanding the general rule in partnership law that each partner has an interest in the proceeds of the division of partnership assets, rather than an interest in specific assets, there is a distinction between the partner's interest in the partnership as a whole and their interest in a specific asset, for example the premises from which the partnership business is carried on. Tax law recognises this distinction, in that BPR is available either on the transfer of land held outside the partnership but used by it, or on the transfer by a partner of his interest in a specific asset owned by the partnership.

Relief on property used in a partnership is set at the lower 50% rate. It is therefore likely that in the transfer of an interest which is an asset of the partnership, the transferor would seek to argue that what is given is in fact a proportionate share of the transferor's interest in the business as a whole, thus bringing the gift within the 100% category of *s 105(1)(a)*. There is scope here for fine argument (see *Featherstonehaugh v IRC)*.

To supplement the 'two-year rule' in *s 105(1)(d)*, *s 112(3)* provides that such an asset will not qualify for relief unless:

- it was used for the purposes of the business throughout the two years immediately preceding the transfer of value; or
- it replaced another asset which was so used, and both it and the other asset (and any asset directly or indirectly replaced by that other asset) were used for the purposes of the business for periods which together made up at least two years falling within the five years immediately preceding the transfer of value.

A common problem for farmers is that buildings may become unsuitable for modern machinery or farming practices but if let on a large scale BPR may be denied because the rent received may tip the balance in favour of an investment activity under the 'wholly or mainly' rule. Consider that, instead of letting the buildings, the farmer enters into a LLP with the proposed occupant of the premises, rather like a farm contracting agreement. The occupier agrees to pay the farmer a fixed profit share, plus bonuses according to the success of the venture. The resulting income to the farmer will be derived from a trade or business and the buildings should qualify for BPR at 50%. The ownership period may already have been satisfied: it is a 'snapshot' test and does not require the buildings to have been occupied for the purposes of the business for two years.

Machinery and plant

5.34 These terms are not specifically defined for IHT and in practice the capital allowances definition is used, but qualified slightly so as to exclude items which would qualify for capital allowances only on the renewals basis. However, that rule is not absolute; some latitude may be given (see

IHTM25227) thus, plant includes anything (see *Yarmouth v France* (1887) 19 QBD 647) which the businessman uses to carry on his business other than his stock in trade. It will include goods and chattels, fixed or movable, live or dead, which he keeps for permanent employment in his business. It does not just mean 'machinery' in the common sense of the word.

Animals

5.35 BPR is available on the value of working animals and production livestock kept for the purposes of a business, and includes live animals kept for permanent employment in farming and which serve to produce saleable products. HMRC Capital Allowances Manual, at CAM21220, accepts that a horse used in a school or show jumping business is plant, as is a guard dog or circus animal; but, in the main, farm animals are not plant unless they work but that would include, for example, a sheepdog or a heavy horse used for ploughing.

Animals which are kept mainly for sale, or to provide a saleable product after slaughter, are outside the description and thus there is a clear distinction between dairy cattle, which do qualify as 'production livestock', and beef cattle, which do not. Laying poultry are plant, but not broilers designated for slaughter whilst young animals not yet ready for use as production livestock do not qualify. Farm animals are normally trading stock such that their cost is a revenue item, thus barring a capital allowances claim.

Intellectual 'tools' and property

5.36 Machinery and plant is not confined to assets such as lathes, lifts and cows. It can include books, as in the case of the law library of a barrister (*Munby v Furlong* [1977] 2 All ER 953), and computer software. Software itself was considered for the purposes of Cases I and II of Schedule D in R1 56 (November 1993). That interpretation summarised the thinking of HMRC at that time on the treatment of expenditure on computer software in the computation of taxable profits from a trade, profession or vocation, noting that most software is acquired under licence and that payments may be regular, similar to a rental, or by way of lump sum. Where the payment for software is a lump sum, it is necessary to decide whether the licence is a capital asset in the trade of the licensee. R1 56 states: 'In broad terms a licence is a capital asset if it has a sufficiently enduring nature'.

Shares Valuation Division of the Valuation Agency have refused BPR on patent royalties, even though, when paid to an inventor, they are earned income, but they have indicated that '… it might be different if the individual invented the product, registered the patent and then engaged in the manufacture …'. For an instance of defective drafting which failed to transfer half the benefit of royalties to a spouse, see *Braithwaite v HMRC* [2008] SpC 674.

Equipment, not 'the setting'

5.37 Plant includes the equipment by which or with which a trade or profession is carried on, but not the place where the business is carried on. This distinction has been argued in many cases and notably in:

- *IRC v Barclay Curle & Co Ltd* [1969] 1 All ER 732: the machinery of the dry dock was plant, but not the cost of excavating the dock itself.
- *ESC 26.2.97*: a caravan provided on a holiday site for holiday lettings, but not on residential sites.
- *Attwood v Anduff Car Wash Ltd* [1997] STC 1167: at a car wash plant, the items of washing equipment were plant, but the wash hall as a whole was not.
- *Bradley v London Electricity* [1996] STC 1054: not an electricity substation itself, although the switch gear was plant.
- *Bestway (Holdings) Ltd v Luff* [1998] STC 357: not the premises in which a company operated cash and carry wholesale supermarkets.
- *Family Golf Centres Ltd v Thorn* [1998] STC (SCD) 106: not three new putting greens, which were part of the golf course.

New categories of 'property'

5.38 Tax law adapts to changing conditions in trade. Even the understanding of what constitutes 'property' may be stretched to include new categories of asset. Is 'potential' taxable? This may be relevant to artists, sportsmen and entertainers.

Example 5.6: Sir Frederick Blaise (1) – assignment of royalties

Sir Frederick Blaise, an eminent musician now in his 78th year, spends two days a week teaching promising young clarinettists. He regularly adjudicates at music festivals, at which some of his compositions are used as test pieces and assigns to his close friend, Eva the recording rights to some of his works which generate royalty payments.

Following the decision in *Nelson Dance*, the PET of the assignment potentially qualifies for BPR under *s 105(1)(a)* because, although Sir Frederick has disposed of a mere asset of the business, this represents a diminution in the value of his estate.

Example 5.7: Sir Frederick Blaise (2) – holding of royalties

Dame Eva, a retired soprano, is the friend to whom Sir Frederick in the previous example had assigned the recording rights. Such rights do not relate to any part of her former business and on her death, they are a mere asset which do not qualify for BPR.

HMRC originally denied relief on the copyright of an author in his books on the death of the author, on the ground that the business had come to an end by virtue of the death or was so personal that it could not be transferred. That was to take too narrow a view of the matter and 'It is only by exploiting the work of his brain and his pen that he can make any professional income … it would be absurd to treat his professional activity as merely turning property to gain' (*Carson v Cheyney's Executor* [1959] AC 412).

How far is the ability of, say, a sportsman to be taxed? He protects his name and endorses products but when he does so, does the endorsement agreement have a capital value to be included in his estate at his death? Is such an agreement an interest in his business as a sportsman qualifying for 100% relief under *s 105(1)(a)*, or is it machinery or plant qualifying for only 50% BPR under *s 105(1)(d)*? For many, the doubt may be resolved by incorporation, so that the asset in the sportsman's estate is a shareholding within *s 105(3)(bb)*.

A soprano so captures the hearts and minds of her public that her endorsement of a product carries a Midas touch, and the mere signature of an agreement either to record a work or to perform at some stadium has a value before the release of the recording or the occasion of the performance. She takes steps not only to develop and protect her voice (and looks) but also her name. She assigns to a friend the benefit of a future recording. Is that Midas touch 'property' within *s 105(1)(a)*, or does it take more of the character of plant as being the 'apparatus' used by the soprano for the carrying on of her business? *Section 110(b)*, which is considered in **Chapter 8**, requires that value to be recognised under the heading of goodwill.

The meaning of 'property' in relation to a share of an unadministered estate has been clarified by the case of *Daffodil (Administrator of Daffodil Deceased) v IRC* [2002] STI Issue 20 (see **Chapter 22**).

Bookmakers' pitches

5.39 This category of property arose from a change in the administration of on-course betting introduced with effect from 8 October 1998 when the National Association of Bookmakers, which had previously allocated pitches, lost control of racecourse betting rings to the National Joint Pitch Council (run by the Horse Race Betting Levy Board). As a result, bookmakers who gave up a pitch could auction their 'seniority' amongst bookmakers wishing to take it over. The right to occupy a pitch is now regarded by HMRC as an asset for CGT purposes and as property to which the deceased is beneficially entitled for IHT within *s 5(1)*. Where the right to occupy the pitch forms part of the deceased's business, its value may qualify for BPR at 100%, subject to compliance with the other statutory conditions.

Where a pitch has been inherited after October 1998, the pitch will have a value so that, on its disposal, *s 62(1)* will apply to provide a book value against which the sale proceeds can be measured. Where that value has been 'ascertained' for IHT purposes it will also be the book value for CGT by virtue of *TCGA 1992, s 274*.

SECTION 105(1)(E): LAND AND BUILDINGS USED IN THE TENANT FOR LIFE'S BUSINESS

5.40 The wording of *s 105(1)(e)* follows quite closely that of *s 105(1)(d)* in allowing BPR on 'any land or building, machinery or plant which, immediately before the transfer, was used wholly or mainly for the purposes of a business carried on by *the transferor and was settled property in which he was then beneficially entitled to an interest in possession*' (italics supplied). It would therefore appear that the whole of the discussion of the meaning 'machinery or plant' set out above would apply with equal force to a claim under this heading as it does to a claim under *s 105(1)(d)*.

The meaning of 'interest in possession'

5.41 The meaning of 'interest in possession' was defined in the House of Lords' decision in *Pearson v IRC* [1980] 2 All ER 479 as:

> 'The first [possible conclusion] is that the power of appointment under clause 2 not having been exercised, the three sisters on reaching that age acquired interest in possession defeasible should the trustees decide to exercise their power to accumulate income. They were then entitled absolutely to the capital and income of the trust fund in equal shares subject to the exercise of that power. The second is that they never secured an interest in possession for they never acquired on reaching that age the right to enjoyment of anything. Their enjoyment of any income from the trust fund depended on the trustees' decision as to the accumulation of income. They would only have the right to any income from the trust fund if the trustees decided it should not be accumulated or if they failed to agree that it should be or if they delayed a decision on this matter for so long that a decision then to accumulate and withhold income from the sisters would have been unreasonable.'

The House of Lords took the second view and as a result a power to accumulate will destroy the existence of an interest in possession. In the light of the *Perpetuities and Accumulations Act 2009*, which removes the limitation on the length of the accumulation period, trusts may postpone for the entire trust period of 125 years the date on which beneficiaries will attain vested interests in income. The test is whether, when income arises, the beneficiary can claim it but the existence of an overriding power which could be exercised to take away future income does not affect the existence of an interest in possession.

History of relief under s 105(1)(d)

5.42 This relief is available only if the business of the life tenant is itself relevant business property and that life tenant holds a qualifying interest as defined. It is subject to exceptions and restrictions. The fundamental changes to the taxation of trusts in *FA 2006* had a direct bearing on this relief as post 21 March 2006 interests in possession (except for a transitional serial interests (TSI)) may only arise as immediate post-death interests, but will not, generally, arise on the creation in lifetime of any settlement.

Timing

5.43 The use of the property for the purposes of a business is to be tested at the time of the transfer. The transferor need not, therefore, have used the land or other asset concerned for business purposes at any other time. Specifically, *s 112(3)* excludes the operation of *s 112(2)* from claims under *s 105(1)(d)*.

Example 5.8: John and Helen – acquisition of established business

John Bailey and Helen Clarke are in partnership as will writers and chartered tax advisers, and work from their respective homes. Each has a fully equipped home office with safe storage facilities for wills and trust deeds, hardware and software, and library (both electronic and hard copy). They had the opportunity to take over the business of an elderly colleague – this included a cherished but elderly library which actually duplicated some of their own books and will soon be out of date and a substantial will bank. It is the latter which was the real attraction of the practice.

The old books fall within the 'excepted asset' exclusion under *s 112(2)*: they will not be used for the business either now or in the future. That does not matter for a claim under *s 105(1)(d)* on, say, a later transfer by either John or Helen, because of the rule in *s 112(2)*, but the effect will soon run off because *s 112(3)* prevents any claim on the value of the old books unless they were used continuously during the two years preceding the transfer. There are succession provisions in *s 112(3)(b)*, but they would not apply in this example.

The meaning of 'assets used in the business'

5.44 The meaning of this phrase was tested in *Finch v IRC* (otherwise *Featherstonehaugh v IRC*) [1985] Ch 1. The life tenant of settled land carried on a farming business as sole trader – he farmed some of the settled land, but all the other assets of the business belonged to him in his personal capacity. On his death on 8 September 1977, the executors and the settled land trustees claimed BPR in respect of the charge to CTT arising on his death. At that time, APR was subject to a limit of 1,000 acres, and in this case that limit had been used up on other land. The only scope for claiming BPR was under the provision now represented by *s 105(1)(a)*. Had the business been carried on in partnership, *s 105(1)(d)* would have applied. The trustees succeeded in the Court of Appeal.

Whilst the main argument concerned valuation of the business on the 'net' basis in relation to liabilities, and is considered in **Chapter 14**, the relevance of the case to situations now falling within *s 105(1)(e)* was the ruling by the Court of Appeal that distinguishes sole traders from partners. In determining the value of the business of a sole trader, every asset in the business is to be taken into account and included in the transfer of value. In valuing the interest of a partner in a business, on the other hand, only partnership assets are to be taken into account. Thus the assets owned by the transferor as his own separate property and used by the partnership business do not count towards the value of the interest in the partnership business. As a result of this, personal assets used by the partnership qualify for 50% relief under *s 105(1)(d)*, and not 100% relief under *s 105(1)(a)*.

The statutory fiction (*s 49(1)*), that the life tenant owns land in which the qualifying interest in possession subsists, has to be applied in deciding the value of the assets of the business for the purposes of BPR. The result of this, the Court of Appeal held, is that the assets of the business included the life interest in the land which was used for the purposes of the business. On that basis, BPR applied to the value of the land.

The decision in *Featherstonehaugh* does create intellectual problems. The test in *s 105(1)(e)* is use 'for the purposes of a business'. That does not sit easily with the expression 'used in the business' in *s 110(b)*, the main section on valuation. Do these two expressions have different meanings? *Featherstonehaugh* left that particular issue open.

The main thrust of *s 110* in relation to liabilities is to set them against the value of the assets used in the business. The changes introduced by *FA 2013, s 176* and *Sch 36* are considered at **8.16** and following. Here the statutory fiction in *s 110* exposes certain impracticalities, some of which are affected by the new rules. Why should liabilities be deducted from assets which are not charged or mortgaged in respect of them? Consider that, as will often be the case, the land is held by the trust and is free of mortgage in the commercial sense, whereas recurrent losses from farming have pushed the life tenant into bankruptcy. In commercial terms, the trustees and the land will usually be safe from the claims of creditors. In *St Barbe Green and Another v CIR* [2005] STC 288, the deceased held two life interests but was himself insolvent – it was held that the personal liabilities reduced the value of the free estate to nil, but the excess could not be set against the value of the settled funds.

The applicability of the lower rate of relief

5.45 Following the decision in *Featherstonehaugh* the question arose whether relief would be available under the more generous provisions of *s 105(1)(a)* or the restrictive provisions of *s 105(1)(e)*. The Financial Secretary to the Treasury wrote, in October 1985, to confirm that the more generous relief was available 'when the circumstances accord with those in the *Finch* case'. Thus the land and the business in such a case are to be included in the same transfer of value. This would normally happen on the death of the life tenant with a qualifying interest.

This leaves the applicability of the lower rate of relief to those situations where the transfer of value does not include the business itself but only the trust asset.

Example 5.9: Barry Bolt – lifetime termination of lifetime interest

Barry Bolt farms partly on land owned by a trust in which he has a qualifying interest in possession and partly on land that he has bought personally. On Barry's death the trust fund passes to his daughter Rachel who works with her father on the farm. The trust also owns a substantial house on the farm land which is occupied by Rachel free of charge.

In December 2017, Barry gave up his life interest in the house so that Rachel could own it (and use it as security for a mortgage to pay for its renovation) and also his life interest in the land bringing about the premature cessation of the trust. That action is treated as a PET made by Barry and if he should die within seven years, it will become a chargeable transfer under *s 52(1)* on the house but not on the business itself.

Other conditions

5.46 The normal restriction that there must be no binding contract for the sale of the asset also applies to claims under this head; neither can the property be an excepted asset. Land or a building of which part only was used for the business may be apportioned between business and non-business use (see **Chapter 7**).

The practitioner may therefore be faced with three somewhat similar situations involving the life tenant with a qualifying interest in possession:

- the life tenant keeps both the land and business right through to death;
- the life tenant gives up the land but keeps the business until death; or
- the life tenant gives up the business but keeps the land until death.

In the first of these situations, a transfer of the settled property and of the business occurs at the same time thus 100% BPR is available. In the second situation, assuming the land is held by the trustees and the business by the life tenant, the release by the latter of his interest in, say, the factory, will qualify for 50% relief only. In the third situation, the life tenant is disposing of an interest in a business which he owns in his own right and the income of which has been taxed on him as his personal income, not as trust income. 100% relief is available on the business. The trustees are left with land which is no longer used for the purposes of a business carried on by the transferor. It is probably, as a result, a mere investment qualifying for no BPR at all.

Chapter 6

Business property: the rates of relief, limitations to the relief

QUICK GUIDE

- Business property must generally be held for two years before BPR can apply but, as with APR, there are rules to extend the actual period of ownership by reference to another holder. There are also rules to allow the effective rolling over from one qualifying asset to another (see **6.8**).
- As with APR, the existence of a contract for sale, or for example provisions in a partnership agreement that amount to a contract for sale, will prevent BPR from being available (see **6.14**).
- The rules as to successive transfers that were seen earlier in connection with APR, have a parallel in BPR (see **6.17**).
- Difficulties can arise when it is not certain whether shares in a company are 'quoted' or where the categories of shares change from time to time by reference to a recognised stock exchange (see **6.20** et seq).

THE RATES: MANNER OF APPLICATION

Interaction with quick succession relief

6.1 The relief applies to the value transferred by a transfer of value which is, wholly or in part, attributable to business property. The reduction in value allowed by *IHTA 1984, s 104* is applied before the claim to any exemption which might make the transfer of value wholly or partly exempt, thus BPR is applied before quick succession relief.

Example 6.1: Clancy brothers (1) – accumulated cash profits

Brothers, Fred and George Clancy ran a successful company specialising in the production of theatrical costumes and fancy goods. Fred died on 30 June

2015 and his estate included shares in the unquoted trading company worth £200,000 which he left to George. Profits have been accumulated in cash within the company, and 25% of the net worth of the company is represented by cash and near-cash assets that are not needed for trading thus representing excluded property under *s 112*.

Assuming an effective estate rate of 20%, the tax charge on Fred's death is as follows:

	£
Value of shares	200,000
Value of shares less excepted assets (say)	(150,000)
Chargeable transfer (200,000 – 150,000)	50,000
IHT thereon at estate rate of 20%	10,000

The business continued to prosper and on George's death on 28 February 2017, the holding that George inherited from Fred was worth £220,000. Of that value, only 20% is now represented by excluded property, with the result that, £176,000 (£220,000 @ 80%) is sheltered by BPR and £44,000 is chargeable, subject to any other relief.

The appropriate percentage under *s 141(3)(b)*, for the time elapsed of more than one year but less than two years, is 80%. The calculation of quick succession relief (QSR) under *s 141(1)(a)* is as follows:

Net increase in George's estate (50,000 – 10,000)	£40,000
Tax charge on the first transfer	£10,000
Reduction, per *s 141(3),* of that tax	80%
Quick succession relief (10,000 × 80% × 40,000/50,000)	£6,400

Interaction with other reliefs

6.2 *Section 104* interacts with *s 165*, which is concerned with the incidence of CGT (note that this is not a reference to *TCGA 1992, s 165* which allows hold over relief from tax on gifts of business assets but, *IHTA 1984, s 165* which also affects a 'gains' situation). Under *IHTA 1984, s 165(1)*, CGT may reduce the value transferred by a transfer – this may occur where a chargeable transfer is, or includes, the disposal of an asset and a chargeable gain arises on that disposal. Commonly, the donor will pay the CGT and, in such a case, there will be no offset relief except that the estate of the donor is further reduced by having to pay that tax. However, where the whole or part of the gain is a chargeable gain and the whole or part of any CGT levied thereon is met by the donee (as defined in *TCGA 1992, s 282*), then the CGT so borne by the donee is treated as reducing the value transferred. BPR, where in point, is applied before applying the reduction available under *IHTA 1984, s 165*.

Example 6.2: Clancy brothers (2) – accumulated cash profits tax thrown away

On 31 December 2015, George, who had made chargeable transfers that exhausted his nil rate band, gave his brother Ron a block of shares in the family trading company worth £200,000. It was agreed that *s 112* excluded from BPR 25% of the value of the shares in the company at that date. The gain on the shares, which attracted no reliefs from CGT, was £80,000. Although a claim could have been made under *TCGA 1992, s 165* on the grounds that the shares qualified for holdover relief, no claim was made immediately and senile dementia, thereafter, prevented the parties from making one. Accordingly, the gain became chargeable. George did not pay the tax on the gain within 12 months, so that tax became recoverable from Ron under *TCGA 1992, s 282*.

The tax to be suffered by Ron of £80,000 would, under *IHTA 1984, s 165(1)* in the case of an asset not qualifying for BPR, reduce the value of the gift from £200,000 to £120,000. However, BPR of £150,000 first reduces the chargeable transfer to £50,000 and that remaining value is wholly eliminated by the tax that Ron has (unfortunately and unnecessarily) paid. There is a right of recovery from the donor (George/George's estate) of such tax under *TCGA 1992, s 282(2)*. (It should be noted that the expression 'close relative' in *IHTA 1984, s 165(3)* does not include a brother. That expression is defined in *FA 1984, Sch 14, para 1(1)* to include only the spouse, children and remoter descendants.)

Grossing up

6.3 The effect of *IHTA 1984, s 104(2)* is also to apply BPR before any grossing up but there are complications in applying BPR where claw-back applies under *s 113(a)*.

Example 6.3: Alice and John – simple grossing up

Siblings, Alice Carr and John Davis were in partnership as motor engineers. On her death in October 2017, Alice, who had already utilised her nil rate band against lifetime transfers, left John, a tax free legacy of her 50% interest in their business, then worth £220,000. Examination of the accounts of the business shows that 25% of that value was represented by investments not in use for the purposes of the business, with the result that BPR was restricted to £165,000 (£220,000 @ 75%). The unrelieved portion of the gift, £55,000 (£220,000 – £165,000), must be grossed up. The grossing up fraction at the death rate is ⅔ thus the tax chargeable is, £55,000 × ⅔ = £36,667 ((£55,000 + £36,667) @ 40%).

BPR is available in appropriate circumstances to trustees of relevant property trusts. *Section 103(1)* includes within the definition of 'transfer of value' an occasion on which tax is chargeable under *Chapter III* of *Part III* of *IHTA 1984*, ie the relevant property trust regime, except for *s 79* (exemption from ten-yearly charge). Thus, *s 103(1)(b)* includes references to the trustees of the settlement concerned within the meaning of 'transferor'. In practical terms, this could give rise to the allowance of an exemption before BPR. This would occur on a partly exempt occasion of charge within the relevant property trust regime but outside the scope of the ten-yearly charge under *s 79*. This will be a rare occurrence.

The rates of relief

6.4 BPR at 100% applies to:

- unincorporated businesses;
- unquoted shares which either by themselves or with such other securities or unquoted shares gave the transferor control; and
- any unquoted shares in a company not listed on a recognised stock exchange (by *Income Tax Act 2007, s 1005* extended to include any investment exchange that the Financial Services Authority so designates) but including those traded on AIM ETHEX but only part of the PLUS (formerly OFEX and now ISDX) market.

The 50% relief applies to:

- shares or securities giving control of a quoted company;
- land, buildings, machinery or plant in a partnership or in a controlled company; or
- land in a settlement (subject to the restrictions as before).

PERIOD OF OWNERSHIP

6.5 *IHTA 1984, s 106* provides that 'property is not relevant business property in relation to a transfer of value unless it was owned by the transferor throughout the two years immediately preceding the transfer'. Ownership for this purpose also includes beneficial entitlement such that if the asset is owned by trustees and there is a qualifying interest in possession (see IHTM25302), property in which the transferor has such an interest can qualify for relief. Where there is no such qualifying interest in possession, it is enough that the trustees have legal ownership.

Where the asset is a shareholding that has been added to from time to time ie accumulation units, HMRC will ask for evidence that the shares actually transferred are derived from a holding that satisfies the two year rule.

Difficulties may arise where, in an unincorporated business, there has been substantial change in the value of a partnership interest shortly before transfer.

Example 6.4: Henry and Ian – purchased interest

Henry Jones and Ian Lang are partners in a printing business which has run for many years. Trade is poor and Henry retired on 31 December 2016, selling his interest to Ian. Ian ran the business until his death on 30 June 2017. BPR is allowable at Ian's death on the whole of the value of his 'old' interest in the business, but on none of that part of the value which is represented by the share purchased from Henry which fails to meet the two year ownership test.

Example 6.5: Donald and Evelyn – organic growth

Donald Blake and Evelyn Wragge trade together in partnership as computer software engineers. In June 2011, the value of their respective business interest was no more than £40,000. Astute marketing in relation to the worldwide credit crunch during the long recession secured lucrative contracts such that, by December 2017, the value of the capital held by each partner had grown to £600,000. There had been no change in profit-sharing ratio over the two-year period and all the growth had been organic. Even though, in real and commercial terms, much of the growth in value has arisen only in spring 2017, on a transfer of an interest in a business by either partner in December 2017 BPR is allowable on the entire value.

For administrative convenience, it is preferable for spouses in a family business to have separate partnership capital accounts, so that ownership of a particular share in the business can be identified in order to prove the requirements of *s 106*.

An attempt to circumvent the two-year rule failed in *Executors of Mary Dugan-Chapman Deceased v HMRC* [2008] SpC 666 (see **6.7**). The aftermath of that case was a claim in negligence by the family against the lawyers: see *Vinton v Fladgate Fielder* [2010] EWHC 904 (Ch), which suggested that, if the scheme had been more precisely executed, relief might have been secured.

Modification of the two-year rule

6.6 The ownership of a mere business asset without business activity does not qualify for BPR; but, once it is established that a business is being carried on, a change in the precise nature of that business will not affect the requirement under *s 106* as to a minimum period of ownership. A change in the use of a business asset from one ('qualifying') trade to another will not prevent the value of that asset, if part of a business, from satisfying *s 106* (see IHTM25303).

REPLACEMENT PROPERTY

6.7 A more fundamental change, from the ownership of one business asset to another, is regulated by *s 107*, under which property satisfies the condition where it replaces other property and both the new property and the old property (and property which directly or indirectly replaces the old property) are owned by the transferor for periods which together amount to at least two years in the last five years before the transfer.

For the replacement provisions of *s 107(1)(a)* to apply, any other property concerned in the exchange must be such that, had the transfer of value taken place just before the replacement of that other property, it would have qualified for BPR (except in so far as the period of ownership of the other property did not satisfy the test in *s 106*). *Section 107* is concerned with the replacement of property which consists of:

- a business or an interest in a business;
- unquoted company securities giving the transferor control;
- unquoted shares;
- quoted shares giving the transferor control;
- land or buildings, machinery or plant used for a business carried on by a company controlled by the transferor by a partnership of which the transferor was a partner; or
- land, buildings, machinery or plant used for a business carried on by the transferor but belonging to a settlement in which he had an interest in possession.

Section 107 does not apply to the replacement of other property, for example mere assets of an unincorporated business and this is consistent with the wording of *s 105(1)*. It is important that the acquisition of the replacement property takes place after the disposal of the original property. The guidance in the HMRC Manual has been revised and updated to reflect recent changes (see IHTM25314 and the flowchart at IHTM25312).

The reasonableness of the approach of the taxpayer was partly in issue in *The Executors of Mrs Mary Dugan-Chapman and Another v Revenue and Customs Commissioners* [2008] SpC 666. It had been hoped by the taxpayer that the use of a rights issue to capitalise a loan might effectively sidestep the two-year ownership rule by deeming the shares acquired to be, in effect, part of an existing shareholding which had been owned for the requisite period. This relied on the reorganisation provisions of *TCGA 1992, s 126* when read with *IHTA 1984, s 107(4)*, but the case turned on its own very special facts which are much compressed here but illustrate several principles.

Counsel had advised the family that a rights issue in the family trading company would cause new shares acquired under it to be identified with shares already held, so that the new shares would be deemed to have been held for as long as the shares in respect of which the rights arose. That would help Mrs Dugan-Chapman, who had a cash balance of £300,000 with the company and who was recently widowed, elderly and might not live two years.

The company had been inactive, which might prejudice a claim to BPR and there were worries that voting balance would be disturbed if only Mrs Dugan-Chapman took up shares. The company sold a painting, easing cash-flow requirements, but was considering fresh purchases and there were issues over pre-emption rights. There was delay, during which the health of Mrs Dugan-Chapman deteriorated. Shares were issued to capitalise her £300,000 loan account but, in addition, documents were prepared for a further investment of £1,000,000. Mrs Dugan-Chapman, through an attorney, was to subscribe for £1 million shares but there was no mention, second time round, of any offer to other shareholders. The taxpayer died only two days after the second issue of shares.

One major factor was that the paperwork did not match the original concept. There were two versions of minutes for the same meeting and two versions of the letters of application for shares, one marked 'wrong' and another 'This one was substituted.' There were also errors in the increase of capital needed.

Section 107(4) allows BPR where shares would be identified with other shares under the 'paper for paper' CGT rules in *TCGA 1992, ss 126–136*. Judith Powell, as Special Commissioner, noted the doubts as to whether BPR would be available in respect of an inactive company and that was relevant to the motivation for the transactions. HMRC agreed that the 300,000 shares did qualify for BPR and it was also accepted by the taxpayer that some shares renounced by a family member and taken up by Mrs Dugan-Chapman did not qualify for BPR because they were not issued in respect of her holding on that date.

Ms Powell held that the final investment no longer had the nature of a rights issue: 'there was confusion about the effect of a rights issue throughout the process'. Actions taken under extreme time pressures did not achieve the effect that had been hoped for. Relief would be available only in respect of new shares that represented a pro rata entitlement based on an existing shareholding, not just any shares issued as part of a rights issue, such as those renounced by other members of the company. Some informality in company dealings can be tolerated, but only where the shareholders are aware of all relevant facts (see the judgment of Neuberger J in *re Duomatic Ltd* [1969] 2 Ch 365). They were not so aware here and relief was denied.

The general principles that practitioners may draw from the case are long established:

- shares issued as part of a rights issue can qualify for BPR immediately; but
- get the paperwork right;
- keep only one (correct) record of the transaction, not two conflicting notes; and
- get on with it.

For an interesting, if provocative and untested, use of *s 107* consider the following example.

Example 6.6: Susanna Green – double BPR

Susanna Green was recently widowed. She has for over two years, held unquoted shares in A Ltd, a trading company, worth £750,000 and also had £1 million on bank deposit. She gave her A Ltd shares into a relevant property trust established for her son Chris and his family, claiming BPR to reduce the chargeable transfer and hold-over relief to avoid an immediate CGT charge.

One week later, Susanna using her cash reserves invested £750,000 in shares in B Ltd, an unquoted trading company.

On her death one month later in December 2017, her executors claim BPR on the retained B Ltd shares on the basis that:

- they replaced other property (the A Ltd shares given away);

- Susanna had owned shares in A Ltd then in B Ltd for periods which together comprised at least two years falling within the five years preceding Susanna's death;
- had Susanna owned the A Ltd shares at her death they would have qualified for BPR.

Is this correct?

Under *s 107(2)* 'relief under this Chapter' shall not exceed what it would have been had the replacement not been made. On Susanna's death, her estate contained cash of £250,000 and B shares worth £750,000 in respect of which BPR is claimed. Can 'relief under this Chapter' mean, not only that claimed in respect of the transfer on death but also claimed in respect of the lifetime chargeable transfer? It is an interesting question, but there is no case directly in point.

IHTM25313 warns against allowing taxpayers to 'trade up' in value shortly before death and sees this as the 'mischief' that *s 107(2)* seeks to avoid but it does not exactly deal with the situation described above. The practical approach recommended in IHTM25313 suggests that the scheme would be challenged, especially if substantial tax was at risk.

Period of ownership of unquoted shares: capital reconstruction

6.8 The deemed period of ownership of unquoted shares may be governed by *IHTA 1984, s 107(4)*. This is without prejudice to *s 107(1)*. Where unquoted shares (falling within *s 105(1)(bb)*) are owned by the transferor immediately before the transfer, and (in a situation such as a paper for paper exchange) those shares would be identified with other shares previously owned by the transferor by virtue of the rules in *TCGA 1992, ss 126–136,* the period of ownership of the 'old shares' counts towards the period of ownership of the 'new shares' for the purposes of satisfying the two-year test in *IHTA 1984, s 106*.

The effect of *s 107* is only to 'save' periods of ownership which have already been earned in respect of the assets which have now been replaced, and not to increase the relief. By *s 107(2)*, the relief is not to exceed what it would have been had the *replacement* of any one or more of the *replacements* not been made. For this purpose, changes which result from the formation, alteration or dissolution of a partnership are to be disregarded. Similarly, changes arising from the acquisition of a business by a company controlled by the former owner of the business may be disregarded. See below as to how *s 107(4)* affects transfers on or after 6 April 1996.

Example 6.7: Roger Lapin – late incorporation

In October 2016, Roger Lapin transferred a business, which he has owned for many years and which was then worth £200,000, to a new company (formed for the purpose) in consideration of all the shares in that company. In December 2017, Roger died at which time the shares were worth £300,000. BPR is available on £300,000, notwithstanding that the shares have been owned for only 16 months.

A difficulty may arise in reconstruction situations where, for example, a taxpayer holds both 60% of the shares in a holding company and 20% of its subsidiary and becomes entitled, through a reconstruction, to extra shares in the holding company. If that augmentation of the shareholding in the holding company increases the total value held by the shareholder in respect of which a claim is made to BPR, HMRC may require evidence of the prior value of each holding so as to prove that (as may well in fact not be the case) the limitation in *s 107(2)* has not been breached.

A similar point may arise in connection with employee share schemes. Consider, the original owner of shares both in a holding company and in its trading subsidiary, at a time when only 50% BPR was available in respect of minority holdings, transferred shares to an employee scheme and thereby lost some of the BPR that would otherwise have been available to him. On a reorganisation, the effect of which is to restore his holding to one qualifying for relief by the time of his death by transferring the shares in the subsidiary up to the top company, it will not be necessary to requalify for BPR by holding the new shares in the holding company for two years from the reconstruction.

Extending the period of ownership

6.9 The period of ownership on all inherited property for the purposes of *ss 106* and *107* is capable of being extended. The legatee, ie the second owner, is treated as having owned the property since the date of the death concerned (the date of death of the person from whom the property is inherited). Where the previous owner was the spouse of the person now inheriting, any period of ownership of the deceased spouse counts towards the period of ownership of the surviving spouse. *Section 108(b)* contains no limitation of this rule relative to the period of the marriage. The provisions of *s 108* are examined further at **6.16** in the context of successive transfers.

Example 6.8: Arthur Nexis – combined ownership period

Arthur Nexis, an elderly bachelor, owned a manufacturing business for many years. In May 2016, he married Bernice his companion of many years, revising his will to make her sole beneficiary of his estate. Arthur died in July 2016. In January 2017, Bernice, having formed a relationship with Carl, made a will in his favour. On her death in October 2017, BPR is available on the business inherited from Arthur notwithstanding that her period of ownership was a mere six months, and that the period of ownership by Arthur, whilst her husband, was also only three months. Carl, whom Bernice never married, died in February 2018. He is deemed, by *s 91*, to have been the owner of the property only since the death of Bernice in October 2017. No relief is due on his death.

New rules, extending the period of ownership, have been introduced by *s 267A* as a result of the enactment of *Limited Liability Partnership Act 2000, s 11* (see **5.21**).

PERIOD OF OWNERSHIP: DETAILED RULES

6.10 *Section 91(1)* provides that, where a person would have been entitled to a qualifying interest in possession (IPDI) in the whole or part of the residue of the estate of a deceased person had the estate been fully wound up, the same consequences follow under *IHTA 1984* as would apply if that person had become entitled to a qualifying interest in possession in the un-administered estate and in the property representing residue (or a corresponding part of it) on the date as from which the whole or part of the income of the residue would have been attributable to his interest, if the residue had been ascertained immediately after the death of the deceased person.

In brief, the legatee is treated as becoming entitled to his inheritance on the day of the testator's death. For the purposes of *s 91*, 'un-administered estate' means all the property in the hands of the personal representatives other than:

- property which passes to them otherwise than as assets for the payment of debts; and
- property which is specifically given to a beneficiary.

Due allowance has to be made for any charges on residue and for any adjustments as between capital and income that have to be made during the administration.

Definitions for s 91

6.11 *Section 91(2)* sets out three definitions for the application of *s 91(1)* thus:

(i) 'Ascertained residue' means residue proper, no longer held by the personal representatives to meet debts.

(ii) 'Charges on residue' means (see *ICTA 1988, s 701(6), (7)*), in relation to the estate of a deceased person, the following liabilities, provided that they are payable out of residue and interest in respect of those liabilities, ie:

- funeral, testamentary and administration expenses and debts;
- general legacies, demonstrative legacies, annuities, and any sum payable of residue to which a person is entitled under the law of intestacy of any part of the United Kingdom or any other country; and
- any other liabilities of the personal representatives 'as such'.

(iii) 'Specific disposition' means (see *ICTA 1988, s 701(5)*) a specific devise or bequest made by a testator. The phrase includes the disposition of personal chattels made by *Administration of Estates Act 1925, s 46* (the widow's legacy), and any disposition which has an effect similar to that of the specific devise or bequest under the law of England and Wales, by virtue of any enactment or by the operation of a law of another country which has a similar effect. In many cases, the provision for a widow at the new rates will 'swamp' the estate. Real estate which is included in a residuary gift, either by a specific description, or by a general description, is deemed to be part of the residue of the estate, and is not deemed to be the subject of a specific disposition.

TRANSFERS AFTER 6 APRIL 1996: SPECIAL RULES

6.12 The effect of *FA 1996, s 184* was to achieve some simplification with conditions imposed for relief on lifetime transfers (and other events occurring) on or after 18 March 1986 and within seven years of the death of the transferor and these were contained in *FA 1986, Sch 19, para 21*, but did not apply to charges under the rules relating to gifts with reservation. It was *para 21* which inserted the claw-back charge of *IHTA 1984, s 113A*.

NO CONTRACT FOR SALE

6.13 The general rule under *IHTA 1984, s 113* is that, even where property would otherwise qualify for BPR in relation to a transfer of value, no BPR is available if a binding contract of sale of the property exists at the time of transfer.

What is 'a binding contract for sale'?

Part performance

6.14 There has been, in *Yaxley v Gotts* [2000] Ch 162, some comment on the doctrine of part performance as it affects the requirement of *Law of Property (Miscellaneous Provisions) Act 1989, s 2(5)* that a contract be 'made in writing' rather than, as was previously the rule, 'evidenced by writing'. *Section 2(5)* does not, it seem, remove the force of all the old cases on the subject, although it might be difficult to show the precise date on which the deemed contract took place.

'Buy and sell' agreements

HMRC do, however, consider that there is a binding contract for sale where partners or shareholder directors of companies enter into an agreement under which, in the event of death or retirement of one of them, the personal representatives of the partner or director so dying are obliged to sell, and the survivors are obliged to purchase, the interest of the deceased in the business or company. These arrangements, known as 'buy and sell' agreements, are specifically referred to in SP12/80. In the view of the Board, such an agreement, requiring as it does a sale and purchase and not merely conferring an option to sell or buy, is a binding contract for sale within *s 113*. As a result, BPR is not available on the interest or the shares, as the case may be. That said, IHTM25292 acknowledges that most arrangements will not constitute 'buy and sell' arrangements: it is therefore necessary to look at each case on its merits.

Examine each case

Further addressing the issue, the ICAEW published memorandum TR557 on 19 September 1984 giving guidance which reproduces correspondence between HMRC and the accountancy bodies concerning the availability of BPR in these circumstances. This memorandum noted the table, published on 6 May 1981 in the *Law Society's Gazette*, summarising the position in various circumstances, which is examined in **Chapter 14**. HMRC regard each situation as separate and distinct, requiring careful scrutiny of the agreement in each case. It was generally considered that *s 113* was intended to deny relief where the transferor, having decided to cease trading, entered into a binding contract for the sale of his business and later died or gave away his interest before completion of the sale. BPR was not to be available on, in effect, the cash realised from the proceeds of sale of a business. The statement of practice extends the application of *s 113*. By inference, it is assumed that BPR will be denied to lifetime transfers, even where there was no intention to cease trading.

Cross-options

On 5 July 1982, HMRC confirmed their understanding that the purpose of *s 113* (as it became) was to limit BPR to transfers of property and to exclude from BPR the transfer of, in effect, the proceeds of sale of a business. HMRC confirmed that, where, for example, on the death in service or retirement of a partner, the surviving partners have an option to buy and the retiring partner (or his personal representatives) have an option to sell, the share in the partnership relief will be available.

That relief will also be available where there is no provision in the partnership deed for the purchase of a partner's share and that partner by his will directs that, if he dies in service, his interest in the partnership is to be sold. The key is whether there is freedom to retain the asset and, for example, sell elsewhere or give it away. Thus, where the provision in the partnership deed is more specific and the surviving partners 'shall purchase' the share of the deceased, no BPR is available. If the contract for sale is not to operate during the lifetime of the partner, such that the partner can give away his interest and that interest continues as an interest in the hands of the donee, BPR will be available.

Accrual/annuity provisions

In the increasingly rare situation where the partnership deed provides that, on the death in service or retirement of a partner, his share is to accrue to the surviving partners who are obliged to pay an annuity to the partner (or to the spouse of the partner as appropriate), BPR will be available on the lifetime transfer of that interest if the accrual/annuity provision does not then come into operation. If that provision does, on the occasion of the transfer, come into operation, then spouse exemption may be due to the extent of the value of the annuity. On death in service, there would be no BPR but exemption would be due on the widow's annuity. In such a case, it is also relevant to consider the Statement of Practice of 17 January 1975, in that any chargeable gains accruing to a retired partner would be computed by comparing the consideration received (including the capital value of the annuity) with the CGT 'cost' or the market value at the date of death for personal representatives of a deceased partner.

Agreements that failed

Further correspondence between the accountancy bodies and HMRC in 1983 highlighted the different views of the interpretation of *s 113* (as it became). *Section 113* excludes from BPR property which is subject to a binding contract for sale 'at the time of the transfer'. The accountancy bodies had argued that the legislation was authoritative for the view that, for the purposes of the tax, the time of a transfer on death is immediately before that death. Building on that argument, it was asserted that a buy and sell agreement in a partnership deed which comes into operation only at the moment of death could not be taken to fetter the interest of a partnership immediately before that death, such that it could not amount to a binding contract for sale at the time of the transfer with the result that the interest of the partner would qualify for BPR.

HMRC did not accept this approach. They considered that the combined effect of what are now *ss 3(4)* and *4(1)* is that the event, namely death, is an occasion of charge and requires a valuation to be made by reference to a hypothetical transfer before death. Effectively, under *s 3(4)*, the death is itself a transfer of value. Thus, when a binding contract for sale exists at the time of death, ie at the time of the statutory transfer of value within *s 3(4)*, BPR is excluded by *s 113*.

Expiry date of cross-options

Whilst cross-options do not constitute a binding contract for sale, care should be taken to ensure that the options do not exist over precisely the same period. This is because HMRC consider that such arrangements do, in effect, between them amount to a contract. The call option should therefore expire before the put option can be exercised.

Exception to the rule in s 113: underlying assets

6.15 The rule in *s 113* applies to the relevant business property, ie that falling within *s 105(1)*. It does not apply to the underlying assets. Thus, where contracts have been exchanged for the sale of an asset which is used for the purpose of a business, and relief is sought in relation to the business as a whole, BPR may still be available on the value of the business, including that portion which is represented by the right to receive the proceeds of sale, if it can be shown that those proceeds are required at the time of the transfer for future use in the business. This will be the case where the vendor was in the process of buying other assets to replace the property sold.

SUCCESSIONS AND SUCCESSIVE TRANSFERS: THE RULES IN SS 108 AND 109

6.16 The period of ownership for the purposes of *ss 106* and *107* may be extended on a 'succession' by operation of *s 108* – see IHTM25332. For succession:

- the person now making the transfer ('owner no 2') must have acquired the property from the original holder of the asset ('owner no 1') on the *death* of owner no 1;
- on the transfer in respect of which relief is now claimed, ie to owner no 3, the period of ownership by owner no 2 starts running from the date of death of owner no 1; and
- if owner no 1 was the spouse or civil partner of owner no 2, the period of ownership by owner no 1 is attributed to owner no 2.

Section 109 extends BPR in certain circumstances where the operation of *s 106* would deny the relief. The requirements are as follows:

- the whole or part of the value transferred by a transfer of value, 'the earlier transfer', must have been eligible for BPR or must have been of a nature which would have been eligible for BPR if that relief could have been given in respect of transfers of value made at the time of that earlier transfer; and
- the whole or part of the property which was relevant business property in relation to the earlier transfer must become, by virtue of that transfer, the property of the person (or of the spouse of the person) who is the transferor on the subsequent occasion, 'the later transfer'; and
- the property, or part of it, or property which directly or indirectly replaces it, must either have been relevant business property in relation to the later transfer, or capable of so being, except that the two-year period required by *s 106* is not satisfied; and
- either the earlier transfer was, or the later transfer is, made on the death of the transferor.

Importantly, and in a parallel to replacement property under *s 107(2)*, *s 109* cannot operate to increase the amount of BPR but only to preserve the BPR which, by virtue of *s 106*, would otherwise be lost. Where, under the earlier of the two transfers, the amount of the value transferred which was attributable to that property was only part of its value that restriction will apply to the claim to BPR on the second transfer. This might arise on a transfer for less than full consideration.

Example 6.9: James Grant – restricted relief (1)

James Grant, a legal aid practitioner, was made redundant in September 2013 and immediately set up in business as a will-writer. He did well and, by the time of his death in March 2016, the business was worth £140,000 analysed as follows:

Cash	£60,000
Debtors	£8,000
Recurring income from storage fees of £4,000, for which an offer has been received of	£40,000
Other goodwill	£32,000

James, who had used his nil rate band against lifetime transfers, left his estate to his common law partner (to whom he was not married) Sally who had also worked in the business and continued to run it with their daughter Gemma whilst Gemma was gaining appropriate qualifications. To encourage Gemma, Sally gave her a 50% share in the business in October 2017 at a time when the business has much the same value as in March 2016. It is considered that ownership of a part of this business is subject to a 90% discount, ie Sally's retained 50% share is worth £63,000, so the gift to Gemma is a transfer of £140,000 – £63,000 = £77,000.

Applying *s 109*: the 'earlier transfer' on James death qualified for BPR as James had owned the business for over two years. However, the business retained £30,000 more cash than it needed and this was 'excluded' such that BPR was restricted to the balance of £110,000, or 78.57% of the full value. The transfer to Gemma took place before Sally has 'earned' BPR by two-year ownership and the extended spousal ownership in *s 108(b)* did not apply but the requirements of *s 109* are met. BPR is allowed on the transfer by Sally of £77,000 × 78.57% = £60,499.

Example 6.10: Jennifer Price – restricted relief (2)

Jennifer Price has owned a business for many years and agreed to sell it to her nephew David in April 2016. The agreed price was £160,000, reflecting all of the value except goodwill, which was worth a further £40,000. David died within 18 months of the purchase.

The sale by Jennifer to David comprised in part a gift of £40,000 (the goodwill) on which BPR was available. When David died, *s 109(3)* operated so as to restrict the relief – if the business (to include the goodwill) had fallen in value to £160,000 BPR will be available only in respect of 1/5, being the fraction of the interest held by David which falls within the requirements of *s 109*. Relief is therefore £32,000.

For a time, specifically between 17 March 1987 and 5 April 1996, an additional rule was imposed by *s 109A*. The rule related to minority holdings of unquoted shares which did not qualify for the higher rate of relief unless the transferor held 25% of the votes throughout the 2 years immediately before the transfer.

LIQUIDATIONS AND RECONSTRUCTIONS

6.17 *IHTA 1984, s 113* specifically excludes from its general rule property in respect of which a binding contract for sale has been made where the property concerned is shares in, or securities of, a company and the sale is made for the purposes of reconstruction or amalgamation.

Under *s 113A*, relief is available only where the relevant business property disposed of by the transferor under a lifetime gift is kept by the transferee as business property in the hands of the transferee and remains so. This application would be harsh where the original property had been replaced. To mitigate this, *s 113A(6)* allows relief where shares owned by the transferee immediately before the death of the transferor would qualify to be identified with the subject matter of the original gift under the CGT rules contained in *TCGA 1992, ss 126–136*, or where those shares were issued to the donee in consideration of the transfer of a business, or of an interest in a business, which in either case represented the original subject matter of the gift (or part of it). For *s 113A* to apply, the replacement must be of the *whole asset*.

A detailed examination of *TCGA 1992, ss 126–136* is outside the scope of this book (and the reader is referred to *Capital Gains Tax 2017/18* (Bloomsbury Professional)) but, in general terms, the relevant provisions concern reorganisation or reduction of share capital. For this purpose, 'reorganisation' means a reorganisation or reduction of a company's share capital; and, in relation to that reorganisation, 'original shares' means shares held before and concerned in the reorganisation, and 'new holdings' means the shares in and debentures of the company as a result of the reorganisation which represent the original shares (*TCGA 1992, s 126(1)*). The effect of *TCGA 1992, s 127*, subject to the general code as to reorganisation, is to treat the original shares (taken as a single asset) and the new holding (likewise taken as a single asset) as being the same asset acquired as the original shares were acquired – see *Executors of Dugan-Chapman deceased v HMRC* [2008] SpC 666.

The liquidation of the company in which shares are held which would otherwise have qualified for BPR has the same effect as the sale of other categories of relevant business property namely that the relief is lost.

The relaxation of *s 113A* where the original property is sold and other qualifying property is purchased, is described in **Chapter 8**.

THE MEANING OF 'QUOTED'

6.18 The meaning of 'quoted' and 'unquoted' is explained in both *ss 105(1ZA)*, inserted by *F(No 2)A 1992, Sch 14, paras 2* and *8* in relation to transfers of value made after 9 March 1992 and *s 13A(3B)*.

Section 105(1ZA) defines the availability of BPR on first consideration of the transfer rather than in relation to claw-back. The key is whether the relevant security is 'listed' rather than whether, from day to day, investors can see the price at which shares are traded; thus, 'quoted' means 'listed on a recognised Stock Exchange' whilst 'unquoted' means 'not listed on a recognised Stock Exchange'.

The existence of a market for unquoted shares does not automatically cause the loss of BPR; thus 'unquoted' includes shares dealt in on the Unlisted Securities Market and the Alternative Investment Market (AIM) and other designated markets. Note that the shares of a company, though not listed in the UK, might also benefit from a secondary listing on a recognised exchange in another country, and that the secondary listing might cause the shares to be 'listed' so as to deprive them of BPR (see **6.21**).

Recognised stock exchange

6.19 The phrase 'recognised stock exchange' occurs throughout the *Taxes Acts*, for example in the definition of a close company in *ICTA 1988, s 415*. The definition itself occurs in *Income Tax Act 2007, s 1005* which is an improvement on the earlier, slightly unhelpful formula and means (with effect from 19 July 2007):

- any market of an investment exchange which is designated as a recognised stock exchange by an order of HMRC; or
- any market outside the UK which is so designated.

'Listed' means:

- admitted to trading in an exchange and included in the official UK list; or
- officially listed on a recognised exchange in a qualifying country outside the UK in conformity with provisions that correspond generally to what applies in EEA states;

and, by a circular definition, a 'qualifying country' is any country in which there is a recognised stock exchange. HMRC issued guidance which whilst not primarily directed to IHT situations does not exclude them. It notes that AIM is part of the London Stock Exchange and that the main exchange is recognised, so AIM shares are traded on a recognised exchange, but they are not 'listed' because they are not on the Official List maintained by the Financial Services Authority, which is the UK listing authority. The guidance draws a distinction between 'recognised investment exchanges' and 'recognised stock exchanges', noting that the investment exchanges will not automatically be designated as recognised stock exchanges; thus such designation will be possible but not a foregone conclusion.

Other trading exchanges

6.20 The present position is thus that BPR is available on a fairly wide range of securities, listed and 'quoted', in the sense that there is in the financial press a regular record of the prices at which deals have been struck. HMRC website contains only an outline of which exchanges are recognised and a link.

It is critical not to assume that all AIM and PLUS (now ISDX) shareholdings qualify for BPR. The 'wholly or mainly' test in *s 105(3)* applies, and the accounts of an AIM or PLUS/ISDX company will be examined to see if it qualifies as a trading company.

Which stock exchanges are actually recognised?

6.21 This is a moveable feast and reference should be made to the HMRC website at www.hmrc.gov.uk/fid/table1-rse.pdf.

Since 2013, HMRC has designated the following as recognised stock exchanges for IHT purposes:

- The Channel Islands Securities Exchange Authority (20 December 2013);
- GXG Main Quote (23 September 2013);
- Global Board of Trade (30 July 2013);
- GXG Official List (16 May 2013);
- ICAP Securities & Derivatives Exchange Ltd (25 April 2013);
- European Wholesale Securities Market (18 January 2013);
- National Stock Exchange of Australia (19 June 2014);
- Singapore Exchange Securities Trading Limited (7 October 2014);
- Singapore Exchange Derivatives Trading Limited (7 October 2014);
- Euronext London (4 February 2015); and
- GSX Ltd (16 August 2016).

Table 6.1

Any stock exchange is designated as a recognised stock exchange in the following countries which is a stock exchange within the meaning of the law of the particular country relating to stock exchanges (or as specified below). Listing is noted separately.

Austria	Vienna Stock Exchange, the Amtlicher Handel (Official market) and the Geregelter Freiverkehr (semi-official market): 22.10.1970
Belgium	NYSE Euronext Brussels, formerly the Brussels Stock Exchange; and Le marché hors bourse des obligations lineares, des titres scindes et des certificats de tresorerie: 22.10.1970; but the NYSE Alternext is unlisted
Canada	Toronto Stock Exchange, Montreal Stock Exchange and Toronto Venture Exchange: 22.10.1970; of these, all of the Toronto and Montreal Stock Exchanges, but only Tiers 1 and 2 of the Venture Exchange, are listed, leaving Tier 3 of the Venture Exchange unlisted
France	NYSE Euronext Paris, formerly the Paris Stock Exchange; MATIF and MONEP: 22.10.1970; all listed, that in Paris as NYSE Euronext France (though doubtless the location of that capital city is actually known to many Americans)
Germany	the regional history of the country lives on in its exchanges, in that, although Frankfurt may dominate, there are 'Amtlicher and Geregelter' exchanges in Berlin, Dusseldorf, Hamburg, Hannover, Munich and Stuttgart; and that in Bremen is allied to Berlin, not (as one might assume) with its closer neighbour Hamburg; all these are known as the 'regulierter' markets and are recognised from 5.8.1971; also recognised are the Risk Management Exchange of Hannover and the European Energy Exchange; the unlisted markets are the Freiverkehr markets and entry markets at these various centres
Guernsey	Channel Islands Stock Exchange: 10.12.2002

Hong Kong	Hong Kong Stock Exchange: 26.2.1971; the Main board is listed, but the Growth and Enterprise Market is unlisted
Italy	The Italian Stock Exchange (Borsa Italiana) is listed as to: its Electronic Share Market, MTAX Market, Expandi Market, MOT, ETF-Plus, SeDex and IDEM; also the MTS, MTS Corporate and the Bond Vision TLX market
Ireland (Republic of)	Irish Stock Exchange, Main market and Alternative Securities Market: 22.10.1970; the Irish Enterprise Exchange is unlisted
Japan	Tokyo Stock Exchange, Main board and Mothers market; the Main market of Nagora and Osaka Stock Exchanges; the Sapporo Securities Exchange and JASDAQ. JASDAQ was originally an over-the-counter market but became recognised from 3.12.2004
Luxembourg	Luxembourg Stock Exchange (Bourse de Luxembourg), Main market and EURO MTF
Netherlands	NYSE Euronext Amsterdam, formerly the Amsterdam Stock Exchange, Endex; MTS Amsterdam: 22.10.1970; the listed markets are NYSE Euronext, Amsterdam Cash Market, Derivatives Market, Endex and MTS; the NYSE Alternext is unlisted
Norway	Oslo Stock Exchange, Nord Pool, Imarex, Fish Pool and Fishex: 22.10.1970; the listed markets are Oslo Bors, Oslo Axess and Oslo Alternative Bond Market
Portugal	NYSE Euronext Lisbon, formerly the Lisbon Stock Exchange; Mercado de Futures e Opcoes; MEDIP: 21.2.1972
Spain	Bolsas de Valores of Barcelona, Madrid and Valencia; Mercados oficiales de Productos Financieros Derivados, MFAO de Futuros del Aciete do Oliva, AIAF Mercado de Renta Fija; Mercados de Deuda Publica en Anotaciones: 5.8.1971

USA	Any exchange registered with the Securities and Exchange Commission of the United States as a national securities exchange: thus the Main markets at the American Stock Exchange, Boston Stock Exchange, Chicago (both Board Options and Stock Exchanges), International Securities Exchange, National (formerly Cincinnati) Stock Exchange, New York Stock Exchange, NYSE Arca (formerly Pacific Exchange), Philadelphia Stock Exchange and NASDAQ. The NASDAQ stock market, as maintained through the facilities of the National Association of Securities Dealers Inc and its subsidiaries, is recognised and, of it, NASDAQ Global, Global Select and Capital Market are listed

The list of recognised stock exchanges is updated periodically: see https://www.gov.uk/guidance/recognised-stock-exchanges.

THE WINDING-UP EXCLUSION

6.22 The concept that BPR will be denied where the business property is subject to a binding contract for sale, is defined in most respects by *IHTA 1984, s 113 (*see **6.13**). It is possible for a person to own an interest in a business and to keep the benefit of BPR, notwithstanding the sale of one of the assets of the business and replacement of that asset by cash or other asset used in the business. However, the situation is different where the asset qualifying for BPR is a shareholding in that, where the shareholding is retained but an order is made to wind up the company, BPR is lost.

Section 105(5) excludes from BPR shares in or securities of a company if, at the time of the transfer:

- a winding-up order has been made in respect of the company; or
- the company has passed a resolution for voluntary winding up; or
- the company is otherwise in the process of liquidation;

unless the business of the company is to continue to be carried on after a reconstruction or amalgamation, and the reconstruction or amalgamation either is:

- the purpose of the winding up or liquidation; or
- takes place not later than one year after the transfer of value.

In effect, therefore, BPR on a shareholding is lost either where the shares themselves are sold, under *s 113*, or where the underlying business is terminated or disposed of under *s 105(5)*.

Chapter 7

Relevant business property: exceptions to the relief: excepted assets

QUICK GUIDE

- Excepted assets can cause problems and particularly so with group companies or a 'hybrid' company (see **7.5**).
- Companies carry an advantage over the sold trader/partnership.
- Where the rule applies, the business entity must be considered as a whole to establish the IHT value then revalued excluding the value attaching to the excepted assets to establish the amount attracting BPR.
- The leading case is still *Barclays Bank Trust Co Limited v IRC*, examined at **7.11**.
- A simple flowchart is set out on page 163.

EXCEPTED ASSETS (*S 112*)

7.1 The impact of *IHTA 1984, s 112* must not be overlooked and its application can give rise to a distortive picture for IHT purposes since assets which are designated 'excepted' are left out of account in calculating BPR. In simple terms, the rule operates to deny relief to assets (to include cash surplus to the business needs) held within a business structure which neither relate to it or its trade. Apart from deliberate act, the problem can often arise over a period of time as the business matures and changes in its method of operation as reflected in a financially strong balance sheet.

In practice, the 'excepted assets' argument is easier to resist when considering a company rather than a partnership or a sole trader. This is since within that corporate wrapper all assets tend to be regarded as held for the purposes of the trade unless, as in the case of a ski lodge in Switzerland, that simply on the face of it cannot be justified – see **7.3**. Basically, if a company passes the 'wholly or mainly' tests examined in **Chapter 9** below, most of its value ('surplus' cash excepted) will usually attract BPR.

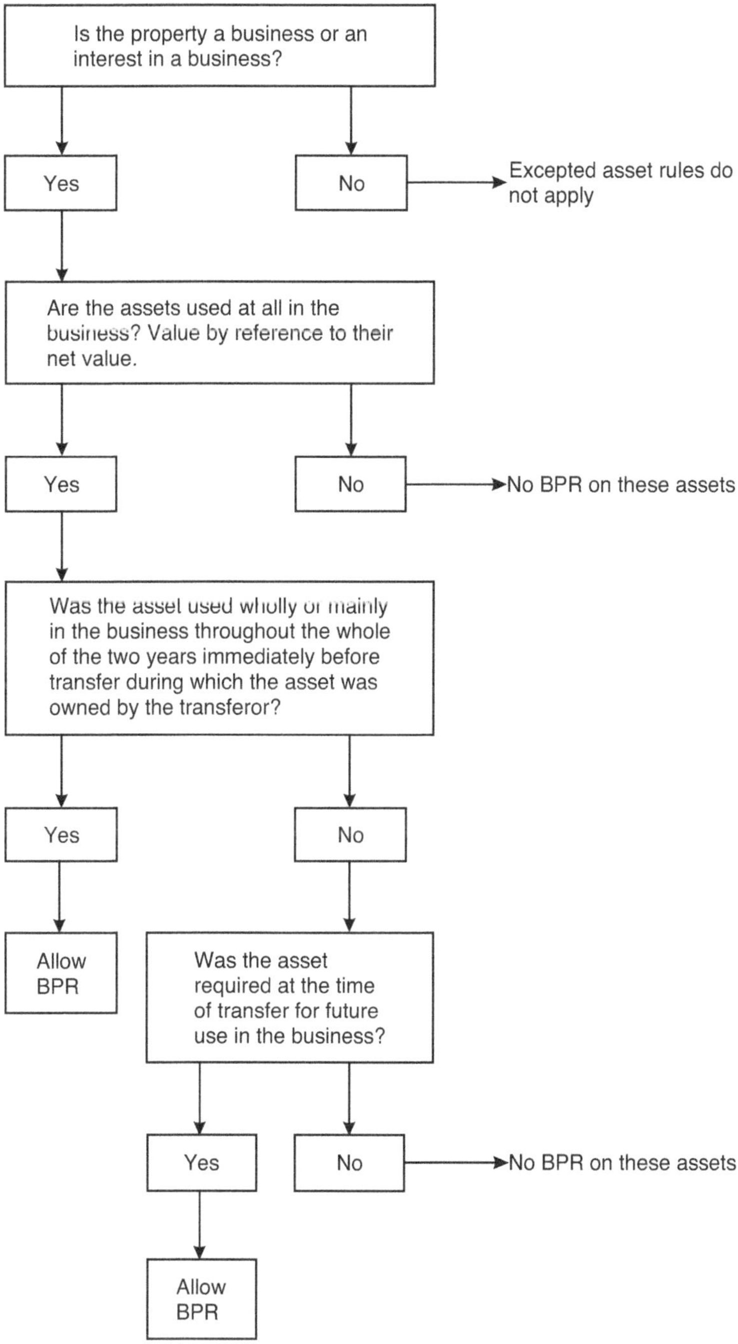
Is the property a business or an interest in a business?
Yes
No
Excepted asset rules do not apply
Are the assets used at all in the business? Value by reference to their net value.
Yes
No
No BPR on these assets
Was the asset used wholly or mainly in the business throughout the whole of the two years immediately before transfer during which the asset was owned by the transferor?
Yes
No
Allow BPR
Was the asset required at the time of transfer for future use in the business?
Yes
No
No BPR on these assets
Allow BPR

Example 7.1: Farley family (1) – organic development of a business

James Farley, a farrier, served the proprietors of canal barges used to haul goods in the early period of the Industrial Revolution. His son Nathaniel expanded the business to include repairs to horse-drawn vehicles, for that purpose buying larger premises with house (in which he lived) and meadowland with the facility for the stabling of his customers' horses whilst awaiting attention and for vehicles to be repaired. His son Alan substituted petrol vehicle repairs for the equine side of the business, which was eventually discontinued.

Operating in partnership, Alan's children now have a transport operator's licence conducted from where the stables and yard used to be, and have residential planning permission on the part of the old paddock which they have not already developed. Grandfather's cottage is vacant, being used only for the storage of old business records and memorabilia. The stabling was demolished when Alan and his children developed the site so as to incorporate the administrative offices of the haulage business, the ground floor being divided between a coffee bar, a small 'heritage' museum devoted to the history of the local canal, a small estate office from which the family's growing portfolio of let properties is managed, and an amusement arcade.

All the freeholds and improvements are still in the partnership accounts at cost. The building work has been paid for. There is now a sum on deposit equal to two years' turnover. The cottage is not used for the purposes of the haulage business and is thus an excepted asset. The estate office is, by analogy with the rule in *Atkinson*, used for the property investment side of the business but that itself may now be so large as to raise the 'wholly or mainly' issue discussed in **Chapter 9**. It is, from the standpoint of the haulage business, an excepted asset. Almost certainly, some of the cash is an excluded asset also: see the case law commentary below. This is a hybrid business: some of the issues raised in *Farmer*, discussed at **9.14**, would be relevant here.

Section 112(2) defines as an excepted asset any asset which is neither:

- used wholly or mainly for the purposes of the business concerned throughout the whole (or the last two years) of the 'relevant period' (see below); nor
- required at the time of the transfer for future use for those purposes;

but with a saving where the business itself is carried on by a company which is a member of a group as discussed below (also see the reference to *s 111* at **8.14**).

Group structures

Relief is preserved where the use of an asset for the purposes of a business was carried on by another company which, at the time of use and immediately before the transfer, was a member of the same group but only provided that the membership of the group of that other company does not fall to be disregarded under *s 111*. The existence of dormant companies within the group does not prejudice BPR on the group: the dormant subsidiaries are simply disregarded under *s 111* such that their assets are excluded from relief.

The test period for excepted assets (*s 112(5)*)

'The relevant period' is defined as the period immediately preceding the transfer of value during which time the asset was owned by the transferor. It is one of the few occasions when the BPR code considers the underlying asset, rather than at the actual interest held.

Where the excepted asset rule does not apply (*s 12(3)*)

The 'excepted assets' exclusion does not apply to an asset which is relevant business property in its own right under *s 105(1)(d)* unless either:

- it was used throughout the two years immediately preceding transfer of value which falls within *s 105(1)(d)*; or
- it replaced another asset which was used in that manner, and both it and the replacement asset or its replacement were used for periods which together comprise at least two years within the five years immediately preceding the transfer of value.

This rule is further qualified by *s 109* (successive transfers) and treated as being satisfied if the assets or the replacement assets were used:

- throughout the period between 'the earlier and the subsequent transfer', as defined in that section; or
- throughout the part of that period during which the assets were owned by the transferor or the spouse of the transferor.

Applying the test

The HMRC approach, as set out in IHTM25341, is to consider the test by means of a two-stage process and this is reflected in the flowchart at IHTM25362 with cross-referencing to other paragraphs in the Manual.

HMRC adopt a relaxed approach where a sole trader business is transferred to a company which he then controls but where the trader retains land or buildings, machinery or plant used in that business. On a strict reading, should the transferor die within the two-year period following the transfer to the company, those assets would not have been used by the company for the minimum two years as demanded by *ss 112(3)* and *105(1)(d)*. However, in practice, the relief is allowed.

Use versus ownership

7.2 The *s 106* two-year ownership period must be satisfied independently of the use test under *s 112*. An asset held for many years, but not used in the business, will not qualify for relief except under the 'future use' test in *s 112(2)(b)* – see the case law commentary below.

Personal use

7.3 An asset does not qualify for BPR where it is used wholly or mainly for the personal benefit of either the transferor or of a person connected with the transferor (see *s 112(6)*). This provision is intended to catch luxury assets such as residential property (virtually no part of which are a business premises), caravans, yachts and similar assets – see **7.1**.

Valuation principles applied where there are excepted assets

7.4 *Section 112(1)* requires that the part of the value of relevant business property which is attributable to the excepted assets must be left out of account when quantifying the value attracting BPR. An asset which is required for *future* use in an unincorporated business will not qualify for relief unless it is being used in the business at the time of the transfer: see the flowchart at **7.1**.

Shares or securities otherwise qualifying for BPR are not immune to *s 112* which operates to exclude from relief the element of that share/security value attributable to the 'excepted assets' held within the company. In practical terms, there may be little difference in value after stripping out the excepted assets in the case of a minority shareholding. However, this can produce a distortive position in the case of a significant shareholding where general valuation principles dictate it is appropriate to consider the net asset value of the company rather than the income those assets produce.

Example 7.2: Farley family (2) – examination of the business

Re-examining the facts in Example 7.1 above, the family carry on separate trades of a haulage business, a coffee bar, the museum, the management of an amusement arcade, and the administration of a portfolio of let properties. Of these, the management of the let properties is an activity within *s 105(3)* and, as such, excepted under *s 112*. Had the family traded through the medium of a company, it would be necessary to determine, by looking at turnover, assets and the main activities of the company, whether it was engaged mainly in haulage or in property management.

A common case of concern is of a company which has a genuine building trade but also retains certain properties as investments. If it can be shown that, on the whole, the company is trading as a builder rather than engaged in property management, all of the assets of the company will qualify for BPR. *Section 112* does not apply so as to exclude the investment properties from the relief; rather it operates to exclude from BPR only those assets which are not used by the company either for its building trade or as investments.

Valuation where assets used in part for business purposes

7.5 Some apportionment is allowed for assets of mixed use (*s 112(4)*); thus where only part of the land/building is used exclusively for business purposes, then only the land/buildings not so used would be regarded as an excepted asset. In such circumstance the division is on a just and reasonable apportionment of the value and this rule can help when considering farmhouses that do not, for some reason, qualify for APR. Where debt is secured on the asset or premises, it must be apportioned and in so far as attributable to the part used for business, that debt reduces the relief otherwise available.

Exclusive use for this purpose will include use by another member of the group, provided that:

- the company using it is a member of the group at the time of the use and immediately before the transfer; and
- that separate company is not excluded by the rule in *s 111*.

There is some interrelation between treatment for income tax and IHT. For example, where a private use proportion of 50% has been agreed for income tax in relation to vehicles used in a business, that same proportion of the value of those vehicles may be disallowed under *s 112*, although it seems that HMRC do not always take this point. There will be cases where this aspect has only a marginal effect on value overall. The value of a building in multiple use can be split and treated as two assets, part being used exclusively for the purposes of the business and qualifying for relief, and part being excepted assets. That would apply, in the case of Alan in Example 7.1, to the part of the new building occupied as administrative offices for the haulage business, and to the parts from which the family carried on the separate trades of a coffee bar and an amusement arcade.

SPECIAL RULES FOR CORPORATE STRUCTURES?

7.6 The rules as to excepted assets can operate more favourably for companies where all assets sit within the protection of a corporate wrapper rather than for partnerships and sole traders – see **7.1**. The issue is whether a company that is otherwise considered to be trading may hold assets that are not used for the main trade but which may be considered to be held for a subsidiary part of the business of the company – see IHTM25342.

The strength of the argument can be highlighted by considering the cases that clearly fall one side of the line or the other. Where the investment activities of the company predominate, the rules as to excepted assets do not apply, because the company is not a trading company within *s 105(3)*. Where, by measuring the turnover, profit, capital and use of human resources, the investment content of a trading company is minimal, say 1% or less, *s 112* will have no effect, because the valuation principles described at **7.4** above are sufficiently imprecise that the value of the shares will be the same, whether or not the company holds assets that are not used for the purposes of its trade.

Example 7.3: Waterfall Ltd – Spanish villa

The trade of Waterfall Ltd is the provision of plumbing and heating installations. It owns two assets not directly related to the company's trade: a villa in Spain that is used by employees who are closely related to the managing director; and a portfolio of quoted investments. The villa will be an excepted asset on the 'personal use' test under *s 112(6)* described at **7.3** above. The share portfolio, to the extent it does not dominate when viewing the company operation as a whole, will not fall to be excluded under the excepted asset rule; rather it is treated as simply forming part of the company's business (albeit not trade).

Provided a trading company is not wholly or mainly an investment company (*s 105(3)*), the fact that it may own an investment property (but not one falling foul of *s 112(6)*) will have no adverse effect on the availability of BPR for the shares – as demonstrated in Example 7.3 above. This treatment does not extend to surplus cash retained in the business and the leading case here is *Barclays Bank Trust Co Ltd v IRC* (see **7.11** below). However, some leading practitioners have challenged the decision in that latter case presenting an argument that by being put on deposit so as to earn interest, the cash was being presently used at the date of the death of the shareholder for the purposes of the business of the company within the decision in *American Leaf Blending Co v Director-General of Inland Revenue (Malaysia)*.

THE MEANING OF 'BUSINESS'

7.7 One practical difficulty is the identification of 'the business' for BPR purposes. The corporation tax case of *Jowett v O'Neill and Brennan Construction Ltd* [1998] STC 482 established that the holding of a bank deposit and receipt of interest did not amount to the running of a business. For the purposes of *CTA 2010*, s *26* (which concerned associated companies and small companies' relief for periods pre 1 April 2015), HMRC considered that a holding company which did not carry on a trade, but which held shares in one or more companies which are its 51% subsidiaries, may or may not be carrying on a business in respect of that holding. Specifically, *CTA 2010, s 26* (see CTM03590) states that a company would not be regarded as carrying on a business if:

- it had no assets other than shares in companies that were 51% subsidiaries;
- it was not entitled to a deduction for management expenses in respect of outgoings;
- it had no income or gains other than dividends which were franked investment income received by the company; and
- the 51% subsidiaries qualified as such under *ICTA 1988, s 13ZA* and related legislation.

The problem is that it is not clear whether the corporation tax definition of carrying on a business can be carried across to the IHT code for BPR purposes. *American Leaf Blending* was considered at **5.7** and must be the starting point for the debate but this sits uneasily with *Jowett v O'Neill*. The VAT case of *C & E Commissioners v Lord Fisher*, also considered at **5.7,** is a helpful summary of pointers towards the carrying on of a business but *Rashid v Garcia* [2003] STC (SCD) 36 goes the other way. That case concerned the liability of the appellant to pay class 2 National Insurance contributions in respect of 'earnings' from property letting. The Special Commissioner held that, whilst it was not free from doubt, the arrangements did not amount to a business rather, it was an investment which, by its nature, required some activity to maintain it. The case of *Manzur v Revenue and Customs Commissioners* [2010] UKFTT 580 (TC) concerned share dealing losses. The taxpayer was a retired surgeon who managed a share portfolio, engaging in rapid trading, but made losses which he wished to set against other income. HMRC argued that:

- the trading volume was low;
- there were no customers;
- the taxpayer had no qualifications and was not regulated by the FSA;
- trading was done within a two hour window in the afternoon;

- some investments were held for months, not days; there were fewer than 300 deals in a year; so
- the activity was consistent with speculative investment.

The First-tier Tribunal agreed and it was held that the taxpayer was not trading; rather the transactions were the management of an investment portfolio.

THE LEADING CASE: *MCCALL V HMRC*

7.8 The nature of a business was considered in *McCall v Revenue and Customs Commissioners* [2009] NICA 12, which considered a claim to BPR on 33 acres of farmland that had significant development value – see **5.6**.

The decision of Mr Hellier as Special Commissioner echoed a recurring theme that a 'business' means more than 'trade'. By a narrow margin, it was accepted that the work done by the relative in *McCall* did amount to a business but it was held to be an investment business. As a general rule, a landowner who derives income from land holds an investment, even though, to earn the income, he may carry out maintenance and management work. The work in this case was within that description. The graziers had occupation that was sufficiently exclusive, even of the landowner, so that the land was an investment. The landowner could not, during the currency of the grazing agreement, use the land for anything else.

The taxpayer failed to secure BPR on development land.

HMRC commentary

7.9 HMRC guidance is direct (see SVM111220) and is targeted at surplus cash, investments and assets used for personal benefit. The specific guidance in relation to a mixed business is that 'the excepted assets rules can only apply to assets which are not used in either part of the hybrid's 'business'. It points out that 'a degree of activity is required to constitute a business, and whether investments involve sufficient activity must depend on their nature and the particular facts' – it emphasises the importance of how matters are viewed by the company directors and shareholders, supported by the business description in the directors report that accompanies the annual accounts.

SVM111220 set out specific guidance on surplus cash and on shares and securities. As to cash, there are five tests to apply, because the holding of cash cannot be regarded as a separate investment business nor as part of the hybrid business: 'It requires no effort and involves no activity.' The tests are:

- Was the cash used for the business?
- Was the cash used to finance the business carried on by the company?
- How much cash did the company use regularly?
- What were its short-term cash requirements?
- Does the amount of cash fluctuate?

Hybrid company

7.10 A practical approach is appropriate.

Example 7.4: James Benjamin Holdings Ltd – mixed development

A BPR claim is made in respect of the shares of James Benjamin Holdings Ltd. The annual report and accounts disclose that the group, of which that company is the holding company, includes the following subsidiaries:

- Jim's Music Ltd – retailer of handheld entertainment devices;
- Benjamin Nursing Homes Ltd – operator of nursing homes;
- Benjamin (Elbingerode) GmbH – property trading and development; and
- Benjamin Heritable Securities Ltd – property investment.

It appears from the accounts that roughly 66% of the profit of the group is earned from selling media devices (Jim's Music Ltd) and from the nursing home (Benjamin Nursing Homes Ltd), and 33% from the property redevelopment activities in the area of the Harz Mountains in Germany (Benjamin GmbH) and the ownership of estates in Scotland (Benjamin Heritable Securities Ltd). Examination of the payroll shows that only 5% of employees are engaged in property development or investment, the remainder work at selling notebooks, headphones and the like or caring for the elderly in the nursing homes.

Seen from the perspective of the employment of capital, the emphasis is rather different. Jim's Music Ltd trades almost exclusively from rented pitches in high street stores and has no freeholds. The nursing home freeholds amount to about 33% of the value of all properties held by the group, 25% of the total being the German redevelopment scheme and the remainder the estates in Scotland. Of this last category, the jewel in the property crown is almost certainly an interest in one of the finest beats on the Tweed. Although this factor is not declared in the accounts of the company, the game book shows that, of the salmon grassed in the previous season, 75% were taken by James Benjamin himself or members of his party.

The issue here is, first, whether the business of James Benjamin Holdings Ltd taken as a whole qualifies for BPR; and, second, if the taxpayer succeeds on that point, whether *IHTA 1984, ss 110(c)* and *112(2)* will operate to restrict relief. Looking at each subsidiary, the operations of Heritable Securities fall entirely within *s 105(3)* and its share value represented in the holding company should therefore be excluded from the claim under *s 110(c)*.

What about the development in the Harz Mountains? The observations of Joel Barnett MP, then Chief Secretary to the Treasury, in 1976 in a reply to a question in Standing Committee E may still be relevant: 'BPR should apply on the transfer of a property dealing business provided that the business includes building construction or land development.' He gave a further assurance that the housing stocks of a building company would qualify for BPR if those stocks were regarded in the accounts of the company as stock in trade. VAT (or its German equivalent MWST) will often be in point here. If, on the facts, it can be shown that Elbingerode is actually developing land and involved in the construction of properties, BPR may be allowable, as it was in *DWP Piercy* examined below.

CASE LAW COMMENTARY

Barclays Bank Trust Co Ltd v IRC

7.11 This case proceeded only as far as a decision of the Special Commissioner, of which the report is refreshingly brief but it is an important case. In *Barclays Bank Trust Co Ltd v IRC* [1998] STC (SCD) 125, at the date of death in November 1990 the deceased held half the shares in a company, and her husband, who had at times been a general dealer, held the remainder. The company dealt, *inter alia*, in bathroom and kitchen fittings, mainly to the trade. The company did not tie up working capital either in premises (which were owned by the husband) or in stock. The company's cash at bank and in hand in 1990 was rising, culminating in a balance of over £450,000. Cash was invested for periods of up to 30 days. The turnover was about £600,000; less in the final year. Although this analysis is not set out in the report, cash relative to turnover was as follows:

Year 1986–87	31.25%
Year 1987–88	42.43%
Year 1988–89	56.96%
Year 1989–90	68.87%
Period 1990–91	92.90%

It was accepted on behalf of the Crown that the company needed a cash reserve of £150,000 which represented 25% of its annual £600,000 turnover. Some eight months before the death of the deceased, the company had approached a similar company with a view to buying the properties of that company but that latter company was later liquidated. In 1997, the company spent over £355,000 purchasing goods which were imported from China.

HMRC issued a notice of determination that £300,000 of the cash deposit at the time of the death was not required for the purposes of the business under *s 112(2)(b)*. The executors appealed.

The Special Commissioner, holding that the case turned on a question of fact, decided on the evidence that the £300,000 was not required for the purposes of a business. It was determined that an asset does not cease to be an excepted asset merely because, at the time for determining the question, that asset might be required at any time in the future. 'Required' in *s 112(2)(b)* did not include the mere possibility that the money might be required should an opportunity arise in two, three or seven years' time for the purposes of the business. The wording of *s 112(2)(b)* implied some imperative that the money would be used on a given project or for some 'palpable business purpose'.

In the particular facts in *Barclays Bank Trust Co v IRC*, there was no evidence that the company would, at the date of death, be the purchaser of the assets of the other company. The money was not 'required'.

Whether an asset is required for future use is a matter of fact, to be decided, as in *Barclays Bank v IRC*, at the time of the transfer. Seasonality in a business will be very relevant, thus the cash balances required by farmers at certain times of year are far greater than at other times, and this argument may be highly persuasive where the date of the transfer happens to occur at a time when cash resources are high. SVM111230 notes that travel agents and insurance brokers may hold high balances at certain times of year to avoid the need for very expensive trading bonds that are otherwise required by airlines and holiday operators. Note should also be made of the HMRC attitude to trading challenges which allows little room for flexibility:

> 'Similarly in times of recession a company may hold significant amounts of cash and/or other investments and in a period when that company is making losses the income stream could be considered as supporting the company's main trading activity.'

HMRC reaffirmed this approach in their published discussions with ICAEW, TAXREP 1/14, where the accountancy body had considered IHTM25352 and IHTM25342 and asked HMRC to recognise that cautious businesses were hoarding cash for good reason. HMRC replied:

> 'We understand that due to the financial circumstances in which businesses find themselves, they may choose to hold more cash in case of a potential downturn in trade. We can also confirm that in recent times we have seen this on a more frequent basis where businesses hold cash in excess of what they would traditionally require.
>
> However, our guidance remains the same, and unless there is evidence which directs us to the fact that the cash is held for an identifiable future purpose, then it is likely it will be treated as an excepted asset. Therefore the holding of funds as an "excess buffer" to weather the economic climate is not a sufficient reason for it not to be classed as an excepted asset …'

IHTM25342 also touches on *Hertford v CIR* (2005) STC (SCD) 177 – see **8.15**.

Salaried Persons Postal Loans Ltd v HMRC

7.12 The case ([2006] EWHC 763 (Ch)) concerned whether an associated company, which had carried on a 'trade' until 1995 but not thereafter, carried on business by receiving rent (its only source of income) from its freehold premises which since 1966 had been let to a third party. The letting had been to the same tenant for over 30 years, involving little activity, so when HMRC restricted the small companies' relief, on the grounds that this was an associated company, the company appealed.

The Special Commissioner held that the company had not carried on a business in the years in point, and that the letting was independent of the trade of the company. That view was upheld on appeal: the company had not bought the premises as an investment. When the company ceased to trade, the rent was the only income left. It merely received rents and authorised review of those rents, but there was no active participation in the management of the property. There was no artificial arrangement to take advantage of the small companies' rate.

THE MEANING OF 'GROUP' FOR BPR

7.13 The case of *Grimwood-Taylor and Mallender v IRC* [2000] STC (SCD) 39 turned on the interpretation of *s 103(3)* (the 'wholly or mainly' exception) in the context of which the Special Commissioner had to determine the meaning of 'group' for BPR and its relevance when considering excepted assets (see IHTM25342). The taxpayer had died in September 1986, holding investments in a number of companies. N Investments Ltd (incorporated in Eire), was wholly owned by the deceased but he only owned 28.99% of

BB Estates Company ('BB') (which was incorporated in England but resident in Jersey) with the remainder held in trust for his children. He also owned NH Farms Ltd and C Ltd (which were registered and resident in England). The reports of the directors showed that N Investments had not traded during the year ended 5 April 1983; BB had as its principal activity the holding and management of property estates; NH Farms Ltd farmed; C Ltd was an investment company; there were accumulated losses in N Investments and in BB, as well as in the other companies. HMRC determined that the holdings in N Investments and BB did not qualify for BPR but the executors argued that N Investments, C Ltd and NH Farms Ltd should all qualify for BPR as part of a group. They acknowledged that C Ltd and N Investments were involved wholly or mainly in holding land and buildings, but these were occupied by NH Farms Ltd, which did qualify for BPR.

The tribunal looked at the report of the directors, which did not help the executor's case. HMRC argued that, as the land had been bought for occupation by the shareholders, the business of the non-trading companies was carried on otherwise than for gain. The businesses therefore fell outside the rules for BPR because of *IHTA 1984, s 103(3)*.

The Special Commissioner would accept that BPR applied only if the companies were a group. 'Group' is not defined in *IHTA 1984*, save that *s 103(2)* indicates that 'for the purposes of this chapter a company and all its subsidiaries are members of a group'. Whilst the deceased owned all the shares in NH Farms Ltd and in C Ltd and in N Investments, the fact that he owned those shares did not make them members of a group. The executors had referred to VAT legislation, but the latest and most helpful version of it was not in force when the taxpayer died. Various arguments were tried including that, since the settlement was for the children of the deceased, the directors 'shared a common purpose' with the deceased.

What was fatal to these arguments was that BB was incorporated in England but resident in Jersey and there was no evidence that it had an established place of business in the United Kingdom. Its investments were run by Chase Bank in Jersey and the company was non-resident.

The Special Commissioner therefore decided that the various companies were not members of a group for the purposes of *IHTA 1984, ss 111* and *112*.

INTERACTION OF EXCEPTED ASSETS AND DEBT

7.14 Subject to the provisions of *FA 2013, Sch 36* in relation to events occurring after 17 July 2013, as described more fully at **8.16** below, an apportionment is required to set debt against taxable assets.

Example 7.5: Donald Drake – treatment of liabilities

Donald Drake practises as an optician and until recently worked from home. The balance sheet of the business comprises:

	£
Premises	300,000
Other business assets	800,000
Gross value	1,100,000
Less liabilities:	
Mortgage*	(100,000)
Unsecured loans*	(400,000)
Net value	600,000

**Both liabilities were established before the FA 2013 provisions came into effect.*

Although the house was formerly part of and used in the business, it is no longer used or needed for the trade – it is merely a residence and thus an excepted asset. The mortgage is set against its value as required by *s 162*, leaving net value not qualifying for BPR of £200,000 (£300,000 – £100,000), which is 20% of the remaining gross value (£200,000/1,100,000), ignoring the unsecured debt. The latter debt reduces the value of the whole business (including excepted assets) under *s 110* and is set against the remaining assets according to value, so debt of £400,000 × 20% = £80,000 allocable against the gross value of the house (£300,000 – £80,000 = £220,000 – mortgage £100,000 = net value £120,000), and the balance of £320,000 (£400,000 – £80,000) reduces the value of the business qualifying for relief. The claim is therefore:

Gross value of the business	£1,100,000
Net value of the business	£600,000
Less relief claimed	(£480,000)
Chargeable estate	£120,000

Chapter 8

Relevant business property: aspects of valuation and the treatment of liabilities

QUICK GUIDE

- The valuation of unquoted shares is always subject to HMRC scrutiny.
- Where the shareholding represents an interest in a 'quasi-partnership' company, there is some argument for suggesting that there should be no discount for a minority holding, although some learned commentators argue to the contrary.
- Where the asset to be valued is an interest in property, there is a strong case for arguing that a discount should apply: see **8.13**.
- The existence of debt within a business can impact on the business valuation for BPR.
- The related property rule in *IHTA 1984, s 161* has a direct bearing on valuation.

BASIC PRINCIPLES

8.1 The general principles of valuation were considered in **Chapter 4.** In particular, it should be noted that the principles in *IRC v Crossman* [1937] AC 26 apply to both business property and to shares in unquoted companies where there may be restrictions that would prevent or hamper a sale.

QUOTED SHARES AND SECURITIES

8.2 The terms 'quoted securities' and 'recognised stock exchange' were considered in **Chapter 6**. However, there is no statutory basis of valuation for quoted securities for IHT purposes, in contrast to the CGT basis as originally set out in *TCGA 1992, s 272(3)* but superseded by the Market Value of Shares, Securities and Strips Regulations 2015 (SI 2015/616) with effect from 6 April

2015. Notwithstanding, the former CGT basis of the 'quarter up' rule in *s 272(3)(a)*, and the alternative rule in *s 272(3)(b)* of halfway between the highest and lowest bargains (other than those at special prices) recorded for the quoted shares on the relevant date is the HMRC accepted IHT method of valuation – see IHTM18093. Where the valuation date falls on a date when the stock exchange is closed, the valuation may be made with reference to the prices ruling on the last day on which the exchange was open or for those on the next day when it is open, whichever produces the better result for the taxpayer.

Markings in the Stock Exchange Official List are observed, such that where the price is quoted ex-dividend, the whole of the net dividend or interest payment must be accounted for separately in arriving at the valuation. Occasionally a security maybe ex-dividend but no longer marked as such; even so the 'rights' must still be valued and in some cases they may carry a separate quotation.

ALTERNATIVE INVESTMENT MARKET (AIM)

8.3 For BPR purposes, AIM shares are treated as unquoted unless importantly those shares have a secondary listing on another recognised exchange (see **6.21**). The AIM is not a very liquid vehicle but such bargains as there are form the starting point for the valuation.

UNQUOTED SHARES

8.4 The intricacies of the valuation process of unquoted shares is a subject in its own right and for that very reason is outside the scope of this book. However, it pivots on *IHTA 1984, s 168* which assumes that a prospective purchaser would have available to him all information which a prudent prospective purchaser might reasonably require if he were proposing to purchase the shares from a willing vendor by private treaty and at arm's length; but an obvious problem is the identification of that available information.

Much of the tension between the taxpayer and HMRC has occurred, not in the field of valuations for IHT but in CGT and notably where a minority shareholding was held at 31 March 1982. The taxpayer would, for CGT purposes, seek a full valuation as at March 1982 but HMRC, would argue that an uninfluential minority holding should be valued on the basis of a very substantial discount thereby resulting in a low base cost. The issue of how much information might be available to the prudent purchaser of a small shareholding was considered in the collective 'Yorkshire Switchgear' cases where, on the same day, the Special Commissioners heard arguments from the point of view of CGT, where the taxpayer urged acceptance of a relatively high value for a minority holding, and contrary arguments for income tax in support of a low value.

Caton's Administrators v Couch

8.5 For IHT, the 'Yorkshire Switchgear' case of note is *Caton's Administrators v Couch* [1995] STC (SCD) 34, which examined several important aspects of valuation and procedure.

Mr Caton died on 7 September 1987, owning 2,495,552 ordinary shares in Yorkshire Switchgear Group Ltd, representing 14.2% of the issued share capital. On 15 April 1988, the company was sold and the executors received £3,269,173. HMRC raised a CGT assessment in the sum of £494,430.25 but the executors appealed disputing the share value. In arriving at the appropriate level of discount, the Special Commissioner noted that: the holding was 14.2% only, thus its owner could not insist on a sale, the company was trading successfully at the relevant date and it was not inevitable that it would be sold. Information about a sale comprised one letter 'requesting discussions leading to the acquisition of an equity interest'. There were so many uncertainties that it was considered that the discount should be 50% or even more.

The Commissioner had taken the worth of the company to be £20 million or £1.12 per share (discounted by 50% to reach a figure of £0.56 per share). That was more correct than the figure of £0.45, which would have been reached on the basis that there no information about a prospective sale. There were actually three issues for determination:

- When determining the value of the shares at the date of death, should it be assumed that unpublished information relating to the trading profits and budget forecasts of the company, and unpublished information relating to a possible sale of the company, would be available to a purchaser?
- What was the value of the shares on 7 September 1987? Was it £0.35 per share, as proposed by HMRC, or £0.50 per share, as proposed by the taxpayers on the basis that no information about a proposed sale of the company was available? Or was it £0.88 per share as proposed by the taxpayers on the basis that information about a proposed sale was available? Or some other figure?
- Were the professional costs incurred by the taxpayer, up to and including the hearing of the appeal, 'incidental costs' of the disposal and thus an allowable deduction in computing the chargeable gain?

The Commissioner referred to *Dymond's Capital Taxes* which stated that the amount of information required will depend on the size of the purchase. If 25% or more of the issued capital is up for sale, the buyer will want to know the figures for sales and profits up to the latest date, with corresponding figures for the previous year and the board's view on trading prospects. He will also want to know the dividend prospects and the directors' dividend policy. If less than

25%, of the issued capital is up for sale, it is likely that the buyer will expect less information.

The size of the company is also important, and a buyer investing £200,000 would obviously be entitled to know more than one investing £2,000. Where the holding was small, say less than £50,000 and less than 5% of the capital, the buyer would not normally be expected to have more than the information which was published or which he could find out without questioning the directors.

Hawkings-Byass v Sassen and related appeals

8.6 The case considered the value at 31 March 1982 of shareholdings in Gonzalez Byass, the sherry and brandy producer but also drew out some interesting concerns pertinent to close run family companies which could easily mirror similar situation in many private companies operating in current times. It was noted that certain family groups could have secured control of the entire company by the purchase of one or more of the holdings to be valued in the case, and that multinational companies might have been interested in a purchase despite the restrictions that would apply to the holding once acquired. The dividend history of the company was patchy and the accounting information somewhat sketchy. Notwithstanding that, the Special Commissioners did take account of the 'political' factors and, in respect of one of the minority holdings being considered, allowed a premium because the acquisition of that holding would have given control to one or other of the families. That premium was fixed at 20%.

Denekamp v Pearce

8.7 A similar issue was considered in *Denekamp v Pearce (Inspector of Taxes)* [1998] STC 1120. The taxpayer owned a minority shareholding in an unquoted company which was liquidated in 1988, giving rise to an assessment to CGT prompting the need to value the holding at 31 March 1982. The argument concerned the appropriate level of discount for that minority shareholding.

The Special Commissioner acknowledged that the question of discount was a difficult one and that a great deal depended on the information available to the hypothetical purchaser at the time. That purchaser must be assumed to hold all the information available to the directors and, in this particular case, the purchaser would have known that dividends had not been paid for many years and were not likely to be paid in the future. A purchaser would also know that the vendor was the 'key man' and that, if he left, profits would probably suffer. The majority shareholders might want to liquidate. A purchaser, therefore, would have been considering whether the company might trade on or be liquidated or sold. The purchaser would know that the company had been up for sale without much success and a purchaser in 1982 might well have concluded that the company would be sold at some point in the future. The Commissioner then reviewed the decision in *Caton's Administrators v Couch* [1995] STC (SCD) 34 in which a holding of 14.02% in a company which was successful and profitable, attracted a discount of 50%. In the present case, the holding amounted to 24% of a company which was unsuccessful and the Special Commissioner concluded that a discount of 55% was appropriate.

CVC/Opportunity Equity Partners Limited and Another v Demarco Almeida

8.8 This case has called into question the basis of valuation of shares in unquoted companies, although commentators are unsure whether it will affect valuations for IHT. Opportunity had engaged four 'deal makers' as employees and directors and each held one share in a company with the remaining 96 shares held by Opportunity. Opportunity carried on business as general manager of a venture capital limited partnership established in the Cayman Islands but the limited partners expelled one of the directors, Mr Demarco, and wanted to exercise a call option over his one share. The issue was what price they should pay.

The case went as far as the Privy Council, where Lord Millett identified as 'quasi partnership companies' those companies where the parties possess rights, expectations and obligations which are not part of the structure of the company. In such companies, the legal, corporate and employment relationships do not tell the whole story: behind them, there is a relationship of trust and confidence similar to that of a partnership. He identified three possible bases on which to value a minority holding:

(1) a value pro rata the total value of the company as a going concern without discount;

(2) a value pro rata the total value of the company, but subject to a discount; or

(3) a value pro rata the net assets of the company at break-up or liquidation value.

The right basis depended on all of the circumstances, but the selection must be fair. He could find no justification for point 3 by adopting a break-up basis of valuation where the purchaser intended to carry on the business but if a going concern basis was appropriate, he had to decide under points 1 and 2 whether or not there should be a discount. He noted that, in the present case, Mr Demarco did not want to dispose of his shares. He was being thrown out. He commented that '… to require Mr Demarco to submit not only to his exclusion from the company, but to the acquisition of his shares at less than their going concern value by a purchaser which intends to carry on the business, is hardly less unfair'.

The Privy Council then decided that, in the case of a quasi partnership company, the majority can exclude the minority only if they offer to buy them out at a fair price. Fairness should be judged by reference to what will happen if the offer is accepted, not if it is refused. In the case of a quasi partnership company, the interest of the shareholder cannot be determined by a sale of his shareholding to his co-venturers unless the price reflects his share in the underlying business: 'The subject matter of the notional sale which forms the basis of valuation is, therefore, not the petitioner's minority holding, but the entire capital of the company.'

As a result, Opportunity had to pay a going concern price without discount.

This decision fits with that in *Hawkings-Bias* and with *Caton's Administrators* and with *Lady Fox* (which was considered in **Chapter 4** in the context of APR rather than BPR). The price to be found in the market is one 'which the evidence shows that various people would have been likely to pay, reflecting, for example, the fact that one person had a particular reason for paying a higher price than others … The valuation is thus a retrospective exercise in probabilities, wholly derived from the real world.'

Phoenix Office Supplies Limited v Larvin

8.9 The contrary view to that in *Demarco* was expressed in *Phoenix Office Supplies Limited v Larvin* [2003] BCC 11. That case examined whether *Companies Act 1985, s 459* gave a member of a quasi partnership company the opportunity to 'put' his shareholding onto the other members at its full undiscounted value, even when he had no contractual right to do so, when he wished to sever his connection with the company for his own reasons?

It was observed that ordinarily a fair value should represent an equivalent proportion of the total issued share capital, that is without discount for it being a minority holding. The offer should provide for equality of arms between parties: both should have the same right of access to information about the company which might bear upon the value of the shares. It would therefore seem that, on the death of a quasi-partner shareholder, the shares may be valued subject to a discount. This is on the basis that, in the absence of a shareholders' agreement to the contrary, there is no means of forcing the other shareholders to pay more.

SO WHICH BASIS APPLIES?

8.10 The determining factor is that there is actually someone in the market who has his own special reasons for buying but the nature of those reasons is irrelevant. The valuer should, with reference to the circumstances of the particular company, consider whether there is another shareholder who would be prepared to make an offer for the shares that are being valued with a view to obtaining an undiscounted value and this might be either in winding-up proceedings or under the terms of a shareholders' agreement. That could make a difference to the valuation, but the difference would be marginal only. The purchaser would not have to offer the full undiscounted value, even though he might hope eventually to realise the shares without the discount but the profit is his. That view is consistent with the comments (see **8.11**) on the case of *Re Courthorpe*.

Commentary

8.11 Even where relatively small sums are in issue, a prudent buyer of an illiquid asset would make fairly searching enquiries. So many businesses keep their financial records electronically that it would be rare not to have management accounts for very recent trading. Any seller of a minority holding would 'set his stall out' by providing all pertinent information that might serve to enhance the desired sale price. It is common for minority holdings to be held by close relatives of majority shareholders and in real life, that minority shareholder is nearly always aware of the company's prospects and the opportunity for sale of the whole company. To ignore these factors discredits the entire system of valuation in the eyes of the lay taxpayer, leading to a search for 'an alternative solution' of some kind, ie probably a solution that a prudent tax adviser would not countenance. The realism shown in *Hawkings-Byass* and *Caton* is to be encouraged.

The inaccuracy of a valuation, whether professional or the opinion of the directors, can have serious implications not least from a compliance perspective. An Inspector will not normally proceed without consulting Shares Valuation Division (SVD) but the difficulty arises where more than one view of the value is tenable and the value agreed is much nearer the SVD figure than the figure put forward by the taxpayer. The use of estimates in preparing accounts for IHT is considered in **Chapter 22**.

There are two main factors which will always be relevant:

- the value of the assets of the company;
- the ability of the company to put those assets to good use.

There is a tendency to regard the value of the assets as the starting point and that approach will be encouraged by HMRC as it is likely to yield a higher valuation than one prepared on an earnings basis. However, *Re Courthorpe* (1928) 7 ATC 538 established the principle that, even in relation to the shares of a company whose assets are all easily valued, a purchaser would apply some discount on the ground that he would not buy the shares except with a view to making some profit on resale or on liquidation. Thus, the liquidation basis of valuation will take into account not only the profit, but also the expenses of running the business down and the tax charges which, having been deferred whilst the company was trading, now 'come home to roost'. For example, on the disposal of assets realising gains, he will wish to provide fully for the tax on those gains, whilst HMRC will resist full provision on the basis that 'it might never happen'. Other additional tax charges may apply on cessation of trade which should be provided for and equally, the valuation on a liquidation basis would have to take account of such matters as redundancy payments, costs of sale and of the liquidation itself.

A basis for discounting?

8.12 Is there any rule of thumb as to the discount that may be applied? It has been suggested that no discount should apply in relation to a 100% holding (which is inconsistent with *Courthorpe*), and that:

- 2.5% is appropriate for a holding of 90% or more;
- 5% to 10% discount for 75% to 90% holding;
- 10% to 15% discount for 51% to 75% holdings; and
- 20% to 25% discount for over 25% to 50% holdings.

SVD will often attempt to value minority holdings, even when no dividends are paid, by reference to price/earnings ratios for comparable companies, applying then a discount because the shares are not as marketable. That approach does indeed seem wrong and certainly where there is no dividend flow, the basis for a price/earnings ratio cannot be supported. Most practitioners will hope to achieve lower valuations than are implied by the discounts set out above.

As commented above, it is generally accepted that the starting point for the valuation of an unquoted shareholding is the value of the company as a whole, taking account of the current market value of the assets, usually on a going concern basis unless, as was observed in *Denekamp*, liquidation can reasonably be expected or assumed. Next, account should be taken of the capital value of the future maintainable profits (assuming a reasonably stable business in a reasonably stable market), by applying a multiple of four. However, this approach is not endorsed by any HMRC statement – valuation of unquoted shares is an art, not a science, as demonstrated in *Demarco* (see **8.8**).

Minority interests in property: what discount?

8.13 It was long considered that a discount between 5% and 15% was appropriate in valuing the interest of a joint owner (but not for an interest valued under the related property rules of *IHTA 1984, s 161*) in investment property. The decision of the Lands Tribunal in *Charkham v IRC* is important and should be seen in the light of the earlier decision of *Wight v IRC* (1982) 264 Estates Gazette 935. In *White*, a discount of 15% was considered allowable for certain residential properties where a view could be taken on how likely it was that a court would exercise its discretion in favour of one of the joint owners to order a sale.

Charkham

In *Charkham*, that principle was taken rather further. It concerned the IHT valuation at various dates of undivided shares in a portfolio of investment properties. The matter was further complicated by the fact that the size of the share in the property portfolio altered at different dates, varying from as little as 6% to over 24%.

The taxpayer's valuer looked at the income produced by the share in question and took the view that a minority owner could not easily obtain an order for the sale of the properties as a whole thus, starting with the income, he applied a multiplier (derived from the investments market) to value the holdings. The tribunal did not approve of that approach but exercised its discretion to consider how likely it was, at any given time, that the minority owner of property might obtain an order for sale. What mattered was not so much the percentage owned by the taxpayer at any given time, but rather the purpose for which the property was held jointly and, if within a trust, whether the purpose of that trust had come to an end. In the light of that principle, the tribunal decided that a purchaser of a minority interest would know that it was not easy to obtain an order for sale and since he would not be able to force a sale transaction he would therefore discount the open market value of the property itself. The tribunal also decided that principle would apply equally to a holding of 6% or 24%.

On the other hand, other factors, such as the obtaining of planning permission, which might affect the likelihood of an early sale, could be taken into account, reducing the discount that might be applicable. On the particular facts of the case this resulted in a discount of 15% in relation to some of the properties and of between 20% and 22.5% on others.

What is clear is that the practitioner should not automatically accept a discount of 10%. He should consider the possibility of redevelopment of the property, and should look at the other co-owners to see whether there is an obvious special purchaser, since the existence of such a purchaser would probably cause the discount to be reduced.

Arkwright and Price

These cases were reviewed in **Chapter 4** and the principles apply equally here.

Effect of partial ownership on reliefs

Beware the situation where a business owner does not hold 100% of an asset used in his trade: it has been argued, for example in relation to farmland, that, where the farmer does not own the entirety, his entitlement to relief may be compromised because others may restrict or interfere with his farming activities.

Chattels

It has been argued that HMRC are wrong to apply a discount to a 95% share of an asset on the ground that the existence of a 5% owner imposes a clog on sale. The point here is that *LPA 1925, s 188(1)* is uncertain of application to assets other than land: it is simply not known if *s 188* gives joint owners of chattels the same rights as it does to joint owners of land.

GENERAL RULES APPLIED IN SS 110 AND 111

8.14 A particular aspect of BPR valuation was considered in the examination of *Finch v IRC* (otherwise *Featherstonehaugh v IRC*) [1985] Ch 1 but the main valuation provision in *IHTA 1984, s 110*, sets out three basic principles:

(1) The value of a business or of an interest in a business should be taken to be its net value.

(2) The net value of a business is the value of the assets used in the business (including goodwill) reduced by the aggregate amount of any liabilities incurred for the purposes of that business.

(3) In ascertaining the net value of an interest in a business, no regard should be had to assets or liabilities other than those by reference to which the net value of the entire business would fall to be ascertained.

These rules are supplemented by additional principles concerning shares and securities contained in *s 111* which provide that, where a company is a member of a group and the business of any group member company is excluded from BPR on the ground that it consists wholly or mainly of dealing in stocks or shares etc or of holding investments, the value of the shares or securities in the company are taken for the purpose of BPR to be what they would be excluding that group member, unless either:

- the business is that of a market maker etc within *s 105(4)*; or
- the business fails to qualify for BPR because it consists wholly or mainly of holding land or buildings, and that land or those buildings are wholly

or mainly occupied by other group companies whose business qualifies for BPR, ie the business of those group companies is not excluded under *s 105(3)* as holding investments or, if it is such a business, it qualifies for BPR under the 'market making' exception in *s 105(4)*.

Section 111 operates where the transferor has control of a company which is a member of the group. 'Holding company' and 'subsidiary', where those expressions are used in relation to BPR, have the meanings given to them by *Companies Act 1985, s 736*. A by-product of the rule in *IHTA 1984, s 111* is that the value of group companies may be affected by the exclusion of one member of the group. This might be the case where a non-qualifying group member gives guarantees which support the trading activities of the qualifying members.

Executors of Eighth Marquess of Hertford, deceased v CIR

8.15 The meaning of *IHTA 1984, s 110* was considered in relation to Ragley Hall for the purposes of a BPR claim in *Seymour Ninth Marquess of Hertford and Others (executors of Eighth Marquess of Hertford, deceased) v CIR* [2004] SpC 444. Ragley Hall is a Grade I listed historic house and in 1991, the whole of the exterior of the building was accessible to the public to view but only 78% by volume of the interior was open with the remainder occupied as living accommodation by the eighth Marquess and later by the then Earl of Yarmouth, (subsequently the ninth Marquess) and their families. In November 1991, the eighth Marquess made a PET (which potentially qualified for BPR at that time) to his son of the business of opening a historic house to the public comprising: Ragley Hall; its contents; the goodwill of the opening business; copyrights in the catalogues and brochures; book debts; cash in hand and at the bank; and the benefit of contracts, motor vehicles, foodstuffs, beverages and all other chattels used in the business. After the gift, part of the hall was let to the eighth Marquess as his living quarters at a rent of £10,000 per annum.

The eighth Marquess died just over six years later causing the PET to fail and requiring a re-examination of the availability of BPR. The executors argued that the part of the value of the business that related to the freehold of Ragley Hall qualified for 100% relief under *s 104*, because the building *as a whole* was one of the assets used in the business and its value was an ingredient of the net value for the purposes of *s 110(b)*. It was agreed that the excepted assets rules in *s 112* were not in point (but HMRC make no reference to that when discussing the case in IHTM25342) and that the case turned on the meaning of *s 110*. HMRC argued that relief should be restricted to the value of the part of the building that was open to the public but the taxpayer argued that where an asset is used mainly for the purposes of a business there is no provision for apportionment in the legislation.

The Special Commissioner noted from the plans of the Hall those areas that were or were not on display to the public, and that the exterior of the building was unaffected visually as a consequence of its occupation by members of the family. Under the claw-back provisions of *s 113A(3)*, the conditions for BPR must be satisfied both at the time of the transfer and at the time of death. At the date of his death, the deceased enjoyed a lease of the living quarters and that part of the Hall could not, without his consent, be used for the business. The asset of the business within *s 110* was the area to which the public had access.

The Special Commissioner considered the arguments in *IRC v Mallender* [2001] STC 514 (discussed at **5.21**) where the court held that what was actually used in the business was a guarantee rather than the underlying asset. The Special Commissioner found the case *Finch v IRC* [1985] Ch 1 unhelpful since although in the assets owned by the trustees included land of which only some was the subject of the BPR claim, the issue of whether the land comprised one or more assets was simply not discussed. It supported the idea that assets used in the business qualify for relief even if owned by a third party, but it did not help to decide the meaning of the term 'asset'.

HMRC argued that the correct approach, when dealing with a sole trader, is to identify the assets in the estate of the trader, and then consider what part of them in each case is used in the business. *Section 112* applies not only to sole traders but also to companies and partnerships, where it is necessary first to identify the assets that belong to the partnership or company, and then consider whether *s 112* applies. Thus, possibly, *s 112* has no relevance to sole traders. An asset might fall within *s 112(2)* because it was not used wholly or mainly for the purposes of the business throughout the whole or the last two years of the relevant period; but, by way of relief, if part of that land or building was used exclusively for the purposes of any business, that part might be treated as a separate asset under *s 112(4)*. *Section 112(4)* would appear at first sight to be directly in point, in that it deals with the situation where part but not the whole of a building is used exclusively for the purposes of a business – see **Chapter 7**. However, *s 112(4)* is confined to the situation where a building would qualify for no relief at all as an excepted asset, except for the part used for the business ie the farm office in part of a farmhouse. *Section 112(4)* would be unnecessary if the 'wholly or mainly' test was irrelevant to identifying the asset for the purposes of *s 110*. That is because the part used in the business would be the *s 110* asset. If, as HMRC argued, the correct approach is first to identify the asset, then to decide which part is used in the business, and treat that part as the asset for the purposes of *s 110*, there would be no need for *s 112(4)*.

The Special Commissioner concluded that there is no procedure for apportioning relief if an asset is wholly or mainly used for the purposes of business. That was enough to decide the case in favour of the taxpayer but consideration was also given to a second argument for the taxpayers, namely that the entire building had a business function. The business consisted of exhibiting a historic house plus most of its contents and the public could see the whole of the exterior plus most of the interior – whilst there was no separate business of showing the exterior, that exterior was still on show and the brochure itself showed the whole building. The lease was merely an exploitation of the asset that was used in the business. The Special Commissioner concluded that the freehold of Ragley Hall was, in the normal sense of the word, a single asset; and that it was simply not possible to divide the Hall in any sensible way so as to exclude part from relief. It was plainly important as a single structure, and the whole building was a vital backdrop to the business carried on. The appeal was allowed.

As noted in IHTM 25342, HMRC do not regard this case as setting a precedent but rather consider its outcome was decided upon the unique and special facts involved. In practice it will be rare that a taxpayer can rely on *s 110* rather than come within *s 112*.

THE TREATMENT OF LIABILITIES: THE NEW RULES

8.16 *FA 2013*, by *s 176* and *Sch 36*, radically changed the way that liabilities are set against property to arrive at the value that qualifies for relief. Schedule 36 also includes provisions relating to excluded property and to discharge of liabilities after death which are not relevant here but passing note should be made to Example 2 in IHTM28024 and the comments in IHTM28027 which refer to *s 175A* and concern BPR. The rules mainly apply to transfers of value made after 16 July 2013 but also to all new borrowing (to include refinancing – see IHTM 28011) taken up after 5 April 2013. In simple terms, the rules require debt relevant to the property qualifying for relief to be set against the property irrespective of where the debt is actually secured, thereby reducing the value on which relief may be claimed.

The liability of direct concern is very widely defined to include any liability which is, wholly or partly, directly or indirectly, attributable to financing one of the following:

- the acquisition of relevant business property: *s 162B(1)(b)(i);* or
- the maintenance of the value of such property: *s 162B(1)(b)(ii);* or
- the enhancement of the value of such property: also *s 162B(1)(b)(ii).*

Section 162B(2) requires the liability to be set against the value of relevant business property before allowing BPR but this carries two qualifications: *first*, the liability must be within *s 162B(1)(b)* and *second,* the liability must not already have served to reduce the value of property under the 'net value' provisions of *s 110(b)*. Thus a liability shown in the balance sheet of the business which has reduced the value of the taxpayer's share in that business cannot be counted again under the new set-off rules in *s 162B*.

HMRC consider (see IHTM28013), that the word 'indirectly' is very wide and does not require a motive; rather it is the effect that counts, however well disguised and however delayed may be the use of the cash.

Example 8.1: Camilla Davies – indirect debt

Camilla Davies has a house worth £2.8 million plus savings of £1.8 million. On the advice of a trusted friend who is well versed in IHT planning, she uses £1 million of her savings to buy shares in qualifying trading companies listed on the AIM which she will hold onto for at least two years. Camilla then takes out a mortgage of £1.5 million on her house – she gifts £1 million to her son and keeps the remaining £500,000 to support her lifestyle following depletion of the cash reserves used to purchase the AIM shares.

Although Camilla has used £1 million of her own assets to acquire the AIM shares which will potentially qualify for BPR (subject to the two-year ownership rule), she has also used £500,000 of the mortgage borrowing to bolster her finances following the cash depletion. So, she has effectively used:

- £500,000 of her own assets directly; and
- £500,000 of the borrowed funds indirectly to acquire the £1 million AIM investment.

£500,000 of borrowing will be repositioned against the value of the AIM shares before applying BPR with the remaining £1 million of the mortgage applied against the value of Camilla's' house.

Example 8.2: John Jones – FA 2013 debt repositioning

John Jones estate comprises:

House, value	£300,000
Investments	£500,000
Total	£800,000

In September 2015 he decides to buy a small engineering business, which includes freehold premises, costing £600,000 inclusive of costs and pays by using his investments and a loan of £100,000 secured on his house rather than on the land. He then runs the business with family help.

At the time of John's death in November 2017 the business has assets of £650,000 but current liabilities of £50,000, so under *s 110(b)* its net value is £600,000. BPR under the old pre *FA 2013* rules described later in this chapter would have been on £600,000 (and the value of the house would have been reduced by the £100,000 loan secured on it) but by *s 162B* the loan is set, against the value of the business, reducing it to £500,000, all of which qualifies for BPR at 100%. The loss of BPR by reason of the repositioned loan under the new rules results in extra tax of £40,000.

As previously stated, where an asset qualifies for both APR and BPR then *s 114* directs that APR is given in priority with BPR applying to any qualifying excess unrelieved value and this same treatment flows through into the alignment of borrowing under *s 162B*. Accordingly, it follows that where such borrowing exceeds the agricultural value of the asset then APR is erased in its entirety – the balance of the borrowing is then set against the non-agricultural business value in order to arrive at the net value attracting BPR.

Example 8.3: Boris Morris – mixed relievable property

Boris Morris, a bachelor who had not utilised his nil rate band, borrowed £1.05 million to buy a farm which he then farmed himself. On his death three years later the estate is worth £3.5 million, of which the farm is worth £1.8 million with an agricultural value of £820,000.

The liability of £1.05 million was incurred to finance the acquisition of assets that are both relevant business assets and agricultural assets. *IHTA 1984, s 114* provides that where both APR and BPR can apply to the same asset APR is given in priority to BPR. Thus under *IHTA 1984, s 162B(4)* and in assuming that the provisions of *s 175A* in relation to the proper discharge of the debt on Boris' death, the position is as follows:

	£
Farm	1,800,000
Other non-relievable assets	1,700,000
Total estate	3,500,000
Agricultural value of farmland	820,000
Borrowing	(820,000)
Balance attracting APR	Nil
Market value of farm	1,800,000
Agricultural value	(820,000)
Balance	980,000
Balance of borrowing (£1.05 million – £820,000)	(230,000)
Balance attracting BPR	750,000
BPR thereon at 100%	(750,000)
Net position	2,750,000
IHT thereon at 40%	1,100,000

8.17 To avoid double deduction, *s 162B(7)* provides that where (and to the extent that) a liability has, under *s 162B,* been set against the business property (so that a chargeable transfer is valued higher than it would have been with the benefit of BPR on the value of the asset without offset of the debt) that same liability is not to set against any later transfer of value by the same transferor. The main purpose of this provision seems to be aimed at 'failed PETs', so that, on a lifetime gift becoming chargeable, the liability cannot be deducted both from the value of the gift and from the estate on death.

Example 8.4: Alan Bennett – liability and BPR offset

In November 2016, Alan Bennett, who had exhausted his nil rate band, made a lifetime gift of his printing business to his nephew Brian. Alan had bought the whole business in October 2014 for £700,000, financing the purchase with cash of £400,000 (leaving himself with bank deposits of £100,000) and a loan of the balance of £300,000 secured on his house, then worth £500,000. At the time of the gift, the business was still worth £700,000. No part of the secured loan was repaid at that time.

Alan died in March 2017, leaving his house to his sister (who lived there to care for him) with residue to his niece. The lifetime gift (in so far as not relieved) is chargeable. His estate comprises:

House, still worth	£500,000
Cash	£100,000
Total	£600,000

but the house is still subject to the loan of £300,000 so the executors control value of only £300,000.

BPR on the gift of the business is calculated:

Value	£700,000
Less loan	(£300,000)
Reduced value before relief	£400,000
Less BPR	(£400,000)

The chargeable value of the gift is therefore reduced to nil.

The taxable estate on death is the value of the gross estate, 600,000 without any deduction for the debt that the executors must now fund, in addition to the IHT charge of £240,000. The instalment option is available but the daughter will probably lose her home and the grandchildren get nothing.

Note:
The residence nil rate band is not in point as the house is not 'closely inherited'.

Note that if seven years elapsed from a gift of relievable property, so that the gift fell out of account, the borrowing that had been 'clogged' by the operation of the new rules would be freed up again and could be used to reduce the chargeable estate, because it had not been used in relation to a chargeable transfer. This is confirmed by IHTM28024.

8.18 The effect of *s 162B(8)* on property held in a 'relevant property' trust is, broadly, that if trustees take a loan facility to finance the purchase of business property, that borrowing will reduce the value on which BPR is available and thus will have an impact when calculating the ten year charge on a relevant property trust.

How will the *FA 2013* rules apply to non-relevant property trusts, for example immediate post-death interests (IPDI)? The life tenant is deemed by *s 49* to be the estate owner and hence regarded as the transferor within *s 162B*. In the situation as described above, it is the trustees that have incurred the liability, not the life tenant, even though for other purposes, but not *s 162B(8),* references to the transferor include the trustees.

8.19 A liability might be attributable partly to buying BPR qualifying property and partly used for other non-qualifying purpose but how is that treated if part of the loan is later repaid? This is addressed by *s 162C* which discharges that liability as follows:

- the first part of the liability to be treated as discharged is that not attributable to the acquisition of relievable property, ie the least helpful treatment for the taxpayer;
- subject to that, discharge of the liability that would have been offset under *s 162B*, for example against business property under *s 162B(1)(b)*, occurs only after all the 'non-attributable' liability has been cleared; and
- subject to that, any remaining debt repayment is allocated to debt incurred in relation to acquiring excluded property which is dealt with under *s 162A* (and which is not within the scope of this work).

As with any anti-avoidance measure, the circumstances make work unfairly against the taxpayer and compliance may prove quite challenging to personal representatives. Taxpayers could make life easier for those who follow them by keeping the borrowing for each purpose separate and distinct thereby side-stepping these penal attribution rules. The *FA 2013* rules are far-reaching in effect: even the clawback rules, discussed in **Chapter 9**, are modified.

8.20 Where a debt is charged on an asset used in the business, the value of the asset is reduced by the amount of the debt irrespective of whether or not that debt was incurred for the purposes of the business. This follows from the simple statement in *IHTA 1984, s 162(4)* that a 'liability which is an encumbrance on any property shall, so far as possible, be taken to reduce the value of that property'. This severely hampers the scope for deathbed tax planning although the repositioning of a 'non-business' debt away from the business asset may not have as great an effect in relation to businesses as a whole as it does in the specific area of farming but even so there are the practical issues of timing and mechanics to consider – see **Chapter 14**.

Prior to the *FA 2013* realignment, HMRC practice was to treat property qualifying for BPR as primarily liable only for those liabilities specifically charged on it (see *s 162(4)*) and any liabilities not so charged, and not taken into account under *s 110*, were deducted against property not qualifying for relief. Only if the value of that other property was not sufficient would the balance of the liability be set rateably against the value of all the property qualifying for relief. There was no statute requirement for the liability to be secured on a particular asset for any particular length of time and repositioning non-business debt away from business assets could increase BPR – that repositioning itself is not a PET. Subject to careful consideration of the extent to which the *FA 2013* rules may apply, this form of planning may be one of the few options available for 'death-bed' situations.

Liabilities of sole traders

8.21 Given the way in which the pre *FA 2013* rules operated the sole trader was often placed at a disadvantage but under the new rules that distinction has disappeared. Where a partner borrowed on the security of his house to introduce capital, that borrowing was incurred not 'for the purposes of' the business but, in effect, to buy a further share in it. Where, however, a sole trader borrows, he in effect *is the business*. That was recognised by Oliver LJ in *Finch v CIR*:

> 'Now, in the case of a partnership, there is not generally very much difficulty in ascertaining what the partnership assets are, for this will be apparent from the accounts, but in the case of a sole trader who may have been carrying on business from his private residence and employing in it property which is not segregated from his other property, there may be considerable difficulty in ascertaining what is and what is not part of "the business". The first part of [paragraph 14(2) Schedule 4 Finance Act 1975, the wording of which is now substantially the same as section 110(b)] is directed … to the case of the decedent who is a sole trader and where what has to be ascertained is not the value of some interest less than the whole but the value of "the business". In such a case there is no such thing as an asset "of" the business. All the assets of the estate, to the value of which the tax is attributable are assets "of" the decedent and in order to identify which of the decedent's assets can be taken into account in assessing the value of the business the legislator adopts the only practicable test – were they assets which were "used in the business" – it being, as [the judge] pointed out, necessarily postulated that the assets are assets, the value of which forms part of the value transferred and therefore which belonged to the decedent. Thus the "assets and liabilities" referred to in the subparagraph are not at this stage identified as assets or liabilities "of" the business. They are simply assets, the value of which is included in the transfer of value, and are to be identified as business assets for the purposes of the ascertainment of the net value by reference to whether they were used by the deceased in his business, just as those of the deceased's liabilities which are to be allowed as a deduction are to be ascertained by asking whether they were incurred by him for the purposes of his business.'

Thus, the debt that is incurred by the sole trader, however and on whatever asset it is secured, operates to reduce BPR, as is indeed now the case for all owners of business property, regardless of the structure employed. The effect of *s 110* with *s 162* as illustrated in **Chapter 14**, is limited to property of the type in the illustration.

The situation under the pre *FA 2013* rules was different where farming interests were concerned because *s 110* was limited by the opening words 'for the purposes of this chapter' to the BPR code and therefore had no relevance to the APR code. APR takes precedence over BPR by virtue of *s 114* and in the absence of the rules contained in *s 110*, the application of debts under *s 162* used to follow in an unrestricted way, ie a liability which was an encumbrance on any property was, as far as possible, taken to reduce the value of that property. That meant that APR was reduced only by liabilities charged on the land, and not by liabilities which were incurred to buy that land but which were secured on other property. That has all changed in the post *FA 2013* era.

Guarantees by business owners

8.22 HMRC practice does not seem to match commercial reality where a business owner has entered into a guarantee, offering perhaps his home as security, to obtain lending for his business – in such circumstance the lender will not release the house from the guarantee until the loan is repaid. HMRC, on the other hand, look at the likelihood that the guarantee will not be called in so unless the lender actually enforces the guarantee, it will not be allowed as a deduction from the value of the estate – the HMRC view is set out at IHTM28353–28355. The statutory basis is set out in *IHTA 1984, ss 5(5)*, which provides that no liability be admitted save where incurred for consideration, and by *162(1)*, which provides that no debt be admitted where there is any right to reimbursement. In such situations the difficulty will lie in valuing the possibility that the guarantor might be reimbursed.

THE RELATED PROPERTY RULES

8.23 One of the more important valuation rules is that contained in *IHTA 1984, s 161*, by which an artificial assumption is made in order to determine the value of property for tax purposes. The rule operates in general to increase the valuation; thus where a person holds a minority shareholding and other shares qualify as 'related property', the minority shareholding falls to be valued as part of an imaginary larger holding comprising both the individual holding and the related property. The same principle will relate to the marriage value of holdings of interest in land (see Example 8.5 below).

Section 161(2) defines related property as property if:

(*a*) it is comprised in the estate of the spouse or civil partner of the transferor; or

(*b*) it is, or has within the preceding five years been,

    (i) the property of a charity and became the property of that charity on a transfer of value made by the transferor, or his spouse or civil partner, after 15 April 1976 which was exempt to the extent that the value transferred was attributable to the property; or

(ii) the property of certain other bodies which became so in the same circumstances as have just been set out in (i) above. Those other bodies are: political parties, housing associations, bodies receiving gifts for national purposes and bodies receiving gifts for the public benefit.

By *s 161(3)*, a two-stage calculation is needed, first fixing the value of each parcel of related property in its own right to establish their relative value and valuing the combined asset, then attributing part of the value of the combined asset to each of the components in the proportions first established. This rule is refined, in relation to shares, so as to exclude value differentials which would be appropriate for different sizes of holding. The proportion which the value of a small number of shares of any class bears to the value of a greater number is to be taken to be the same as the number of shares, despite the fact that valuation of each holding in isolation might produce a very different result. That rule applies equally to shares, stock, debentures 'and units of any other description of property'. This rule applies where the shares are listed on a recognised stock exchange as being of the same class or would be if they were listed see *Arkwright v CIR* (discussed in **Chapter 4**).

The related property rules are of wide application and the impact must be considered in the context of IHT planning. The relief afforded under s 176 permitting a revaluation of shares sold within three years of death dis-applying the related property rules, may, at first sight seem quite attractive. However, in reality the relief based on the probate substitution of the lower sale value may often simply not be worth claiming because the separation of the holding from the larger imaginary holding under the related property rules may cause the loss of other reliefs.

Example 8.5: St John's Retreat – ransom strip

The Diocese agreed that estate known as St John's Retreat should form the centre of a team ministry and that the Rectory lying immediately to the north of the parish church together with the 'glebe' land, should be sold to finance the building of a vicarage that would be more suitable for the ministry. Andrew and his wife Helen bought the whole of the property, inserting into the name of the house 'Old' in accordance with the requirements of the Diocese. The northern portion of the Old Rectory garden was laid out to orchard, and adjoined a separate eight foot driveway to the meadow, which lay behind old developed property facing Church Road but gave an open view of countryside to the west. A plan of the Retreat and of the property appears below.

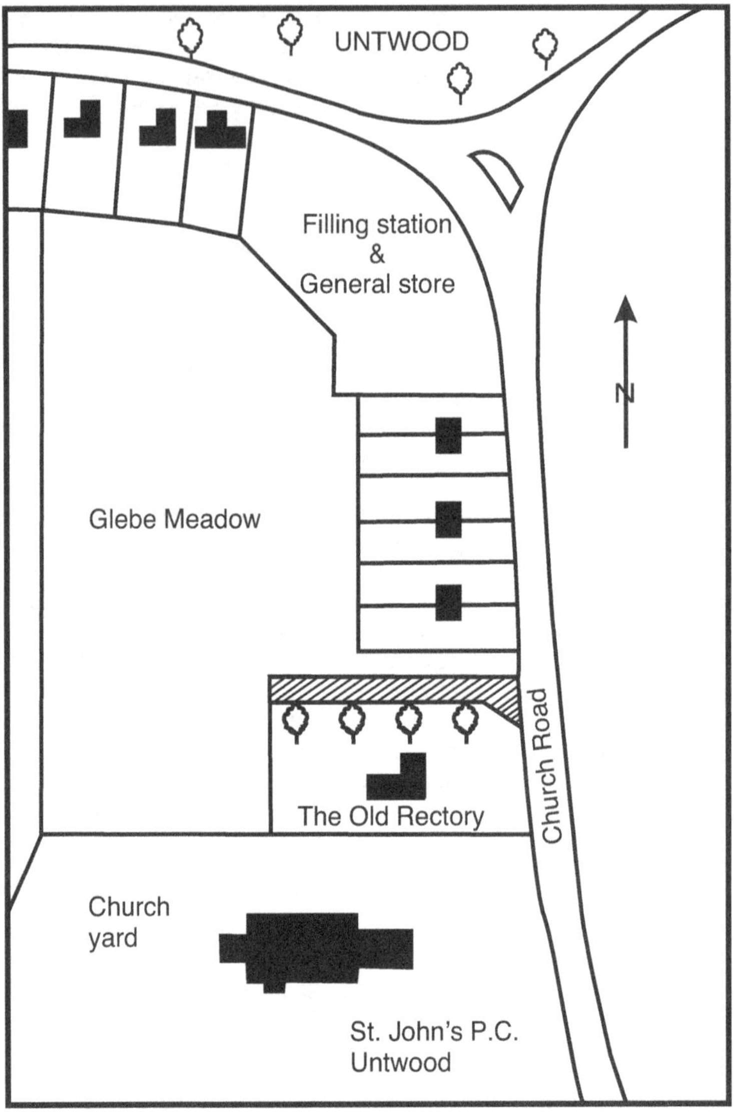

In January 2005, Andrew and Helen jointly transferred to (what was then) an accumulation and maintenance trust ('the 2005 Trust') a strip of land along the northern boundary of the Old Rectory garden about 10 feet in width (shown hatched on the plan). In 2007, with the benefit of local knowledge that the land to the west of Glebe Meadow would be developed, Andrew and Helen transferred the Glebe Meadow to a discretionary trust for their children ('the 2007 Trust') and severed their joint tenancy in the Old Rectory and the remainder of the land enjoyed with it.

On Andrew's death in January 2017, valuation advice was obtained. Glebe Meadow extends to 3.2 acres. By now, it is surrounded by developed land, but the eight-foot driveway is not sufficient for access to Church Road for a development site. Without planning permission, Glebe Meadow is worth £25,000. The 'ransom strip' that was given to the 2005 Trust is sufficient to create an access from Church Road to Glebe Meadow and sight-splay, and thus holds the key to the development value of Glebe Meadow which would be of the order of £880,000. The ransom strip is worth 1/3 of the development value, ie £290,000. The Old Rectory, minus Glebe Meadow and without the ransom strip, would be worth £250,000. If the title to the Old Rectory still included the ransom strip, the Old Rectory would be worth £600,000. The share in the Old Rectory passes to Helen under Andrew's will.

There is evidence that, since 2007, the trustees of the 2007 Trust have let Glebe Meadow on a commercial basis to the daughter of the proprietor of the filling station (see plan) to graze and exercise her horses. There is also evidence that the ransom strip was never fenced off. There is no evidence that the trustees of the 2005 Trust *bona fide* assumed possession and enjoyment of the ransom strip within the meaning of *FA 1986, s 102(1)*. Rather unfortunately, an eagle-eyed District Valuer, when inspecting the inside of the Old Rectory to establish its probate value, noticed a large-scale aerial photograph of the Old Rectory and its orchard, which appears to show the deceased and his family relaxing in front of a summer house which is clearly on part of the ransom strip. Even less fortunately for the family, it is clear that the photograph was taken at or about the time of Andrew's 80th birthday in June 2014, judging from the caption to the aerial photograph incorporated within its frame.

Section 161 requires any valuation of Andrew's half share in the Old Rectory to be made without discount for a half share, because the other half share is owned by Helen. Whilst there is no need to consider the value of Glebe Meadow in relation to the Old Rectory, because it was fully and effectively given away 12 years before Andrew's death, a problem does arise in relation to the ransom strip, because the working of the gift with reservation rules has the result that, under the *Inheritance Tax (Double Charges Relief) Regulations 1987*, HMRC have the opportunity of disregarding the gift in 2004 and treating the ransom strip as if it were still part of the estate of Andrew (and of Helen) as at the death of either of them.

Andrew's estate, therefore, includes half the value of the Old Rectory, valued without discount for a half share, because the other half share is owned by Helen, plus half the value of the ransom strip. The £600,000 value is split between house and strip – £250,000: £290,000 – giving a value for the house of £277,800 and for the land retained within the 2005 Trust of £322,200. Thus, the exempt transfer to Helen is £138,900, ie slightly greater than the actual value that she receives. The trustees of the 2005 Trust cannot set against the transfer to them of £161,100, such part of the nil rate band as was available to Andrew in 2005 because, for the purposes of the *Inheritance Tax (Double Charges Relief) Regulations 1987*, the gift in 2005 is ignored (see example 1 in *reg 5* thereof).

Valuation of shares and loan notes

8.24 In *McArthur's Executors v HMRC* [2008] SpC 700, the issue was whether certain option rights must be valued or whether, as the executors argued, they had little or no value. The shareholder had lent money to a family company on terms that gave him a right to subscribe for more shares. The Special Commissioner found that the terms of the documents (the 'writs of the debtor') gave the shareholder rights to repayment and to conversion into shares. Those rights must be valued. The actual unit of valuation was the shareholding in the company, plus the relevant conversion rights.

Chapter 9

Relevant business property: exceptions to the relief: 'wholly or mainly holding investments'

> QUICK GUIDE
>
> This chapter is concerned with case law. The dispute with HMRC concerns, mainly, the argument that certain land-based activities demand so much time and effort from their owners that they should be regarded as trading businesses whilst, in the view of HMRC, the income of the entity really only results from the exploiting of interests in land. Arguments turn on the nature of income in each example, so this chapter reviews the cases in considerable detail.

INTRODUCTION

9.1 There is a distinction between what the layman might regard as a business and what will qualify for BPR under the legislation. That distinction is sometimes academic and may take relatively little or no notice of the amount of effort put in by the taxpayer over the years to generate the wealth that is now to be taxed.

The all-important BPR exclusion of certain activities that might otherwise be considered to be a business is contained in *IHTA 1984, s 105(3)*. In essence, BPR will not apply to a business, or an interest in the business, or shares in, or securities of, a company where that business:

> 'consists wholly or mainly of one or more of the following, that is to say, dealing with securities, stocks or shares, land or buildings or making or holding investments.'

In practice, the majority of cases turn on whether a given activity falls within the latter heading of 'the making or holding of investments'.

Whilst there is some similarity of legislative purpose between the exclusion of certain assets from BPR under the excepted assets rule described in **Chapter7**, and the exclusion of businesses from BPR which are in the nature of investments, there is usually more for the adviser to play for where the 'wholly or mainly' argument is advanced by HMRC. This is because, in the case of an excepted asset, the remainder of the business will normally qualify for relief.

The application of the 'wholly or mainly' test is an 'all or nothing' test, as was recognised in the First-tier Tribunal decision in *Best v HMRC* [2014] UKFTT 77, discussed at **9.20** below. Where the business is incorporated, the entire value of the shareholding is excluded from relief but for the sole trader, the application of the rules can be less harsh.

Example 9.1: Percy and Maureen – mixed trade

Percy and his wife, Maureen own a long established tailoring business in the West Midlands carried on from freehold shop premises which they own as partnership property. One flat over the shop is let at a nominal rent to a longstanding employee as part of the terms of his employment. A second flat is let to students. Percy and Maureen share a mutual interest in restoring and upholstering period furniture which they carry out from a converted lean-to/large shed adjoining the cutting room. This is a hobby which is not carried out for financial profit and they mainly supply family and friends.

The accounts of the separate elements of these activities show that:

- The tailoring business is very profitable.
- The income from the student lettings is, owing to the location of the premises, only moderate.
- The furniture renovation activity just about breaks even.

On taking on Percy and Maureen as new clients, their adviser, Cerys, reviews the business from the standpoint of all taxes and comes to the following conclusions:

The furniture restoration should be regarded as entirely separate from the other activities. There is no likelihood of a charge to income tax, because the activities are not carried on for gain, they currently yield no profit, and they have yielded no great profit in the past. The lean-to/shed from which that activity is carried on is inconvenient, ill-appointed and of little value only. As a result, it matters little that no BPR will be available in respect of it.

The student lettings should be treated as part of the overall business activity of the tailoring shop. The rents received are small, both in relation to overall turnover and profit. On examination, it appears that the accommodation available to a student at any one time is small, both in area and in value, by comparison both with the accommodation occupied by the employee and with the premises occupied and used directly in the tailoring business. Seen in the round, the undertaking carried on at the premises is mainly that of a very successful tailoring business: the rental income is merely incidental.

As a consequence the receipt of rents from the intermittent student lettings should be included within the turnover of the tailoring business, though naturally identified as rental income. By this means, Percy and Maureen stand a more than reasonable chance of securing BPR on the whole of that undertaking, including the value of its premises, relying in part on the important ruling in *Farmer (Farmer's executors) v IRC* mentioned at **9.14** below.

THE *MARTIN* CASE

9.2 In *Martin and Another (Executors of Moore Deceased) v IRC* [1995] STC (SCD) 5, Mr and Mrs Moore had run nine industrial units in Leicester in partnership, mainly on three-year leases. Mr Moore died in November 1989 and Mrs Moore carried on, though at a lower level of activity, until her death in June 1991. In considering the availability of BPR to the letting business, it was necessary to look closely at the nature of the work undertaken by Mrs Moore actually in relation to the properties. There were three aspects to that work:

- finding tenants, negotiating rents, granting leases, and deciding whether to accept surrenders or allow assignments of leases;
- keeping the outside of properties painted and in good repair and generally complying with her obligations as a landlord; and
- managing the estate as a whole, keeping properties tidy, secure and in good repair.

The Special Commissioner held that the first two of these categories were clearly attributable to the holding or making of investments. The third category were incidents of the business of holding investments – the money that resulted from these activities was the net rents and nothing else. The properties as a whole were an investment in the estate of Mrs Moore for the purposes of *s 105(3)* and wholly failed to qualify for BPR.

THE *BURKINYOUNG* CASE

9.3 *Burkinyoung v IRC* [1995] STC (SCD) 29 was heard with the *Moore* case. Mrs Burkinyoung bought a house in west London in the 1940s, converting it some 30 years later into four flats, all of which, at her death in December 1989, were let furnished on assured shorthold tenancies. Mrs

Burkinyoung did not deal with day-to-day management herself but entrusted that to managing agents, however, she did control and direct the way in which the property was let. The managing agents dealt with maintenance, repairs, looking after common areas, arranging and supervising the lettings, collecting rents, insurance and paying rates. All of these aspects of management were considered by the Special Commissioner to be natural and necessary incidents of holding property as an investment – Mrs Burkinyoung simply received the rent from the property. The business was wholly one of making or holding investments for the purposes of *s 105(3)* and BPR was denied.

PRINCIPLES EMERGING FROM THE COMMENTS OF THE SPECIAL COMMISSIONER

9.4 Although *Martin* and *Burkinyoung* were decisions of the Special Commissioner they carry considerable authority. Argument for the taxpayer in *Burkinyoung* included reference to VAT cases which drew a distinction between business and 'mere investment' but that line of authority was rejected, as throwing no useful light on the application of *s 105(3)*. In argument for the taxpayer in *Moore*, there was reference to judicial recognition of the difference between an intermediate landlord, who sublets 'passively', and one who, by virtue of what he does, can properly be described as an occupier entitled to protection under the *Landlord and Tenant Act 1954*. Again, the point was rejected by the Special Commissioner as not being an appropriate test for *s 105(3)*.

Is a business an investment business? The Commissioner considered the tests as to existence of a business put forward by Ralph Gibson J in *C & E Comrs v Lord Fisher* [1981] STC 238 (see **Chapter 5**).

Over what period is the matter to be tested? The essential nature of the business is to be decided by reviewing the activities of the landlord over 'a sufficient period'. In *Moore* this included the time when Mr Moore was alive and the level of business activities was higher than after his death. In *Burkinyoung* the Special Commissioner referred to 'the length of her ownership of the property, and the way she has been letting out the property for a rental income'.

THE LEVEL OF ACTIVITY

9.5 It is important to appreciate that an investment business does not cease to be such simply because the landlord is actively involved – the late Mr Moore was at the property on most days from 7am. What decides the availability of the relief is not merely how much work is done but the nature of that work. Activities which are no more than the natural and necessary incidents of holding the property as an investment in order to derive rental income from it are the activities of 'holding investments'. This principle has proved decisive in later cases.

Are income tax cases any help to the taxpayer? Not at all. In *Fry v Salisbury House Estate Ltd* (1930) 15 TC 266, a distinction was drawn between, on the one hand, what the landlord did under the terms of the lease such as repairs and improvements, and on the other hand, business activities of the landlord not required under the lease and not connected with nor incidental to the holding of that property as an investment. The second category included heating or cleaning services for which a separate charge was made. Where activities are required by the lease for which no separate consideration is paid, those activities concern 'holding investments' and that test was fatal in both *Moore* and *Burkinyoung*.

Is *Hansard* any help to the taxpayer? Not at all. *Moore* and *Burkinyoung* are among the earliest recorded authority on interpretation of *s 105(3)*. The Special Commissioner thought that the relevant part of that subsection was sufficiently obscure that he should look at *Hansard* (in reliance on satisfaction of the other tests in *Pepper v Hart* [1993] AC 593) and he considered the statement by the Chief Secretary to the Treasury, that BPR would not apply to the letting of land. That was enough for the Special Commissioner.

Zetland

9.6 In the case of *The Trustees of David Zetland Settlement v HMRC* [2013] UKFTT 284 (TC) the taxpayer claimed that the level of services carried out to the tenants of Zetland House meant that it ought to be classified as a business for the purposes of BPR. Access to the relief was very relevant since the Settlement was approaching its ten-year anniversary charge. Zetland House was originally a multi-storey factory occupied by a number of printing firms but the property had outlived its original purpose. There were a number of long-term leases and by 1997 25% of the building was unoccupied – gross rents were £510,000 per annum.

The taxpayer acquired the premises and transformed them by offering flexible space for computer, media and high-technology businesses all of which required major changes to both the physical appearance and the usage/occupation of the tenant. He made office space more available, offered attractive short-term leases and made the individual offices smaller. The gross rents and service charges in the year ended 5 April 2007 were approximately £2.4 million. The taxpayer hired more staff to provide a wider range of services and facilities to the tenants. The building had a restaurant, gym, cycle arch, Wi-Fi, portage, 24-hour access, meeting rooms, media events, outdoor television screens and an art gallery – it was run on the basis of a community with regular barbecues and social gatherings.

HMRC argued that the purpose of the additional services was to increase occupancy and hence increase the rental receipts – the services were no more than incidental to the core business of collecting rent and service charges. The tribunal looked closely at the income and profit of the various activities which clearly indicated that the bulk of income, well in excess of 50%, derived from the rent and service charge. It was held that the activities were predominantly investment activities or related to investment. The services provided were mainly of a standard nature aimed at maximising income through short-term tenancies. The non-investment side was incidental to the core business and the services were insufficient to make the business one that was mainly non-investment. The purpose of the activities was largely to improve the building and its fabric, to secure the tenancy and to keep the occupancy rates high.

SUBSEQUENT DECISIONS ON S 105(3)

Hall v CIR

9.7 There were a series of cases which concerned caravan parks of which the first was *Hall (Executors of Hall deceased) v CIR* [1997] STC (SCD) 126.

In 1967 the deceased commenced trading as a caravan park proprietor and was later joined in partnership by her son. The partnership owned 18 acres, on nine of which stood nearly 100 caravans. There was a separate 'zone' of about one acre on which there stood 11 wooden chalets let on 45-year leases. The rent was reviewable every five years. The lessors covenanted to keep the site clean and tidy, maintain the road, car parks, hard standings and footpaths and were obliged to supply water, electricity and drainage. This was not a touring caravan park; rather the caravans were static but they were not owned by the business. The season ran from 1 March to 31 October and there was no right of occupation during the winter.

The deceased ran the business whilst her son did the physical work – there were no other employees. The terms of the site licence required them to provide numerous facilities and services. Caravan tenants were not allowed to sell their vans, save through the agency of the business, which charged 10% commission. The taxpayer provided an analysis of the total income of the partnership for the year ended 31 October 1993:

Rents and standing charges:	£30,754
Mains supplies were collected on a non-profit basis.	
Sales of bottled gas produced:	£216
Fishing rents	£205
Takings from coin boxes (not all profit)	£294
Commission from caravan sales	£5,492

The Special Commissioner swiftly decided that the chalets let on long leases were investments but the caravan pitches were a greater problem. HMRC had agreed that the business of the caravan park did not consist *wholly* of the making or holding of investments but that was not the point at issue. The Commissioner had to decide whether that business consisted *mainly* of the making or holding of investments and as 84% of the total income came from rents and standing charges that was sufficient to deny BPR.

The *Pearce* case

9.8 In *Powell (personal representatives of Pearce deceased) v IRC* [1997] STC (SCD) 181 the caravan park concerned had pitches both for long-term and short-term lettings. The deceased worked the site and lived there. Someone was always in attendance to ensure continuous supplies of gas (through bottles), electricity etc. Many of the residents were long term, many were retired and social and medical visits were organised for which no charge was made. Maintenance of the park included meter reading, grass cutting, hedging, ditching, security, painting, cleaning and maintaining the caravans.

Some 23 of the 33 units were owned privately, with the owners paying a pitch fee and three were on short-term lets. Income tax was assessed under Schedule D but the Special Commissioners ruled that the tax status of the income of the business was irrelevant. All the income came from pitch fees and all fell on the 'holding investments' side of the line drawn in *Burkinyoung* and *Martin*. The work done was required under the terms either of the caravan licence or of the lettings whilst the social and medical visits, which might have fallen on the other side of the line, were made for no charge.

A distinction was drawn between the position of proprietors of caravan sites and of hoteliers, on the ground that hoteliers offered 'a high standard of service' to their clients and did not merely rent rooms. The services of hoteliers were business activities, not 'holding investments'. It was not relevant that the business was carried on by workers 'in occupation' of the site, nor that the deceased herself worked there. The facts were very similar to those in *Hall*, perhaps even less favourable to the personal representatives. There was little difference between the situation in *Pearce* and the position of the owner of long leases who received ground rents. The deceased was carrying on a business consisting wholly or mainly of the holding of investments and BPR was denied.

Furness: a victory for the taxpayer

9.9 The taxpayer fared better in *Brendan Peter James Furness* SpC 202. The caravan park in that case was licensed for static caravans and touring caravans and at weekends the park held rallies. No permanent residence was permitted at the park. The appellant considered he spent 80% of his time looking after the welfare of the residents at the caravan park, maintaining it and its structures. He employed three people full time to help and there was much evidence of how much work, and what kind of work, they did.

There was an argument that less than half of the net profit of the business came from pitch fees for static caravans. Alternatively, the sheer quantity of work involved in running the site was enough to take the business outside the scope of *s 105(3)*. The Special Commissioner examined the evidence of the source of net profit – caravan sales for the years 1993 to 1996 inclusive produced more profit than pitch fees and even if the net profit in respect of caravan sales were to be reduced in respect of tours and caravan rally charges, the sales still exceeded the pitch rentals for each of the years in question.

The source of the profit may have been sufficient to decide the matter in any event, but the Special Commissioner took into account the very considerable amount of work undertaken both by the appellant and by his employees. He considered that that level of activity did not correspond to what one would normally find in a business concerned wholly or mainly with the holding of investments, and therefore BPR was allowed.

Weston v IRC

9.10 In *Weston (Executor of Weston, deceased) v IRC* [2000] STC 1064 the income from caravan sales was highly relevant. The company in which Mrs Weston held shares at her death in 1993 bought and sold caravans on commission and granted rights to pitches on the park run by it. It maintained the park and supplied electricity and bottled gas.

This park lay near the M25 and was residential for people aged 50 plus. There was no shop or social club, only a laundry and storeroom. The caravans were owned by their occupiers, purchased from the company or, if from another resident, through the company's agency. Pitch fees usually exceeded fees from sales which were few and far between. Much staff time went on maintenance of the park – three employees between them spent 111 hours per week on site maintenance and 37 hours on sales. In every year from 1988 to 1994 (except 1989), profits from pitch fees were greater than those from sales.

The Special Commissioner decided that, on balance, the company ran a park, rather than a dealership in caravans and sales were ancillary to pitch fees. BPR was denied on grounds that the activity fell within *s 105(3)*. On appeal to the High Court the taxpayer's extra argument was that caravan owners paid for services by the company, not merely for rights of occupation but the argument was rejected for lack of new evidence. The moral must be to take particular care in presenting the appeal at first instance.

Stedman's Executors v IRC; CIR v George and Another

9.11 The leading caravan park case is *Stedman's Executors v IRC*, first reported at [2002] STI issue 33 when it came before the Special Commissioner. Evidence was produced of the various business activities carried on. The issue was whether:

- the holding of land so as to produce a profitable return was incidental to the carrying on of some other business; or
- it was the very business that was carried on by the company.

It was argued that each business activity must be looked at in isolation before considering the business as a whole. There were eight elements to the business, of which three in particular were relevant.

(1) The residential homes park

The caravans were owned by the residents, not by the company. Site fees were paid that yielded the company a profit on the supply of services. There was lighting of the common parts. The roads were maintained, as were fire hydrants and a car park for visitors. Rubbish was collected weekly. The company took a commission on the sale of caravans on site and sold some caravans itself.

The Special Commissioner considered that the site fees were like rent: they were a return on investment. However, they were also similar to a service charge, reflecting work done by the company for the benefit of the occupiers. Much of the site fees went on the expenses of running the site. That level of expense suggested a non-investment type of business. The right approach was to say that the investment business was that of making a return on the licensing of the site and its facilities. The expenditure relating to the provision of services did not relate to the business of holding investments and 72% of the site fees went in overheads. The service element was greater than the investment element.

The provision of gas and electricity did not relate to the holding of land and was quite unconnected with making or holding investments. Equally, the earning of commission on sales of caravans was not related to the holding of land but was part of a business.

(2) The club for residents

It was agreed between the parties that running the club that the company had established was not an investment activity.

(3) Caravan storage

An area was set aside for winter storage, subject to agreements for periods of six months to a year. There was a security fence and a security guard. A good deal of staff time related to this part of the business. The main element of the supply was the security. Nevertheless, even taking account of these overheads of security, the nature of the agreement relating to a plot showed that it was mainly to generate income from holding land. Thus, in relation to this part of the enterprise, the business that was carried on by the company was the holding of land as an investment.

(4)–(8) Other property and sources of income

The company maintained an office from which the park was administered. It owned and let a warehouse and a shop. There were fields let on grazing licences. The company enjoyed income from insurance agency commissions and interest on cash balances. The Special Commissioner found that the income from letting the warehouse and the grazing land was clearly investment income. Interest was income from investing money belonging to the company. The commission, on the other hand, was not income from holding investments. The cash arose from all of the activities of the company together. The earning of interest was not a business in itself, so this income did not arise from the holding of investments.

Taking the business of the company in the round, 'investment-type' income for 1998 represented 40% of turnover, 20% of gross profit and 16% of net profit before the fees of directors. Those figures were not outweighed by other factors. They gave a good reflection of the nature of the business. The business of the company was not, in the opinion of the Special Commissioner, mainly that of making or holding investments.

The arguments of the taxpayer failed to persuade Laddie J in the Chancery Division, who considered the issue from the point of view of a businessman and concluded that the core investment was in land and that the other activities were merely incidental to it. The various trading activities and the receipts from them amounted to the holding of an investment that included the site fees from the park, the right to place vehicles on company land and the storage facility. The supply of water was ancillary to the investment business. The core activity of the company was that of holding investments.

The taxpayer appealed successfully to the Court of Appeal, which held that the overall approach of the Commissioner had been correct in law. *Section 105(3)* did not require the opening of an investment 'bag' into which might be put the activities that were linked to the caravan park simply because they might be regarded as 'ancillary' to the investment business. The court did not need to decide whether or not investment was 'the very business' of the company. That was not required by the statute. This was a hybrid business. The holding of property as an investment was only one part of the business and, on the findings of fact of the Commissioner, not the main part. It was difficult to see why an active family business, such as that comprised in the present case, should fail to qualify for BPR simply because one necessary component of making the profits was the use of land.

Review of the law in the light of *George*

9.12 The result in *George* may perhaps lead to consideration of a 'special skills' test. Are any particular skills needed to run the enterprise? What is the level of service provided? Consider that a 'pure investment' yields a return on capital *per se* but often the capital is in the form of land. On the other hand, the business of, say, a restaurant, whilst it may include freehold or valuable leasehold premises, demands culinary expertise if it is to thrive. The mere receipt of rents demands clerical skill, in asking for the money, but relatively little entrepreneurial skill.

HMRC have reviewed their approach since the George decision (see IHTM25279), but taxpayers may prefer to rely on the decisions of the Upper Tribunal and the higher courts. The taxpayer will have a better chance of securing the relief if the facts are thoroughly researched beforehand and the case is well presented; better still if the family get good advice long before the chargeable transfer takes place and run the business with an eye to qualifying for the relief. The adviser should visit the park site and assess the extent to which it offers space in one or more of the following categories:

- chalets;
- mobile homes;
- pitches for touring caravans that are taken for the season; and
- pitches for 'pull-on tourers', ie holidaymakers who are just passing through.

Of all these categories, it is customers in the last group who make the most work for the site owner. A pitch may be taken for only one night; people may arrive late but still expect the site shop to be open and to stock fresh (local) produce. Undoubtedly, the letting of a pitch for a whole season is attractive, both in providing reliable cash flow and in securing goodwill for eventual sale of the site. To enhance the BPR claim, however, the cases show that a site owner should cherish his 'pull on' customers and put up with the work they cause him.

Holiday lettings

9.13 Great care is needed when dealing with holiday lettings but there is no doubt that the ability to secure BPR against that activity has been severely reduced following several high-profile cases all found in HMRC's favour. Original HMRC guidance indicated that BPR would generally be allowed if the lettings were short term (ie weekly or fortnightly) *and* there was substantial involvement with the holidaymakers both on and off the premises. This applied even if the lettings were for only part of the year.

HMRC guidance (IHTM25278) was amended in late 2008, and at time of writing states:

> 'HMRC's view is that furnished holiday lets will in general not qualify for business property relief. The income derived from such businesses will largely consist of rent in return for the occupation of property. *There may however be cases where the level of additional services provided is so high that the activity can be considered as non-investment, and each case needs to be treated on its own facts*' (emphasis added).

The guidance goes on to indicate that cases involving claims for business property relief on holiday lettings should be referred to HMRC's Technical Team (litigation) – it is clear that there is a great deal of unease within HMRC.

Although, following a period of consultation, *FA 2010* has restored in part the favourable income tax treatment of furnished holiday lettings, this did not impact on the IHT position which has always been dealt with quite separately under the IHT code. Nonetheless, it must be appreciated that any claim to BPR in respect of furnished holiday lets will always be carefully scrutinised by HMRC and it must be said, with more than a jaundiced approach. It is clear that HMRC do take a strong stance and thus the practitioner must be prepared to mount a robust defence.

Holiday lettings: *Pawson*, *Green* and *Ross*

9.14 The narrowness of the HMRC BPR position on holiday lettings was challenged in *Pawson (deceased) v Revenue and Customs Commissioners* [2012] UKFTT 51(TC). In this case the deceased had 25% interest (with the remaining ownership held by other family members) in a picturesque large seaside holiday letting cottage in respect of which the executors claimed BPR. The income from the property in recent years had steadily increased producing a small taxable profit in all but one year due to expenditure on necessary repairs etc. The property was fully furnished with heating and hot water turned on before the arrival of visitors, the well-equipped kitchen was kept in good working order, the cleaner attended the property between each letting and the gardener was hired to ensure that the grounds were kept in good order at all times during the letting season – the family owners dealt with the letting arrangements personally and attended the property on a regular basis. In a surprising decision at the First-tier Tribunal it was held that the activities amounted to a business and the claim to BPR was allowed. However, as expected, HMRC appealed to the Upper Tribunal and were successful in their bid to overturn the earlier decision.

This effectively restores the position to that determined under *Stedman's Executors v IRC*; *CIR v George and another* namely that the rental exploitation of land with a view to profit is still fundamentally a business of holding the property as an investment. It is generally accepted that the *Pawson* case was rather weak such that the appeal decision was probably the correct one on the facts as they stood. However, the concern is that the Upper Tribunal commentary is sufficiently damming as to be likely to prevent a BPR claim in any furnished letting case.

The case of *ACC Green v HMRC* [2015] UKFTT 0236 (TC) which considered the availability of BPR on transfer of 85% of the appellant's interest in an established self-contained holiday business. The appellant's team argued that a wide spectrum of businesses involved the use of property which at one end involved the granting of a tenancy and at the other end the running of a hotel or a shop but the tribunal were not swayed that the business in question fell on the hotel side. The Tribunal placed heavy reliance on the authorities set out in the cases of *Pawson v HMRC* [2012] UKFTT (TC), *McCall v HMRC* [2009] NICA 12 STC 990 and *Best v HMRC* [2014] UKFTT 077 (TC) and concluded that 'the extra services provided are insufficient to demonstrate that the business is other than mainly one of holding the property as an investment'.

What must be considered as the final death knell was sounded in the case of *Marjorie Ross (dec'd) v HMRC* [2017] UKFTT 507 which considered a business of holiday cottage lettings. At the time of her death, Mrs Ross owned a share in the Green Doors Cottages Partnership, which owned a portfolio of properties comprising eight cottages and two flats. The cottages were all holiday lets, one flat was on a long term let to a local hotel which used it for staff accommodation, and the other was rented out to the handyman, and sole employee, of the partnership. The estate claimed BPR which HMRC denied on grounds that the business was wholly or mainly one of holding investments. The estate appealed.

The FTT quoted extensively from uncontested witness statements which gave a rounded view of the services provided by the partnership. These services included:

- regular cleaning;

- the ability to take meals at the hotel (which was originally owned by the deceased);
- local tourist advice;
- an onsite handyman to fix any problems;
- a laundry room;
- provision of dog bowls and creation of a pet friendly environment; and
- personal guest services provided at the hotel such as a left luggage facility and newspaper delivery.

One guest concluded by saying that: 'A stay at one of Green Door Cottages provided the complete holiday experience as it was similar to checking into a hotel but with the ability to self-cater …'. However, that was not enough to convince the tribunal that BPR was due – self-catering cottages were fundamentally different from a hotel business, no matter how much their visitors made the comparison:

> 'However high the standard of services which were provided, and whatever the level of expenditure incurred on those services, what guests at Green Door Cottages really wanted was access to a property to call their own in a beautiful part of Cornwall to enjoy for a specific period. The essence of that is the right to rent land in the form of one of the Green Door Cottages for a specific period. That is an activity which consists mainly of the investment in property …'

The appeal was refused.

Livery business: *Vigne v HMRC*

9.15 The case of *Maureen Vigne (dec'd) v HMRC* [2017] UKFTT 632 provide a useful insight into the livery business which traditional operation sits on the very margins of *s 105(3)*.

The late Mrs Vigne owned 30 acres of land from which she provided access to livery services to the horse owners. The horse had the right to reside in a field and was provided with a stable but the day-to-day care was wholly provided by the horse's owner. However, ancillary services were provided as part of the livery business which included:

- the provision of worming products and where an owner was unable and/or unwilling so to do, the administering of such products on a quarterly basis;
- a hay crop was grown on a part of the land referred to as the hayfield and this was used to provide feed during the winter months at a time when the grass in the field might not provide a sufficient food source;

- collecting and removing horse manure from the fields in which the horses spent most of their time; and
- undertaking a daily check of the general health of each horse.

On Mrs Vigne's death, the estate claimed BPR on the livery business and not expectantly. HMRC refused on the grounds that the business was wholly or mainly one of making investments – namely, holding land from which an income was generated from licence fees. The estate appealed.

The FTT looked carefully at the factual evidence, which was not significantly disputed, and considered case law and in particular paid close attention to the arguments considered in *Stedman's Executors v IRC* [2003] EWCA Civ 1763, [2004] STC 147 (see **9.11**)and *McCall and Keenan v Revenue and Customs Commissioners* [2009] NICA 12 (see **5.6**). HMRC placed heavy reliance on the 'minimal' time spent per horse on services over and above those of providing the land and stabling but the FTT rejected this approach: picking up manure, for example, was not carried out on a 'per horse' basis but to clear the whole field. In total, the yard manager worked around 20 hours per week.

The tribunal appeared to indicate that a typical 'DIY' livery would likely be a business of holding investments, but described what was being offered in this case as 'enhanced livery'. The tribunal considered the livery business was 'providing a level of valuable services to the various horse owners, which prevents it being properly asserted that the business was mainly one of holding investments'. The taxpayer's appeal was allowed.

Mixed use estate: *Farmer v IRC*

9.16 One of the most decisive leading cases is and remains *Farmer (Farmer's Executors) v IRC* [1999] STC (SCD) 321, which considered an estate run as one business where the majority of income came from lettings or licence fees but the main capital value was held in the farmland, farmhouse and farm buildings. The anonymous 'Mr Farmer' both farmed and let surplus properties. In the accounts of the business, there was no distinction between the two business activities. For most of the period of years up to Mr Farmer's death in 1997, the turnover from farming was greater than that of rents, although the net rental profit was greater than the net farming profit.

Out of 449 acres, farmland and woodland accounted for 441 acres. The balance of eight acres was tracks and let properties. There were 23 tenancies, mainly shorthold or licences. The landlord was responsible for repairs and supplied water for which the tenants paid. For probate, the farmhouse, farm buildings and farm land were valued at £2,250,000 and the let property at £1,250,000. HMRC considered that the business was not relevant business property. It was mainly making or holding investments within *s 105(3)*. One argument, resting on *s 103(3)*, was that a business must be carried on 'for gain', and that profits are therefore an essential test of the value of a business enterprise.

The Special Commissioner eventually found in favour of the executors. All relevant factors must be considered. Profits were not the only test. The business was a landed estate. Most of the land was used for farming. The let properties were subsidiary in two ways: they occupied a relatively small proportion of the total area and, being sited towards the centre of the land, would not have existed but for the connection with the farm.

The Special Commissioner concluded that the business consisted mainly of farming. Importantly, the let property was not excluded property under *s 112*: once it had been decided that there was a single business for the purposes of *s 105(3)*, all the relevant assets qualified for that relief.

The *Earl of Balfour's* case

9.17 The Scottish case of *Brander (Representative of Earl of Balfour deceased) v HMRC* [2009] TC 00069, although slightly weaker, reinforced the *Farmer* decision. The Earl of Balfour entered into a farming partnership with his nephew in November 2002. In June 2003, he died. HMRC considered that BPR was not due, and the Earl's personal representative appealed. Two issues arose. The first was whether the requirements of *s 107* as to replacement property were satisfied. They were: this issue was considered in general terms at **6.7**.

The second issue was whether the business failed the 'wholly or mainly' test of *s 105(3)*. By a settlement created under the 1923 will of the first Lord Balfour (who had been Prime Minister during the Great War), the family estate was tied up with a view to preserving it intact. The taxpayer, in applying the various tests, noted in relation to capital value that it was not intended by the first Earl, nor by his successors, that the land should ever be sold. Although the deployment of capital is a relevant factor, here its value was less important than it might otherwise have been.

The First-tier Tribunal considered that Lord Balfour had managed the estate the way most of his generation did: without a specific plan but as a unified business, where each part contributed to and supported the whole, being a traditional mix of: agriculture, both in hand and let; woodland and forestry management; sporting interests; and letting of cottages and other properties, either to estate workers or others. Where cottages were not let to estate workers, a prospective tenant who had skills that might benefit the estate would be favoured over one without, even if the rent commanded might be less. HMRC appealed, arguing that some particular circumstances had been ignored, especially the value of capital involved.

The Upper Tribunal disallowed the appeal. Various factors, none of which might be conclusive on its own, could be considered to decide whether a business was mainly investment. The Upper Tribunal might have allocated some of the values differently (this was a large and complicated estate), but the lower tribunal was right in law and was entitled to attribute little worth to the capital value of investment properties.

Moneylending: the *Phillips* case

9.18 In *Phillips and Others (Executors of Rhoda Phillips Deceased) v HMRC* [2006] SpC 555, there was a company which made informal, unsecured loans to related family companies. That was the company's business, which, it was held, did not consist wholly or mainly of making or holding investments. The shares in the company were held to be relevant business property within *s 105(3)*.

Building and developing: *Piercy*

9.19 *DWC Piercy's Executors v HMRC* [2008] SpC 687 concerned a property development company that owned land in Islington on which it had built workshops for letting. The executors of a major shareholder claimed BPR, but HMRC denied it on the ground that the company received substantial investment income and its business was therefore mainly making or holding investments. The executors claimed that the company was still trading: it still held undeveloped land that it wished to develop for housing, but had to wait for uncertainty to be resolved concerning proposals for a new railway line.

The Commissioner found that the land was still held as trading stock, not as an investment. A land dealing company will not qualify for BPR if it is a speculative trader or dealer, but all other land-dealing companies that actively develop land or build on land are outside the exclusion. For a company to be wholly or mainly holding investments, the Commissioner held that it must actually have investments; thus, holding the land as trading stock was highly relevant here. The Commissioner allowed the appeal.

Carrying on a business 'for gain'

9.20 In *Grimwood-Taylor and Mallender v IRC* [2000] STC (SCD) 39 (see **Chapter 7**), the taxpayer failed on the preliminary issue over whether the companies concerned were even carrying on a business 'for gain'. IHTM25253 notes the exclusion of businesses carried on otherwise than for gain, noting that stud farms may fall into this category.

HMRC will strongly resist the argument, which was almost advanced in *Clark and Another (Executors of Clark Deceased) v HMRC* [2005] SpC 502, considered at **9.19**, that an activity, merely by being vigorously pursued, can become a business, even though its true nature is the holding of investments.

It should be noted that the principles that may serve to deny BPR are applied separately from the test as to whether a business exists at all. This has the important result that, where it is accepted that a business exists but it is argued that *s 105(3)* denies relief, the instalment option will still be available, as will the benefit of relief from interest on instalments unless the business interest is a company.

Property management as a business: *Clark*

9.21 In *Clark and Another (Executors of Clark Deceased) v HMRC* [2005] SpC 502 an 1895 partnership, later incorporated, bought a large site and built homes. Later it bought back many of the properties and leased them to the former occupiers. The company had its own workforce to deal with the building and maintenance. At the date of death of C, the company owned 92 dwellings, four shops, four offices, two industrial units and 20 lock-up garages. The Clark family owned 141 other properties. All of them required very active management. The directors spent 75% of their time on managing the rental activity and 25% of their time on building works. The company owned 42% of the total property portfolio.

In support of a claim to BPR on the shares of the company, the executors argued: that letting property was no longer an investment activity; that the accounts of the company did not show just how much work and effort was required by the management and staff; that the two elements of the portfolio should be treated the same, ie as non-investment activity; that it was irrelevant that rental income is investment income for income tax; that 'investment' could be something done by a trading company, for example 'investing' in plant and machinery; that the rental income was part of the business activities of the company and was not investment; and that the attitude of HMRC discriminated against letting compared to other active businesses.

HMRC argued that activities that produce a rent are necessarily investment. They also wished to distinguish some of the cases relied on by the taxpayer as being based on other legislation rather than the most recent Court of Appeal cases on IHT. This was a mixed business which included, on the one hand, investment and, on the other hand, building. Mainly, the business consisted of holding investments.

The Special Commissioner could see a distinction between managing properties owned by third parties and managing properties owned by the company. A management fee was not the same as a rent. The rents came from leases of company assets rather than from an active building trade. The balance sheet showed the land as 'fixed assets – investments', which was correct. The rents were income from the ownership of that property. They would still be rents, even if someone else managed the properties and maintained them. The maintenance work was inherent in ownership of the property. Thus the rental activity was itself a business of holding investments.

The test was whether the business of the company consisted 'mainly' of holding investments, for which the Commissioner must look at the business in the round, reviewing profit, the overall context, capital employed, time spent and turnover. He therefore analysed the profit and loss account according to the different activities of the company but applying a consistent basis. An estimate of time spent on investment activity was used to apportion administration and establishment expenses, being mainly salaries. However, the profit on building and management activities was too small to bear the proportion of management time that was spent on it. It therefore looked as if the rental activity was subsidising the other activity.

Investment activity produced more turnover and profit, but took less time. Apportioning the time spent on the basis of turnover and profit resulted in a large loss for the non-investment activities, so that basis of apportionment could not be justified. The main asset shown in the accounts related to investment activity. Looking at the position in the round, the Commissioner concluded that the business carried on by the company consisted mainly of holding investments; thus BPR was denied.

Property management: *Best v HMRC*

9.22 The case of *Best v HMRC* [2014] UKFTT 077 (TC) considered the issue of 'wholly or mainly' in the context of *s 105(3)*. Bullick Developments Ltd owned a business centre in County Antrim. The development was financed in part by the Local Enterprise Development Unit and it was argued that such finance was not available for mere investments. The centre included offices, showrooms, industrial and warehouse space, from very small to quite large units. It was agreed that one self-contained area, bringing in 15% of the income, was an investment. Of the rest, 10% was offices, the rest light industrial or warehousing. 400 people worked there.

The standard form licence agreements did not give exclusive possession of the land. Services were listed in two schedules, but it was agreed that most of those in the first schedule were of a type commonly provided by a landlord. Further services were listed in the second schedule and much was made of the availability of a forklift driver.

It was a weakness of the case that no copy invoices were provided. The company employed a site maintenance person, receptionist and a forklift driver full time; two part-time people; and three security guards. Most of the units had no separate landline, so calls were put through by the receptionist. Many units were accessible only to small vehicles, making the forklift useful, though the extent of that use was not proved.

Income of £611,472 broke down to:

	£
Licence fees	447,068
Service charges	53,335
Sundry	4,517
Equipment hire	4,452
Heat and light	31,284
Telephone	45,568
Postage	20,056
Parking	5,192

Applying the tests in *George, McCall* and *Pawson*, the tribunal considered that the provision of office-like facilities and of the forklift did not save the business from its investment nature. Those services did not predominate when considering the business as a whole. The business was 'well towards the investment end of the spectrum' and BPR was denied.

OVERVIEW

9.23 Taxpayers will try to show that theirs is a 'hybrid' business, not relying only on property-based income. If rural, they will try to show that the rules for a 'landed estate' apply. The questions for decision in *Farmer,* which still serve as a good guide thus:

(1) Was the business of the company run on a business-like basis with the help of agricultural consultants, farm managers, business plans and budgets?

(2) Who administered the lettings and what proportion of total management time was taken up with them?

(3) What was the allocation of turnover between different activities of the business?

(4) What was the allocation of profit from the business?

(5) How is capital employed between farming and lettings?

(6) Is the business of the company carried on 'for gain'?

(7) The standing back test.

In the light of *George*, there are three additional questions:

(8) Is there investment activity?

(9) Apart from pure profit rental, is there any income which the company claims to be trading income which is, in truth, disguised investment income?

(10) In the view of the reasonable businessman, is this mainly an investment business?

Practitioners may find it difficult to explain to clients that the 'wholly or mainly' test is applied in isolation, disregarding such matters as the amount (rather than the nature) of work done; or the treatment for other taxes whether national or local; or the treatment for other purposes, for example planning law.

One particular structure that can cause difficulty is the group of companies that participates in joint venture agreements, notably overseas where often, local law will require the holding company to accept less than 50% of the joint venture. The relevant tests may be those in *s 105(1)(cc)* or *s 105(4)*. Arguably, however, there is comfort in such a situation from the result in *Farmer*: one looks at the business entity as a whole, and overlooks the fact that some subsidiaries are investments rather than group companies. In the farming context, where the claim is to BPR rather than to APR, one may perhaps argue for a distinction between 'original' farm buildings, now used for holiday lets, and 'new' units, such as those bought in by the business as investments over the years. The new properties may be treated less favourably than the original units. Where does a barn conversion stand in this analysis? The shell was perhaps original, but it now has a value and a purpose that together bear no resemblance to a farm building as one commonly thinks of one. It will be a matter of fact and degree; but, when the developing of farm assets goes too far, the claim to BPR is endangered.

Chapter 10

Clawback

QUICK GUIDE

- The clawback rules are complex.
- There is a seven year risk period following any lifetime gift of property initially qualifying for APR and BPR.
- The clawback provisions are considered first in relation to BPR and then in relation to APR.
- Clawback of the relief may be partial or total, depending on the circumstances: see the rules in **10.6** and following.
- Clawback affects both PETs and CLTs but in different ways.
- A flow chart to help the practitioner to follow a course through the legislation is included at **10.21** followed by an extended example of the rules.
- The most important section of this chapter is a discussion of steps to be taken to avoid clawback and this begins at **10.24**. A significant problem is that the practitioner may be acting for the donor rather than the donee, and yet it is the donee who needs to be aware of the clawback rules and how they operate.

INTRODUCTION

10.1 The BPR and APR clawback provisions were originally introduced by *FA 1986* and later 'simplified' by the changes: (a) to a single 40% death rate; and (b) to the allowance of 100% BPR on all unquoted shareholdings.

CLAWBACK OF BPR: THE BASIC RULE

10.2 BPR may be subject to clawback where following a transfer which potentially (PET) or actually (chargeable lifetime transfer (CLT)) qualified for BPR, certain conditions are not satisfied at point of retest. The rule, which is contained in *IHTA 1984, s 113A*, operates to claw back BPR which would otherwise have sheltered a lifetime transfer of value thus it only applies where that transfer is made within seven years of the death of the transferor. To achieve this, *s 113A* introduces the concept of a 'notional transfer' made by the *donee* (note: not donor) of property immediately before the death of either:

- the transferor; or
- if earlier, the donee.

Thus, under *s 113A(2)*, if the conditions set out below are not satisfied, BPR due on the original transfer of value will be clawed back where that transfer, being made within seven years of the death of the transferor, becomes chargeable.

Subject to special rules relating to replacement property (see **6.67**), the conditions set out in *s 113A(3)* are that:

- the original property was owned by the transferee throughout the period beginning with the date of the chargeable transfer and ending with the death of the transferor; and
- subject to an exception discussed below, in relation to a notional transfer of value made by the transferee immediately before the death, the property would qualify for BPR but for the operation of *s 106* (ownership period).

Example 10.1: Oscar – takeover intervening

In January 2013, Oscar (a bachelor), having made no prior gifts, gave his 20% shareholding in Hughes Fabrication Ltd valued at £200,000 to his nephew Roger. The shares potentially qualified for BPR at the time of gift. In July 2013, the company was taken over and Roger received £325,000 for his shareholding.

Oscar died in December 2017. His estate, left to his niece Rose, comprised his house worth £400,000 and cash of £100,000. The now failed PET made to Roger must be retested for its eligibility for BPR which was potentially in point when the gift was first made. However, as Roger no longer held the shares as at Oscar's death there is nothing to test and as a consequence no BPR will be available. The tax on Oscar's estate is calculated as follows:

	£
House	400,000
Cash	100,000
Estate in hands of executors	500,000

		£
Nil rate band (2017/18)	325,000	
Failed PET (200,000 – (3,000+3,000))	(194,000)	
Balance of nil rate band		(131,000)
Chargeable estate		369,000
IHT thereon at 40% and met from the residue		147,600

Note: The residence nil rate band is not available as the property is not 'closely inherited'.

COMPUTATIONAL PROBLEMS RELATING TO CLAWBACK

10.3 Where a gift of business property was a PET and, by reason of the donor's death within seven years of the transfer, that PET fails, clawback operates to increase the donor's cumulative 'clock' by the amount of that failed PET for the purpose of calculating the IHT due on the donor's estate. However, where the gift of business property was a CLT when first made, clawback operates only to charge extra tax on the original CLT gift, on the basis that no BPR is available but there is no change to the donor's cumulative 'clock'.

Where clawback applies, the additional IHT chargeable on death must be calculated without the benefit of BPR, unless the conditions set out above are complied with. By *IHTA 1984, s 113A(3A)*, special rules apply to shares or securities which were quoted at the time of the chargeable transfer, or which were unquoted and qualified for BPR under *s 105(1)(b)* or *(bb)*, and remained unquoted throughout the period from the transfer to the death. For the purposes of *s 113A(3A)*, 'quoted' means listed on a recognised stock exchange, and 'unquoted' means not so listed.

Example 10.2: Stanley – control shareholding

Stanley had a controlling shareholding in a quoted company. In December 2012, he gifted part of that shareholding to his son Arthur and the shares potentially qualified for BPR under *s 105(1)(b)*. Arthur still owned the gifted minority shareholding when Stanley died in December 2017.

The original gift was of quoted shares which therefore fell within *s 113A(3A)(a)* rather than within *s 113A(3A)(b)*. As a result the exception in the first part of *s 113A(3)(b)* comes into play so that, in relation to the deemed transfer by the transferee, it is not necessary to satisfy the BPR test in *s113A(3)(b)*. It is enough that Arthur retains the shares so as to satisfy *s 113A(3)(a)*. Arthur therefore succeeds in claiming BPR even though his holding is a minority one.

If the donee has died before the transferor, then the time for satisfying the conditions is the date of death of the donee. As will be seen, this has implications for the rules by which a taxpayer can 'recover' relief.

Additional tax: changes to the rules on computation introduced in 1992

10.4 *F(No 2)A 1992, Sch 14* applies to transfers of value (and other events) occurring on or after 10 March 1992. *Paragraph 9(1)* is concerned with two situations occurring in relation to a death on or after 10 March 1992:

- the failure of a PET; and
- clawback in respect of a CLT;

and requires *ss 113A* and *113B* to be interpreted on the basis that the amendments in *Sch 14* came into effect at the time the transfer was originally made, and that so much of the value transferred as would have been reduced in accordance with Chapter 1 of Part V (ie as would have qualified for BPR), as amended by *Sch 14*, was in fact so reduced.

PARTIAL CLAWBACK: THE OPERATION OF S 113A(5)

10.5 Situations can arise where the conditions of *s 113A* apply to only part of the gift of the original property and where that happens:

- in relation to a PET, only a proportionate part of the value transferred as is attributable to the original property qualifies for BPR; and
- in relation to a CLT, the extra tax is calculated as if only a proportionate part of the value transferred (as was attributable to the original property) had qualified for the relief.

Example 10.3: Sarah (1) – partial clawback

At Christmas 2011, Sarah Brown, who always used her annual exemption on 6 April each year, made a gift to Tom of her unquoted holding of 300 shares in a trading company value at £480,000. The shareholding is within *s 105(1)(bb)*, so the gift potentially qualified for BPR.

Sarah died in December 2017 causing the PET of the gifted shareholding to fail and requiring consideration of the availability of BPR under the clawback rules. However, at the time of Sarah's death Tom only holds 150 shares (having sold 150 shares five years ago) – the retained shares meet the conditions for BPR. As a consequence BPR is only available to shelter half the value of the original gifted shareholding with the remaining PET value of £220,000 remaining chargeable.

Example 10.4: Sarah (2) – partial clawback trust

In April 2011, Sarah Green, who always used her annual exemption on 6 April each year, created a discretionary trust for her adult children and transferred to it 100 shares in the family trading company valued at £600,000. Whilst the gift into trust was a CLT (not PET) the shares qualified for BPR at that time such that no IHT entry charge was due. In November 2016, the trustees accepted a cash offer for 40% (40) of their holding but kept the remainder (60).

Sarah died in December 2017 requiring consideration of the availability of BPR against the original changeable transfer under the clawback rules. The retained shareholding meets the conditions for BPR so relief from the 'top up' death rate IHT charge under *s 7(4)* is restricted to £360,000 (£600,000 × 60/100).

Interaction of reliefs: clawback and 'pre-death loss relief'

10.6 An oft overlooked but valuable relief is allowed under *s 131* in respect of any property where, broadly:

- a PET fails and the subject matter of the transfer has been earlier sold at arm's length for less than it was worth when originally gifted; or
- IHT at 20% on a CLT falls to be 'topped up' to a maximum of 40% because death occurs within seven years of the chargeable transfer.

Where the property transferred is still held by the donee (or his spouse or civil partner) but has fallen in value by the date of death, that fall in value may reduce the transfer. Where the property is no longer held, but has been sold under a 'qualifying sale' (one at arm's length to an unconnected person where there is no 'buy back' term in the sale), the same reduction may apply.

Disregard of BPR: the rule in s 113A(7A)

10.7 A difficult technical point arises from the operation of *s 113A(7A)*, which requires the provisions of BPR to be disregarded when deciding, for the purposes of clawback, whether there is a PET or CLT in any particular case. The difficulty of this provision could arise where, for example, the nature of property has changed between the time of transfer and the later death, such as unquoted shares becoming quoted.

Example 10.5: John – clawback after flotation

In April 2014, John gifted his shareholding in a fledgling trading company listed on AIM (thus unquoted) valued at £600,000 to his daughter, Iris – the shares potentially qualified for BPR at time of gift. In June 2016, this niche company was floated successfully on the Stock Exchange. Iris still owned the shares at the time of her father's death in December 2017.

By virtue of *s 113A(7A)*, the gift was treated as a PET within the requirements of *s 113A(1)* and this is so, whether or not the value transferred in April 2014 would have been covered by BPR. The conditions of *s 113A* apply to trigger an IHT charge on John's death. This is because, in relation to a notional transfer of value made by the donee (Iris) immediately before the donor's death (John), the shares would no longer be relevant business property within *s 113A(3B)*.

ADDITIONAL CONDITIONS

10.8 *IHTA 1984, s 113A* has imposed additional conditions to the general BPR conditions which require the donee to show that:

- the property transferred was relevant business property at the time of the transfer, if that transfer is a PET which later fails; or
- if that transfer was a CLT, it qualified for the relief at that time.

Where this cannot be shown, BPR that may have been available through satisfaction of the asset status on the death of the donor will be denied. Thus if the transfer was a CLT, additional tax is payable (see *s 113A(2)*) without the benefit of the relief and the additional IHT payable by reason of the donor's death is the difference between:

- IHT at the death rate without the benefit of BPR; and
- IHT paid at the lifetime rate on the value, but with the benefit of BPR.

Where the gift is of property within *s 105(1)(d)*, such as machinery used for a business in which the transferor is a partner, *s 113A(3)* requires that the property requalify for BPR in relation to the notional transfer of value by the donee. This can create a problem where the asset comprising the gift does not so qualify by reference to the donee (rather than donor) because the donee is not a partner or (in a corporate example) does not control the company concerned.

Where there is a family company, relief will be lost if, for example, there is a gift of the land, from which the family business trades, not to a single person but, say, to two children equally. If neither of the children controls the company, the land is not, in relation to either of them, a qualifying asset for BPR; so clawback may apply. The better solution may be a gift of the shares and the land together to a discretionary trust, where the land itself, after 50% BPR, is worth no more than £325,000, ie within the trust nil rate band overall. Then the trustees will control the company, and clawback may be avoided.

Definitions employed in s 113A: 'the original property' and 'the transferee'

10.9 *Section 113A(8)* refers to 'the original property', as property owned by the 'transferee' (donee) in the period beginning with the chargeable transfer and ending with the death of the transferor which qualified for BPR in relation to the chargeable transfer. For this purpose the 'transferee' means either:

- the person whose property the original property became on the chargeable transfer; or
- the trustees of a settlement where, on the transfer of the original property, it became settled on relevant property trusts (as will apply to almost lifetime trusts created after *FA 2006*).

The definitions also apply to replacement property (see *s 113B(8)*).

The result of this can be to trigger clawback even where the trustees no longer own the asset having made an absolute appointment to a trust beneficiary, but see **10.28** below.

RELIEF FROM CLAWBACK

Problems of identification: the rule in s 113A(6)

10.10 It would be very harsh to apply the clawback rules where, in substance, the gifted asset has been retained but (perhaps through matters beyond the donee's control) has changed its identity whilst remaining within the category of property for which BPR was designed. This is allowed for in *s 113A(6)* and applies in two situations:

(1) Where the shares comprised in the original transfer are no longer held by the donee but through a reorganisation of share capital or a conversion of securities, new shares are identified with the old shares under the rules set out in *TCGA 1992*, *ss 126–136* (reorganisation or reduction of share capital), those shares are treated as if they were the original property.

(2) Where the subject matter of the original gift was a business or an interest in a business, and that interest is no longer held by the transferee but shares were issued to him in consideration of the transfer by him of the business, again those shares are treated as if they represented the original property. BPR is given pro rata where the subject matter of the original gift is only partly represented by the new securities.

Replacement property for the purposes of clawback (BPR)

10.11 There are provisions allowing for the requirements of *s 113A* to be satisfied, even though the donee may have disposed of all or part of the original subject matter of the gift before the death of the transferor. Importantly, any replacement must (see *s 113B(2)(b)*) be by way of a transaction at arm's length or, if it is not at arm's length, must be on terms such as might be expected to be included in a transaction at arm's length and the donee must apply all of the proceeds of the disposal in buying a replacement property within the 'allowed period' of the disposal. Nothing in *s 113B(1)(b)* would seem to allow partial relief where only part of the proceeds are reinvested.

Example 10.6: Fred – partial reinvestment

In April 2013, Fred gave to his niece Victoria a small industrial property potentially qualifying for 50% BPR under *s 105(1)(d)*. The property is subject to a mortgage and Victoria agrees to take the property subject to that charge. In April 2017, Victoria sold the property intending to buy similar premises at a different more prosperous location. The lending requirements of the bank are such that, on the occasion of the new purchase, a portion of the proceeds of sale of the first property must be applied in discharging Victoria's personal debts. No BPR under *s 113B(1)(b)* will be available on Fred's death within seven years of the gift because Victoria has not applied all of the proceeds of sale of the first property in purchasing the second.

For replacement property to qualify, such that relief is not clawed back, it must be purchased within 'the allowed period' of the disposal. That period is (see *s 113B*) three years or such longer period as HMRC may allow.

Further conditions: s 113B

10.12 *Section 113B(3)* sets out further conditions to preserve BPR from clawback:

- the replacement property must be owned by the donee immediately before the death of the transferor; and
- throughout the period beginning with the date of the chargeable transfer and ending with the death (disregarding any period between the disposal and acquisition), either the original property or the replacement property was owned by the donee; and
- in relation to a notional transfer of value made by the donee immediately before the death, the replacement property would (apart from the requirements of *s 106* as to ownership for two years) be relevant business property.

Example 10.7: Swift purchase of replacement assets

Consider that:

- a donor made a lifetime gift (PET) potentially qualifying for BPR at that time; and
- the donor died within seven years of that gift; and
- that death occurred before the donee's death; and
- all or any part of the original gift had been disposed of before the donor's death (or failed to qualify for BPR because it was subject to a binding contract for sale); and
- the donee acquired the replacement property (or contracts to do so) after the death of the donor but within the 'allowed period' of three years after the disposal of the original subject matter of the gift (or part of it, as the case may be).

At first sight, the primary condition in *s 113B(3)(a)* is not satisfied because the donee does not own the property at the date of death so BPR would be lost. However, this may be salvaged by the overriding application of *s 113B(5)*, which states that for this purpose a reference to a time immediately before the death is instead regarded as a reference to the time when the replacement property is acquired with the result that BPR in this example is preserved.

In just the same way that, under the main clawback provisions of *s 113A*, the death of the donee during the 'clawback risk period' could cause the clawback conditions to be retested so, by *s 113B(4)*, if the donee has died before the transferor, a reference to the death of the transferor within *s 113B* has effect as a reference instead to the death of the donee.

Other provisions parallel with those in *s 113A* are the definitions of 'the original property' and 'the transferee', which have been noted above. The timing of transactions is also important as by *s 113B(7)* a disposal is regarded for the purposes of *s 113B(2)(a)* and *(5)(b)* as taking place at the time of a binding contract for sale thus, it is the contract date which determines applicability of clawback, whether under *s 113B(5)(a)* on disposal or *s 113B(5)(b)* on acquisition.

Again, in a provision which is parallel in *s 113B* to that in *s 113A* in relation to identification of property, *s 113B(6)* provides that *s 113A(6)* shall apply in relation to replacement property as it does to the original property.

In just the same way as the reconstruction and amalgamation provisions of *TCGA 1992, ss 126–136* apply under the main clawback rule to save BPR from loss, they apply by virtue of *IHTA 1984, s 113B(6)* to replacement property to save the relief.

CLAWBACK OF APR: THE BASIC RULE

10.13 APR may be lost where, following a transfer which would otherwise qualify for that relief, certain conditions are not satisfied with the rule operating in a similar manner to BPR. The rule is found in *ss 124A* and *124B* and as with BPR has effect in relation to transfers on or after 16 March 1986. APR is lost in two situations where certain conditions are not satisfied:

- any part of the value transferred by a PET qualifying for APR fails; and
- part of the value transferred by a chargeable transfer qualifies for APR and the transfer is made within seven years of the death of the transferor.

In the second situation, the tax concerned is the 'top up' from lifetime rates to death rates but the value of the chargeable transfer does not impact on the donor's cumulative 'clock' which remains the same as that at the time of the chargeable transfer.

The conditions which must be satisfied for avoid a clawback of APR are set out in *s 124A(3)*:

- The original property must be owned by the donee throughout the period from the chargeable transfer to the death of the transferor and must not, at the time of that death, be subject to a binding contract for sale.

- Generally, the original property given must be agricultural property immediately before the death and must have been occupied for the purposes of agriculture throughout the relevant period, either by the donee or by someone else.
- The exception to the rule just set out arises where the original property consists of shares in or securities of a company. Here the condition is that, throughout the period from the gift to the death of the donor, the agricultural property qualifying for APR by virtue of the 'farming company' rules of *s 122(1)* was owned by that company and was occupied for the purpose of agriculture, either by that company or by someone else.

Example 10.8: Mary – costly health problems

In January 2012, Mary gave to her adult farmer son William her farm at Low Street, together with the farmhouse adjoining which William then lived in – the gift (PET) potentially qualified for APR. In January 2017, William is forced by ill health to give up farming – he sold the land, bought an insurance bond with the proceeds and continued to live at Low Street Farmhouse. Mary died in December 2017 causing the PET to William to fail requiring the availability of APR to be retested under the clawback provisions.

APR is no longer available on the farmhouse because it has been separated from the land; nor can it be claimed on the investments now representing the proceeds of sale.

By *s 124A(4)*, if the donee dies before the transferor, the date at which retention of agricultural property is tested for the purposes of escaping clawback under *s 124(3)* is that of the death of the donee.

Example 10.9: Andrew – donee death before clawback event

In June 2013, Andrew gave his brother James a parcel of agricultural land to add to James' existing farmland holding – the land potentially qualified for APR at time of gift. James died in October 2015 still owning the gifted land, which was sold by his executors in September 2016.

Andrew died in December 2017 causing the PET to James to fail requiring the availability of APR to be retested under the clawback provisions. Even though the gifted land had been sold at the time of Andrew's death, APR will still be available because it satisfied the conditions at the time of James death.

PARTIAL CLAWBACK: THE OPERATION OF S 124A(5)

10.14 As was noted in relation to BPR, APR clawback can arise where the conditions of *s 124A* apply to only part of the property; thus:

- in relation to a PET, only a proportionate part of the value transferred as is attributable to the original property qualifies for APR; and
- in relation to a CLT, the extra tax is worked out as if only a proportionate part of the value transferred (as was attributable to the original property) qualified for APR.

Example 10.10: Jennifer – partial clawback

Jennifer owned the Kingsnorth Estate, comprising a sheep farming operation, an adjoining moor with sporting rights, and a number of estate cottages commercially let. In June 2013, she gifted the entire estate to her brother Tom. At time of gift, APR was potentially available in respect of the lowland sheep farm and, with some difficulty, on the moorland but not on the cottages. Tom kept the lowland farm which he continues to operate and also the cottages but sold the moor.

Jennifer died in December 2017 causing the PET to Tom to fail requiring the availability of APR to be retested under the clawback provisions. APR is available in respect of the retained farm but is clawed back in respect of the moorland as it is no longer owned by Tom. The conditions of *s 124A(3)* are not satisfied in relation to the cottages, in respect of which APR was never available.

DISREGARD OF APR: THE RULE IN S 124A(7A)

10.15 This rule (applicable to transfers of value on or after 28 November 1995) finds a parallel for BPR in *s 113A(7A)* by disregarding APR in calculating the transferred value of a PET or a CLT. Importantly, the rule does not affect the rate of relief, nor does it affect the amount in respect of which the relief is given.

Definitions employed in s 124A: 'the original property' and 'the transferee'

10.16 *Section 124A(8)* defines 'the original property'. It is the property which, in relation to the chargeable transfer for clawback was either agricultural property qualifying for APR under *s 116* or shares in a farming company qualifying through *s 122(1)* for APR under *s 116*.

'The transferee' (donee) means either:

- the person whose property the original property became on the chargeable transfer; or
- the trustees of a settlement where, on the transfer of the original property, it became settled on relevant property trusts (as will apply to almost lifetime trusts created after *FA 2006*).

The definitions also apply to replacement property (see *s 113B(8)*).

The result of this can be to trigger clawback even where the trustees no longer own the asset having made an absolute appointment to an object of the trust. The effect of this is examined in relation to trusts later in this chapter. Further, a seemingly innocent gift may have disastrous consequences.

Example 10.11: Terence – transferee ownership

In December 2011, Terence, aged 75, made an outright gift to his son Jeremy of the farmhouse in which at that time Jeremy was living with Susan, his wife, and their two children. It had been the family home but Terence had moved out after his wife had died a few years before handing it over in, frankly, a fairly dismal condition. Susan had used her accumulated earnings to refurbish the house, installing a modern kitchen, bathroom and central heating.

During an estate planning exercise carried out for Jeremy by his solicitor, Jack, it was proposed that, to recognise Susan's contribution, the farmhouse should be transferred by Jeremy into their joint names. By this time, Terence was aged 81 and in poor health.

Jeremy mentioned the proposal in passing to his accountant, Cerys who identified that a transfer by Jeremy would cause him, as 'the transferee' within *s 124A(8)*, to no longer be the owner of 'the original property' as required by *s 124A(3)(a)* although it might well be argued that, after the transfer, Jeremy was still the owner of half the farmhouse. It could not be ruled out, however, that HMRC might maintain that Jeremy's interest in the proceeds of sale of the farmhouse was materially different from his original interest, ie that, if Terence were to die within the next year or so, clawback might apply at the very least to one half, and possibly to the whole, of the value of the farmhouse. Either way, with six years having already elapsed since the gift, it would not be prudent to risk clawback and as a result, the transfer to Susan did not take place.

The effect of clawback

10.17 Clawback operates differently according to whether:

- the clawback charge arises on the CLT or failed PET caused by the death of the transferor; or
- the clawback charge arises on the failure of a PET, that PET being the release of the reserved benefit (as to which see **10.23**).

Example 10.12: Paul – APR exchanged for BPR

In September 2011, Paul gave Phillip a small farm with potentially qualified for APR at the time of gift. Phillip continued to farm but developed part of the land as a trout fishery, which he conducted on business-like principles (see *C & E Comrs v Lord Fisher* – **Chapter 5**)).

Paul died in December 2017 causing the PET to Phillip to fail requiring the availability of APR to be retested under the clawback provisions. APR is available in respect of the undeveloped portion of the farm land but not to the trout fishery, although that may qualify instead for BPR by reference to Phillip's activities.

If the transfer when made was a CLT, rather than a PET, and qualified at that time only for APR with IHT being payable at the lifetime rate, on the death of the transferor within seven years the 'top up' tax charge is calculated as if APR had not been available at the outset. There is no effect on the original tax paid at the time of transfer and the donor's 'clock' is unaffected. As a separate matter, BPR is not available in the alternative to relieve the 'top up' charge to tax; this is the effect of *s 114(1)*.

Example 10.13: Cedric – APR rate change

In January 2013, Cedric transferred a parcel of let agricultural land into a discretionary trust established for the benefit of his adult children. Although qualifying for APR at time of gift, as the tenancy had commenced prior to 1 September 1995, the rate of APR was set at 50%. After the tenant's death in January 2014, the trustees commenced to farm the land in-hand and at the time of Cedric's death in December 2017, the trustees still owned and occupied the land.

Under the notional transfer deemed to arise immediately before Cedric's death, the land qualifies for APR by reference to the trustees (as donees) and thus potentially it would seem at the higher 100% rate. However, the original transfer to the trustees cannot benefit from this increased rate of relief rather it must remain at 50%, which was the rate applicable for such property at the death of Cedric.

An exception to clawback

10.18 Where the value transferred exceeds *the value of the property transferred*, relief from clawback operates in an interesting way, as explained in the IHT Manual at IHTM24174 and the first example there set out, which is paraphrased below.

Example 10.14: Jessica – separation values

In January 2012, Jessica gave to her son Tom all of the farmland but retained ownership of the farmhouse in which she continued to live. The value of the gift (established on the estate before less estate after basis) is (*a*) the value of the gifted land plus (*b*) the drop in the value of the house once separated from its land.

Tom had kept and continued to farm the land, all of which qualifies for APR at the time of Jessica's death in December 2017. Assuming all conditions are met, APR is available on the entire 'transfer value' of the land, ie including the element representing the depreciation in value of the farmhouse even though, being now separated from its land, it no longer qualifies for APR.

This interpretation follows from the definition of 'the original property' in *s 124A(8)*: the conditions are satisfied in relation to that property. In current times it is debateable whether a farmhouse would be any less valuable if shorn of its land.

RELIEF FROM CLAWBACK

Problems of identification: the rule in s 124A(6)

10.19 As was noted in relation to BPR, it would be harsh to apply the clawback rules where the substance of the asset given continues to be owned by the donee although its identity has changed. By *s 124A(6)*, clawback is avoided where shares (in a farming company) now held by the transferee can be identified with the shares originally given because they are effectively the same company by virtue of the rules in *TCGA 1992, ss 126–136* (reorganisation or reduction of share capital). The replacement shares are simply treated as if they were the original property.

Furthermore, where:

- the subject matter of the original gift was agricultural property; and
- on incorporation of the farming business, the donee now holds shares which were issued to him in consideration of the transfer to the company of the agricultural property;

the legislation operates to treat the shares, which in effect represent the interest in the original property, as if they were the original property.

The specific wording of the legislation is that the period of ownership of the transferee of the original property is to be treated as including his period of ownership of the shares. This wording is intended to get round the conceptual problem of the original legislation, which provided that the shares should be treated as if they were the original property (or part of it). It was not clear how the shares in such a circumstance could qualify for relief from clawback under *s 124A(3)* but the deeming provision as enacted by *FA 1987, s 58* resolves this problem.

Replacement property: the conditions

10.20 Further relief is provided under *s 124B* which concerns the situation where the original property has been sold and replaced by other qualifying property. There is a four-stage test before replacement property relief may be given:

- *Disposal triggering clawback*: The donee of the property must have made a disposal, before the death of the transferor, of part or all of the original subject matter of the gift.
- *Purchase of new property*: Whether that disposal was of part or of all of the original subject matter, the entire proceeds from that disposal must have been applied in purchasing the replacement qualifying property. *Section 124B(1)(b)* refers to 'the whole of the consideration received by him for the disposal', which is generally regarded as being the proceeds net of both costs of sale and CGT.
- *Time window*: The replacement property must be acquired, or at least there must be a binding contract for that acquisition, within the 'allowed period' of three years following the initial disposal, or such longer period as HMRC may allow.

- *Everything at arm's length*: The disposal of the subject matter of the original gift and the purchase of the replacement qualifying property must be by transaction at arm's length, or must be on terms such as might be expected to be included in a transaction at arm's length.

The effect of *s 124B* is to avoid clawback by deeming the conditions in *s 124A(3)* to have been satisfied with regard to the replacement property but only provided four further conditions are met, namely:

- *Donee retains asset*: the replacement property is still owned by the transferee immediately before the death of the transferor, and is not at that time subject to a binding contract for sale;
- *Old land always farmed*: throughout the period from the original transfer to the exchange of property, the original property was owned by the transferee and either he or someone else occupied it for the purposes of agriculture;
- *New land always farmed*: throughout the period from the exchange of property to the death, the replacement property was also occupied for the purposes of agriculture, either by the transferee or by someone else; and
- *New land still farmland*: the replacement property is, immediately before the death, still agricultural property.

As with BPR, if the donee has died before the transferor, *s 124B(4)* provides that it is the position at the date of death of the donee which determines whether the provisions as to replacement property avoid clawback.

The following flowcharts, which assume that 'T' is the transferor of the asset and that 'B' is the beneficiary/transferee, act as a broad guide to the replacement provision process.

Clawback flow chart

10.21

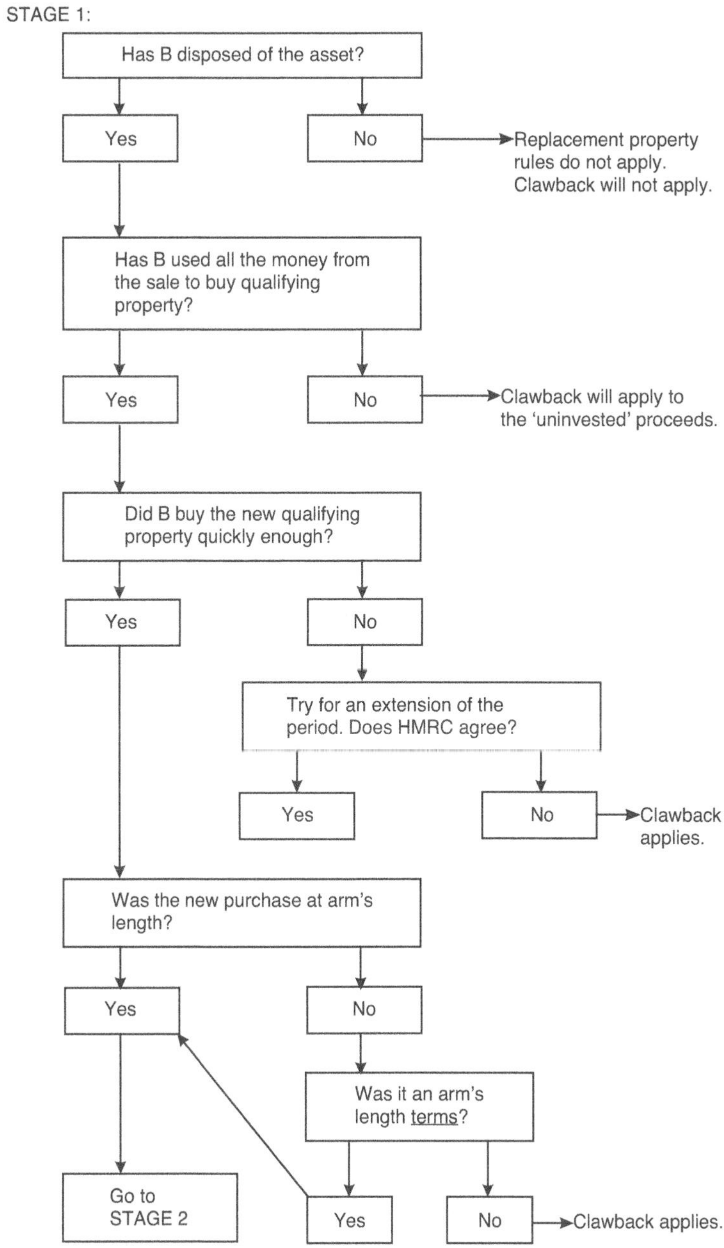

10.21 *Clawback*

STAGE 2:

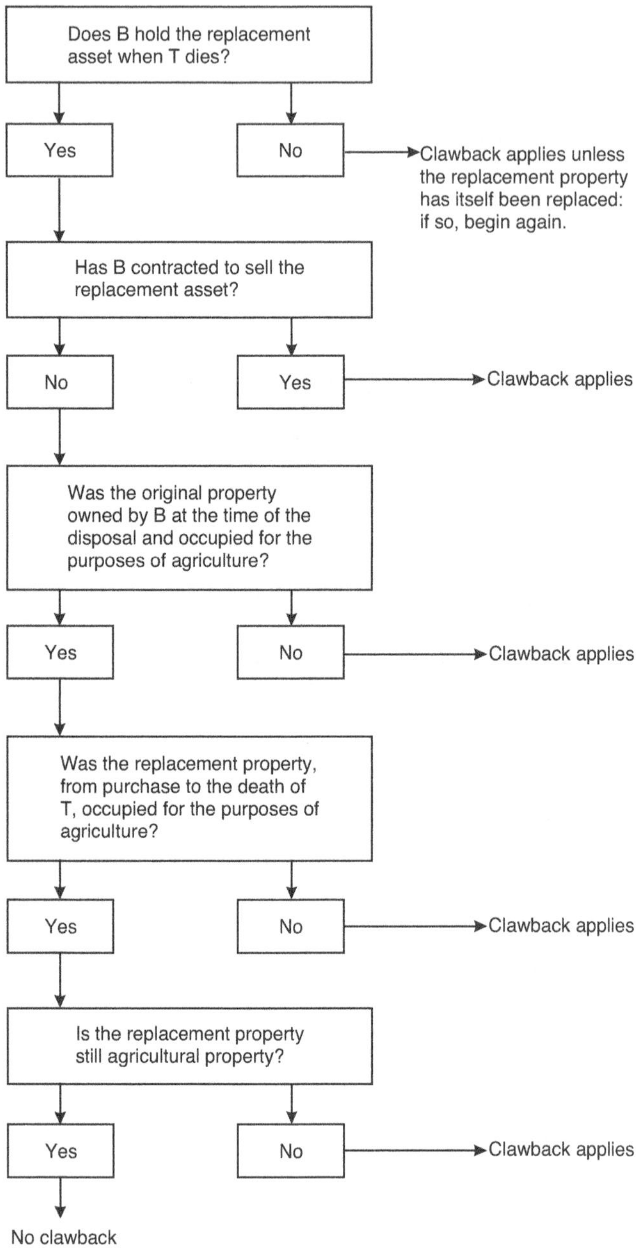

Replacement property: how the rules work

10.22 The clawback provisions could prove restrictive where, for good farming or other reasons, the sale of the gifted land has taken place but the contemplated purchase of replacement land has not yet been made. Accordingly, relief is afforded under *s 124B(5)* which provides that clawback may be avoided where:

- all or part of the original property has been disposed of before the death of the transferor; or
- is subject to a binding contract for sale at the death of the transferor; and
- the replacement property is acquired; or
- there is in place a binding contract for the acquisition of replacement property;
- after the death of the transferor but within 'the allowed period' after the original disposal; and
- the transferor dies before the transferee.

For timing purposes, it is the date of the contract that matters (see *s 124B(7)*) but the purchase of the replacement property must be made within three years or such longer period as HMRC may allow.

The replacement property provisions of *s 124B* are generally wide enough to include shares in a farming company, by allowing *s 124A(6)* to apply to the replacement property as it did to the original property.

Example 10.15: Church Farm – replacement property

Jim bought Church Farm near Bovey Combe in North Devon many years ago and farmed it until his death in December 2013, when the land passed under his will to his wife Sarah. Although she had been a great help on the farm all her life, much of the work was done to a greater or lesser extent by her five children (Andrew, Barbara, Celia, David and Edward). Andrew was now farming on his own account nearby at Birkham; Barbara had married a local builder and lived nearby; Celia worked in London as a hotel manager; David was farming at Biddleton; Edward had stayed at home and in April 2014 entered into partnership with his mother, Sarah, to run the home farm.

In April 2016, Sarah made gifts in order substantially to reduce her estate and all potentially qualified for APR at that time:

To Andrew she gave 12 acres near the main road. To Barbara, Sarah gave some redundant farm buildings which she thought Barbara and her husband might be able to make something of. There was a row of pretty but redundant cottages which seemed an appropriate gift for Celia, who could perhaps develop them into a holiday cottage business. David received a parcel of off-lying farmland amounting to nine acres. To Edward, Sarah gave a half interest in the remaining farmland.

In May 2016, the local council eventually undertook steps on a long proposed a road widening scheme and a strip of land was taken across the middle of the land which Andrew had received, leaving him with eight out of the gifted 12 acres and to one side of the land that David had received leaving him with six out of the gifted nine acres. In both cases, the remaining acreage was no longer an economic unit. Meanwhile, Barbara and her husband, having carried out a feasibility study on the barns and having obtained planning permission for a development, decided that they would sell the barns in an undeveloped state but with the benefit of that planning permission.

Jack, the family solicitor, advised Andrew, Barbara and David that given Sarah's age and poor health she was unlikely to survive the April 2016 gifts by seven years, with the risk that clawback might apply to deny them APR. He explained that the compulsory land purchase made by the local council, could trigger the clawback and that they should make the situation good by putting the money back into farmland before Sarah's death. By April 2017, Andrew, worried by Jack's advice and having found no land that was particularly suitable, made a hasty purchase of land at a slight overvalue. Barbara did not act so quickly – the property market had taken a downturn, so she and her husband kept the proceeds of the barns on deposit.

David did not want to ignore Jack's advice, but there was no land near him available for sale. However, he knew that a neighbouring farmer was planning to retire within a couple of years and that extra land immediately adjoining David's land would likely come onto the market. He mentioned the problem to his accountant, Cerys who advised David to wait explaining that under *s 124B(5)* he had up to May 2019 to make a purchase without risk of clawback (on the assumption there was no reason for HMRC to extend the three-year period).

Sarah died in December 2017 which left the residue of the estate, greatly diminished in value following the April 2016 gifts, to her five children.

Jack was struggling with the administration of the estate – in terms of value, the 2016 gifts were substantial in relation to Sarah's remaining estate. Jack was concerned that he would have to tell the children not only was there nothing more to inherit but that they would also have to fund from their own resources any tax that might be payable in respect of the lifetime gifts.

Having missed the point that Cerys had grasped, he was concerned that clawback would apply to Andrew's gift, and that, since it was the first of the five gifts, it would fully utilise Sarah's combined nil rate band/transferable nil rate band.

He considered, correctly, that the barns given to Barbara would not qualify for APR, because they had not been used for agriculture for many years. Similarly, the cottages, although of a character appropriate, had not been occupied by farm workers or retired farm workers or the dependants of farm workers and did not qualify for APR. He was, therefore, very glad to learn that Andrew had made a purchase that would protect him from clawback. Concern then focused on David, until Cerys explained he had until May 2019 to make his valid purchase. Jack had never been worried about the gift to Edward: he had stayed at home and was still farming the land which he had occupied in partnership with his mother since April 2014.

It is important to note that unlike *s 124B(4)* which quite specifically allows substitution of the death of the donee for that of the transferor in relation to the replacement property provisions of *s 124B(1)*, *(2)* and *(3)*, the more narrow *s 124B(5)* is only potentially in point where the transferor dies before the transferee thus clawback could apply if the deaths 'are in the wrong order'.

It is not essential that the property (in the hands of the donee) at the death of the transferor must qualify for APR at the same rate as the subject matter of the original gift. It is simply sufficient that the property qualifies for relief, even if at a different rate, for the clawback to be avoided albeit relief is awarded at the rate that was appropriate at the time of the original transfer. For example, the gift might be of farmland which, considered in the light of the circumstances at the time, qualified for relief at 50%. Whilst the land might be retained by the donee, circumstances might change such that, at the later death within seven years of the transferor, relief might be available at 100% only because there had been a change in the occupation of the land. In such a circumstance clawback would not apply and relief would continue to be available at the full rate.

Gifts with reservation and the application of the clawback rules

10.23 The interaction of the codes for clawback and reservation of benefit only serves to complicate an already complex subject – see **Chapter 14**. The IHT application will pivot on whether a benefit was initially reserved but later released, or where death intervened whilst the benefit was still enjoyed – see IHTM24201 and IHTM24202.

COUNTERING CLAWBACK IN RELATION TO BPR AND APR

10.24 The sheer complexity of the clawback provisions make it difficult to easily explain the rules to the client but managing clawback is a combination of not only avoiding automatic loss of APR or BPR but also ensuring that the transferor/donee have a reasonable grasp of the rules that apply and receive sound advice when circumstances change which could jeopardise the relief. It would seem, particularly in the light of comments in the judgment of Arnold J in *Swain Mason v Mills & Reeve* [2011] EWHC 410 (Ch) at para 197 as to the framing of letters of engagement, that it is not reasonable merely to expect the client to sign a letter acknowledging that he has studied and understands the effects of the clawback provisions. It will be prudent, though, for the adviser to have on his file some record that he advised not only the donor of the gift but that he advised the donee or urged the donee to take separate advice to avoid the risk of clawback.

Earlier in this chapter it was noted that clawback may be triggered where trustees make an absolute appointment to a trust beneficiary – this will apply to all trusts within the relevant property regime. However, where the interest of a life tenant with a qualifying interest in possession comes to an end during their lifetime in circumstances where the life tenant becomes the absolute owner of the BPR qualifying property, there is no clawback. This is because, by virtue of the deeming provisions, the life tenant is effectively treated as the owner of the trust property.

Replacement of agricultural property by business property

10.25 RI 95 remains current. It sets out the HMRC view as to the availability of relief where agricultural property is replaced by business property (or vice versa) and considers the availability of the 'new' relief to the replacement asset but be aware of the mismatch in the provisions concerning the operation of *s 124A(1)* – **see Chapter 12**.

Surrender of farm tenancies and FBTs

10.26 If a tenancy itself is the subject matter of a gift attracting APR and that tenancy is surrendered, the death of the transferor within seven years of the original gift must trigger clawback under *s 124A* because the asset which was the subject matter of the gift no longer exists. The parties cannot substitute a FBT in place of the old tenancy in the hope of relying on *s 124B* because there is no 'consideration received for the disposal therefore no consideration applied in acquiring other property' within *s 124B(1)(b)*.

However, consider the following:

A gift of the qualifying tenancy is followed by a further indirect gift of the freehold reversion by the landlord to a nominee for the tenant. That would ensure that the initial gifted asset, the tenancy, remained in existence in so far as it is possible (which might require careful arrangement) for the tenants to be their own landlords. If, on examination of the values, there is no transfer of value, clawback will not be a problem but a word of caution – it may be necessary to re-examine the profit-sharing arrangements within the farming partnership to ensure that there is, commercially, no transfer of value.

The mention of surrender of a tenancy prompts the adviser to consider the possible impact of CGT on the disposal, but ESC D39 may be in point if the surrender is part of an arrangement that results in extension of the lease.

FURTHER HAZARD: DISCRETIONARY TRUSTS

10.27 Different considerations affect the practitioner in relation to discretionary or accumulation trusts, where the trustees may decide to advance trust capital. Clawback can be triggered under *s 124A(2)* in respect of the 'top up' charge should the settlor die within the seven-year period post trust creation. The trustees should, therefore, be warned of the risk of clawback, so that their decision to advance funds is an informed one.

Example 10.16: Jonathan – BPR asset no longer held by trust

In May 2015, Jonathan who always used his annual exemption in April each year, transferred cash of £325,000 into a discretionary trust (the 'May 2015 Trust').

In September 2015, Jonathan transferred a minority holding of unquoted shares in a trading company valued at £150,000 and cash of £150,000 to a discretionary trust ('the September 2015 Trust') – the shares qualified for BPR at that time. Jonathan paid the IHT on the transfer.

In November 2015, Jonathan set up a third discretionary trust ('the November 2015 Trust'), having made no other transfers in between. He transferred a 29% unquoted holding in the same trading company valued at £87,500 and cash of £87,500 into the trust. The 'estate before less estate after' calculation on the November 2015 Trust produced a transfer value of £175,000. Jonathan paid the IHT on the transfer.

10.27 *Clawback*

In February 2016, the trustees of the September 2015 Trust appointed the entire fund to Rachel absolutely.

In February 2017, Jonathan died. The November 2015 trust was still in existence and still held the shares which qualified for BPR at that time.

(1) IHT due on the original creation of the May 2015 Trust was as follows:

	£	£
Transfer into trust		325,000
Nil rate band		(325,000)
		Nil
Cum. IHT thereon @ 20%		Nil

(2) IHT originally due on creation of the September 2015 Trust was as follows:

	£	£
Total of previous transfers (the May 2015 Trust)		325,000
Transfer into trust	300,000	
Less BPR on £150,000 @ 100%	(150,000)	
Net transfer 150,000	150,000	
Tax on net transfer (gross rate 25%)	37,500	
Gross transfer		187,500
Cum. total		512,500
Nil rate band		(325,000)
		187,500
Cum. IHT thereon @ 20%		37,500
IHT due on previous transfer		Nil
IHT due on the current transfer		37,500

(3) Tax originally due on the November 2015 Trust was as follows:

	£	£
Total of previous transfers brought forward		512,500
Transfer into trust '175,000' ('B')	175,000	
Less BPR £87,500 @ 100%	(87,500)	
Net transfer	87,500	
Tax on net transfer (gross rate 25%)	21,875	
Gross transfer		109,375
Cum. total		621,875
Nil rate band		(325,000)
		296,875
Cum. IHT thereon @ 20%		59,375
IHT due on previous transfers		(37,500)
IHT due on November trust		21,875

(4) Recalculation of IHT on death as a result of clawback

	£	£
The September 2015 Trust transfer		
Total of previous transfers (May 2015 trust)		325,000
The net transfer (BPR no longer due)		300,000
Tax paid by the donor on the net transfer		37,500
Gross transfer		662,500
Nil rate band		(325,000)
		337,500
Cum. IHT thereon @ 40%		135,000
IHT due on previous transfer		Nil
Revised IHT due on September 2015 trust		135,000
Tax originally paid		(37,500)
Top up clawback IHT		97,500*

** Reconciled as taxable amount due to loss of BPR of £150,000 @ 40% plus 20% top up charge on original chargeable transfer of £187,500*

	£	£
The November 2015 Trust transfer		
Total of original lifetime transfers brought forward		512,500
Transfer into trust '175,000' ('B')	175,000	
Less BPR £87,500 @ 100%	(87,500)	
Net transfer	87,500	
Tax on net transfer (gross rate 25%)	21,875	
Gross transfer		109,375
Cum. total		621,875
Nil rate band		(325,000)
		296,875
Cum. IHT thereon @ 40%		118,750
IHT due on previous transfers @ 40%		(75,000)
IHT due on November trust		43,750
Tax due originally paid		(21,875)
Top up IHT		21,875

As has been noted at **10.18**, the clawback charge for a CLT does not alter the settlor's cumulative clock so when calculating the top up charge on the November 2015 trust no account is taken of the increase in the cumulative total by reason of the clawback of BPR which remains at £621,875. Nevertheless, the complexity of this calculation would have been avoided if funds had not been released to Rachel prior to Jonathan's death.

Further difficulty: value-shifting

10.28 A transfer of value by a close company which is apportioned (under *s 94*) to the participators always gives rise to problems and this is no less so in the context of clawback. The cause of the difficulty is that the charge in such a situation may well arise because the relative value of shareholdings has been affected by the transfer of rights in them but no one person is actually the 'donee' in such a situation. This is so since the person who benefits is the one who already holds shares which, by virtue of the waiver or other action, are more valuable than they had been previously. Accordingly, there cannot be a notional transfer by the donee and thus under strict interpretation of the rules avoidance clawback is impossible.

The beneficiary of the value-shifting cannot identify the additional asset acquired by virtue of the operation, so the notional transfer rules required for clawback and its avoidance cannot work in his favour. As a general rule, therefore, where there is a lifetime transfer by means of the shifting of value in close company shares, there is always a risk of clawback of BPR on the death of the participator within seven years.

COMPARISON OF POTENTIALLY EXEMPT TRANSFERS (PET) WITH IMMEDIATELY CHARGEABLE TRANSFERS (CLT)

10.29 The treatment of a PET is almost the same as that of a CLT where the subject matter qualifies for 100% BPR/or APR. However, in certain circumstances, there can be an advantage in making the CLT of property rather than making a PET and this is set out in the following table:

Table 10.1 Clawback: effect on transfer of relevant business property

Property entitled to 100% APR/BPR: comparison of PETs with CLTs.

	PET Treatment	*CLT Treatment*
1. Lifetime gift	No immediate charge – BPR/APR qualification merely a comfort	No IHT charge if application of APR/BPR reduces chargeable value to nil
2. Death within seven years of transfer: assets retained by donee and continue to meet BPR/ APR requirements	No IHT charge if APR/ BPR potentially due on initial gift.	No IHT charge if APR/ BPR due on initial gift which reduced chargeable value to nil
3. Death within seven years of transfer where clawback applies:	IHT charge (subject to nil rate band)	IHT charge (subject to nil rate band)
(*a*) as to the transfer itself	On the full amount	On the full amount
(*b*) as to aggregation with the later transfers, eg on death	Full value is aggregated	Value is nil, so there is an advantage in making the transfer by CLT
4. Transferor lives seven years from gift.	No IHT: transfer has become fully exempt	No IHT: falls out of cumulation

CLAWBACK: A LOOPHOLE?

10.30 Clawback is not limited to PETs but PETs do bear the full brunt of the clawback knock on impact.

FA 1996, s 184(4) amended the clawback rules in respect of the definition of shares or securities in *IHTA 1984, s 113A(3A)* so as to include not only securities of a company giving the transferor control but also the more general category in *s 105(1)(bb)*, 'any unquoted shares in a company'.

Commentators have argued that this opens up a specific loophole analysed as follows:

Section 113A(3A) is stated to apply to two classes of shares or securities namely:

- shares that were quoted at the time of the PET that fails; or
- shares that were quoted at the time of a CLT which, by virtue of *s 113A(2)*, qualified for BPR and was made within seven years of the death of the transferor.

The shares must then fall within *s 105(1)(b)* or *(bb)* in relation to that initial transfer and must remain unquoted thereafter on the death of the transferor. It has been argued that the planning opportunity arises because it does not matter that, at the time of the 'clawback event', the shares are no longer relevant business property.

Example 10.17: Hetty – imaginative use of the rules

Hetty has a terminal illness likely to result in her death in three years' time and has consulted her solicitor, Jack, for advice on steps that could be put in place to minimise the IHT charge on her estate which will pass to her two sons. She owns 85% of an unquoted trading company which operates a care home but neither of her sons work in the business nor hold any shares in it. The premises from which the care home is run is owned by Hetty personally and has a freehold value of £900,000 thus whilst BPR is available it is restricted to 50%.

Hetty would like to transfer the property to the company in exchange for shares.

In so far as the value of the issue of shares should not equal the value of the property, there would be a chargeable transfer made by Hetty because it would be a transfer at undervalue to the company. However, that chargeable transfer calculated based on the reduction in value of Hetty estate post gift would not be great because Hetty already owns 85% of the company. The transfer would be within Hetty's unused nil rate band but, effectively, it would be chargeable accepting that Hetty would not live seven years.

Hetty's plan is to transfer the property to the company in exchange for shares, then to transfer her shares to her sons on the basis that the transfer of shares would be sheltered by 100% BPR. The intention would be that post gift and prior to Hetty's death the sons would then sell the trade but keep the company as a mere investment vehicle. Jack thought clawback would be a problem, so asked his assistant Leslie to research it who concluded that there would be no clawback provided the sons kept the shares until after Hetty's death. It was Leslie's interpretation that the retained property need not continue to qualify for BPR in the hands of the donee, and he summarised his reasoning as follows:

- *Section 113A(1)* sets out the rule for clawback.
- *Section 113A(2)* provides that the IHT has to be calculated without the benefit of BPR where clawback applies.
- *Section 113A(3)(a)* imposes a condition that the original property must be owned by the transferee from the original transfer to the death of the transferor.
- *Section 113A(3)(b)* provides the general rule that on the death of transferor the property is retested for BPR by reference to a notional transfer made by the transferee.
- *Section 113A(3)(b)* may be disapplied in so far as testing for BPR by reference to the transferee by *s 113A(3A)*.
- *Section 113A(3A)* has two limbs:
  - Clawback will not apply to an original transfer of quoted shares that qualified for BPR at that time as part of a control holding held by the transferor.
  - Clawback will not apply to shares that were part of the old category of unquoted shares which, with other securities, gave control and which remained unquoted through to the death, or unquoted shares under the general rule in *s 105(1)(bb)* and which remained unquoted.

Jack was not convinced. He considered that all the categories of property in *s 105(1)* are subject to the opening words of that section which in turn makes them subject to the rest of the section as a whole. He considered that the 'wholly or mainly' provision of *s 105(3)* must 'colour' the subject matter of a gift demanding the requirement that the application of *s 105(1)(bb)* is still subject to base BPR conditions set out in *s 105(3)*. Jack concluded that the proposed strategy would not preserve the gifted shares from BPR clawback on Hetty's death.

Chapter 11

Woodlands

QUICK GUIDE

- This is a 'Cinderella' relief and one that is used almost as a last resort. Essentially, it operates by way of tax deferral rather than absolute tax relief and it is not extinguished on death; accordingly, it must be recognised that at some later point the deferred tax must and will spring back into charge.
- APR or BPR should always considered as a priority option to woodlands relief.
- The conditions for the relief to apply are set out at **11.5** and summarised with the aid of a flow chart.
- The tax charge itself is examined at **11.9** onwards and specific reference is made to the 50% reduction under *IHTA 1984, s 127*, noted at **11.9** and illustrated at **11.13**.

COMMERCIAL OCCUPATION OF WOODLANDS: OVERVIEW OF TAX TREATMENT

11.1 Although operating as a tax deferral mechanism, this book would not be complete without reference to woodlands relief. The taxation of woodlands generally operates on a favoured basis and it is desirable if not necessary to have at the very least some basic knowledge of the special application for taxes other than IHT.

Historically, profits derived from the occupation of woodlands managed on a commercial basis was charged under Schedule B. However, that charge was abolished with effect from 6 April 1988 by *FA 1988, Sch 6, para 2(1)* which states 'The charge to tax under Schedule B in respect of the occupation of commercial woodlands is hereby abolished…'. As a result, the proceeds of felling and selling the timber have since that date been free from income tax, and the value of growing timber is free from both income tax and CGT.

The corollary of the *FA 1988* measures is that losses arising from the occupation of woodlands are not trading losses and they cannot be offset against any other income. By the same token there is no relief for revenue expenditure, capital allowances are not available and there is no relief for interest paid. Where that interest is paid by a company whose business consists of the occupation of woodlands, that interest is not deductible as a charge on income but there is a proportionate restriction where the company is engaged partly in the management of woodlands and partly in other trades – that proportion will (see *FA 1988, Sch 6, para 3(5)(b)*) relate to the business of the whole of a group of companies where a woodland management company is part of the 51% group.

Is there a trade?

11.2 It is still necessary to identify the activities which amount to no more than the commercial occupation of woodlands and those which go a step further by effectively amounting to the carrying on of a trade. There are three key cases which still reflect the law in this area:

(1) In *Christie v Davis* [1945] 1 All ER 370 the owner and occupier of a woodland estate set up a sawmill. Timber from the estate was used to produce fences and gates and the bark was sold on for other manufacturing use. Timber was also bought in from other estates and applied for the same purpose. The owner appealed against the HMRC decision to assess the profits to tax under Schedule D, Case I and it was held that working the sawmill was an ordinary method of making the timber marketable and that, to the extent that the timber came from the owner's estate (but not that timber bought in), the profit was covered by the Schedule B assessment.

(2) In *IRC v Williamson Bros* 1950 SC 391, timber merchants wished to ensure future supplies of timber and bought woodlands which were some distance from their sawmills. They cut the timber and carried it to their sawmills, where it was sold either to other commercial sawmills or direct to other businesses. It was held that the Special Commissioners were bound to follow the decision of the High Court in *Christie v Davis* (above) such that the profits from the felled timber were properly excluded from an assessment under Schedule D.

    HMRC pursued an appeal to the Court of Session who returned the case to Special Commissioners to decide whether the operations of felling and sawing timber were part of the right of ownership/occupation or should be treated as separate trades in their own right. It was decided that the timber was not cut up into planks before it was sent out to other sawmills but rather was sent in its raw state. HMRC withdrew their appeal.

(3) *Collins v Fraser* [1969] 3 All ER 524 required a detailed analysis of the process undertaken. The taxpayer owned woodland and also a sawmill

some 12 miles away. The timber was felled, stripped, trimmed, and reduced to cross-cut timber at the wood or was cut to five-foot lengths at the wood or at the mill. Once it reached the mill, the timber, if over five feet long, was reduced to that length and then turned into planking one inch or one and a half inches thick, converted into boards and made into boxes.

The issue was how much of this entire operation could be said to be done so as to make the wood marketable in some adjusted shape or form such that the profit could properly fall within Schedule B. On appeal, the case was sent back to the Commissioners for them to decide what was done with the timber, up to and including the planking stage, which might extend beyond what could legitimately be regarded as normal preparation of timber for the market 'and disposing of the fruits of the taxpayer's land'.

Are woodlands occupied for commercial purposes? The reality is that many woods are retained mainly for amenity and sporting value and there is very little profit for the grower, only for the woodland and timber contractors. Whilst the question is no longer relevant to the income tax treatment, it is still relevant for CGT when considering the availability or not of rollover relief applicable to the gain arising on the value of the land itself (but not on the value of timber which is exempt). Similarly, VAT may be chargeable on the sale of timber or on the grant of the right to fell timber if the woodlands are managed on a commercial basis. BPR should be available (see **11.7** below) on the basis that the woodlands constitute a business, if it can be shown that separate accounts are maintained for those woodlands which are commercial rather than those which are not.

Case commentary on timber: whether income or capital

11.3 Cases, examined in earlier editions of this work, centred on the deductibility in the computation of profits of a payment for the right to obtain trading stock, particularly where that stock was timber or growing crops. The key cases are:

- *Kauri Timber Co Ltd v New Zealand Commissioner of Taxes* [1913] AC 771, where the Privy Council upheld the HMRC view that the transactions by the taxpayer under which the rights were acquired gave them an interest in possession of the land rather than the mere purchase of the timber.
- *Murray v IRC* (1951) 32 TC 238, where the Court of Session confirmed the view of the General Commissioners that the profit on re-sale of the right to cut timber was part of the taxpayer's normal trade.
- *Hood Barrs v IRC* [1957] 1 All ER 832, where the amounts that the taxpayer had paid for two lots of standing timber were capital and not deductible from his Schedule D, Case I trading profit.

- *Hopwood v C N Spencer Ltd* (1964) 42 TC 169, where expenditure was incurred on the purchase of a wood to secure future supplies, pivoted on whether a transaction represented the purchase of stock in trade, rather than of a capital asset. The company provided the purchase price but the conveyance was recorded in the name of the controlling shareholders. The only 'formal' agreement between the shareholders and the company comprised a minute of a company meeting that the company was the owner of the timber in the wood and that the wood had been purchased to provide a reserve. The company expected the wood to be cleared within seven to ten years. It was held that the expenditure was incurred not on the purchase of a capital asset, but on buying stock in trade.
- *Chu Lip Kong v Director General of Inland Revenue* [1981] STC 653 where a timber merchant bought land and sold it without taking any timber. The purchaser specialised in felling timber that belonged to other people. He was assessed on the basis that the profit from the sale of land was a trading profit. Although the Special Commissioners regarded the immediate sale as a fictitious transaction, the Privy Council held that the taxpayer had sold the land at a profit without separate exploitation of the timber on it. There was no evidence that the taxpayer had previously dealt in land. This was an isolated transaction. As a result, the land was purchased as a capital asset, and the subsequent sale gave rise to a gain of a capital, and not of an income, nature.

The common theme running through these cases is the failure to fell timber makes it easier to regard the transaction as a capital one.

The former woodland relief

11.4 Effective for deaths on or before 11 March 1975, *Finance (1909–10) Act 1910, s 61(5)* provided estate duty relief which operated to exempt the value of growing timber until its later sale. The rules contained one key anomaly worth mentioning:

FA 1986, Sch 1,9 para 46 denied PET treatment to any transfer that included any element (albeit small) of woodland that was subject to the deferred estate duty charge. By ESC F15, the scope of the paragraph is restricted solely to that part of the value transferred which is attributable to the woodlands which are the subject of the deferred charge. The overall effect of this provision is that no liability would arise to estate duty after 12 March 1975, unless money was received:

- from the sale of timber felled or cut after 12 March 1975; and
- at or before the time of the first transfer of the woodland concerned for IHT purposes (other than a transfer to a spouse).

RELIEF ON DEATH

11.5 An election, which will not apply where APR is in point, may be made within two years of a death (or longer, see below) for the value of any trees or underwood (in this chapter called 'timber') on any land situated in the UK or, any EEA state, to be left out of account in determining the notional value transferred on death.

The election must be made by the person liable for any part of the IHT charge (ie executors, trustees etc) and the effect of the claim is to defer tax until a later disposal.

Under *IHTA 1984, s 125(2)(a)*, the value of timber is left out of account in fixing the notional value transferred on a death (subject to the other terms of the code contained in *ss 125–130*). The circumstances for the relief to apply are as follows:

- it is available on the occasion of death only (*s 125(1)(a)*) and (see IHTM04374) cannot apply to a failed PET;
- the estate of the deceased person includes land in the UK or any EEA state;
- there must be timber on that land;
- the circumstances must be such that APR does not apply;
- the deceased must have owned the land for five years before death else if not owned for that period, must have been acquired otherwise than for consideration in money or money's worth ie by inheritance on death;
- there must be an election made by the person liable for the whole or part of the IHT; and
- that election must be made within two years of the death or within such longer time as HMRC may allow.

The above conditions are straightforward but again it must be recognised that first, the relief only relates to the timber and not to the land on which it stands; thus it is necessary to establish the value of the underlying land excluding the timber; and second, the (deferred) IHT becomes chargeable on a later disposal of the timber.

The flow chart (Availability of woodland relief) sets out the steps to be satisfied. The relief is not particularly common, since the representatives of the taxpayer will prefer, wherever possible, to claim the full relief of APR or BPR.

Availability of Woodland Relief

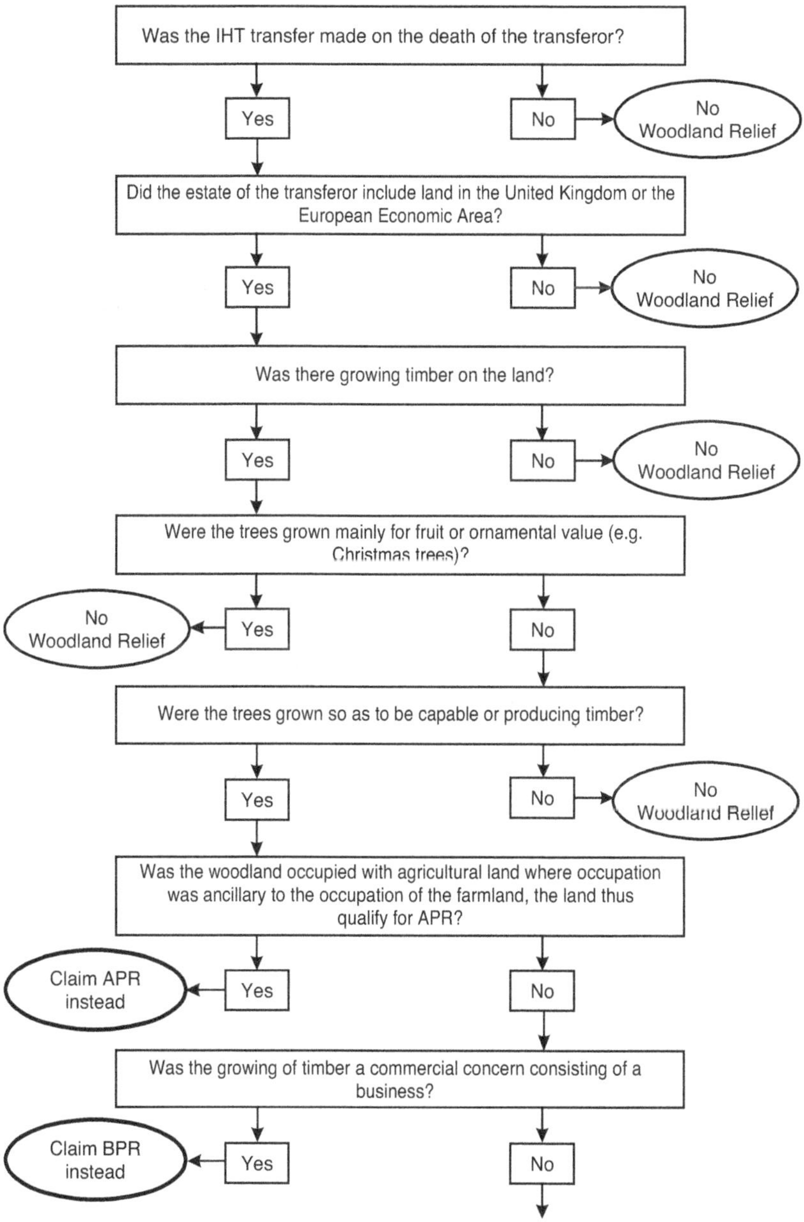

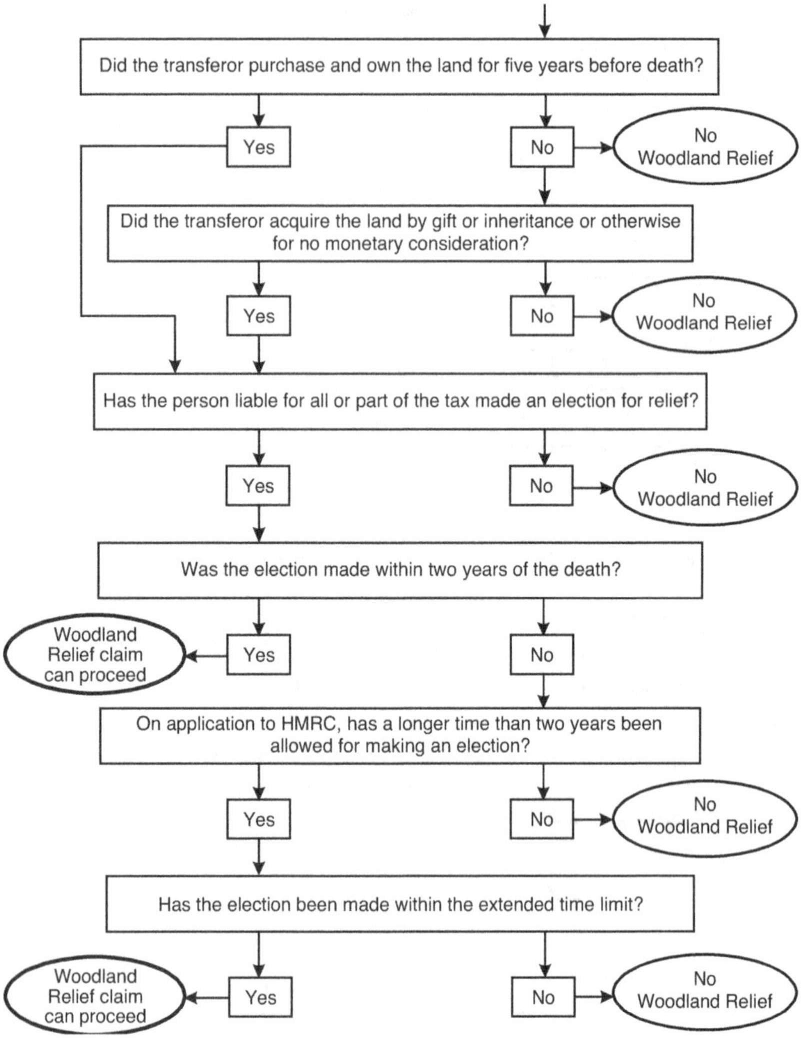

Note the required link between ownership of the timber and the bare land – see IHMT04376. Once that link is severed then even if the trees still stand, access to the relief is lost. Accordingly, when creating business structures such as partnerships or syndicates it is vital to ensure that each participant directly owns the land on which he is growing trees. By the same token, relief may be available where ownership is deemed to persist through the operation of the 'gift with reservation' rules – see IHTM04374.

An election need not be for the whole of an area qualifying for relief but if it is for part only that part must be identified – see IHTM04375.

Example 11.1: David – GWR and 'deemed' ownership

David owned woodland which formed part of the family estate. Anxious to minimise his IHT exposure, in October 2016 he transferred to his daughter Jennifer by separate conveyance, first the main house and parkland, and second a large area of woodland nearby.

Notwithstanding advice to the contrary, David continued to enjoy sporting rights over the estate. David did not merely ride, or just take a gun when walking his dogs, which might (just) have been *de minimis* within RI 95; rather he would occasionally organise events with a group of friends. On such days the woodland workers were unable to carry out timber operations for obvious safety reasons.

Even though David made no money from his hobby, the shooting did (albeit to a minor extent) restrict Jennifer's use of the land by interrupting the business of the timber management and that amounted to reservation of benefit by David, at least as far as the woodland was concerned. That reservation of benefit made him and not Jennifer the 'deemed' owner of the woodland in the context of any subsequent claim to woodland relief.

Prior relief as agricultural property

11.6 Where the woodland is ancillary to the occupation of agricultural land or pasture (see **Chapter 2**) an APR claim should always be considered first.

Prior claim to BPR

11.7 Having eliminated the availability of APR, consideration should be given to the availability of BPR which may apply if the woodland is part of a business which itself qualifies as relevant business property. Under *s 127(2),* BPR will be available in respect of the value of trees and underwood but not the land. It may be difficult to support the business connection where the woodlands are not part of a syndicate but merely part of a rural landholding so small in area as to be barely viable from a commercial point of view, but yet have high amenity or sporting value.

There is a natural overlap between woodlands relief and BPR. Where, for example, the estate of the deceased includes growing trees and underwood, woodlands relief will enable the value of that timber (but not the land) to be left out of account. A subsequent disposal of the timber before any transfer of the land on which it is growing, will give rise to a charge on the timber; but, if that timber qualifies for BPR, then the latter relief will be applied after deduction of the deferred IHT which became payable in respect of the earlier death. An example of how the section works appears at Example 11.3.

The amount on which tax is charged under *s 126* (the postponement provision) is reduced by 50% where, had the value of the timber not been left out of account under *s 126*, it would:

- have been taken into account for the purposes of a claim to BPR; or
- in relation to early cases, would have been taken into account for BPR if the current legislation had then been in force.

This later alternative will apply only where the relevant death occurred before 7 April 1976.

PROPERTY QUALIFYING FOR RELIEF

11.8 *IHTA 1984, s 125* applies only to woodlands held by the *deceased* in one capacity or another, but not to woodlands owned by a *company* in which the deceased held shares. Importantly, woodlands relief has no parallel to the APR provision in *s 119* by which occupation of land by a company controlled by the transferor is treated as occupation by that transferor himself.

The definition of woodlands, for the purposes of woodlands relief, excludes orchards (note the failed attempt to claim APR in *Dixon v IRC*, considered at **2.6** above) and nurseries. The latter was tested (for income tax purposes) in *Jaggers (t/a Shide Trees) v Ellis* [1997] STC 1417 and whilst that case has little relevance for IHT (see below), it does contain some pertinent definitions which are so relevant. Young conifers were planted on nine acres of land which the Special Commissioner held did not comprise woodland and was not in any sense being prepared for use for forestry purposes; rather it was a Christmas tree plantation. On appeal the High Court held that the terms 'woodland' and 'forestry' were not clearly defined in *ICTA 1988* and reliance was placed on expert evidence. It was held that the growing of Christmas trees did not amount to the production of timber. By reference to *Brutus v Cozens* [1973] AC 854, the court held that the right approach is that:

> 'the meaning of an ordinary word of the English language is not a question of law … it is for the tribunal which decides the case to consider, not as law but as fact, whether in the whole circumstances the words do or do not

> as a matter of ordinary usage of the English language cover or apply to the facts which have to be proved. If it is alleged that the tribunal reached the wrong decision … the question would normally be whether their decision was unreasonable in the sense that no tribunal acquainted with the ordinary use of language could reasonably reach that decision.'

The Commissioner had been right in his decision. 'Woodlands' in *ICTA 1988, s 53(4)* meant 'land used for forestry purposes' and the Oxford English Dictionary defined 'woodland' as 'land covered with wood, ie trees; a wooded region or piece of ground', and 'forestry' as 'the science and art of forming and cultivating forests and the management of timber'.

The taxpayer had argued that the site was covered with trees and was therefore woodland but to accept that was to disregard age, size, type or appearance of trees, the way they were spaced or pruned, the age at which they were cut or removed, and the general appearance of the trees themselves. 'Woodland' meant a wood, a sizeable area of land to a significant extent covered by growing trees of some maturity, height and size. As a rule of thumb, the trees should be capable of being used as timber. It was a matter of impression and personal judgement. In the instant case, the site had not struck the Commissioner as woodlands and would not strike Lightman J for a moment as wood or woodlands. The trees looked more like bushes than timber trees, and the site looked more like a nursery than a wood. The appeal was dismissed.

This case is more relevant for APR, by virtue of the exposition of the meaning of 'woodlands', than to woodland relief. That is because *s 125(1)* refers to land 'on which trees and underwood are growing', whereas the value of the Christmas tree is in the cut article. A Christmas tree plantation will usually qualify for APR as a nursery and likewise, a grower would benefit more from 100% BPR than from woodland relief.

THE CHARGE TO TAX ON DISPOSAL

11.9 Where:

- the value of timber has been left out of account by virtue of an election under *s 125*; and
- part of that timber is later disposed of; and
- that disposal occurs before any part of the value transferred on the death of any other person 'is attributable to' the value of the land on which that timber grows;
- IHT is, by *s 126(1)*, chargeable under *ss 127* and *128* unless that disposal is made by any person to their spouse (*s 126(1)*).

Where IHT has been charged by *s 126* on the disposal of timber, IHT is not charged again in relation to the same death on a further disposal of the same timber but the person entitled to the proceeds of sale (or who would be entitled if the disposal of the timber had been by sale) is liable for the tax. Form IHT100f must now be used in conjunction with form IHT100. By virtue of *s 216(7)*, the account must be delivered by the person liable within six months of the end of the month in which the disposal occurs.

The basis of charge

11.10 Where there is a chargeable event, the value on which IHT is charged is the net proceeds of sale (where that sale is for full consideration) else the net market value of the timber at the time of disposal. However, in both cases that value is reduced by 50% (relief under *s 127(2)*) if BPR would have been in point (or would have qualified for BPR had the relief been available at that time) had the woodlands relief election claim not been on the occasion of the first death.

The rate of tax

11.11 The rate of IHT is determined (by *s 128*) as the rate at which it would have been charged on the initial death (that in respect of which the original claim was made) by treating it as the top slice of the estate as increased by:

- the amount (net value or net proceeds of sale) fixed by *s 127*; and
- any amount on which IHT was previously charged under *s 126* in relation to that death.

Where deferred IHT becomes chargeable following a reduction in IHT rates by substitution of a new table, the new rates will be used even though the death occurred when an old scale applied on which the rates were higher.

Net value: deduction of expenses

11.12 The following expenditure may be claimed as a deduction to arrive at the net proceeds of sale/ net market value of the timber:

- sale costs of disposal of the timber;
- costs of replanting to replace the trees within three years of a disposal (or such longer time as HMRC may allow); and
- costs of replanting, so far as not allowed on the previous disposal (see *s 130*).

The HMRC manual has long since amended the original guidance which directed the District Valuer to ignore any allowable expenses/obligations to replant, and to value the 'net value' as the higher of:

- the value of the trees for sale for felling, including any underwood; and
- the value of the standing trees or underwood as part of the estate of the transferor.

Rather current practice is on the basis of standing trees where such estates contain young trees or trees with high value as amenity rather than simply for forestry.

Credit for tax charged

11.13 Where a disposal occurs giving rise to a chargeable event on which IHT is chargeable under *s 126* and that disposal is itself a chargeable transfer for IHT purposes (ie not a sale at market value) the value of that chargeable transfer is calculated (under *s 129*) as if the value of the timber had been reduced by the *s 126* charge.

Example 11.2: Algernon – credit relief

Algernon, a bachelor, who had a full nil rate band available to him having made no chargeable transfers in his lifetime, died in December 2015 leaving an estate comprising:

	£	£
House, cash and investments		690,000
Woodlands – Bare land	60,000	
Growing trees	140,000	
		200,000
Total estate		890,000

Algernon left the whole of his estate to his sister Julia who made a woodlands election under *s 125*.

In June 2016, Julia had some of the trees cut down and incurred felling costs of £25,000 and replanting costs of £18,000. The remaining trees were then worth £175,000 and she gifted them to her nephew, Jim. At the time of the gift Julia had already utilised both her nil rate band and annual exemptions.

The original IHT position on Algernon's death was as follows:

	£
Gross estate	890,000
Less s 125 claim value	(140,000)
Taxable estate	750,000
Less nil rate band	(325,000)
Chargeable transfer	425,000
Tax at 40% thereon	170,000

By giving the trees to Jim, Julia has released the deferred charge to IHT but calculated with reference to the position on Algernon's death. The basis of the transfer is:

	£
Gross value of gift of trees	175,000
Less costs to fell	(25,000)
Less costs to replant	(18,000)
Net value of trees	132,000

IHT is payable on £132,000, which is regarded as being part of Algernon's increased estate of £882,000 (£750,000 + £132,000) calculated as follows:

	£
Gross estate	890,000
Less original s 125 claim value	(140,000)
Taxable estate	750,000
Add deferred amount	132,000
Adjusted total	882,000
Less nil rate band	(325,000)
Chargeable transfer	557,000
Tax at 40% thereon	222,800
Less original tax paid	(170,000)
Deferred IHT charge	52,800

Julia died in December 2017 and this caused the PET of the gift of trees to Jim to fail. The PET is calculated as follows:

	£
The value of the timber on gift	175,000
Less the deferred charge on Algernon's estate	(52,800)
Failed PET value	122,200
Tax due thereon at 40% payable by Jim	48,880

The deferred tax charge in respect of Algernon's death (£52,800) is payable immediately but by *s 229*, the instalment option is available in respect of the tax due on the failed PET of the gifted woodland to Jim.

50% REDUCTION UNDER S 127

11.14 The operation of the 50% reduction under *IHTA 1984, s 127* is illustrated as follows:

Example 11.3: Fred – woodlands and BPR

Fred Turnbull died in March 2015 leaving a gross estate of £1 million. This included a forestry business (all covered by BPR) the timber of which was valued at £125,000 and land on which the timber grew valued at £35,000, all of which passed to his daughter Susan who made a claim under *s 125* to defer the IHT charge on the timber. Susan did not carry on the forestry business and in December 2017 having already used her nil rate band and annual exemptions, she transferred the timber and the land on which it stood to a discretionary trust – at that time the land was worth £45,000 and the timber was worth £180,000.

This was a CLT, but the gift also triggered release of the deferred IHT charge on Susan's £125,000 inheritance of the timber from Fred's estate. The timber would have qualified for BPR on Fred's death and thus the 50% reduction is still available in calculating the deferred tax charge as follows:

	£
Original value of timber	125,000
Less s 127(2) relief at 50%	(62,500)
Chargeable	62,500
IHT thereon at 40% due by Susan	25,000

The gift into trust was also a CLT and Susan agreed to meet the IHT due thereon. The chargeable transfer is calculated as follows:

	£
The value of the gift (£180,000 + £45,000)	225,000
Less the deferred charge on Fred's estate	(25,000)
Net chargeable transfer	200,000**
Gross chargeable transfer*	250,000
Tax due thereon at 20% met by Susan	50,000

**As Susan is meeting the tax due the value of the transfer must be grossed using the formula net transfer x100/80.*

*** Had Susan continued the forestry business the transfer of the timber to the trust would qualify for 100% BPR.*

RELIEF FOR LIFETIME GIFTS

11.15 It is worth restating that woodlands relief is only applicable on death and may not be claimed against a gift made during lifetime. Such a life time gift of woodlands must instead rely on APR and/or BPR as illustrated in the flow chart above.

THE TREATMENT OF LIABILITIES: THE NEW RULES

11.16 *FA 2013*, by *s 176* and *Sch 36*, realigned the basis on which liabilities are set against property in order to arrive at the value which qualifies for relief. The rules apply to transfers of value made on and after 17 July 2013 but with effect to fresh/restructured liabilities incurred on or after 6 April 2013. In simple terms, the new rules require debt to be set against the property qualifying for relief (even if the debt is secured against other property – see IHTM28023) thus reducing the value on which relief may be claimed – see **8.16** and IHTM28011.

The liability is very widely defined, see *s 162B(5),* as any which is, wholly or partly, directly or indirectly, attributable to financing:

- the acquisition of the land or of the trees or underwood: *s 162B(5)(c)(i)*; or
- the planting of the trees or underwood: *s 162B(5)(c)(ii)*; or
- the maintenance or enhancement of the value of the trees or underwood: *s 162B(5)(c)(iii).*

HMRC consider (see IHTM28013) that the word 'indirectly' is very wide and that there need be no intention to convert cash into relievable property: it is the effect that counts, however well disguised and however delayed may be the use of the cash. This will present practitioners with difficult compliance duties.

11.17 To avoid a double deduction, *s 162B(7)* provides that where (and to the extent that) a liability has been set against the woodland property (so that a chargeable transfer is valued higher than it would have been with the benefit of woodland relief on the value of the asset without offset of the debt) that same liability is not set against any later transfer of value by the same transferor. This provision is aimed at 'failed PETs' preventing the deduction of a liability from both the value of the gift and the estate on death and as such it does not sit easily with the regime for woodland relief.

11.18 A liability might be attributable partly to buying woodland and partly for other purposes but what if it is partly repaid? *Section 162C* addresses this issue by approaching the matter in the following order:

- the first part of the liability to be treated as discharged is that not attributable to the acquisition of relievable property, ie the least helpful treatment for the taxpayer;
- subject to that, discharge of the liability that would have been offset under *s 162B*, for example against trees and underwood under *s162B(5)(c)*, occurs only after all the 'non attributable' liability has been cleared; and
- subject to that, any remaining debt repayment is allocated to debt incurred in relation to acquiring excluded property which is dealt with under *s 162A* (and which is not within the scope of this work).

As with all anti-avoidance legislation, there may well arise situations that work unfairly against the taxpayer and compliance in particular may prove quite challenging to personal representatives. Taxpayers could make life easier for those who follow them by keeping the borrowing for each purpose separate, thus avoiding these attribution rules.

Regrettably, what is rather more likely is that borrowing (and the security offered) will be pooled; that old borrowing will be mixed with funds added after 5 April 2013; the estate owner may forget, and pay a planting contractor out of the wrong account, increasing the borrowing; or that part of the property may qualify for relief and another part not; or not at the same rate.

Chapter 12

Overlap of APR and BPR

QUICK GUIDE

- It is common for both reliefs to apply to the same asset but priority is always given to APR.
- As APR is only applied to the agricultural value of land and property, access to BPR is critical not only for those assets to which APR cannot apply ie plant and machinery but also more importantly to mop up any excess market premium value.
- A short list of the relative merits of each relief is set out in **12.6** and **12.7**.

HISTORICAL PERSPECTIVE

12.1 *FA 1981, s 96* and *Sch 14* introduced measures intended to both remove anomalies and simplify the previous law. There is a similarity in the drafting of several sections and this is particularly evident in the following areas:

- the manner in which APR/BPR is applied to assets held within relevant property trusts;
- the fact that APR/BPR works by reducing the value transferred by a transfer of value rather than by reducing the chargeable transfer itself;
- the manner in which replacement property is treated;
- the succession provisions;
- the denial of APR/BPR where the property is subject to a binding contract for sale;
- clawback provisions;
- the manner in which property is treated which is subject to a reservation; and
- the rates of APR/BPR.

PRIORITY/ELECTION

12.2 The main statutory provision regulating the interaction of APR and BPR is contained in *IHTA 1984, s 114,* which provides in *sub-s (1)* that where:

- any part of the value transferred by a transfer qualifies for agricultural relief; or
- would qualify for APR except for the rules as to successive transfers and in particular *s 121(3)*;

then no BPR is allowable in respect of that part of the value.

Similarly, *s 114(2* avoids double relief in relation to woodland relief. The value to which BPR may apply under *s 104* is the net value after taking the trees and underwood out of account. To this extent, therefore, there is no election for relief: if the relief is available, it will be given in the order in which the statute allows.

TERRITORIALITY

12.3 There is a clear distinction in the territorial scope of application for the two reliefs. By *s 115(5)*, APR is restricted to agricultural property in the UK, Channel Islands, Isle of Man or an EEA state. However, for BPR there is no corresponding situs restriction within *s 103* which is available on worldwide qualifying property.

AMOUNT

12.4 Neither BPR nor APR is limited in amount. That having been said, APR carries a natural limitation since the relief is restricted to the agricultural value of the property whereas BPR applies to market value which may and often does command a premium.

Example 12.1: Wilfred – market value premium

Wilfred owned land which he farmed for many years until his retirement in 2002 and since then the land together with the bungalow has been let under a farm business tenancy (FBT) for the purposes of agriculture. The holding adjoins gravel workings which are now nearly exhausted. It is believed that further gravel deposits lie under Wilfred's land, and the operator of the adjoining quarry has expressed interest in purchase of the land for a sum which is significantly in excess of the agricultural value.

Wilfred does not enjoy good health and no longer lives on the farm, preferring to live with his niece who also has a lasting power of attorney for him. In his present state of health, Wilfred cannot realistically set up a new business so as to exploit the potential of the mineral deposits himself:

On Wilfred's death, APR should be available on the land and the bungalow by 'piggy backing' on the underlying farming activities carried out by the tenant farmer, but it is unlikely that any claim be made for BPR in respect of the land premium in excess of the agricultural value. The niece might, in her capacity as attorney, begin to take active steps to exploit the gravel deposits on her uncle's behalf, by arranging for soil tests etc and she might be able to enter into a joint venture agreement with the gravel operator to exploit the land in the hope that her uncle will survive by two years the date on which it can be shown that there is in being a business which might be recognised as such for BPR. During that two-year period, Wilfred would be the (long-term) owner of an asset used in a (new) business, so BPR at 50% might, just might become available immediately under *s 105(1)(d)*.

It might help the niece to terminate the farm business tenancy so as to recover possession of the land and be recognised as the occupier of it, thereafter entering into farm contracting agreements over such part of the acreage as is not immediately required for gravel working. However, all this is really clutching at straws: HMRC can be expected to deny BPR unless it can be shown that there really is a business in existence and not merely the taking of preliminary steps to establish a business. Merely walking the land, cleaning drinking troughs etc did, just, amount to a business in *McCall v HMRC* [2009] NICA 12; but it was not a farming business, rather an investment, business so did not qualify for BPR.

The issue of what constitutes a legal partnership was considered in *Khan v Miah* [1998] 1 WLR 477, where the Court of Appeal had to establish what, if any, business the parties had agreed to conduct. The case concerned a restaurant. The court held that an agreement to set up and carry on a business as a partnership was not proof that a partnership existed at that stage, any more than a mere statement by the parties that they were partners would on its own suffice to establish a partnership. If the court could establish what business the parties had agreed to conduct, it must then decide whether the business was being carried out by the partners at the material time.

REPLACEMENT ASSETS: SWITCHING

12.5 RI 95 sets out the HMRC view in relation to the operation of *ss 113B* and *124B* (see **Chapter 10**) where agricultural property is replaced by business property (or vice versa). After preliminary observations as to the definition of a PET and the operation of BPR and APR, HMRC noted qualifying conditions for relief and the rules which allow for the sale and replacement of qualifying property.

RI95 states that where agricultural property (the original property) which is part of a farming business is replaced by non-agricultural business property (the replacement property), the period of ownership of the original property will be relevant for applying the minimum ownership condition to the replacement property. BPR will be available on the replacement property if all the conditions for that relief are satisfied. Where non-agricultural business property is replaced by a farming business, and that farming business is not eligible for APR, it may still be possible to claim BPR under *s 114(1)* if conditions for that relief are so satisfied.

In this respect HMRC determine that where the original property of agricultural land is not part of a farming business, any replacement property could only qualify for BPR if it satisfied the minimum ownership conditions in its own right. HMRC accept that such situations are likely to be rare.

APR is denied on the value transferred by a PET of a farming business which the donee later sells and replaces with a non-agricultural business (*s 124A(1)*) but by virtue of *s 114(1)* BPR is not so excluded if the conditions for BPR are satisfied. In the reverse situation, the farming business acquired by the donee can be 'relevant business property' for the purposes of *s 113B(3)(c)*.

Where the gift is of an APR qualifying tenanted farm which is later sold and replaced by another APR qualifying farmland (irrespective of whether tenanted or owner-occupied) clearly such replacement will qualify as 'APR property'. If the land originally given is also occupied by the donee the gift may qualify for BPR or for APR or both, giving the donee more scope on replacement.

WHERE APR IS AVAILABLE BUT BPR IS NOT

12.6 As a general rule, BPR is far more favourable than APR but there are some instances where APR is more generous and these include:

- land which the transferor does not farm but is farmed by the tenant;
- land belonging to a sole trader, which is transferred by itself without the business carried on there;
- land with the benefit of vacant possession used by a partnership where the transferor is a partner but where the land itself does not belong to the partnership;

- land farmed by a company, but not owned by it, where the transferor is a minority shareholder only;
- let land which is owned by a company which is controlled by the transferor; and
- a farmhouse occupied by the transferee who farms the land.

THE SITUATIONS WHERE BPR IS MORE FAVOURABLE THAN APR

12.7 BPR is typically more favourable and examples include:

- assets of farming businesses other than land and buildings;
- non-controlling unquoted shareholdings in farming companies; and
- businesses or shares attributable to agricultural property where the ownership requirement is not yet sufficient for APR but which replace non-agricultural business assets (so that the assets qualify for BPR under the replacement property rules).

THE FARM: APPLICABILITY OF BPR

12.8 Where the owner of the farm is also the person carrying on the farming activities the difficulties noted in Example 12.1 may be avoided. The farming activity will constitute a business within *s 105(1)(a)* and the value thereof not already relieved by APR may attract the benefit of BPR. It is exceedingly common for a part of the farm to carry a premium value which exceeds its agricultural value ie the land with development value and the farmhouse by reason of its location. It is the accepted approach that the excess premium value of the farmhouse will not qualify for BPR but where, in exceptional cases, it can be demonstrated that a specific part of a farmhouse is used solely for the purposes of the business, BPR may be in point. However, that would not apply to the whole farmhouse but rather would be restricted to the business part and this would need to be very robustly supported (ie business usage reflected in the business accounts).

Sheltering proceeds of sale of a business through agricultural property

12.9 The entrepreneurial taxpayer faces a new problem following the successful sale of his business (the gain on which is likely sheltered by CGT entrepreneurs' relief). The former business value is now represented by cash and unless this is promptly invested in qualifying property, it will no longer have access to the protection of BPR in the event of premature death. This was precisely the situation in *Swain Mason v Mills & Reeve* [2011] EWHC 410 (Ch) and *Swain Mason v Mills & Reeve (a firm)* [2012] EWCA CIV 498.

An attractive option might seem to reinvest the sale proceeds in the purchase of a landed estate to include: a private residence, land farmed in hand, some let farmland, some let properties and local facilities such as part of a village. The estate would be run as a single venture but would that single venture qualify for APR and BPR as a single composite business or would it be necessary to look at each individual component? APR may be allowable on the agricultural properties/land but the let properties would clearly not be occupied for the purpose of agriculture and thus would not attract APR. Depending on the estate location, such properties could command a premium value but this could not be franked by BPR if, taking the estate as a whole, it should be treated as consisting wholly or mainly of making or holding investments.

This specific aspects were considered in the cases of *Farmer (Executors of Farmer Deceased) v IRC* and *Brander (Executor of Balfour Deceased) v HMRC* [2009] TC 00069 – see **Chapter 9**. The correct approach requires a review of the business in the round leading to a consideration of five individual components:

- Profit
- Turnover
- Time spent
- Value of the assets
- The fundamental nature of the business.

The entrepreneur who 'turns his hand' to farming

12.10 For the entrepreneur businessman who having made his wealth 'turns his hand to farming' there are perhaps several obvious questions:

- Why is he in the business of running a landed estate?
- Is the business carried on otherwise than for gain, disqualifying it for relief under *s 103(3)*?
- Is there an element of the 'vanity' that was described in *Higginson*?
- How far can it be said that he pursues a business as opposed to indulging in a mere hobby?

It could be argued that a composite enterprise, such as the ownership of a landed estate, should be separated into individual parts – see the *Hall* case in **Chapter 9**. Those parts of the enterprise which, taking one year with another, consistently make losses could then be disregarded for BPR purposes without tainting the remainder.

Where the life expectancy of the taxpayer is not great then putting steps in place to secure the APR/BPR protection for a working estate has a greater advantage than reducing the estate by the making of lifetime gifts. This is because of the lower two-year APR/BPR qualification compared to the longer seven-year run off period for the lifetime gift. However, it should be noted whereas BPR (at the lower 50% rate) is available in respect of a building which is occupied for business purposes by a partnership to which the owner of that building belongs (whether or not a rent is charged) in the identical position for the corporate shareholder BPR is denied on that asset to the extent the shareholder holds 50% or less of the shares.

Use a corporate 'wrapper'?

12.11 As noted in Example 7.4, the APR/BPR tainting that might apply to a landed estate can be minimised where the assets are all owned through a company. The two tests to be applied are:

(1) whether a company with mixed activities is mainly trading; and

(2) whether the 'excepted assets' rules will apply.

In a corporate scenario an asset is an 'excepted asset' if it is not used in any of the activities of a company and by implication, that suggests a company can carry on a number of activities which together make up a single business. However, the obvious risk with such a mixed operation is that a progressive increase in peripheral investments may tip the balance in favour of an investment company which fails the wholly or mainly rule (*s105(3)*) – see **Chapter 9**.

FAMILY LIMITED PARTNERSHIPS

12.12 Following the trust reforms introduced in *FA 2006*, the Family Limited Partnership (FLP) provides a viable alternate structure to the traditional family trust.

Example 12.2: The Vittles family – family limited partnership

The Vittles family has substantial wealth sheltered neither by BPR nor APR, where:

- land is merely grazed but has development value; or
- the investment content of a business now swamps the original business; or

- an existing business has been sold, converting sheltered assets into vulnerable cash; or
- relief is restricted to 50% or to the agricultural value only.

The value is large enough that a gift into trust of even a modest part would exceed the nil rate band and trigger an immediate IHT charge at 20% (or at 25% if the Trust has no liquid cash and relies upon the settlor to settle the IHT entry charge).

To address this, the Vittles family sets up a family limited partnership in which one person is appointed the general partner who runs the business. Cash is given to other relatives who merely reinvest it in the venture. The arrangement is transparent: it is a contract and may therefore appeal to those from civil law jurisdictions who have such trouble with the trust concept. The partnership is not an entity in itself and has no legal personality.

The main purpose of this type of structure is to protect assets and to trickle value down the generations. The partners could include minor children (although on reaching the age of 18 they could decide for themselves whether or not to repudiate the contract) who at the age of 18, could become more involved in the venture as an encouragement to leave value in, rather than dissipating it. This would be less of a problem where the children's money is already held by trustees.

Treatment of FLPs for law and tax

12.13 The FPL is transparent thus if the funding of a child's interest is from a parent, the income profit share is assessable on that parent; but, if from a grandparent or acquired through inheritance, the income is that of the child. The child (or his trustee) owns the right to a share in the assets of the partnership but not to specific assets over which there is no control. The limited partner cannot interfere in the management of the fund, save by withdrawing capital.

A gift into the partnership for the benefit of the child is a PET to that child and on their death, the partnership share forms part of their estate. This is not a trust arrangement and the relevant property regime of periodic or exit charges does not apply. However, the gift of an interest in the FLP may trigger a CGT charge (excluding the cash constituent of the share and assets qualifying for holdover relief under *TCGA 1992, s 165*). Even so this provides an advantage over a trust interest since creation of the latter will be a CLT for IHT and in the family context, the trust will likely be settlor interested preventing access to CGT holdover relief.

The general partner runs the FLP which must be operated for profit in order to be regarded as a genuine partnership under general principles – the mere holding of property or investments is not enough.

A well-drawn up FLP will set out a defined plan which controls the capital withdrawal thus:

- no withdrawal for any purpose for five years;
- no withdrawal until the partner has reached the age of 22 (and has completed his education);
- staged withdrawal thereafter for any purpose and (with consent of other partners) lump sum withdrawal for the purchase of a residence or of another business to be run separately from the FLP; and
- early withdrawal for other purposes and at other times only after deduction of penalties.

The general partner

12.14 The general partner may be an individual or even a corporate body which might help to limit liability and provide cover in the event of death. If the general partner is a wealthy individual they may feel vulnerable to a financial claim and would naturally look to secure indemnities. The dividing lines of responsibility should be clear, and this is particularly important if the FLP benefits an extended family class.

If the general partner is a corporate then control may be an important issue as the person who funds gifts to their children will want to own shares in the corporate body. On death, the founder's shares are important strategically and these should not pass to the limited partners as a group as that could make them owners of the whole structure, which might weaken asset protection.

The general partner may be paid for their work but there is no clear guideline as to the level of that remuneration. However, care must be taken to prevent the structure from degenerating into a vehicle for the founder to milk the assets to supplement his pension, whilst hoping that capital value has left his estate. It can be seen how control may be reserved, but over that control hangs the spectre of reservation of benefit from gifts made to children if it appears that the money was, from the beginning, destined to be invested in an FLP.

Regulation and protection

12.15 The management of investments is regulated by the Financial Conduct Authority and as the general partner will be promoting investment, it follows they must be duly authorised. The FLP is a 'collective investment scheme' and there will be costs associated with the registration as well as delays. The values involved should be sufficient to justify this combined regulatory and financial burden.

It must be stressed that each partner owns a share in the FLP, not a direct share in the underlying assets and it may prove difficult to realise that share in order to meet liabilities eg in the case of divorce proceedings. The FLP share will be a financial 'resource' to which the court may look to provide for the other party to a marriage and this may be where restrictions on the withdrawal of capital requiring the consent of other partners will be helpful. Much will turn on recent management of the FLP – if it produces a steady income, an order might be made to use it to fund maintenance.

The FLP offers a level of protection not so different from a discretionary trust by providing a pattern of income or of capital distributions that a court may take into account in deciding to what extent to regard the FLP share as a financial resource. Some protection may perhaps be achieved by taking the FLP offshore as a foreign court might disregard an English court order and/or refuse to enforce it. However, the off shore compliance costs could be significant and the FLP should be of sufficient value to justify such a course of action. In practice, it will usually be the location of the defendant who owns the FLP share which is of paramount concern.

SHARE FARMING OR CONTRACT FARMING

12.16 These arrangements were described in **Chapter 3** – very broadly, both the landowner and the operator farm the land where the operator will provide the working machinery and labour and will pay a proportion of the variable input costs.

APR is restricted to the agricultural value of the farmland and thus where there is a marked difference between that and market value it will be very important for the landowner to show that he qualifies for BPR to mop up the excess premium value – this was the point at issue in *McCall v HMRC* [2009] NICA 12, examined in detail at **5.6**. The Country Land and Business Association (CLA) model agreement may be a useful guide here. As a side issue, it should be noted that farmers' averaging relief is not available in respect of contracting income of the contractor (who might well be related to the landowner). This may be a material consideration where contracting income forms a significant proportion of the income of a farming business and particularly so where the agriculture activity concerned gives rise to cyclical variations.

Care should be taken in drawing up the agreement. It is easy for it to slip into the nature of either a tenancy on the one hand or a partnership on the other, particularly since a partnership can be created (for tax purposes at least) without the need for any actual deed. Where there is no deed, it may be argued that the landowner retains possession of the land, but that may not be consistent with what the parties have actually agreed. HMRC consider that it is the availability of vacant possession, rather than the existence of any tenancy, that will decide whether relief is at 50% or at 100% – see IHTM24081. If, as is common, the agreement has a limited time period a careful note should be made of the expiry date.

To achieve the desired purpose of leaving the landowner with CGT reliefs whilst allowing the contractor to undertake the risk of the farming operation, it is crucial for that agreement to include an obligation that the landowner must provide the variable inputs and that he actually pays for them. The crops in the ground should remain the property of the landlord throughout the farming cycle and for that reason it is not recommended that the contractor be given the right to sell any crops before harvesting. HMRC might interpret that as giving the contractor occupation of the land which would be fatal to the claim that the landowner was the farmer and would deny access to BPR on any premium value – also see IHTM24082.

The same risk of slippage into partnership or tenancy exists in contract farming but the tenancy problem is avoided if neither the landowner nor the operator has exclusive possession. The landowner needs to retain occupation of the land and of any buildings for CGT purposes whilst the operator should have a mere licence to occupy. The agreement should provide that the landowner can terminate the agreement within 12 months, subject to paying the contractor compensation. A difficulty will arise where, for example, animals on the land are owned by one party or the other to the agreement rather than in shares as the occupation of the land by the animals might lead to their owner being regarded as the occupier of the land.,

The risk of inadvertent entry into partnership can be avoided by a careful definition of what each party receives from the agreement. Each will conduct a separate business, keep their own books and should maintain separate bank accounts except perhaps for one into which gross takings are paid which are then divided into agreed shares. However, as with so many arrangements which push the boundaries of available reliefs it will be necessary to underpin the interpretation with a solid base. Proper accounting procedures should be in place and a record made of meetings held to discuss/decide farm policy. Without such minutes, it may well be very difficult to show that the landowner has taken any active part in the farming operation at all.

Two factors in particular create difficulties – too much bad paperwork and not enough good machinery. The farmer may have a great deal of practical farming knowledge, but the lack of expensive modern machinery may prevent him from personally carrying out many of the traditional farming operations. That leaves him only the care of hedges, and looking after livestock, which may also represent someone else's investment. Meanwhile, the greater part of many farming incomes arises not from the tending of any crops or animals, but from the completion of such forms as will secure the best government subsidies. It is very easy for the farmer in later life to delegate more and more of the work to the point where he is in truth no more than an investor, as if landlord of his own farm. If family leave home and cannot really work in partnership with their parent and take on what the latter can no longer do, there is every prospect that relief may be endangered.

A further word of caution – entering into a farming contract may be the right commercial decision but it may also jeopardise APR on the farmhouse. Farmhouse cases favour the farmer who can show that he is involved, in a meaningful way, with farming operations on a day-to-day basis – see IHTM24082.

GRANTING AGRICULTURAL TENANCIES

12.17 By *IHTA 1984, s 16* the grant of a tenancy of agricultural property (in the UK, Channel Islands or the Isle of Man) for agricultural purposes is not a transfer of value if made for full consideration in money or money's worth. The term 'grant' will include the coming into existence of the tenancy by any means and will include the creation of a farm business tenancy (FBT).

HMRC have argued that 'full consideration' should be a 'tender rent' rather than a mere 'rack rent', ie in areas where agricultural land is keenly sought after, it should not merely be the amount of rent passing on the recent review of the rent of similar land, but should be the premium rent that would be offered by a farmer in that area to secure the land in question. It may also be necessary to look at the personal circumstances of the tenant, including their agricultural experience or formal training or financial status. HMRC will not accept as evidence new rents charged to existing tenants; rather the yardstick is the rent which would be expected to be obtained on the grant of the tenancy on the open market by a willing grantor to a willing grantee, but on the same terms as the actual tenancy, including provision as to rent reviews and obligations to repair and the like.

Security of tenure

12.18 By virtue of *Agricultural Tenancies Act 1995, s 4*, agricultural tenancies beginning after 31 August 1995 generally do not have security of tenure.

Agricultural Holdings Act 1986, s 1 gave security of tenure to agricultural tenancies where the whole of the land was let for use as agricultural land. Certain exceptions were allowed which might not, however, substantially affect the character of the tenancy. Tenancies or leases which covered the use of land, only part of which was agricultural, might not enjoy the statutory protection. Where the lease does not grant that protection and where it can be assigned, it can have a value and, in so far as the property is agricultural, APR may be available in respect of it. The valuation will not be straightforward, since it must be on the basis that the lease is assignable but that the assignee will be restricted to agricultural use of those parts which are so used for the assignment.

The discount, which seems to range from 20% to 50%, will be the subject for negotiation with the District Valuer. Even where the tenancy has been created since 1 September 1995, such that the tenant does not have security, some discount should be allowable because of the delay in gaining possession (which can be up to two years). The valuation should also allow for any compensation payable to the tenant on leaving. Valuation principles in general are considered in **Chapter 8**.

BPR ON AN AGRICULTURAL PROPERTY COMPANY

12.19 Agricultural property may be the underlying asset of a trading company and it may then qualify as relevant business property. The operation of *IHTA 1984, s 114(1)* will always ensure that APR is applied first with BPR applied against the excess unrelieved value of the underlying asset. APR will not be available in respect of a minority holding of shares in a farming company but 100% BPR will (subject to the observation below) be allowable where the holding is of unquoted shares.

Similarly, a company may have owned business property which is disposed of and replaced by agricultural property. Where the agricultural property has not been owned long enough to qualify for APR, the combined period of ownership may still be sufficient for BPR. This was examined at **12.5** in the context of RI 95.

A significant limitation on BPR is that, where a company owns a tenanted farm, BPR will not be available because the company is not trading; it is 'wholly or mainly holding investments' within *s 105(3)*. A minority shareholding in such a company will not benefit from either BPR or APR, although the controlling shareholder will secure 50% APR if it can be shown that the company has owned the farm for seven years, and if the controlling shareholder has likewise owned the shares for seven years.

FARMERS DIVERSIFYING INTO NON-AGRICULTURAL ENTERPRISES

12.20 Many farming families are reluctant to give up their way of life, even though the acreage held is no longer economically viable. The change in farming methods can frequently mean that buildings, both cottages and farm buildings, are no longer occupied for the purposes of agriculture. The height of farm buildings, which in the last century was perfectly adequate to house farm equipment is totally inadequate for the size of machinery now in use. Yet the farmer may not wish to dismantle attractive old farm buildings (particularly where they lie close to the farmhouse) to replace them with large modern and more functional structures.

A common alternative is to divide them into small workshops and holiday lets which may provide the farmer with a more predictable and regular return of cash than that derived from his crop and other stock. The difficulty is, of course, that, from an APR perspective, the buildings will no longer be occupied for the purposes of agriculture; and, from a BPR perspective, no relief (unless *Farmer v IRC* [1999] STC (SCD) 321 (reviewed at **9.13**) is in point) will be available in respect of the part of the business which is wholly or mainly holding investments, however active the farmer may be in managing the land and buildings producing the leisure-based or craft-based income.

Other difficulties arise from a planning point of view, in that the development of redundant farm buildings into a profitable portfolio of investment properties may render those properties far more valuable than their cost plus the cost of improvements. Accordingly, CGT is likely to be a significant factor in deciding whether the portfolio of 'leisure-based' buildings should be the subject of a gift. Commonly, the planning recommendation will be the creation of a lifetime trust (limited to amounts within the nil rate band) typically discretionary in nature to which the assets can be transferred, with the gain held over under *TCGA 1992, s 260*.

However set up, the trust arrangement must be genuine and robust in construction, notwithstanding the cost to the family in terms of professional time. The trustees should insist on meeting from time to time to consider the exercise of their discretion, should be seen to exercise it, and should minute that exercise.

The problem of diversification might be considered within the context of incorporating the farm business and has been noted elsewhere in this chapter and in **Chapter 7** that a 'hybrid' company, ie one with activities which encompass both trading and non-trading activities, may secure limited protection through BPR for activities which, taken by themselves, would not qualify for BPR but which do so within the context of being used for part of the trade of the company as a whole. One reason for not incorporating the farming business in order to secure relief for a small part of the operation, such as the ownership of two holiday cottages, is the restriction on relief in respect of minority holdings of farming companies, which has been noted earlier in this chapter. Where, for example, it is intended to allow members of the family a share in the farming business, even though they are not actively engaged in the conduct of the farm, the tax adviser may find, on looking at all the circumstances, that partnership is a better vehicle for carving out an interest for a non-farming member of the family. Alternatively, farming assets may be the subject of a trust whose trustees farm the land in partnership with the other owners of farm assets.

Holiday lettings

12.21 One area of confusion for clients and of contention for the tax practitioner concerns holiday lettings which are treated as a trade for income tax by virtue of *ITTOIA 2005, s 323(2)* and *(3)*, subject to fulfilment of certain base conditions namely:

- the property must be let on a commercial basis;
- with a view to the realisation of profits; and
- the tenant must be entitled to the use of furniture.

To be 'qualifying holiday accommodation' *ITTOIA 2005, ss 325* and *326* imposes further conditions namely:

- it must be available generally as holiday accommodation for periods which amount to not less than 210 days in the year (the 'availability condition');
- it must actually be let for at least 105 days per year (the 'letting condition'); and
- for a period of at least 155 days in the year, it is not normally in the same occupation for a continuous period exceeding 31 days ('period of longer-term occupation').

ITTOIA 2005, s 326 sets out averaging rules and an election to deal with accommodation ('under-used accommodation') which would qualify as holiday accommodation if actually let 70 days per year.

For CGT purposes, holiday cottages may be a trade for the purpose of securing certain 'business' reliefs such as entrepreneurs' relief, roll-over relief etc. However, irrespective of the income tax/CGT status the business of managing holiday cottages does not immediately qualify for either APR or BPR in its own right. In the past, HMRC could consider, for BPR, the inclusion of the value of holiday cottages within a business; but, where the cottages were not part of the business, they did not qualify for any relief. The issue of BPR on holiday property is now largely determined by reference to the *Pawson* decision supported by the later cases of *ACC Green v HMRC* [2015] UKFTT 0236 (TC) and *Estate of M Ross (deceased) v HMRC* [2017] UKFTT 507 discussed at **9.13**.

IS THERE A NEED FOR TWO RELIEFS?

12.22 For many, farming is no longer a viable commercial activity and the uncertainty of BREXIT (at time of publication) does little to improve the position. Farmers are now in competition with producers of crops from other countries and there are clear difficulties for agriculture as a sustainable activity.

The official line for the continuing need for both APR and BPR is as follows:

> 'The purpose and scope of the two reliefs differ. Business relief is designed to encourage and protect investment in risk-taking family trading ventures generally whereas agricultural relief is focused more specifically on farmland and buildings and includes letting of farmland. But the practical outcome is that most trading companies and farms get 100% exemption from Inheritance Tax.'

However, the two reliefs sit uneasily side by side, both from the point of view of policy and of application – they are not good bedfellows. There is, for example, no clear and logical distinction between the premises from which, say, a farrier carries on his trade where he also happens to live, and the premises enjoyed by his farmer customer, but yet the farmer gets relief from IHT on his residence (albeit limited to the agricultural value), whereas the farrier does not.

It is in this context that the case of *Farmer v IRC* [1999] STC (SCD) 321 could offer the means of a very significant simplification of the law if extended into a statutory principle. Farming is a business that qualifies for BPR. The letting of, say, warehouses is not such a business, but developing land is. If, following the principle in *Farmer*, the management of land on an active basis could be regarded as a business for BPR, many of the difficult cases would be removed but as always the problem is where to draw the line. Under the law as it stands, farmers are leaving the land, or allowing it to lie fallow, because farming is not profitable. The existence of APR has not prevented that; it has merely persuaded that farmer to retain the asset, but not to put it to any good commercial use. To that extent, therefore, it could be argued that APR is not working.

Almost certainly, the existence of APR made the purchase of farmland attractive to people who had wealth to invest and who were seeking a safe haven in the form of an IHT shelter. That distorted the market and, for as long as it lasted, placed on agricultural land an artificial value that was so high that farming operations ceased to be commercially viable.

Chapter 13

Tax shelters

QUICK GUIDE

- APR/BPR can shelter assets from IHT and, where a tenant farmer can buy the freehold reversion, exposure to very considerable IHT can be saved overnight.
- Shares listed on AIM and similar unlisted markets can secure valuable relief from IHT provided the investor accepts the risks inherent in those markets.
- Whilst investment properties will not normally qualify for BPR, there are group structures that may command relief: see **13.16**.

INTRODUCTION

13.1 Quite apart from traditional tax planning and compliance adherence, the duty of the adviser might be thought to include an obligation to the client to consider what actions might legitimately be taken to mitigate the burden of tax, even by the use of special schemes: see for example the controversial case of *Mehjoo v Barber* [2013] EWHC 1500 which concerned the sale of a family business which realised a gain of £8.5 million triggering a CGT liability of approximately £850,000. It was held at first instance that the accountant should have referred the client to a tax adviser specialising in advising non-UK domiciled taxpayers. However, on appeal, the Court of Appeal, as reported at [2014] EWCA Civ 358, saw a distinction between fairly routine tax planning and the highly specialised scheme that might have helped the taxpayer in the particular circumstances of the case.

A well planned and executed scheme, in respect of which full disclosure is made, brings with it little, if any, risk of prosecution under the offence of tax fraud introduced by *FA 2000, s 144* – see **Chapter 22**. The emphasis should be on good execution, here meaning not merely good documents but careful adherence in practice to what those documents put in place. It is the sloppily executed transaction that is the easiest prey for the diligent HMRC Inspector – see *Dugan-Chapman* and *Vinton* referred to in **Chapter 6**.

THE GENERAL ANTI-ABUSE RULE (GAAR)

13.2 A full discussion of GAAR is outside the scope of this book but it is necessary to make reference to it in the context of APR and BPR.

GAAR is enacted in *FA 2013, ss 206–215, Sch 43* and supported by *HMRC's GAAR Guidance* – it came into force on 17 July 2013. The prime focus of the legislation is to counteract a tax advantage (to include IHT) arising from 'tax arrangements' that are perceived as 'abusive' – for this purpose the definition of 'tax advantage ' is very wide (*FA 2013, s 207(1)*) and covers essentially anything that was put in place where tax saving was the 'main purpose'.

It prompts consideration of the following in relation to any transaction:

- Are the results consistent with any principles on which the relevant tax provisions are based?
- Are the arrangements contrived or abnormal?
- Are the arrangements intended to exploit any shortcomings in the relevant tax provisions?
- Do the arrangements accord with established practice?

(1) (i) 'Arrangement' includes 'any arrangement, understanding, scheme, transaction or series of transactions (whether or not legally enforceable)' (*FA 2013, s 214*).

(2) 'Abusive' (*FA 2013, s 207*) describes any arrangement where it cannot reasonably be regarded as a reasonable course of action but where the arrangement follows established practice it is not considered abusive. It is necessary to consider whether the results achieved are consistent with the principles of the legislation else:

 – Are those results achieved using one or more contrived steps?

 or

 – Does the operation exploits loopholes in the law?

(3) 'Tax advantage' includes securing tax relief ie APR/BPR (*FA 2013, s 208(a)*).

The legislation works to nullify the tax advantage in so far as that is just and reasonable. The Guidance asserts that the effect of the GAAR is to reverse the rules established in such cases as *Fisher's Executors v CIR* [1926] AC 395, *Duke of Westminster v CIR* [1936] AC 1 and *Ayrshire Pullman v CIR* (1929) 14 TC 754 and in particular the Guidance states:

> 'Taxation is not to be treated as a game where taxpayers can indulge in any ingenious scheme in order to reduce their tax liability … as far as the

operation of *GAAR* is concerned, Parliament has decisively rejected this approach, and has imposed an overriding statutory limit on the extent to which taxpayers can go in trying to reduce their tax bill. That limit is reached when the arrangements put in place by the taxpayer to achieve that purpose go beyond anything which could reasonably be regarded as a reasonable course of action.'

GAAR is not targeted at straightforward tax planning, such as investing in assets that qualify for BPR but taxpayers must bear GAAR in mind when satisfying any compliance requirement – see **Chapter 22**.

FARMS

As landlord

13.3 The purchase of let farms may provide an IHT tax shelter by providing access to APR – the land is agricultural land or pasture within *IHTA 1984, s 115(2)* and occupied for the purposes of agriculture; after ownership for seven years (*s 117(b)*) APR is available in respect of the agricultural value either 100% (or 50% if the tenancy was established before 1 September 1995).

Share-farming and contract farming

13.4 The nature of these farming arrangements and some of the difficulties encountered is considered in **Chapters 3** and **7**. Properly structured, the existence of a share farming or contract farming agreement may enable the owner of a farm to show that he is the 'paramount occupier' with the result that only the minimum two-year period of occupation/ownership is required to gain access to APR rather than seven years. The benefit of APR rewards the effort required to ensure compliance with the statutory requirements and thus the success of use of farmland as a tax shelter lies largely in the care with which arrangements are executed.

Tenant buying out landlord

13.5 An established tenancy (namely one created before 1 September 1995) carrying security of tenure may have considerable value, part of which reflects the possibility that the landlord may be willing to sell to the tenant on terms which give that tenant part of the 'marriage value' arising when they become the owner of the property with the benefit of vacant possession. In this context, the minimum period of occupation or ownership, does not require the transferor both to own and occupy the land to qualify (*s 117(a)*).

Example 13.1: Ben – tenant to owner

Ben, a tenant farmer, has occupied Low Farm for more than two years and should he purchase the freehold, perhaps using cash assets which would not qualify either for APR or BPR, he would immediately satisfy the test in *s 117(a)* and without the requirement to wait for a further seven years. Should Ben be given the opportunity to buy the freehold of the farm on his deathbed he could still secure immediate access to100% APR.

However, there may be practical issues where, for example, Ben wishes to provide for family members who are not engaged in the family farm – the purchase locks up the capital that could have been given to non-farmers.

These rules may, however, be part of an exercise to help non-farming members of the family.

Example 13.2: Boris – protecting the farmer son

Boris, in partnership with one of his sons, grants a tenancy which gives the son security under the *Agricultural Tenancies Act 1995*, perhaps until retirement age. He leaves the freehold to all his children equally, with the result that after his death they receive rent only until sale. On that eventual sale, the son's interest in the land as tenant is bought out (unless his lease has long since expired). CGT entrepreneurs' relief may shelter the farming son's gain.

APR will be available to the farming son after two years, either by virtue of his occupation as a partner before the tenancy or under the tenancy itself, but the other children must wait seven years from date of their inheritance.

If all the children were willing to become partners with the farming son, they might earn the CGT benefits of roll-over relief and (subject to meeting all the tests) entrepreneurs' relief. In these circumstances, it may be better to keep the land outside the partnership but to grant the partnership a tenancy which itself would be a partnership asset. A disposal could then fall within *TCGA 1992, s 169I(8)(c)* as a material disposal of business assets.

A difficulty would arise in the use of a company structure with multiple shareholders as only a control shareholding would qualify for APR. There is also the risk that the company might be no more than an investment vehicle: this is considered at **12.17**. The absence of control would also dent access to 50% BPR under *IHTA 1984, s 105(1)(d)*.

BUSINESSES

Listed unquoted trading companies

13.6 The chief and most accessible category of property qualifying for BPR (see *IHTA 1984, s 105(1)(bb)*), is 'any unquoted shares in a trading (or holding) company' (but see **13.10** for a more detailed discussion on excluded activities). Other categories are mentioned below. For this purpose, 'shares' means any kind of share – this can include a preference share, but tax practitioners should take great care to draft the company documents in such a way as to ensure that the share has all the attributes of a share and is not really a loan to the company disguised as a share. There is no case directly in point but the drafting must be robust. Furthermore, not all such shares will qualify for BPR since *s 105(3)* (wholly or mainly holding investments) also applies to such companies. The majority of AIM companies, particularly those engaged in high-technology industries, will have the characteristics of relatively high risk and relatively low cash resources that will qualify them as typical trading companies, investment in which BPR is intended to encourage.

'Unquoted' for this purpose has a defined meaning (see *IHTA 1984, s 105(1ZA)*) and whilst in relation to UK companies this will usually mean not quoted on the full list of the London Stock Exchange it will also require a non-listing on *any* recognised exchange. This has given rise to considerable investor interest in small companies post *FA 2006* which now bring most lifetime trusts within the relevant property regime.

Income Tax Act 2007, s 1005 was amended with effect from 19 July 2007 so as to allow HMRC to regard as a 'recognised stock exchange' any investment exchange that the Financial Services Authority designates as a 'recognised investment exchange'. This power was exercised, for example, in relation to part, but not all, of the market formerly known as OFEX, now ISDX. The result is confusing: some shares on ISDX qualify for BPR, while others do not and as a consequence great care should be taken in this area.

Since 2013, HMRC has designated the following as recognised stock exchanges for IHT purposes:

- The Channel Islands Securities Exchange Authority (20 December 2013);
- GXG Main Quote (23 September 2013);
- Global Board of Trade (30 July 2013);
- GXG Official List (16 May 2013);
- ICAP Securities & Derivatives Exchange Ltd (25 April 2013);
- European Wholesale Securities Market (18 January 2013);

- National Stock Exchange of Australia (19 June 2014);
- Singapore Exchange Securities Trading Limited (7 October 2014);
- Singapore Exchange Derivatives Trading Limited (7 October 2014);
- Euronext London (4 February 2015); and
- GSX Ltd (16 August 2016).

There are three practical difficulties in dealing with such shares:

First, by the very nature of the market it is illiquid. Dealings in a number of the companies proceed only on a 'matched bargain' basis. It may, therefore, take some time to assemble a substantial shareholding and, following a death, the executors may find that there is a delay in liquidating the holding.

Second, identifying stockbrokers willing to advise on unlisted securities. This is a specialist market where some of the stocks are fledgling companies with only a very short established track record. Among ISDX securities there are, however, some very well kept secrets, in the form of companies of considerable quality with an established record, which would be an asset to a share portfolio even without the specific advantage of BPR. Most financial advisers are wary of direct investment in individual companies, preferring 'packaged' investments offering the taxpayer an investment in a discretionary fund comprising a selection of companies the shares in which qualify for BPR once the holding period has been satisfied.

Third, the commitment to hold the portfolio until death or until the relevant chargeable transfer.

Caveat: secondary market listings

13.7 Trading in shares is conducted on an international stage and stock exchanges are often not the property of governments but are themselves freely traded. Exchanges compete and companies 'shop around' for the regulation regime and the pool of investors that suits them best. As a result, even a small company may be traded on several exchanges and, whilst it may be unquoted in the UK, it may have a listing in another country. This should be checked, since a listing anywhere in the world will deprive shares of BPR.

Fully private companies

13.8 The lack of or absence of a ready market in the shares of fully private companies may be sufficient to discourage most investors apart from family members of the family company and this restrictive market place was discussed in **Chapter 8**. Nevertheless, where there is sound commercial justification for an investment, or where for example a minority holding in a trading company is inherited and enjoys the benefit of an income stream from dividends, there can be good reasons for its retention. The unfettered operation of *s 105(1)(bb)* affording 100% relief to unquoted shares regardless of the size of the shareholding, has greatly increased the attraction of keeping a minority holding.

Conversion of debt into equity

13.9 Debt is not part of the value of an unincorporated business (see **Chapter 14**) and as such the transfer of a debt owed by a business from one person to another does not attract relief. Sound commercial reasons may dictate that on the formation of a private limited company, an investment is made by way of loan rather than by way of shareholding and the existence of a loan account offers the founder shareholder the opportunity of withdrawing funds from the company without suffering a tax charge. However, from an IHT perspective, an equity holding is infinitely more attractive as it will qualify for 100% BPR under *s 105(1)(bb)* (assuming of course that the nature of the business carried on by the company is one which is not disqualified from BPR by *s 105(3)*).

The conversion of debt into equity will not in normal circumstances amount to a transfer, although it can do.

Example 13.3: Susan – exchange of debt for shares

Susan had graduated from Exeter University with a degree in engineering and, although in full-time employment, in her private time she was working on an invention that she hoped might eventually be marketable. A company was formed for the purpose and, when orders materialised, Susan lent the company money for development. The company quickly spent the cash in the development stage and the bank refused finance unless Susan agreed to convert half her loan account into shares, which she did. Contracts were placed but the pricing was wrong, and the company came close to liquidation. A year later under pressure from the bank, Susan converted the rest of her loan account into shares. Susan died in a driving accident within a year of the second conversion.

Her executors successfully claimed BPR on both her original shareholding and on the shares resulting from the first conversion of loan account, which had been more than two years before death. BPR was denied on the shares resulting from the second conversion on the grounds that these had not been held for two years prior to her death. The executors claimed the shares were of negligible value but HMRC countered that, when Susan agreed to the second conversion, she was throwing good money after bad and was, in effect, making a gift to the company.

A conversion will seldom give rise to a chargeable gain and whilst there will be expenses on the issue of shares those costs will usually be small by comparison with the potential IHT saving. The conversion of debt into equity is, therefore, a useful device in suitable circumstances, particularly where the holder of the debt is considered likely to outlive the event by at least two years even if not the full seven years.

Conversion of cash into equity

13.10 A simple purchase of additional shares in an unquoted trading company will only secure access to relief after satisfying the minimum two years ownership period. As originally envisaged in *The Executors of Mrs Mary Dugan-Chapman and Another v Revenue and Customs Commissioners* [2008] SpC 666 a two-year period may be abridged where the company makes a rights issue which is taken up. This is in reliance on *IHTA 1984, s 107(4)* when read with the 'paper for paper' provisions of *TCGA 1992, ss 126–136.*

Buildings or machinery

13.11 A problem may arise where buildings or machinery used in the trade of the family company are neither owned by that company or its controlling shareholder. If the latter settles all his shares on the owner of the buildings or machinery, the new owner of the buildings/machinery may make a gift of those assets with the benefit of 50% BPR even if the shares ultimately revert to the settlor (but only provided the shares continue to qualify for BPR in their own right). This may of course trigger a chargeable event for CGT purposes but if the buildings/machinery constitute business assets, *TCGA 1992, s 165* holdover relief may apply to defer recognition of the gain.

Nursing home businesses

13.12 Such businesses, which deliver both accommodation and care, merit separate mention. The operating of a nursing home business is not necessarily tainted by the former activity so as to fall within the BPR exclusion under the 'wholly or mainly one of making or holding investments' rule. The case of *Powell (Personal Representatives of Pearce deceased) v IRC* [1997] STC (SCD) 181 distinguished the position of the services provided by the proprietor of the caravan site (passive activity) and the services provided by the hotelier (business activity). This proposition is consistent with the earlier case of *Griffiths (Inspector of Taxes) v Jackson* (1983) 56 TC 583, on which HMRC rely (see IHTM 25277). This considered the failed argument that income derived from the management of a portfolio of 180 student lettings in Bristol amounted to a trading income.

The care of the elderly involves much more than merely providing accommodation for which a rent is charged. Where the investment content of an unquoted company which operates a nursing home is not excessive, ie that it can be shown that the business does not carry liquid resources which are far beyond the regular needs of the business, a nursing home business may be a very acceptable investment as a shelter securing 100% BPR. The attraction of this kind of business has in the past been that generally it was low risk, being substantially property-based, but reference should be made to the HMRC position set out in IHTM25277 which is much less encouraging, at least in relation to 'residential homes'. Access to BPR will be very much dependent on the level of services provided.

Lloyd's interests

13.13 Investment in Lloyds is no longer an obvious choice for an effective IHT shelter. Those who remain underwriters enjoy substantial relief because the investments or other assets which support the underwriting qualify for BPR at 100 (see **Chapter 5**) but an underwriter wishing to retire faces a difficult choice as the release of assets from his deposit increases his exposure to IHT overnight. Immediate planning, perhaps including the making of substantial gifts, may well be appropriate.

INVESTMENT PROPERTIES

13.14 Simply put, the direct holding of mere investments in whatever format does not qualify for BPR and the long-term solution may be the creation of a nil rate band discretionary trust of the investments, holding over any gain into the trust (assuming the settlor-interested rules are not contravened). However, the holding of investments through the medium of a subsidiary of a group of companies could provide access to relief, but this stratagem is not without risk and the interaction of *IHTA 1984, ss 110* and *112* must be carefully observed (see **Chapter 8**).

Chapter 14

General planning points

QUICK GUIDE

The sheer complexity of the APR and BPR rules can create hazards and bear traps for the unwary practitioner as illustrated in this chapter.

- As a general rule, a reservation of benefit may be less of a problem and in some cases may provide a significant advantage where the asset qualifies for APR or BPR. The reserved asset is still treated as part of the estate of the donor and is tested for relief by reference to them – see the detailed rules at **14.3** onwards.
- It is still possible to achieve immediate IHT savings where an existing non business debt within a structure has the effect of limiting BPR or APR. Similarly, the 'cleansing' of an asset that has already been held for two years, so that obstacles to a claim for relief are removed, brings immediate reliefs: see **14.29**.
- Generally, the existence of a contract for the sale of an asset denies access to APR or BPR. The specific rules and 'buy and sell' agreements and options are considered at **14.6** onwards.
- Retirement from and sale of the business will lead to a loss of APR/BPR where perhaps an alternate structure could be maintained to reserve access to the relief – see *Swain Mason and Others v Mills & Reeve* [2011] EWHC 410 (Ch), examined at **14.14**. In particular, there is immediate loss of relief where on retirement a partnership interest is converted to a mere loan to the partnership: see especially the *Beckman* case discussed at **14.17**.
- Clawback problems may arise where the donee of gifted APR/BPR qualifying property wishes to dispose of it.

UNFORTUNATE GIFTS

14.1 Traditionally, the planning process often starts with the recommendation that the owner of business or agricultural property should make lifetime gifts. However, in certain circumstances such outright gifting could lead to a greater IHT exposure than had the assets retained within the donor's estate (for example see **Chapter 15** which discusses the double use of 100% relief in respect of property passing from transferor to children and repurchased by the spouse of the transferor).

Example 14.1: Hugh – lifetime gifting and death

Hugh, who always utilised his annual exemption in April each year, made the following transactions in the period commencing 31 July 2003:

- On 30 June 2010, a cash gift of £200,000 to a discretionary trust.
- On 31 August 2014, a cash gift of £360,000 to his son Ian.
- On 31 October 2016, the sale of his farm to an unrelated third party.

On 31 July 2017 Hugh died leaving a chargeable estate was worth £3 million (mainly comprising the proceeds of the farm sale) but it is also necessary to re-examine the chargeability of gifts made in the seven years prior to his death:

(1) The CLT to the discretionary trust took place more than seven years before Hugh's death and therefore falls out of accumulation. No additional IHT is due.

(2) The now failed PET to Ian was made less than three years before Hugh's death and is therefore outside the scope of taper relief. To calculate IHT thereon, it is necessary to recognise CLTs (but not PETs) made in the seven year period counting back from the date of the gift which brings into account the transfer into trust made in June 2010. The position is therefore as follows:

Lifetime gift:	£
Cum. Transfer (June 2010)	200,000*
Cash gift (August 2014)	360,000
Cum. Total	560,000
Nil rate band	(325,000)
	235,000
IHT thereon @ 40%	94,000
IHT paid on initial gift	Nil
IHT due on cash gift payable by Ian	94,000*

** As the value of the July 2010 transfer into trust fell within the nil rate band all of the IHT now due is attributable to the failed PET*

(3) The knock on effect to the estate is as follows:

Death estate:	£	£
Estate		3,000,000
Nil rat band (2017/18)	325,000	
Failed PET	(360,000)	
Balance of nil rate band		Nil
Chargeable		3,000,000
IHT thereon @ 40%		1,200,000

(4) Ian's position may be summarised as follows:

	£	£
Lifetime gift	360,000	
IHT due thereon	(94,000)	
Net gift to Ian		266,000
Estate	3,000,000	
IHT due thereon	(1,200,000)	
Net estate to Ian		1,800,000
Total received by Ian		2,066,000

(5) Had Hugh not given Ian the cash gift but had instead made equivalent provision for Ian by will, the position would have been as follows:

Death estate:	£
Estate	3,360,000
Nil rate band (2017/18)	(325,000)
Balance of nil rate band	Nil
Chargeable	3,035,000
IHT thereon @ 40%	1,214,000

(6) Ian's position may be summarised as follows:

	£
Estate	3,360,000
IHT due thereon	(1,214,000)
Total received by Ian	2,146,000

The tax saving had the gift to Ian not be made would have been £80,000 (£2,146,000 – £2,066,000). The moral is that someone in Hugh's position, who has made a substantial chargeable transfer, must weigh up quite carefully the advantages of early giving and the risk that a charge by virtue of *IHTA 1984, s 7(1)* may apply.

THE TAXATION OF RELEVANT PROPERTY TRUSTS OF RELIEVED PROPERTY

14.2 A full discussion of the taxation of relevant property trusts is outside the scope of this book and reference should be made to chapters 6 and 8 of *Inheritance Tax 2017/18* (Bloomsbury Professional), chapter 10 and *Trusts and Estates 2017/18* (Bloomsbury Professional). Nevertheless, the practitioner should be aware of the interaction of APR/BPR with relevant property trusts.

Since 22 March 2006, the creation of a lifetime trust (other than for a disabled person or for charity) is now a CLT albeit APR/BPR may still apply to frank the chargeable value to the extent that the property so qualifies for relief. However, by *IHTA 1984, s 68(5)(a)* such relief is dis-applied in calculating the exit charge on any capital distribution made within the first ten years post trust creation, unless the asset itself qualifies for relief in its own right.

In the past, to avoid the impact of the related settlement rule which requires aggregation for the purposes of sharing the trust nil rate band (*IHTA 1984, s 62*) a common device involved the creation of a number of pilot trusts all established on consecutive days to which relievable property or substantive property was later added—see the *Reynaud* and *Rysaffe* cases (see **14.23** and **14.24**). This devise has now been superseded by measures introduced in *F(No2)A 2015* reflected in *IHTA 1984, s 62* (effective for all chargeable events taking place on or after 18 November 2015) which operates to ensure that where property is added to more than one relevant property trust on the same day (either on or after commencement of those trusts) that added value together with the original value of that trust when first settled (to the extent not already recognised as a related settlement) will be brought into account in calculating the rate of tax for the purpose of the ten-year and exit charges. There is a *de miminis* exemption where the same-day addition is £5,000 or less (*IHTA 1984, s 62(2)–(4)*) nor does the provision apply to a will executed before 10 December 2014 where the death occurred before 6 April 2017.

The above anti-avoidance legislation does not apply if either (or both) trust A and trust B is a 'protected settlement' (*s 62C*) namely a trust which commenced before 10 December 2014 and either of two conditions (A or B) is satisfied:

- Condition A is broadly that the settlor has not added value to the settlement since 10 December 2014.
- Alternatively, Condition B is that the settlor makes a transfer on or after 10 December 2014 resulting in the value of the property being increased, where the transfer arises (under *s 4*) on the settlor's death before 6 April 2017 by reason of a 'protected testamentary disposition' (ie under the settlor's will, where its provisions are substantially the same as they were immediately before 10 December 2014).

GIFTS WITH RESERVATION

14.3 Where on or after 9 March 1999 an individual gifts an interest in land the gifts with reservation rules set out in *IHTA 1984, s 102(2)–(4)* will apply if:

- at any time in the relevant period,
- the donor or the spouse of the donor enjoys 'a significant right or interest' in relation to the land, or
- is party to 'a significant arrangement' in relation to the land.

(1) The 'relevant period' means a period ending on the date of the death of the donor and beginning seven years before that date or, if it is later, on the date of the gift (see *s 102(1)*).

(2) A 'significant' right, interest or arrangement (see *s 102A(3)*) is one which entitles or enables the donor to occupy all or part of the land or to enjoy some right in relation to all or part of the land otherwise than for full consideration in money or money's worth. The rule does not apply if that right or interest granted was acquired more than seven years before the gift of the property (*s 102A(5)*). However, an exception relief is found in *FA 1986, s 102B(4)* (which is carried into the Pre Owned Asset code by *FA 2004, Sch 15, para 11(5)(b)*) which provides protection where the donor and donee together occupy the land and the donor receives no benefit, other than a negligible one, provided by or at the expense of the donee himself for some reason connected with the gift.

Gifts with reservation: de minimis

14.4 In the case of *Personal representatives of Lyon deceased v HMRC* [2007] SpC 616, it was held that the receipt of only £15,965 from a trust of £2.7 million could be the reservation of a benefit so as to taint the gift. The settlor retained a right of revocation and was a potential beneficiary of the trust, and it was in that capacity that the benefits were received. The way the trust had been operated showed that possession and enjoyment of the property had not been assumed at the beginning of the relevant period.

Gifts with reservation: treatment where BPR is available

14.5 An IHT charge may arise on property which is subject to a reservation on the death of the donor, or which is at that date no longer subject to reservation but was subject thereto at sometime within seven years preceding their death. BPR may be available but special rules apply which operate as if the occasion, by reference to which the tax is charged, is a transfer of value by the donee rather than one made by the donor.

Outline of the rules

The main rules are set out in *FA 1986, Sch 20, para 8* and apply in any of the three following circumstances:

- there is a disposal by gift of property which qualifies for BPR in relation to the donor; or
- there is a disposal by way of gift of property which qualifies for APR under *IHTA 1984, s 116*; or
- there is a disposal of shares or securities which qualify for APR under *s 122(1)*, ie agricultural property of companies.

The test as to whether shares (which originally qualified for BPR at date of gift) qualify for BPR at date of death is determined as if they were owned by the donor and had continued to be owned by him since the gift. However, the percentage at which APR/BPR may be applied is determined as if the transfer of value were one made by the donee (*FA 1986, Sch 20, para 1A*). In all other cases, the requirement is that the gifted property in which the benefit was reserved:

- qualified for APR/BPR when the original gift was made; and
- qualifies for APR/BPR by reference to the donee.

Tracing the subject matter of the gift

This is one area where there is no complete 'mix and match' between agricultural property and business property, contrary to the more general rule noted at **10.23** and considered in more detail at **12.5**.

The tracing provisions would appear to operate more favourably where the original property qualified for APR and the later property for BPR but the rule is otherwise where the original property was shares qualifying for APR and the later property qualified for BPR only. The authority for this is the combined effects of *FA 1986, Sch 20, para 8(1)* and *(3)*.

The effect of these relieving provisions is logical. They treat the donor as making the hypothetical transfer rather than the donee, since the donee would not normally be able to show that he had owned the relevant asset for two years before the deemed transfer to him. The result (see *para 8(2)*) is that the donee steps into the shoes of the donor and counts the donor's ownership of the asset prior to the gift as part of the ownership requirement of the donee. Equally, occupation by the donor of property prior to disposal, and occupation by the donor after that disposal, is treated as occupation by the donee.

The position of the donee

The donee may receive the benefit of the donor's period of ownership in the situation described above, but in all other respects the clawback legislation in relation to gifts with a reservation of benefit, still consider the notional transfer as if made by the donee. Thus, if the requirements of BPR are not met, perhaps because the donee does not control the company that uses an asset he has been given, clawback may apply to a gift that otherwise satisfies the spirit or policy of the legislation.

Further substitution rules affecting gifts with reservation

14.6 There are further substitution rules (see **Chapter 10**) where the donee has since died. As far as any period up to the death of the donee is concerned, any reference in *FA 1986, Sch 20, para 8* to the donee is construed as being a reference also to the personal representatives of the donee or to the beneficiary of any will or intestacy affecting the donee.

Further rules relate to shareholdings which are the subject of gifts with reservation which apply where control is in issue, for example under *IHTA 1984, s 105(1)(b), 105(1)(cc)* or *122(1)(b)*. Who must show control? The issue is decided on the hypothesis that the donor owned the shares even after the making of the gift and this will generally facilitate the claim to BPR.

Example 14.2: Owen – gift with reservation

Owen owned 66% of Intellectual Resources Ltd (a non-farming unquoted company), from which he drew a salary which far exceeded the commercial value of any work that he did for that company. In June 2015, he gave 16% of the company to his friend Colin but insisted that his salary remained the same, notwithstanding the gift. Colin still owned the 16% shareholding in the company at the date of Owen's death in December 2017.

The failure by Owen to reduce his salary to commercial levels following the gift of shares to Colin was a reservation of benefit which continued until Owen's death. As at that latter date Colin had only held the shares for 18 months thus less than the *s 106* minimum two-year requirement to secure BPR. However, by reason of *FA 1986, Sch 20, para 8(2)* Colin is credited with Owen's period of ownership by treating the shares as if still held by Owen.

Example14.3: Denis – assets changing character

In May 2011, Denis formed a company to carry out farming activities. He kept 79% of the issued shares in the company, which the following month bought farmland. By October 2016, Denis' son Sam had joined him in the farming operation and Denis was considering a gift of shares to Sam. He consulted his solicitor, Jack, for advice, explaining that he was terminally ill and that his doctors had given him at most 10 months to live at the most. He wished to make a gift of the shares but still retain his (artificially high) salary as a director of the company. He was considering a gift of between 28% and 51% of the company, though he made it clear to Jack that despite his poor health he did not really wish to part with control of the family business.

Jack advised on the IHT effect of the minimum gift by Denis, of 28% of the company to Sam noting that if Denis retained a high salary this would make the gift one with reservation. Jack considered that this might not be fatal since, if Sam retained the shares, the rules would operate to credit Sam with his father's period of ownership and to treat the transfer as being made out of Denis' controlling shareholding.

Jack considered that BPR would apply to the gift with reservation and that, even if Denis were to die as early as October 2017, provided the shares had been retained by Sam, BPR would still be available on the transfer.

Fortunately, before posting the letter, Jack ran it past his assistant, Leslie. He observed that, although the situation as described by Jack might be considered to be similar to that of a gift in a non-farming company, there was actually a distinction between the rules for BPR and those for APR. Critically, there was no provision for APR which exactly paralleled the BPR treatment under *FA 1986, Sch 20, para 8(1A)*, which relates only to securities qualifying for BPR under *IHTA 1984, s 105(1)*. Leslie therefore considered that although BPR would be available on the gift, there would be no BPR if Denis were to die, because the test would be based on the availability of the relief on a hypothetical transfer *by Sam* of his shares in August 2017. Since the gift was proposed to be of 28% of the company only, it would not qualify for APR since minority shareholdings in farming companies do not so qualify. It would not be possible to 'borrow' control in this circumstance, as it would have been in relation to a non-farming company.

Leslie considered that BPR, rather than APR, would be available, in so far as the shares qualified in their own right as being shares in an unquoted trading company. However, this would require Sam to demonstrate that the company was indeed a trading company and that there were no excepted assets. Leslie observed that often the trade of farming companies is carried on, not by the companies themselves, but by managers for the company or even tenants. Thus, the availability of BPR may be in doubt should the APR test not be satisfied, because either the company is not carrying on a trade, or it is but not to such an extent as to be wholly or mainly trading by reason of holding land that is farmed by others.

Leslie suggested that the gift by Denis to Sam should be of not less than 51% of the company and since Denis was reluctant to abandon any of his present salary, he would not lose any income by making the larger gift. Leslie considered that, in the event of Denis' early death, BPR would be available in respect of the gift but only provided Sam retained the shares given to him. This was because:

- BPR was available on the original gift with reservation;
- Sam kept the subject matter of the gift;
- viewing the matter as at August 2017, it could be shown that Sam had a shareholding which controlled the company, thus satisfying the requirements of *s 123*; and
- the minimum period of ownership would be treated as satisfied by virtue of the rule in *FA 1986, Sch 20, para 8(3)* by reference to *para 8(1)(c)*.

Example 14.4: Alan – clawback and reserved benefit

Alan farmed in a long established partnership with his son Barry and Clare (Barry's common law partner). Alan had owned the freehold of the farm for many years but did not charge the partnership a rent. When Alan's wife died he moved to a bungalow on the edge of farmland and Barry and Clare moved into the farmhouse. On 30 June 2012, Alan gave the farmhouse and the adjacent farm to Barry together with 15 acres of pasture adjoining – Alan retained 300 acres of farmland. There was no adjustment to the profit-sharing ratio at the time of the gift.

On 31 January 2014, there was an opportunity to buy 75 acres of land close to the farmhouse and, as part of the financing arrangement, Barry transferred the farmhouse and land that had been given to him by Alan into joint names with Clare, so that they could between them raise the funds to buy the extra 75 acres.

On 31 March 2017, Alan died and it became necessary to consider the rules both as to clawback and gifts with reservation.

In this situation, there have been two transfers, each of which will be relevant to the IHT analysis.

Alan's gift on 30 June 2012

Alan has reserved a benefit out of the gift, even though he moved out of the farmhouse. This is so since part of the house may have served as the farm office, and the 15 acres may have contributed in some way to farm profits but there were no changes in the partnership shares of profit to recognise the fact that Alan had less capital in the business post the gift. (*Note: There is also a contrary but weaker argument that Alan did not reserve any benefit out of the gift because there was no direct benefit to him and because his profit share remained unchanged and was a fair share for the work that he did; but assume for the purposes of this example that a benefit was reserved.*)

The HMRC approach (see IHTM24201) is to test this gift for relief in June 2012 and to test it again at the time of clawback, ie Alan's death in March 2017. HMRC procedure requires (see IHTM24202) different treatment according to whether the clawback charge arises on the death of the original transferor or on a failed PET on the earlier release of the benefit.

Where (as in this example) the clawback is triggered by death, HMRC consider whether BPR would have been available had the gifted house and 15 acres still been owned by Alan at the date of his death. Here it may be assumed that the house and 15 acres, if considered in isolation, might not have qualified for BPR but when placed in the context of Alan's estate which included the retained 300 acres it would likely have been allowable.

However, as noted in the review of relief from clawback at **10.20** it is also necessary to satisfy all four additional conditions for lifetime transfers (see IHTM24173), namely that:

(1) The transferee still owns the replacement asset at date of the transferor's death.

(2) Throughout the period from date of lifetime gift to the sale of that original property it was owned by the transferee and occupied (by him or another) for the purposes of agriculture.

(3) Throughout the period from date of the sale of the property to the date of the transferor's death the replacement property was owned by the transferee and occupied (by him or another) for the purposes of agriculture.

(4) The replacement property is agricultural property as at the transferor's death.

In applying the above criteria it can be seen:

- Barry has not owned all the property continuously from 2012 to 2017;
- the original property has not being sold;
- it still qualifies for APR; and
- it has been occupied for agricultural purposes from the transfer to the date of death.

Whilst tests (2)–(4) are met, test (1) is not and thus the clawback would apply.

There is some qualification to the rules, because the test as to whether the house and 15 acres still qualify for APR is by reference to Alan rather than by reference to Barry. However, following *Hanson*, the house would still be seen in the context of all the land farmed.

Barry's gift to Clare on 31 January 2014

Again, there has been no change in profit-sharing ratios and still no rent is charged, so it is also considered that Barry has reserved a benefit noting that this is not a gift between spouses. In the event of Barry's death within seven years, *FA 1986, Sch 20, para 8* would be applied to Barry's reserved benefit. However, at first sight this could give rise to a quirky difficulty when dealing with a reserved benefit:

In noting that BPR and APR serve only to reduce the value transferred it follows there is a requirement for the transfer to take place. No clawback charge arises on the death of Alan in relation to the gift made by Barry to Clare in January 2014 because as at March 2017, Barry is still alive. On that basis, clawback itself is not relevant to Barry's onward gift but it is still relevant to the original June 2012 gift from Alan to Barry. Therefore, even though *Sch 20, para 8* would operate to treat the farmhouse and 15 acres as forming part of Barry's estate on Alan's death in March 2017 had Barry also died on that date, those provisions do not deem them to be owned entirely by Barry for the purpose of preserving that June 2012 gift from clawback.

The position is salvaged by *FA 1986, Sch 20, para 2(4)* which operates to preserve the relief where the donee makes a gift of property that was comprised in the original gift to him and in which the donor has reserved a benefit – see IHTM14374. Paragraph *2(4)* states that where the gift is not back to the original donor (Alan), the donee (Barry) is treated as continuing to have possession and enjoyment of all of the house and 15 acres and not merely the retained half interest in it. Reading that back to Alan's gift, we can deem Barry still to own the entirety of what Alan gave him, so the 'four tests' are satisfied and clawback does not apply to Alan's gift.

DEDUCTIBILITY OF DEBTS

14.7 The treatment of liabilities was considered in **Chapter 8**. The repositioning of debt which may have been beneficial for BPR/APR has been thwarted by *FA 2013* (see **8.16**).

The repositioning of a debt is not a PET and thus there is no need for seven year run off nor is there a requirement to satisfy the two year ownership rule for BPR; rather the effect is immediate. There may be expense in connection with the advice and the disposals, but there need not be any other substantial tax costs; however, it is important to examine the availability of liquid resources which can be used to discharge 'unnecessary' debts. Given the 'discrimination' against sole traders described in **Chapter 8**, repayment of debt can be particularly effective for the latter.

It was also noted in **Chapter 8** in the absence of the rules contained in *IHTA 1984, s 110*, for APR purposes *s 162* applies debts in an unrestricted way. A liability which is an encumbrance on a specific item of property is taken, as far as possible, to reduce the value of that property. APR will therefore be lost only in respect of qualifying assets subject to specific charges.

Example 14.5: Brendan – debt allocation (1)

Brendan and his wife, Mary, have farmed successfully for many years but in recent times economic pressures have forced them to diversify. The farm, in addition to its own well stocked farm shop selling its own produce, now has a virtual mini out-of-town shopping centre with profitable franchises for the sale of goods which have no connection with farming. The farmland as a whole has been used as security to obtain bank lending to purchase of further farmland and for the development of the shopping centre.

Brendan and Mary are concerned about the impact of the diversification on the trading status on the farm as a whole and consult their accountant, Cerys. After considering the position in the round, Cerys concludes the level of family involvement in the sales of goods from the shopping centre is now so slight by comparison with the work done by others, and the revenue from the farm shop is so small by comparison with the franchise income, that even after considering the extensive time management of the enterprise, there is a very high risk that the venture will be regarded as 'wholly or mainly holding investments' and will lose its BPR entitlement.

In order to preserve APR, Cerys advises that the bank loans secured on the farmland must be apportioned between the part used to acquire farmland and the part that has financed the shopping centre. Once the loans are apportioned the farmland will be subject to less debt under *s 162*, with an immediate increase in the overall amount of APR available to the family. The part of the debt attributable to the purchase of farmland cannot now be shifted away from that land, as was formerly possible prior to *FA 2013*: it must be set against the farmland under *s 162B*.

Example 14.6: Gerald – debt allocation (2)

Gerald, a bachelor, has traded successfully as a printer for many years. He has learnt that he has a terminal illness and has a life expectancy of only six months. He asks his accountant, Cerys to conduct a review of his estate which holds the following assets some of which are connected with his business:

	£
Unit 3, Welsford Trading Estate (print shop)	340,000
Unit 1a, Welsford Trading Estate (store)	70,000
17 Lyhart Street (shop premises)	140,000
The Chestnuts, Swardeston (residence)	600,000
Investments (mainly cash)	150,000
Heidelberg Printing Press	550,000
Other stock, plant and machinery	200,000

Gerald has never kept his non-trading activities entirely separate from the printing trade but rather the assets of each have supported the other. There are two loans outstanding: loan A, of £60,000; and loan B, of £200,000. Investigation shows that loan A, secured on Unit 3 (from which the business trades), was in fact applied in the purchase of the shop premises at 17 Lyhart Street, which have never been used for the purposes of the trade and in respect of which there was never any intention of such use. On the other hand, loan B, secured on the residence, is applied specifically to the purchase of the Heidelberg Press.

A pro forma account for IHT planning purposes suggests the following calculation:

	£	£
Personal or private assets (no BPR):		
Investments	150,000	
The Chestnuts	600,000	
17 Lyhart Street	140,000	
		890,000
Business assets:		
Unit 3	340,000	
Unit 1a	70,000	
Heidelberg Press	550,000	
Stock, plant and machinery	200,000	
Less loans	(260,000)	
		900,000
Estate before reliefs		1,790,000
BPR at 100%		(900,000)
Nil rate band (2017/18)		(325,000)
		565,000
Tax at 40%		226,000

Cerys negotiates with the lenders for loan A to be rescheduled and charged specifically on the shop premises. It was not used to acquire relievable assets so falls outside the restrictions of *s 162B*. Loan B is in part discharged from the surplus cash, with the balance being secured specifically on the Heidelberg Press. That reduction in debt falls within *s 162C*. No part of it was used for non-business purposes so none is attributable to non-business assets under *s 162C(2)*. As a result, the cash and the non-business assets are reduced in value, but the amount of loan A, which could otherwise have reduced BPR, no longer does so. This results in a revised calculation:

	£	£
Investments	90,000	
The Chestnuts	600,000	
17 Lyhart Street	140,000	
		830,000

	£	£
Unit 3	340,000	
Unit 1a	70,000	
Heidelberg Press	550,000	
Stock, plant and machinery	200,000	
	1,160,000	
Less loan B	(200,000)	
		960,000
Estate before relief		1,790,000
BPR at 100%		(960,000)
Nil rate band		(325,000)
Estate after reliefs etc		505,000
Tax at 40%		202,000
Net saving: reduction in tax	24,000	

The effect of *IHTA 1984, ss 110* and *162* illustrated above is limited to property of the type in the illustration. For holders of other kinds of property, the scope for reducing IHT by moving debt may be slightly greater.

Example 14.7: Quentin – debt allocation (3)

Quentin, an elderly bachelor, holds the following assets:

	£
Shares in Neptune Ltd, traded on AIM	150,000
Debt on purchase of shares	(50,000)
Balance	100,000
Partnership share: Marine Chandlers	100,000
Unit 6 Quayside Trading Estate	150,000
Liability incurred on purchase	(60,000)
	290,000
Residence (free of mortgage)	700,000
Bank deposits	80,000
Quoted investments, ISAs	100,000
	1,170,000

14.7 *General planning points*

(1) Quentin consults his solicitor, Jack, who later produces the following pro forma account for IHT planning purposes:

	£	£
Residence		700,000
Cash		80,000
Investments		100,000
Unquoted shares	150,000	
Less liability	(50,000)	
Net value for BPR		100,000
Partnership share		100,000
Business premises	150,000	
Less liability	(60,000)	
Net value for BPR		90,000
Estate		1,170,000
BPR claim at 100%:		
AIM shares	100,000	
Partnership share	100,000	
Premises	90,000*	
Total claim		(290,000)
Estate after reliefs		880,000
Nil rate band 2017/18		(325,000)
Chargeable estate		555,000
Tax at 40%		222,000

An analysis of Jack's pro forma account indicates the following:

(i) Jack has overlooked the point that BPR on the business premises falls within *s 105(1)(d)* and is therefore restricted to 50%. The chargeable estate is thus £600,000 (£555,000 + (£90,000 @ 50%)) and the true liability is therefore £240,000.

(ii) Jack's advice to reposition the debt away from the relevant business property, would have been sound except that critically he has not yet assimilated the effect of *FA 2013* which automatically allocates the debt for BPR purposes.

(2) Based on Jack's incorrect advice Quentin renegotiates the loans. The £50,000 secured on the unquoted shares is reduced, as to £20,000 from cash resources with the balance being secured on the quoted investments,

which Quentin is reluctant to sell in a bad market. The loan of £60,000 is re-secured on the residence. Jack recalculates the adjusted IHT calculation as follows:

	£	£
Residence	700,000	
Loan secured on house	(60,000)	
		640,000
Cash		60,000
Investments	100,000	
Less loan	(30,000)	
		70,000
Unquoted shares		150,000
Partnership share		100,000
Business premises		150,000
Estate, gross		1,170,000
BPR claim:		
Shares	150,000	
Partnership share	100,000	
Premises	150,000*	
Total claim		(400,000)
Estate after reliefs		770,000
Nil rate band 2017/18		(325,000)
Chargeable estate		445,000
Tax at 40%		178,000
Tax saving, on Jack's figures (222,000 – 178,000):		44,000

An analysis of Jack's adjusted account indicates the following:

(i) *Jack's figures are wrong – he has made the same mistake as before in that BPR on the premises could only ever be claimed at 50%.

(ii) The effect of *FA 2013, s 162C* is that whilst the repayment of £20,000 of debt frees up an equivalent amount of BPR, the balance of that debt that has been repositioned must still be set against the value of the shares.

(iii) Similarly, moving the debt away from the industrial unit has achieved no saving. There is still a net saving on the exercise, but it is only £8,000 (£20,000 @ 40%).

PLANNING SINCE *FA 2013*

14.8 The *FA 2013* measures were introduced to counter schemes that sought to secure relief not otherwise intended under the legislation but how should practitioners now proceed?

- *First*, great care should be taken not to 'taint' any existing borrowing that is 'effective' by increasing or renegotiating it after 17 July 2013.
- *Second*, do not mix new non-business borrowing with borrowing that relates to relievable property, thereby avoiding the complex allocation rules demanded by *s 162C*.
- *Third*, analyse existing borrowing to establish whether it relates both to relievable and other property. Consider how any repayment would be treated under *s 162C*. *FA 2013* now dictates that repayment of a mixed loan must be set first against non-relievable property: over a period of years it may become possible to show that new debt relates to the purchase of new non-relievable assets, and gradually to pay down debt on relievable assets.

Consider the following situations:

Example 14.8: Matthew – roll-over

In October 2015, Matthew purchases Top Farm for £1 million, financed by £200,000 from savings and a loan of £800,000 secured on other non BPR property. In mid-April 2016 the borrowing is increased to cover the cost of drainage works at Top Farm. Shortly afterwards a long-planned road scheme is signed off by the local council and a compulsory purchase order takes most of Top Farm. Matthew sells off the remainder of the farm to his neighbour and repays the debt from the proceeds.

Anxious to roll over at least part of the gain on the sales, Matthew then buys Middle Plantation for £800,000 financed by a bank loan of £200,000. That loan must be set against the value of the new land but no adjustment is required for the old borrowing which has indirectly financed the second purchase.

Example 14.9: Nigel – serial investments

Nigel has a long-standing arrangement with his bank that allows some gearing of his investment portfolio of AIM shares, with a limit of £300,000. The amount outstanding at any time is secured on the deeds of his residence.

In July 2017 the portfolio is worth £1.4 million and the borrowing is at its limit. Under the old rules, BPR would have been available on the full value of the shares (if held for two years or replacing others so held). The borrowing would have been set against the house under *s 162(4)*.

In December 2017 a different bank, keen for Nigel's business offers him slightly better banking terms and agrees that he can switch to them in other respects on the same basis as before. The new bank clears the £300,000 facility with the old bank but how will the new facility be treated under the new rules?

The new loan has not facilitated the purchase of any of the shares, either directly or indirectly. However, it is clear that the new loan is linked to the old one, which may be a close enough connection for HMRC.

How would the GAAR impact on this? It is clearly intended to protect 'loose' legislation and to prevent exploitation of defects in anti-avoidance legislation: see *para B7.2* of the *GAAR Guidance*. However, in its base terms all that has happened here is that Nigel has simply switched banks for commercial reasons – tax was not a factor in that decision.

14.9 The simplest planning will involve a review of all borrowing ensuring where possible that no part sits against relievable property which does not relate to such property. Where the taxpayer owns several businesses some of which hold surplus cash which would fall foul of the excepted asset rule it would make sense (subject to income tax and similar considerations) to release that money to pay down debt elsewhere.

RESTRICTION ON OFFSET OF DEBT

14.10 The case of *St Barbe Green v CIR* [2005] EWHC 14 (Ch) illustrates the peril of a debt sitting in the wrong place. The deceased had a qualifying interest in possession in a substantial trust fund which fell to be aggregated with his estate but the latter was insolvent. The trustees sought to set the unrelieved personal debts against the value of the trust funds to arrive at the IHT payable but HMRC rejected this treatment and the trustees' appeal to the High Court failed. Mann J held that the words 'except as otherwise provided by this Act' in *s 5(3)* qualified the principle that debts were deductible. The personal estate must be seen in isolation: once debts had reduced it to nil, there was no asset against which further debt could be set. The trust funds gained no benefit by way of offset.

Probably few settlors would want trust funds to be dissipated by advances to cover the debts of beneficiaries. However, the *St Barbe Green* case shows that it could be useful for trustees to have a power to advance capital to a life tenant; and that, if that power were exercised seven years before death, the funds to be taxed would be less and (for those who care about such things) family honour might be saved by avoiding insolvency.

BUY AND SELL

14.11 The issue of whether or not there is a contract for the sale of property in relation to business property was considered in some detail in **Chapter 6**, where reference was made not only to SP 12/80 but to the ICAEW memorandum TR557 which considered 'buy and sell' agreements in some detail. TR557 specifically identified a number of provisions which might be found in partnership agreements and suggested the 'apparent effect' on which HMRC commented on 5 July 1982. The result of this can be seen in several situations:

- The partnership determines on the death or retirement of the partner and his share remains his asset: BPR is available both in respect of lifetime transfers and on death.
- On the death of a partner, the partnership continues but the share of the partner dying or retiring is free to be sold and, if not sold, passes to the personal representatives or a beneficiary: BPR is available both in respect of lifetime transfers and on death.
- The situation is as just described and the personal representatives have an option to sell and the continuing partners an option, but no obligation, to buy: BPR is available both in respect of lifetime transfers and on death.
- The partnership continues notwithstanding death or retirement, but the share of the outgoing partner automatically accrues to the continuing partners who must then pay for that share (usually according to a predetermined formula): no BPR is available in respect of the share, whether on a lifetime transfer or on death.
- The partnership continues as before and the partnership share passes to the personal representatives but those representatives are under an obligation to sell. Here the situation is finely judged and will depend on the particular circumstances. Even though there is, within the context of the partnership agreement, a contract to sell the partnership share on the happening of an event which has not yet occurred, BPR may be available in respect of a lifetime transfer if that transfer itself, within the terms of the partnership deed, does not trigger the automatic purchase provisions. In the event of death, however, no BPR will be available, because the automatic purchase provisions have, as a result, been triggered.

- The partnership continues on death or retirement and where there is no obligation within the partnership agreement to sell. Here, BPR will be available both in respect of lifetime transfers and on death, even though the outgoing partner may have executed a will directing the executors to sell. This is logical: the will itself could be revoked at any time and, in the absence of compulsion within the partnership deed, there is no reason, for the mere intention of the outgoing partner that the property should be sold, to deny him relief during his lifetime.

A way out, perhaps?

14.12 It is not uncommon for the practitioner to be faced with a situation of a partnership deed drawn up to include a binding contract for the sale of a partnership share thereby operating to deny BPR to the outgoing partner. The practitioner should look carefully at the present constitution of the partnership as not infrequently, it may be discovered that the partnership deed is outdated and that there have been partnership changes since it was first signed. If, as is equally likely, there has been insufficient formality in the accession of later partners, there may well in fact be a partnership at will only. All the present partners have informally adopted the partnership agreement and the accounts have been prepared in a manner which is entirely consistent with it, without realising that any one of them at any time could retire from the partnership at will without being bound either as to any provisions as to notice or in relation to the offending obligations respectively to sell and to buy.

OPTIONS IN RELATION TO SHARES

14.13 Shares may have variable rights whereby entitlement to the distribution of profits or to assets in a winding up may differ in a future accounting period from the present one. The percentage of profits or assets for the current accounting period is the lower of:

- the percentage to which the parent company is entitled in the current accounting period; and
- the percentage to which it would be entitled in any later accounting period in which the rights had changed (see *ICTA 1988, Sch 18, para 5*).

Where the shares or securities are held within arrangements which might result in a change to entitlement, either to distributions or to assets at some future time, *ICTA 1988, Sch 18* assumes that effect would be given to those arrangements. Those arrangements are, however, confined to the variation of rights attaching to shares and securities; they do not include arrangements which could affect the ownership of shares. In the case of *J Sainsbury plc v O'Connor* [1990] STC 516) which concerned group relief rather than IHT, the

Court of Appeal held that the mere existence of an option to purchase shares did not affect the beneficial ownership of those shares by the taxpayer company.

RETIREMENT

Consider alternatives to an immediate sale, especially where ill health may be an issue?

14.14 The sale of a business may be a stressful process of which tax is only one of the considerations on the table. The main shareholder in the family company may feel that it is his duty to see the sale through and to provide his family with lasting security but it may be better to let others resolve the problem whilst he retains a significant interest in the business through to his death. That way, his estate may save both CGT, through the uplift on death, and IHT, by retaining assets qualifying for BPR.

These very issues were raised in *Swain Mason and Others v Mills & Reeve* [2011] EWHC 410 (Ch), [2012] EWCA Civ 498 (CA). Mr Swain, a company director who was 61, overweight and who had previously suffered a heart attack, consulted solicitors in June 2006 about selling his controlling shareholding in the family company. He was by then spending most of his time in Thailand and had agreed in principle to sell the company to the current management. The solicitors advised in a letter in January 2007 about the tax consequences of the sale. A letter applying for certain tax clearances referred to Mr Swain's 'continuing weak health', to his heart attack and diabetes and to the fact that he would not return to active work. That letter also showed that Mr Swain, though still domiciled, resident and ordinarily resident in the UK, might cease to be so resident as soon as the 2007/08 tax year started, ie within a few weeks.

Mr Swain made an appointment for a hospital procedure to take place in Thailand on 17 February 2007 and told all relevant parties, including the solicitors (though in a somewhat oblique way), that this was his priority and that he would not be attending board meetings. The sale went ahead on 31 January 2007. Mr Swain died during the hospital procedure.

IHT on his estate was greater than it would have been had the sale been deferred primarily due to the loss of BPR but CGT was also incurred that might have otherwise been avoided – that combined tax liability amounted to approximately one-third of the sale proceeds. The daughters sued the solicitors, arguing that they should have advised delaying the sale until after the operation. The solicitors argued that they were not instructed to advise on what might happen in the event of death soon after completion; the risk of death had been negligible: there was evidence that it was less than a one in one thousand chance; Mr Swain had not asked their advice on the issue that had caused the loss.

The pleadings were amended so as to claim that the January advice letter should have extended to advice on IHT and that, had it done, Mr Swain would have sought further advice before proceeding. Expert evidence suggested several courses of action that might have mitigated the burden of taxes.

A central issue was whether any solicitor dealing with this type of case must advise on the tax consequences of death intervening and possibilities for mitigation of that tax. On retrial as ordered by the Court of Appeal Arnold J examined carefully the extent of the solicitors' retainer and whether the solicitors had excluded tax advice as to the effect of the disposal. Arnold J held that they had, but by only a narrow margin, because the letter could have been more specific about IHT and CGT. He dismissed the claim. That decision was upheld by the Court of Appeal, which ordered the family to pay 60% of the costs of the solicitors. It is the postscript that reminds practitioners of the value of good advice:

> 'The trial, and its outcome, will have been disastrous for the four daughters of Mr Swain; not only in costs terms but also (it is not difficult to apprehend) in terms of anxiety, pressure and emotional upset.'

What lessons can be learnt from Swain Mason?

14.15 Lessons learned from tax cases can be more fruitful than a passive study of the statute and all the 'Jack' examples are designed to support that theory (consider that the terms of engagement were the most important single factor in *Mehjoo* (see **13.1**)). The solicitors in *Swain Mason* eventually succeeded but did not recover all their costs. In that latter case there was much difficulty over the issue of disclosure of documents (which was costly but achieved little) and an observer might feel that the lawyers won but only on a technicality.

The case highlights the following precepts:

- the engagement letter must state what areas of advice are *not* covered, as well as what areas are;
- the need to keep in touch with the client: do not 'steamroll' a transaction through;
- in a commercial case, remember the 'private client' issues: have regard to all taxes; but
- if sued, do not throw in the towel but use all means allowed by the court rules to defend the position.

That will be small comfort to clients, who may have hoped that judges like Peter Smith J, the original trial judge in *Swain Mason*, would uphold their rights. They may remember that he heard the case of *RSPCA v Sharp*

[2010] EWHC 268 at first instance and there made observations about another equally eminent firm of solicitors, but was overruled on appeal.

The following principles should be adopted by both clients and practitioners alike:

- establish what advice is covered, and what is not;
- get advice on all aspects of a proposed transaction;
- give full instructions;
- don't rush; and
- if things go wrong, research carefully before embarking on litigation and get all relevant evidence ready at the beginning.

The expert in the case considered that a competent solicitor would have recommended a short delay not least since (under the then rules) Mr Swain could have claimed to be non-resident and outside the scope of CGT from 6 April 2007. It would, however, take three years to lose UK domicile but Mr Swain could have invested the proceeds in Free of Tax for Residents Abroad (FOTRA) securities, being gilt-edged securities excluded from IHT (*IHTA 1984, s 6(2)*). In so far as Mr Swain received payment in loan notes, FOTRA would be no help, but perhaps insurance cover for three years, until UK domicile had been shed, might have been possible.

All of this prompts the question of 'How good at their job are practitioners expected to be?'. This point was raised in the case hearing and in referring to *Midland Bank Trust Co Ltd v Hett, Stubbs & Kemp* [1979] Ch 384 at 402, it was considered that the test is 'what the reasonably competent practitioner would do having regard to the standards normally adopted in his profession'. The standard by which specialist or 'above average' practitioners will be judged in these circumstances is illustrated in *Swain Mason,* where the defendants, a national firm whom many will consider to be of outstanding repute, were to be judged by the standard of competence of firms with specialist tax departments.

Inactive partner still a partner

14.16 As noted in **Chapter 5**, *IHTA 1984, s 105(1)(a)* allows relief in respect of property consisting of an interest in a business. There is no requirement that the owner of that interest should be actively concerned in the furtherance of its trade. It has been noted that *s 103(3)* excludes from the definition of 'business' one which is 'carried on otherwise than for gain', but that is not taken to mean that the partner who by reason of age or infirmity contributes very little to the running of the business (and may even obstruct or frustrate the efforts of younger partners, perhaps his own children) is carrying on the business otherwise than for gain so as to deny himself BPR.

In practical terms, there may be difficult decisions to make. If in a partnership scenario the elderly partner retires from the business, the younger partners may be able to drive the business forward more successfully than before. On the other hand, the elderly partner will often have substantial capital in the business far exceeding that of the younger partners and withdrawal of the same may cause problems for the remaining partners.

Leaving money in

14.17 In reality, it will often be financially impossible for the younger partners to introduce fresh business capital equal to what the elderly partner holds. If the latter is to take a less active role, the question arises as to how his interest should be treated in the accounts. Loans to a partnership are not part of a business and it is therefore important that the interest of the elderly partner should not be converted from capital account to loan account losing access to BPR in the process.

This issue was examined in *Beckman v IRC* [2000] STC (SCD) 59. Mrs H ('mother') was in partnership with her daughter Mrs B but retired in 1993. There had been no partnership agreement but profits had been shared equally and mother's capital account was £169,185 at the point of her retirement but had reduced to £112,811 at the time of her death. The daughter argued that the value of the capital account represented her mother's interest in the business to which BPR ought to apply. Whilst HMRC agreed that mother had had an interest in the business up to the time of her retirement, they argued that her relationship with the business changed fundamentally when she retired. The money due to her was an unpaid debt representing the capital and undrawn share of profit but post retirement she no longer had an interest in the business within *s 105(1)(a)*. The Special Commissioner agreed with HMRC. Before her retirement, mother had enjoyed the rights of management and suffered the liabilities of a partner. After mother retired, her rights were like those of a creditor. The capital account did not represent an interest in the business within *s 105(1)(a)* after the retirement.

A gift of such a loan account, made by increasing the loan account of one partner and reducing that of another, does not attract BPR. On the other hand, BPR may be available to shelter a transfer between the capital account of one partner (but see the observations noted in Example 5.2).

The partnership structure is well adapted to succession within the family provided that arrangements are put in hand early enough. Where the younger generation are admitted to partnership with a fair share of profit but not all of which is drawn, there is the opportunity over many years for substantial capital accounts to be established by that generation and for the older partner/partners to draw down fully on their shares of profits – any capital requirements of the business could then be provided by the younger partners. In practice, difficulties arise where succession arrangements are only put in place when the older partners are within a few years of retirement. In the case of farming businesses where the return on capital may have been low, the older partners will be quite happy to part with substantial assets as long as their income is not diminished. An appropriate structure may be to divide profits in tranches, first recognising the time and other skills which each partner brings to the business, then the capital which each makes available to it.

All too often the older partner having perhaps made a gift of farmland, will wish to continue enjoy the share of profits which should in commercial terms go to the donee of the gifted land. If the profit sharing ratio is not adjusted there is a reservation of benefit by the older partner, in that the drawings policy has been fixed artificially so as to benefit him. A better structure is for drawings to be on a similar basis for all partners. If that puts the business under strain financially, and if specific capital expenditure is contemplated, then the younger partners may decide, as a transaction entirely separate from regular drawings, to introduce capital as and when it is needed.

MONEY BOX

What is possible?

14.18 Clearly, there is an advantage in holding non trading assets within the favourable tax environment of a business, subject only to the normal commercial risks of investment. The limit of what may reasonably be held in such a business was examined in some detail in **Chapter 7** in relation to excepted assets and as noted the test can be quite strict. Apart from the guidance contained in *IHTA 1984, s 112*, there is little authority on the subject other than observations in the HMRC Manual and a decision of the Special Commissioner in *Barclays Bank Trust Co Ltd v IRC* [1998] STC (SCD) 125, examined in some detail at **7.11**.

Where money is left in a business, steps should be taken to support that the business either had an immediate need for that money or will have a need of it in the reasonably near future for, as was noted in *Barclays Bank Trust Co Ltd v IRC*, some 'palpable business purpose'. The hallmark of money held within a business which is surplus to the business needs (and which as a consequence will not qualify for BPR) is that it lies in an account undisturbed from one year to another (save for the addition of interest and further capital).

In the case of *Brown, Ralph Louis (executors of) v IRC* [1996] STC (SCD) 277 the Special Commissioner had to consider whether the nature of a company's business had changed after the sale of the main asset of the business. The deceased owned 99% of the issued share capital of an unquoted company, Gaslight

Entertainments Ltd Company and in January 1985, an offer was accepted for the nightclub run by the business. The proceeds of sale went onto short-term deposit whilst Mr Brown searched for an alternative nightclub for the company to purchase but he died in November 1986, at which time most of the proceeds of sale were still on deposit. It appeared to be agreed between the parties that the company was carrying on a business during 1985 and 1986 but the question was the exact nature of the business. Mere intention was held insufficient; rather it was necessary to look at what the company was actually doing.

HMRC argued that, following the sale of the nightclub, effectively the business of the company was no more than holding investments and thus disqualified from BPR by *s 105(3)* but the Special Commissioner did not agree. Whilst acknowledging that the proceeds of sale were held in an account which produced income, such funds were available at short notice so that, when Mr Brown found a suitable alternative nightclub, he could buy it. There was evidence to show the efforts made to find an alternative venture and but for his illness and death, he might well have made a purchase.

Brown v CIR is a case which turns on its own particular facts but it does show how, in suitable circumstances, money may be held within a business on deposit and yet qualify for BPR. Short-term deposits are more likely to qualify for BPR than long-term fixed deposits.

VARIOUS STRUCTURES

Wholly or mainly

14.19 **Chapter 9** examined the test in *IHTA 1984, s 105(3)* as to whether a business consisted wholly or mainly of dealing in securities, stocks or shares, land or buildings; or making or holding investments. However, it should be recognised that this problem can and often does arise by stealth as the balance sheet of the business strengthens. Apart from *Brendan Peter James Furness* SpC 202 (noted at **9.9**), most of the decisions in this area have gone against the taxpayer and this should be noted.

For the unincorporated business, there may come a point at which the holding of investment assets can threaten the BPR eligibility of the business as a whole. The question is one of degree, as was noted in the discussion of excepted assets in **Chapter 7** and the relevant factors are:

- the value of assets employed in each part of the business;
- the source of turnover;
- the source of profit; and
- the way the directors or partners spend their time.

It is necessary to look at the business as a whole and where the tipping point between trading and investments is borderline, it may be wise to remove the latter from the balance sheet so as to strengthen the argument that the remainder of the assets do qualify.

Dual/multiple company structure: the hybrid company

14.20 **Chapter 7** also included an examination of the valuation principles which are applied where there are excepted assets, the view of HMRC as set out in the Shares Valuation Division Manual, and an illustration, see SVM111220, of a hybrid company. The term 'hybrid company' appears in the Manual, rather than in decided cases, but is a useful concept. It should be accepted that *IRC v George*, though containing many helpful observations, was essentially a case which turned on its own facts and specifically where the Court of Appeal refused to upset the findings of the Special Commissioner.

Each company structure must be examined on its own and the importance of *ss 110(c)* and *112(2)* must be understood. A trading company can support a certain level of 'non-trading' investment without compromising its BPR status but the difficulty lies in deciding how far to push that argument. There may be a point (which will be different for each trade) at which it has to be accepted that the bulk of the assets, income and activity of the company no longer relate to trade and do not justify the protection of BPR. Before that tainted point is reached, it may be helpful to remove from the group structure (if this can be done in a tax effective manner) the investments or group companies which clearly are wholly or mainly holding investments.

USE OF RELEVANT PROPERTY TRUSTS TO SECURE 100% RELIEF

14.21 Several factors may encourage the owner of business or agricultural property to make a lifetime gift into a trust whilst accepting that irrespective of trust type following *FA 2006*, this will assume the form of a chargeable transfer. A discretionary trust would be appropriate for the donor who cannot make up his mind as to whom to give the property to but has in mind a group of potential beneficiaries such as children or grandchildren. Such a trust type would also avoid potential dissipation by younger family members or claims in matrimonial proceedings of younger family. This may also serve the reluctance of the settlor to part with the asset save on terms that allow him to retain a degree of control or even, put bluntly, power over the asset and, by implication, over the beneficiaries.

In relation to non-business assets, there can be very good reasons to make a gift by way of a lifetime trust from which importantly the settlor is excluded since, as a chargeable transfer, it would allow access to general CGT holdover relief (*TCGA 1992, s 260*). That argument is not as strong in relation to business assets, where *TCGA 1992, s 165* also gives (in relation to assets other than shares) relief by way of holdover from an immediate charge to CGT (provided the settlor retains no interest under the trust, which he should not do anyway, to avoid reservation of benefit).

Given that a transfer to a relevant property trust is a chargeable one rather than a PET, the availability of 100% APR or BPR becomes highly relevant to the decision whether or not to make the gift at all. Frequently, a viable plan to mitigate the prospective burden of IHT is not implemented because it is recognised that there will be an immediate charge to tax, even though the payment of tax on the occasion of the gift can, in certain circumstances, earn for the transferor and transferee significant saving in tax at a later date.

One important aspect of such transfers is that the clawback rules are not relevant to fixing the IHT payable at lifetime rates on chargeable lifetime transfers. If the transferor dies within seven years post transfer, the 'top up' charge will apply unless the APR/BPR conditions are met to save the transfer from clawback, for example in relation to agricultural property that the property remains agricultural property and is used as such – see **Chapter 10**. The availability of BPR, even at 50%, can greatly increase the value of the assets which a transferor is willing to settle on lifetime trust and this particular point features strongly in the tax-planning scheme.

Associated operations: Reynaud v IRC

14.22 The structure adopted in *Reynaud v IRC* [1999] STC (SCD) 185 illustrates a particular form of planning which, when it came before the Special Commissioner, was successful. However, as discussed in **Chapter 13**, its use has been largely eroded by the *FA 2015* insertion of *s 62* into IHTA 1984 which applies to chargeable events arising on or after 18 November 2015. The principle upon which the *Reynaud* scheme rested was that the availability of relief depends upon the facts at the date of the transfer and the previous history, but not upon what happens afterwards. Thus the existence of a contract before a transfer causes relief to be lost, but a contract immediately after the transfer does not change the relief available at the instant of transfer itself.

Mr Reynaud and his three brothers established discretionary settlements for, in each case, the benefit of family and charity (but excluding, in each case, the settlor and his wife). This took place on 27 April 1995 and shares in C Ltd, a company wholly owned by the four brothers, were transferred to the trustees. On 28 April 1995, C Ltd purchased some of its own shares with the aid of a bank loan. The remainder of the shares were then sold to M Ltd. On the day when the discretionary settlements were made, there was a real possibility that the sale to M Ltd would not proceed. C Ltd would not have purchased its own shares from the trustees if it was not certain that the sale of the remaining shares to M Ltd would take place, since C Limited did not have enough funds of its own to satisfy the share purchase.

HMRC argued first, that the transfer by Mr Reynaud of his shares to the discretionary trustees was a transfer of value under *IHTA 1984, s 3* and second, that the transfer of shares and the own share purchase by C Ltd were associated operations within *s 268*. On that basis, those operations took place at the time of the latest of them, ie the date on which C Ltd bought the shares and on that date no BPR was available. As an alternative, HMRC argued that, under the principle in *Ramsay v IRC* [1981] 1 All ER 865, the transaction should be treated as a gift of cash to the trustees and in this respect it may have been the fact that each of the brothers entered into similar transactions. The taxpayers argued that the gift of shares to the discretionary trust was a transaction which stood alone, being complete at that stage and that the purchase of the shares from the trustees by C Ltd was not a 'relative associated operation'. That purchase did not contribute to any transfer of value represented by the initial gift to the trustees by virtue of which the estate of the taxpayer was reduced. BPR should therefore be available.

The Special Commissioner held there was only a single disposition, namely the operation which reduced the estate of each taxpayer arose solely through the transfer of the shares into the discretionary settlement. The disposition could not be treated as having taken place later than the date on which it had actually occurred and on that basis, BPR was available.

The Special Commissioner considered the principle of fiscal nullity under *Ramsay* which required a pre-ordained series of transactions or a single composite transaction with steps inserted which had no commercial and no family or non-tax purpose. In this case there were two transactions, namely the settlement of the shares and the purchase of those shares by the company from the trustees and it was held that those transactions were not part of a single composite transaction. The sale to the third party, M Ltd, took place after a day of negotiations and there must have been some reasonable likelihood that the negotiations might fail.

HMRC instruction at IHTM14821–14836 shows how closely associated transactions are scrutinised – all the facts will be examined in a search for any formal linkage which may show that the subject matter of the gift is, in reality, not the favoured, relievable property but the proceeds of its sale.

Associated operations: the Rysaffe case: now or never?

14.23 The issue of associated operations was tested again in *Rysaffe Trustee Co (Channel Islands) Ltd v CIR* [2002] STI Issue 23, where initially the

Special Commissioner and eventually the Court of Appeal had to consider ten discretionary settlements, five made by each of two settlors in exactly the same form apart from the date which, it appeared, had been inserted by the settlor's solicitors after execution. Each settlement was of £10, satisfied by a single payment from the settlor to the solicitors of £50 each. The intention of the scheme had been that shares in a private company (of which the settlors were shareholders and directors) would be transferred to the trustee, in equal proportions to each settlement. It had been intended that deferred shares be issued, but unfortunately that issue proved to be invalid; so, instead, the settlors took some existing ordinary shares, re-designated them as deferred shares, and transferred them to the trustee.

The issue of 'associated operations' arose in connection with the ten-year periodic charge on the settlements. Under the law that charge is determined, among other things, by the size of the transfer. The settlors had wanted each settlement to stand alone for the purposes of the computation. However, HMRC argued that there was, in relation to each settlor, a single settlement and as a result, tax should be charged under *s 64* at the rate applicable to the total value, in relation to each settlor, of all five settlements. Alternatively, the five settlements in each case were 'dispositions of property' within the meaning of *s 43* and resulted in one settlement. The trustee argued that the 'associated operations' rules did not apply or, even if they did, the settlements were not 'dispositions of property' as a result of which the property was held in trust, within the meaning of *s 43(2)*. The trustee required each settlement to be considered separately, because they were separate settlements under trust law. Even if there were a single disposition by associated operations, it could result in property being held on the trusts of five separate settlements.

The Special Commissioner had found in favour of HMRC, though that decision was overturned on appeal. In an examination of the transaction, it was clear that the transfer of £10 was part of the establishment of each settlement and that each had been created to facilitate the later transfer of shares to the trustees. It did not matter that the original transfer failed because the issue of shares was invalid and that matters had to be put right later. The creation of each settlement and the later transfer to each of them of the shares were together associated operations.

The Special Commissioner found that all of the settlements were treated in exactly the same way by both the settlor and the trustee. It was clear from *s 268(1)* and *(3)* that the operations might be carried out at different times but it could not be clearly established when the sum of £50 was paid over and thus when each settlement commenced. For IHT, therefore, the settlements all began on the same date, regardless of the care with which they had all been dated differently. What was more important was not the date that the lawyers had put on the deeds but the fact that the 'pump priming' sum of £10 for each settlement was paid as part of a block payment for the group of trusts and that the deferred shares transferred to each of the settlements all came from one holding of shares. That was enough, in the view of the Special Commissioner, to link the five documents. The associated operations thus found within the scope of *s 268* comprised the establishment of the settlements, the transfers of £10, and the later transfers of deferred shares.

Next, the Special Commissioner considered the interpretation of *s 43(2)* and the definition of 'settlement'. She held that the associated operations were the establishment of the five settlements in each case and the later transfers of the shares. They were all within the definition, namely 'dispositions … whereby the property is for the time being … held in trust'. The trusts were the same. The property was comprised in 'a settlement' by means of a disposition which included all the dispositions which had been affected by the associated operations. That was enough to bring it within *s 43(2)*.

She further considered whether the five settlements comprised 'any disposition or dispositions of property' within *s 43(2)* in order to decide what property was referred to as property comprised in 'a settlement', ie in a single settlement rather than all five settlements. Since each settlement was identical in that, on the basis of the previous findings, each had:

- the same settlor;
- the same trustee;
- the same type of trust property;
- the same beneficiaries;
- the same commencement date;
- each was a disposition; and
- each was affected by reference to the other;

there was one settlement within the meaning of *s 43(2)*.

This had serious consequences for the original tax plan as it had been part of the structure of the shares that, at the time of settlement, they carried very limited immediate rights, though they were to have equal rights with ordinary shares 12 years later. If the plan had worked, the value of what was settled would have been low and, even at the time of the ten year anniversary, the value would be moderate although, with the prospect of increased rights only two years later, there would doubtless be by then some 'hope value'.

On appeal to the High Court, it was held that it was actually not relevant that the five settlements were identical – there were, in fact, separate documents with separate dates which evidenced that the settlor had intended to create five separate settlements. They could not be artificially amalgamated merely because their terms were similar. On its true construction, *s 268* came into effect only in so far as the expression 'associated operations' was used elsewhere. It was not an anti-avoidance provision that could be invoked to nullify the tax advantages of any scheme. In the present case, the making of the five settlements was, for IHT purposes, a disposition. There was no reason to consider the associated operations rule. The associated operations rule had been designed to identify a disposition that had been made by several transactions and not by just one. Here, the making of each settlement was a single disposition and there was therefore no need for the associated operations provisions to apply. Each settlement was separate. The ten-yearly charge applied to each settlement separately. They were not to be amalgamated to calculate the ten-year charge.

HMRC's appeal to the Court of Appeal also failed, on a strict interpretation of the charging section, ie *s 43*. Each of the five trusts was a 'settlement' within *s 43(2)(b)* and having determined that, the charge applied to the assets within that settlement for the purposes of *s 43(1)*, so that the trusts were not aggregated. HMRC arguments based on *s 268* were rejected: the court held that the IHT charge was arrived at on the plain words of *s 43*, without any need to ask whether a disposition was 'associated operations'. *Section 268* came into play in those cases where a dispute arose as to whether there was a 'disposition' at all for IHT purposes.

There are, however, lessons to be learned here:

First, a scheme that had some intellectual attraction was very nearly spoilt through the lack of meticulous care in execution. This could have been avoided by variation of one settlement from another, not merely in respect of date, but by the choice of different beneficiaries or different trustees, and by establishing quite clearly different and distinct transfers of property to each.

Second, there is some intellectual difficulty in the argument that five transfers of shares from the same holding are linked, as if the shareholding itself could be a single asset. No authorities were quoted either for or against the argument. The Special Commissioner did draw a distinction between transfers of shares and transfers of cash. She was not willing to accept that the rules as to associated operations could extend to the payment of cash from the same bank account, but for some reason did not have the same difficulty in treating the shares as a single asset rather than as being fungible. For the time being, the issue is closed.

Transfer between settlements

14.24 The effect of *IHTA 1984, s 81* in relation to transfer of assets from one settlement to another, should be carefully noted. Consider that:

- Relevant property trust A, just over two years away from its ten-year anniversary charge, holds cash.

- Relevant property trust B, which provides for similar beneficiaries, holds BPR qualifying property.

(1) What is the effect of acquisition by the A trustees of BPR assets from the B trustees, who would not otherwise be interested in transferring to anyone?

and

(2) Will that give the A trustees relief from the ten-year charge?

The answer is that irrespective of where the property sits *s 81* treats the property as being still comprised in trust B, unless a person becomes entitled to the property itself.

The maiden aunt

14.25 In the same way that excluded assets may accumulate late in a business as it matures, so holdings of shares in a private family company can become aggregated in the hands of a single person which transfer on their death may not benefit from either the spouse exemption or the transferable nil rate band. On the one hand, the family as a whole may greatly benefit, since such a person may make few demands on the business and may build up capital which the younger (and sometimes not so young) members of the family can cheerfully dissipate later. On the other hand, in planning transfers, whether by lifetime gift or by will, the practitioner should have regard to the possibility that a controlling shareholding may build up in the hands of the 'wrong' member of the family for IHT purposes.

CGT planning: exploiting ss 191 and 179

14.26 The effect of an election under either *IHTA 1984, s 191* or *s 179* is to revalue assets on sale for IHT purposes. The revised lower, value will have been 'ascertained' for IHT purposes, which triggers revaluation for CGT under *TCGA 1992, s 274*. CGT will be the more attractive option levied at either of 10%/18%/20%/28% with relief given for expenses of sale instead of IHT at 40% chargeable on the gross value. The advantage arises where relief from IHT is available either because the asset is within the nil rate band or it is covered by the spouse exemption or it attracts APR or BPR at 100%.

Stonor (executors of Dickinson deceased) v IRC

14.27 The case considers the interaction between the IHT rebased probate value and the ongoing CGT base cost. Mrs Dickinson held freehold properties, valued at the time of her death in 1992 at £582,000 and by her will left some chargeable gifts which were all within the nil rate band with the residue left to charity. Within the time limits prescribed by *s 191*, the executors sold the properties for £918,457, and sought to substitute the gross sale prices as the amended probate values.

As IHT was not in point, HMRC declined to investigate the values and the executors then made a formal claim under *s 191*, hoping thereby to use the higher values at sale as the base values for CGT. HMRC issued a notice of determination on grounds that there was no 'appropriate person' to make a claim. The executors on the other hand considered that, had the residuary beneficiaries not all been charities, they would have been liable for tax on the estate or would have been so liable if the chargeable transfers had exceeded the nil rate band.

The Special Commissioner dismissed the appeal. The purpose of *s 191* was to grant relief from IHT, not to reduce liability to CGT on a later disposal. Whilst the section did not state that it could not apply where there had been an increase in values after the death, any claim must be made by the 'appropriate person', ie one liable for IHT. That tax was charged on a chargeable transfer, one which was not an exempt transfer. Here, the transfers were exempt, and so there was no tax chargeable. Since no tax was payable, there was no liability for tax. There was no person liable to pay the tax.

More specifically, *s 190(1)* provided that, if there is more than one person liable for the tax and one of them is in fact paying the tax, it is that person who is the 'appropriate person'. In the light of this, there was in the instant case clearly no 'appropriate person', so the claim must fail.

DEATHBED PLANNING

14.28 *IHTA 1984, s 106* requires ownership of relevant business property for at least two years before the transfer of value in order to secure access to BPR but curiously and quite critically, the legislation does not require that the property must have been relevant business property throughout that two year period. In the case of land or buildings, machinery or plant that qualifies for BPR under *s 105*, the use must continue throughout the two years that immediately precede the transfer of value, subject to replacement property rules. However, a different rule can apply to shares in a company.

Example 14.10: Investment into trading company

An elderly shareholder, or perhaps one suffering from a terminal illness, might own shares in an investment company. It may be possible to change the character of the business carried on by the company such that, at the time of death, it is in truth a trading company the shares in which are then capable of benefiting from BPR. The actual mechanics of achieving such a change would be difficult to arrange and might involve the disposal of investments and the acquisition of trading assets, so the idea may be more academic than practical; but, if it could be achieved there would be a substantial saving.

Example 14.11: Asset switch to earn relief

There could be a transfer of assets to a trading company which the trading company could then use in its business. Subject to the discussion of this point at **7.5** above, it might be possible by this method to shelter from IHT the assets so transferred. However, given the purposive advice in the Shares and Assets Valuation Manual reviewed at **7.9**, such transfers might well be regarded as provocative. There could easily be issues of capital gains tax on the transfer to the company.

Chapter 15

Will planning

> QUICK GUIDE
>
> - The attribution rules can have the effect of wasting APR or BPR on gifts that would be otherwise exempt.
> - Much of this chapter is devoted to the difficult subject of the allocation of reliefs and the complexities of double grossing. The simplest and probably best advice must be to avoid these complexities if at all possible.
> - Successful will planning involves the full use of the reliefs and making informed decisions on whether to use the nil rate band on the first death or to leave intact and available for use on the second death. This is illustrated in an extended example at **15.9**, and in a review of the use of the nil rate band at **15.10–15.12**.
> - This chapter considers the use of APR and BPR in the context of IHT planning but the use of wider will planning to mitigate the burden of IHT is outside the scope of this book and for a more general approach, reference should be made to *Ray & McLaughlin's Practical Inheritance Tax Planning* 14th Edn (Bloomsbury Professional) or *Tax Planning 2017/18* (Bloomsbury Professional) or *Inheritance Tax 2017/18* (Bloomsbury Professional).

THE NIL RATE BAND

Infinite nil rate band

15.1 This heading is a misnomer since only where all of the property qualifies for 100% APR or BPR can there be said to be a nil rate band which, in suitable circumstances, may encompass the whole estate. As demonstrated in Example 6.4, regard must always be given to the distinction between those assets which qualify for the full relief and those which do not. Nevertheless, one of the cardinal principles of IHT will planning used to be the utilisation of the nil rate band to avoid or to reduce the aggregation of estates between spouses on the second death. Specific family circumstances often thwart the traditional tax routines relied upon by the practitioner but in general tax planning terms, assets which do not qualify for relief are most suited for transfer to exempt beneficiaries such as the surviving spouse or a charity, whilst assets qualifying for APR and BPR should be directed to non-exempt beneficiaries.

Double nil rate band

15.2 This is not a reference to the transferable nil rate band but rather the process by which the same asset may, by agreement between the family and with suitable planning, be used to earn relief twice over. Where the finances of the family permit, assets qualifying for APR or BPR are left to the children with other not qualifying assets passing to the surviving spouse. The latter applies their inheritance in the purchase of the business assets from the children. If the spouse survives two years, APR or BPR (assuming the qualifications are still met) will be earned for a second time.

Partly exempt transfers: interaction with APR and BPR

15.3 The notional transfer on death is treated for IHT as a single gift of the entire estate which may be wholly taxable, partly taxable or non-taxable dependent on the size and destination of the estate (for residence nil rate band purposes), availability of reliefs and the nil rate band/transferable nil rate band Problems arise, however, where the estate passes partly to taxable beneficiaries and partly to those who are exempt. The rules in *IHTA 1984, ss 36–42* operate to attribute the available exemptions/reliefs among the various gifts, making distinctions between, on one hand, the residuary estate and, on the other hand, all other gifts, which are called 'specific gifts'.

The specific gifts may or may not be exempt but, where they are not exempt, it is necessary to establish whether they bear their own tax and this will in turn depend on the terms of the will (if there is one) of the deceased. Gifts of residue may likewise, of course, be exempt or not dependant on the beneficiary but, where a gift of residue is not exempt, it will always bear its own tax.

Non-exempt 'gross' legacy

15.4 The value of the non-exempt specific gift and whether or not that value is subject to gross up is dependent on whether it bears its own tax or not, and the value of other specific gifts.

Example 15.1: Joanna – gift bearing its own tax

Joanna, a widow whose late husband had used his nil rate band, died on 31 December 2017 leaving an estate of £2.4 million, having made no lifetime chargeable transfers. She left her son £400,000 to bear its own tax, and divided the rest of her estate between the RNLI (a registered charity) and her daughter.

The residue is £2 million (£2.4 million – £400,000) thus the daughter's 50% share is £1 million. The tax on death is charged on £1.4 million (£2.4 million – £1 million exempt share of residue to charity). The IHT thereon, calculated at the lower 36% rate by virtue of *IHTA 1984, Sch 1A,* is £387,000 ((£1.4 million – nil rate band £325,000) @ 36%) but for the provisions of *IHTA 1984.*

The IHT is allocable *pro rata* as follows:

By the son £110,571 (387,000 × 400,000/1,400,000)

By the daughter £276,429 (£387,000 × 1,000,000/1,400,000)

Note: There is no RNRB as the estate value exceeds the threshold.

Non-exempt 'free of tax' legacy

15.5 Where a legacy is left, as will very often be the case, free of tax, single or double grossing may apply.

Example 15.2: Kylie – single grossing

Kylie died on 31 December 2017 having made no prior chargeable lifetime transfers. Her estate is £1 million of which she has left a legacy of £500,000 to her sister Joyce and the residue to her husband.

She is treated as having made a net legacy transfer on death of £175,000 over and above the nil rate band of £325,000. IHT on the gross equivalent of the excess net £175,000 is £116,666 ((£175,000 × 100/60) @ 40%), making the gross legacy for Joyce £616,666 (proof: (£616,666 – (£616,666 – £325,000) @ 40%)) = £500,000). That leaves exempt residue for the husband of £333,334 (£1 million – £616,666).

This example illustrates the effect of *s 38(3)*. The grossing up attributes to the specific gift:

- the amount of the gift itself; and
- the tax that would be chargeable if the value of the gift equalled the gift plus its tax.

A mixture of legacies, with exempt and non-exempt shares of residue

15.6 The IHT calculation becomes far more complicated where there is a mix of non-exempt specific gifts which do not bear their own tax and some which do and/or where there are gifts of residue which are partly taxable and partly exempt. This situation brings into play, under *s 38(5)*, 'the assumed rate' which is found by dividing the assumed amount of tax by a portion of the value transferred which would be charged on the transfer after making a hypothesis, namely that:

- the amount which corresponds to the value of the specific gifts not bearing their own tax is equal to the aggregate already identified in *s 38(3)*, which has been illustrated above; and
- the parts of the value transferred attributable to specific gifts and to gifts of residue or shares in residue are determined accordingly.

The assumed amount of IHT is the amount that would be charged on the value transferred on that hypothesis. The following worked example breaks the calculation into its constituent parts for ease of understanding:

Example 15.3: Tanya – double grossing

Tanya died on 31 December 2017 leaving an estate worth £2,850,000. In the seven years before her death she made (now failed) PETs of £280,000. Her will gives a free of tax legacy of £270,000 to her son Wayne, and the residue to her daughter Cherie and to her husband Jason in equal shares.

After offset against the failed PETs of £280,000, the balance of the nil rate band available to the estate is £45,000 (£325,000 – £280,000).

Stage 1: calculating the gift to Wayne and residue

Of the £270,000 tax-free legacy to Wayne, £45,000 is covered by the nil rate band, leaving a chargeable net amount of £225,000 which equates to a gross of £375,000 on which the IHT is £150,000 (£375,000 @ 40%). The equivalent of Wayne's net tax free legacy is £420,000 (£270,000 + £150,000). That is the basic 'stage one' grossed-up value of the legacy, used to identify the residue of the estate, part of which is exempt and part chargeable.

The residue is found by taking:

	£
The estate	2,850,000
Less the grossed up legacy	(420,000)
Balance	2,430,000
Of that, Cherie's half share is chargeable	1,215,000

Taking the provision for Wayne and Cherie together, the gross chargeable transfer is thus:

	£
Legacy, grossed up	420,000
Plus share of taxable residue	1,215,000
	1,635,000

Stage 2: calculating the 'assumed rate'

What is the IHT chargeable on £1,635,000? That will produce the 'assumed rate' for revising the grossing up of the legacy calculated as follows:

	£
Gross chargeable transfer on death (as established above)	1,635,000
Deduct the nil rate band (325,000 – 280,000)	(45,000)
	1,590,000
Tax on that at 40% is	636,000

This produces 'the assumed rate', ie the proportion which the tax on death bears to the chargeable transfers on death. It is:

$$\frac{636{,}000}{1{,}635{,}000} = 38.90\%$$

Stage 3: re-grossing in the context of the whole estate

We must now review the legacy to Wayne. The sum is:

$$\frac{£270{,}000 \times 100}{(100 - 38.90)} = £441{,}898$$

The new grossed-up taxable value of Wayne's legacy is greater because it is now seen in the context of an estate half of which is subject to tax.

Stage 4: recalculating residue after re-grossing Wayne's legacy

It is then necessary to re-examine the transfer on death as a whole and recalculate the residue:

	£
The gross estate at death	2,850,000
Deduct the new grossed-up legacy of	(441,898)
Adjusted residue	2,408,102
Cherie's half share	1,204,051

Thus, the value passing to chargeable beneficiaries is now:

	£
Legacy to Wayne	441,898
Share for Cherie	1,204,051
Total	1,645,949

This final total is the one that is subject to the tax. The calculation is thus:

	£
Chargeable transfers on death	1,645,949
Nil rate band (325,000 – 280,000)	(45,000)
Chargeable	1,600,949
Tax at 40%	640,379

Who actually pays the tax?

So far, the computation has served only to establish the IHT to be charged. The nature of a gift of residue is such that both Cherie and Jason must, from their inheritance, fund both the legacy for Wayne and the IHT thereon. Once that has been established, Jason takes the rest of his inheritance tax free (by virtue of the spousal exemption) whilst Cherie suffers IHT on her share. At this point, it is also necessary to consider what administration costs there may have been. In the present example, the costs and disbursements amount to £17,000, attracting VAT at 20%, so a total bill of £20,400.

First, calculate the residue:

	£
Estate	2,850,000
Deduct legacy	(270,000)
Deduct tax	(640,379)
Pay the costs and VAT	(20,400)
Residue	1,919,221

The way that the IHT is calculated on the legacy is actually non-statutory. The basis commonly adopted is to assume that the proportion of tax attributable to Wayne's legacy is calculated by reference to the gross value. Returning, therefore, to 'stage 4' of the double grossing calculation, the gross legacy was £441,898 and the total transfer on death £1,645,949. The proportion of tax attributable to the gross legacy is:

$$\frac{441{,}898}{1{,}645{,}949} \times 640{,}379 = 171{,}926$$

The total that may be distributed has already been established at £1,919,221.

The tax payable on death of £640,379 may be apportioned:

	£	£
Attributable to Wayne	171,926	
Attributable to residue	468,453	
Residue as calculated		1,919,221
Add back the IHT attributable to residue		468,453
Estate available for division between Cherie and Jason		2,387,674
Of this, Jason receives half free of any deduction, ie		1,193,837
From her half, Cherie must suffer the tax of £468,453 leaving		725,384

Summary:

	£
Husband	1,193,837
Daughter	725,384
Son	270,000
Lawyers	20,400
HMRC	640,379
Estate	2,850,000

It is not uncommon for the estate to be of insufficient value to satisfy all the legacies intended under the will with the result that the legacies must be abated. This is recognised by *s 37(1)*, which operates to apply the 'double-grossing' process on the legacies treated as reduced as far as is necessary to fit within the size of the estate. That reduction is made in the order in which it would apply under the terms of the will, or under any rule of law which determines abatement in the absence of a specific term of the will.

Difficulties arising under s 41(b)

15.7 In the above Example 15.3 it was provided that none of the IHT attributable to the value of the property comprised in residue should fall on any gift of a share of residue if, or to the extent that, the transfer was exempt with respect to that gift (*IHTA 1984, s 41(b)*). That provision was examined in *Lockhart v Harker (Re Benham's Will Trusts)* [1995] STC 210, where the court had to consider the effect of the apparent intention of the testator to override *s 41(b)*. For a time, this decision caused concern since the rules tended to work in favour of the chargeable beneficiaries and against charitable beneficiaries. Not surprisingly, charities took a very strong line even though, in some respects, the decision in *Re Benham*, inconvenient though it may have been, probably did reflect what the average testator would want, ie to achieve the same net (post tax) inheritance for exempt and non-exempt beneficiaries alike. However, the real difficulty was that the calculation which emerged out of *Re Benham* actually increased the tax payable overall.

This is no longer such an issue for two reasons:

(1) HMRC published the view that:

> 'generally speaking the court is concerned in such cases to establish the intention of the testator or testatrix from the wording of the will and admissible extrinsic evidence. If the will is drafted in common form with the direction to ascertain residue after payment of funeral and testamentary expenses and debts followed by a bequest of that residue then it is focusing on the ascertainment and division of disposable residue rather than on what each residuary beneficiary is to receive. Accordingly wills so drafted would not appear to involve *Benham*-style grossing up computations …'.

(2) The situation was, for most practitioners, put beyond doubt by the decision of the High Court in *Holmes v McMullen (Re Ratcliffe deceased)* [1999] STC 262. There, the testatrix left one half of her estate to relatives and the other half to four charities. Mindful of the decision in *Re Benham's Will Trusts*, the executors issued a summons for a declaration as to the correct method of administering the net residuary estate.

The court held that the respective half shares were to be calculated *before* the payment of IHT due in respect of the shares passing to the relatives such that the net amount received by the relatives would be less than the net amount received by the charities. An equal division of the residue between the charities and the relatives was bound to mean that the IHT attributable to the share of the relatives had been borne by that share. To subject the share of the charities to any part of the burden was prohibited by *s 41(b)*.

The court naturally considered the decision in *Re Benham's Will Trusts* but did not follow it, finding that it did not establish a principle which needed be followed. By implication, the *Re Benham* decision was not approved but nevertheless, practitioners need to be aware of the case.

The impact of additional charity relief

15.8 *FA 2012, s 209* and *Sch 33* introduced a reduced rate of IHT where at least 10% of a net estate is given to charity. Where applicable, the rate of IHT applied to the estate is reduced from 40% to 36% and applies for deaths occurring after 5 April 2012.

The gift of a share of residue to charity might seem the obvious solution, but for two issues:

The first is that it may commit the executors to the double grossing calculation (see Example 15.3).

The second concerns the worries that many practitioners have about giving shares of residue to charities in any circumstances. Charities can be quite demanding beneficiaries but a professional executor should be willing to:

- answer questions as to the administration of the estate;
- produce timeous reports and accounts; and
- manage the estate so that, as far as consistent with the executors' duty to all beneficiaries, charities are not disadvantaged by tax rules.

However, for the lay executor charities can seem intrusive, even hectoring, in their approach. Further, as was illustrated in *RSPCA v Sharp and Others* [2010] EWCA Civ 1474 which found in favour of the charity, any error or ambiguity in the terms of the will is likely to be litigated where the charity beneficiary has deep pockets.

There is certainly a widely held view that any gift to a charity is best made by way of fixed legacy and that would certainly have avoided the difficulties in the *RSPCA* case. Where the gift to an exempt beneficiary, be it spouse, civil partner or charity, is fixed, calculation of relief is simple: the amount given is exempt. The rest of the estate (passing to a non-exempt beneficiary) is chargeable, so grossing up and double grossing are avoided. A testator (and this will include most) who is uncertain how much they will leave at death, but who wishes to take advantage of the reduced 36% rate, will be able to leave a fixed amount to charities (with residue to chargeable beneficiaries) but only after also including a discretionary trust of a sum that effectively 'tops up' the charitable legacy to at least 10% of the estate at death. The following example of a simple estate, ie one comprising only the 'general component', illustrates the principle:

Example 15.4: Diana – substantial charitable gift

Diana is widowed. Her late husband did not use his nil rate band and she has made no chargeable lifetime gifts. She has a son, Fergus, now well set up in life, and two grandchildren. She is a keen supporter of her local church, which is about to undergo major restoration, and she heads the appeal committee.

Her estate comprises:

	£
Residence	1,300,000
Shares in family trading company	400,000
Quoted shares	200,000
Let properties	640,000
Total	2,540,000

Diana is aware that the nil rate band is fixed at its present level of £325,000 until 5 April 2021 and understands that the value of her estate is above the RNRB threshold denying access to the RNRB/TRNRB, so there will be a combined nil rate band of £650,000 available to her executors. The family shares qualify for BPR at 100%, so may be inherited free of tax and as Fergus runs the company, Diana wishes to leave him the shares. The let properties have been valued recently such that the figure of £640,000 may be relied upon but Diana wants her grandchildren to have them tax free. Diana therefore makes her will as follows:

- specific bequest of shares to Fergus;
- specific devise of let properties, free of tax, to a trust for the grandchildren;
- legacy to the church restoration fund of £150,000;
- legacy to a discretionary trust of £100,000 to be held for the grandchildren and the church, with a letter of wishes explaining that Diana would like

the total gift to the church to be at least 10% of her net chargeable estate; and

- residue to Fergus.

Assuming that Diana dies on 31 December 2017, at a time when all the values are the same except that the quoted shares are worth £250,000, the IHT calculation is:

	£
Residence	1,300,000
Shares in family trading company	400,000
Quoted shares	250,000
Let properties	640,000
Total	2,590,000
Deduct BPR	(400,000)
Deduct legacy to church	(150,000)
Chargeable estate	2,040,000
Deduct nil/transferable rate band (T/NRB)	(650,000)
Chargeable estate	1,390,000
Primary charge to IHT at 40%	556,000
Test for the 36% rate:	
Chargeable estate as above (net BPR)	2,040,000
Add back charitable legacy (but not BPR)	150,000
Adjusted Estate	2,190,000
Deduct NRB	(650,000)
'Baseline amount'	1,540,000
10% of baseline amount	154,000

On the figures above, the charitable gift is slightly lower than 10% of the baseline amount, so the IHT charge remains at 40%. If the trustees exercise their discretion more than three months but not later than 24 months after Diana's death to increase the legacy to the church, that will be read back under *IHTA 1984, s 144*. If they act promptly, as soon as the extent of the estate is known, the deed of appointment of extra funds to the church can be lodged with Form IHT400 to secure immediate relief. There is still no grossing up: it is a simple increase from £150,000 to, say, £155,000.

That gift would then exceed 10% of the baseline amount, so the 36% rate can apply. The revised calculation is:

	£
Total estate as above	2,590,000
Less BPR and revised charitable legacy	(555,000)
Estate	2,035,000
Less NRB	(650,000)
Chargeable estate	1,385,000
IHT at 36%	498,600

The chargeable estate is reduced. The family have lost £5,000 to the church but this is more than compensated for by the IHT saving of £57,400 (£556,000 – £498,600).

STEP has produced a 'model' form of legacy that is designed to give as much of the estate to charity as will secure the 36% rate. That form is noted and accepted by HMRC as being effective. However, the model form is not really appropriate where BPR or APR is in point because the rules as to attribution discussed in this chapter may produce unexpected results. In such circumstance the use of a discretionary will trust is recommended: just what Diana and her executors found useful in this example.

Will planning for an elderly couple

15.9 An extended example.

Example 15.5: Jeff and Jill – BPR maximised

Jeff and his wife, Jill, both aged 73, made their money in farming and, a few years ago, secured planning permission on part of the farmland for a golf course. They sold the remaining farm but retained the farmhouse in which they continue to live and just under two acres enjoyed with it. Jill's share of the proceeds was invested in a share portfolio currently worth £1.6 million of which £1.2 million is an unquoted shareholding in a regional brewery. Jeff, a golfer himself, invested his share of the proceeds in Driver Ltd, an unquoted company which now operates the golf course built on the former farmland. That investment is worth £1.4 million. Jeff has other investments of £300,000.

Jeff and Jill have two adult children and five grandchildren and wish to make tax-efficient provision for the whole family. Driver Ltd pays Jeff fees as a non-executive director which he regards as part of his pension, and dividends of £20,000 per year. Their present wills provide for the entire estate to pass to the surviving spouse, with the usual gifts over to the children and grandchildren. Cerys, a chartered tax advisor, is a member of the club and they consult her.

Cerys's pro forma IHT computation of the prospective liability to IHT is based on Jill's comment that, with all the golf he plays, Jeff is reckoned to be the better life of the two. Neither has made any lifetime gifts of any substance, although Jeff was wondering if he should give away some of his Driver Ltd shares because the club is doing well and the company might soon develop land that is not needed. The computation is as follows.

Jill (assumed to die first):

	£
Portfolio	1,600,000
Share of house	1,200,000
Contents (share)	25,000
Personal effects	10,000
	2,835,000

Under the present will, all is exempt, passing to Jeff.

Jeff (assumed to survive Jill):

	£
Driver Ltd	1,400,000
Portfolio	300,000
Share of house and contents	1,225,000
	2,925,000
Inheritance from Jill	2,835,000
Less probate costs on Jill's estate	(15,000)
	5,745,000
Less BPR on Driver Ltd	(1,400,000)
Less BPR on brewery shares	(1,200,000)
	3,145,000
Less nil rate band/transferable nil rate band	(650,000)
	2,495,000
Tax at 40%	998,000

Note: There is no RNRB/TRNRB as the estate threshold is breached.

Cerys advises Jill that no APR is available on the farmhouse, now shorn of its farmland. Cerys advises that it is no longer necessary to use the nil rate band on the first death, but that Jill should consider leaving her brewery shares to her children on the basis that that gift will be sheltered by BPR and that, after her death, Jeff may buy the shares back from the children (who probably need cash more than a long-term investment) funded from the sale of his share portfolio. If Jeff survives that purchase by two years and still holds the shares at his death, they will qualify for BPR a second time. There is little advantage in causing an immediate charge to IHT by making legacies to the children or grandchildren (of assets not qualifying for relief) any greater than the nil rate band, unless it is considered that the asset will appreciate very quickly and it is desired that that appreciation be taken in the names of the younger generation.

Cerys advises that, in the light of *IRC v Lloyds Private Banking* [1998] STC 559, there is little to be achieved by attempting to pass a half share in the house to the next generation for so long, at least, as Jeff as the survivor will need to live there.

Jeff should not make an *inter vivos* gift of the Driver Ltd shares, since: (*a*) BPR will almost certainly be available on his death (if the law remains unchanged); and (*b*) the shares yield useful income, leaving Jeff free, if he wishes, to make gifts from other funds. Cerys considers that the land on which Driver Ltd might get planning permission is, for the time being, still used by the club for parking and for a practice range; so, although the club might sell it, it is not an 'excepted asset' within *IHTA 1984, s 112*, and so BPR is not at risk. To allow for the possibility that he might die first, Jeff should in his will leave the Driver Ltd shares to his children on the basis that Jill will have enough, with her own securities and what she inherits from him, to 'recycle' BPR in the same way, by buying his Driver Ltd shares back so as to give the children cash and hold the BPR assets until the second death.

On this basis of Jill's instructions, the revised IHT computation is:

	£
Jill's estate (as before)	2,835,000
Distributed to children	(1,200,000)
Residue (to Jeff)	1,635,000
No IHT payable	

Jeff's estate (assuming he survives Jill by two years):

	£
Portfolio of 300,000 (less cost of brewery shares and, say, CGT of 10,000)	90,000
Driver Ltd	1,400,000
Brewery shares	1,200,000
Share of house and contents	225,000
Inheritance from Jill (net of probate costs of 15,000)	635,000
	3,550,000
Less BPR on Driver Ltd and brewery shares	(2,600,000)
Less nil rate band/transferable nil rate band	(650,000)
	300,000
Tax at 40%:	120,000

Cerys shows Jeff that, although he will have incurred some CGT on the sale of the share portfolio used to raise the cash to buy the brewery shares, he may achieve a useful saving of IHT with these plans.

Jeff tells Cerys that he is happy with her advice, except for two points. He had understood that a discretionary trust may be used so as to avoid the aggregation of the estates of husband and wife whilst still allowing the surviving spouse access to all the income in the discretionary trust fund. That seems to him to give him the best of both worlds, in that Jill need not actually leave anything to the children at this stage. Could he not sever the joint tenancy in the house, so that Jill's share went into a trust? The other point is that he has met a (very persuasive) financial adviser in the bar who recommends the use of borrowing to fund lifestyle and lifetime gifts. If the Driver Ltd shares are doing well, might they be good security for such a scheme?

Cerys urged caution. How did Jeff imagine that such a discretionary trust might work? Would he want to be one of the trustees? Would he want to regard the income of the trust fund as being his, in all but name? What if, for example, following Jill's death, he were to sell the house as being too large for him on his own: would he expect the trustees to buy a smaller house and, in the exercise of their discretion, allow him to live in it? Did he think that, if he lived in a house belonging to a discretionary trust, the proceeds of sale might later be realised free of CGT?

Jeff nodded approval to all these propositions until Cerys pointed out that, specifically, there would be a serious risk that Jeff would be regarded as the tenant for life of a trust fund if he did indeed have significant control over the destination of the income, and particularly if the income were, for reasons of convenience, to be mandated to his bank account. If Jeff really felt that he could not live in retirement without the benefit of the income from Jill's securities, then he must accept that, on his death, IHT would be chargeable on the fund that had supported his lifestyle.

Further, Cerys pointed out the effect of SP10/79, by which the exercise by the discretionary trustees of a power to allow a beneficiary to occupy a dwelling house might well cause that house to be regarded as subject to an interest in possession in favour of the occupier.

If Jeff wanted the proceeds of resale of such a property to be tax free, as would be the case if the trustees could bring themselves within *TCGA 1992, s 225* (principle private residence relief), then it would follow that, on Jeff 's death, there would be a charge to IHT on the fund; whereas, if Jeff could show that he was not the only occupier of the property, had no actual right to live there, and that the trustees from time to time went so far as to take away from him the benefit of occupation and in their discretion gave it to someone else, such that he was not regarded as having a life interest, then on resale *s 225* would not apply and a CGT charge might be levied.

Cerys advised Jeff that discretionary trust arrangements which are intended to give a surviving spouse benefits from a fund, without aggregation of that fund with the free estate of the surviving spouse, must be transparent in their operation. The trustees must have full discretion to make provision for beneficiaries other than the surviving spouse, and for such an arrangement to stand any chance of success the trustees must, as a matter of course, no matter how tedious and expensive the operation, meet from time to time for the purpose of exercising their discretion. It greatly helps in such a situation if the surviving spouse is not one of the trustees. If Jeff lived in a 'trust house', payment of a notional rent to the trustees for the use of any part of it that he did not own would tend to show that there was no life interest in that part of the property but such payment is not actually needed to secure relief under *s 225*.

Cerys was not keen on the idea of borrowing, and definitely not on the security of the shares, since that would cut down BPR which was so valuable here. With the inescapable, if unpalatable, logic of an accountant, she said that, if the family home was too big, they should 'downsize' and release equity that way. Jeff had had enough of a lecture by this time. He graciously observed that clubroom chat made no mention of SP10/79 or of BPR, and accepted Cerys's advice.

DRAFTING POINTS – THE USE OF A CAP

15.10 Estate planning for many married couples and civil partnerships was simplified following changes introduced by *FA 2008* and contained in *IHTA 1984, ss 8A–8C*. The transferable nil rate band provisions broadly allow a claim for all or part of an unused nil rate band on the death of a spouse or civil partner to be transferred to a surviving spouse or civil partner (who dies on or after 9 October 2007) – the rules apply in the same way whether the survivor leaves a will, or dies intestate. The facility to make claims for the transfer of unused nil rate band remains very useful despite the nil rate band initially remaining frozen at its 2009/10 level of £325,000 for the tax years 2010/11 to 2020/21 inclusive. Transferable nil rate band claims cannot be made by cohabiting couples who are not legally married or civil partners (see *Executor of Holland deceased v IRC* [2003] SpC 350), or by family members who occupy the same property, such as siblings (see *Burden and another v United Kingdom* [2008] All ER (D) 391 (Apr)).

Nevertheless, the nil rate band is still useful for 'extended' families, where each spouse has children from an earlier marriage and they wish to find a balance between providing for each other and satisfying the reasonable capital expectations of their own children. The simple solution is to take a will precedent and limit the definition of 'the nil rate sum' to an amount which both husband and wife agree is appropriate in relation to the overall assets of the family. There is clearly a risk that the adviser will be found to be negligent if he fails to point out the (relatively) unlimited nature of a simple gift of a nil rate band legacy. Particular care should be taken where the estate includes assets qualifying for APR or BPR, since a 'slack' draft could result in a gift of all such property within the nil rate sum.

A newly introduced residence nil rate band (RNRB) is available on death from 6 April 2017 if an interest in the home passes to a lineal descendant (*IHTA 1984, ss 8D–8M*) and as with the general nil rate the RNRB may generally be transferred to a surviving spouse to the extent that it remains unused on the first death, subject to a successful claim being made. However, the RNRB (and the TRNRB) is subject to tapered withdrawal for estates valued at more than £2 million on death and importantly that estate value whilst calculated net of liabilities is before application of reliefs.

DRAFTING POINTS: DEFINING THE LEGACY

15.11 As explained at **15.4**, the ambiguity of the will had a detrimental impact in *RSPCA v Sharp* [2010] EWCA Civ 1474. Clear will drafting is critical in order to achieved the intended result and the gift of the nil rate band should be specified as of a sum after taking account of other gifts, in lifetime or within the will itself and where desired excluding property qualifying for APR or BPR.

ATTRIBUTION RULES

15.12 *IHTA 1984, s 39A (*introduced by *FA 1986, s 106)* has a particularly harsh effect and operates to determine the way in which value should be attributed to specific and to residuary gifts with respect to transfers of value made after 17 March 1986. It was considered that an anomaly arose where an individual made a PET, some or all of the value transferred by which related to property qualifying for APR or BPR and thus *s 39A* sought to clarify the situation.

Example 15.6: Julienne (1) – BPR position pre FA 1986

Julienne, who had made no previous chargeable transfers, died on 28 February 1986 leaving an estate which included:

	£
Residence	100,000
Personal effects	20,000
Shares qualifying for 50% BPR*	600,000
* *The rate then in force*	720,000

Julienne's will gave to her husband Bernard the house, the contents and half the shares. The residue was given to their son Alphonse.

It could be argued that the estate was partly exempt because some of the property passed to Bernard (and was therefore exempt) while the rest passed to Alphonse (this transfer was potentially chargeable). Special provisions in *IHTA 1984, Part II, Ch III* (ie *ss 36–42*) applied whereby BPR reduced the value transferred by the transfer of value in question. There was nothing in the legislation requiring a specific gift made out of property qualifying for, in this case, BPR to be taken into account in fixing the value of the gift. The application of the rules could work either for or against the taxpayer. In this example, the total value transferred by Julienne after BPR was:

Property without reliefs	£	£
Residence	100,000	
Chattels	20,000	
		120,000
Shares	600,000	
Less BPR	(300,000)	
Balance		300,000
Adjusted Estate		420,000

It could be argued that the value of specific gifts by Julienne to Bernard was £420,000, ie £120,000 non-business property and £300,000 drawn from the value of the shareholding. It could also be argued that all of the value being transferred was attributable to Bernard and therefore exempt. Alphonse therefore received, free of (in those days) capital transfer tax (CTT), property worth £300,000.

Had Julienne made a specific gift to Alphonse of the non-relievable non-business assets and given the remainder (including all the qualifying shares) to Bernard, the rules would have worked the other way. The business relief would have been wasted, because the transfer to Bernard was exempt anyway.

THE EFFECT OF S 39A

15.13 The relevance and therefore impact of *s 39A* is often overlooked in will planning and this can lead to unintended results. *Section 39A(2)* requires the value of any specific gifts of relevant business property or of agricultural property to be taken to be their value *after* relief (the legislation is written to ensure the relief reduces the value transferred by a transfer, rather than the value of the property).

Section 39A(3) requires the value of any specific gifts which do not qualify for APR or BPR to be 'the appropriate fraction' of their value. 'The appropriate fraction' is defined in *s 39A(4)* as being one where:

- the numerator is the difference between the value transferred and the net value of any gifts qualifying for relief as reduced; and
- the denominator is the difference between the unreduced value transferred and the gross value of gifts without the value of relief.

The effect of *s 39A(3)* is to attribute part of the APR or BPR to an exempt transfer thus leading to a loss of relief.

Example 15.7: Julienne (2) – BPR position post FA 1986

Value of specific gifts to Bernard:	£
Shares	300,000
House and contents: £120,000 multiplied by:	
$\frac{(420{,}000 - 150{,}000)}{(720{,}000 - 430{,}000)}$ ie	
$\frac{270{,}000}{290{,}000} \times 120{,}000$	111,724
Total	411,724

As a result, the value transferred at Julienne's death of £720,000 is attributed as follows:

Bernard (as above)	£411,724
Alphonse	£308,276

The gift to Bernard is exempt but the gift to Alphonse is chargeable in so far as it exceeds the nil rate band.

There is a trap for the will draftsman in these provisions.

Example 15.8: Jessica – gift of NRB legacy

Jessica's farm is worth £1.2 million and qualifies for APR at 100%. She has made no chargeable lifetime gifts and has life policies amounting to £320,000 which, by virtue of bad management, will fall into her estate. Her present will gives the residue to her husband Jim, subject to a legacy of the nil rate band to her friend Joss. However, the will has been drafted in such a way that the legacy consists of the value of the farm after relief, such that Joss inherits the entire farm (having a nil chargeable value after APR) and all the policy proceeds. Jim gets nothing.

Consider instead that, according to the way the will is drafted, the farm assets are not specifically allocated to the legacy. The legacy is treated as coming partly from the value of the farm and partly from the policy proceeds and, to that extent, BPR may be wasted. The sum is:

	£
The legacy	325,000
Divided by the estate of	1,520,000
Multiplied by the legacy of	325,000
To produce a figure of	69,490

Joss's legacy of £325,000 is treated for IHT as representing £69,490, from the value of the farm and, as to the balance, from the policy proceeds.

Avoiding the difficulties of s 39A

15.14 As far as possible the complexities of *s 39A* and indeed double grossing should be recognised and avoided by making clear identification of the intended gifts to be passed to exempt beneficiaries. If a chargeable specific gift is made of an asset that qualifies for relief, that fixes the availability of the relief.

Example 15.9: Kenneth – BPR legacy

Kenneth, who has made no chargeable transfers nor PETs in the last seven years, owns the following assets:

	£
Residence	800,000
Savings: quoted shares	400,000
Business property qualifying for 50% relief	200,000
	1,400,000

His current will leaves the house and the business assets to his widow Leonie and the quoted shares to his daughter Marie. That is inefficient, triggering a charge to IHT on death of £30,000 ((£400,000 – £325,000) @ 40%). His will should be redrafted (or a deed of variation entered into following his death) to give Marie the business assets and (possibly) £200,000, with residue to the widow, Leonie. That way, for IHT purposes, £400,000 will still be attributed to the specific gift to Marie but, after deducting BPR of £100,000 (£200,000 @ 50%), all the value will be within the nil rate band. Later, mother and daughter might think about a second transaction, to re-use BPR on the business assets in the way already described earlier in this chapter.

As an alternative, the will might have been drawn, or might after death be varied, so as to give Marie sufficient quoted shares equal to the nil rate band of £325,000. There will be no charge on Kenneth's death, though there will be no further nil rate band, at least from Kenneth, to transfer on the widow's death. Leonie, taking the residue, might consider a simple gift to Marie of the business property. There would be no charge on Kenneth's death and if done promptly such that there is little or no movement in value, there might be no gain on the shares, so no CGT either. If Leonie made a lifetime gift of the business property but failed to survive that gift by seven years then, subject to clawback, BPR would be available by virtue of the succession provisions of *IHTA 1984, s 108*. Clearly, care would be required to avoid, in the case of a variation, any understanding that Leonie would make the gift to avoid the suggestion within *s 142(3)* that there had been any consideration for the making of the variation.

Sometimes, a more conscious effort is needed to avoid *s 39A* in order to give appropriate advice when taking will instructions.

Example 15.10: Maud – dilution of BPR

Maud lived in modest circumstances with her friend Winifred. The house belonged to Maud, who also held stocks and shares worth £600,000, including an unquoted shareholding worth £150,000 in the family trading company.

In consulting her solicitor, Jack, about a new will, Maud made it clear that, in the event of her death, she wanted Winifred to be able to continue living in the house for as long as she might want. In fact, Winifred was not in terribly good health and it seemed quite likely that, in the event of Maud's death, Winifred would move into sheltered housing. To tide Winifred over, Maud wanted to ensure that there was a suitable sum of money available tax free. Subject to that, Maud wished her estate to pass, as far as possible, in equal shares to three charities and to a niece and nephew.

Jack was very concerned to avoid the '*Re Benham*' problem (see **15.7**), not being aware of the later decision in *Re Ratcliffe* which might have afforded him some reassurance. By concentrating exclusively on avoiding wording which might fall precisely within the difficulties of *Re Benham*, Jack overlooked one or two other points.

On Maud's death, it appeared that the house was rather more valuable than Jack had appreciated, having the benefit of an area of land to the rear with planning potential. For a time, it seemed that Winifred would be happy to continue to live there and, in his capacity as executor, Jack saw no reason to encourage her to move on. In accordance with the power given to him by the will, Jack earmarked £30,000 (the limit allowed by the will) as the 'cushion' to tide Winifred over.

Jack was both surprised and concerned to receive a letter from HMRC concerning the operation not only of *s 39A* but also of SP10/79 (see Example 15.5), of whose existence he had previously been blissfully unaware. The HMRC argument was that the power in the will to allow Winifred to continue to occupy the dwelling house amounted to an immediate post death interest even though she had no formal interest in possession, with the result that, in calculating IHT, the value of the house (and the land to the rear which Winifred was also, in effect, occupying) fell to be regarded as a transfer to a non-exempt beneficiary.

Further, the clause in the will allowing up to £30,000 to be made available for Winifred was a gift free of tax. Since there was no chargeable specific gift of an asset qualifying for relief, the gift of the £30,000 must be treated as 'spread' across the property qualifying for BPR, the shareholding in the family company, and the remainder of the estate. Clearly, as a result, some of the available BPR was thus wasted. It also became necessary, because of the distribution of residue between exempt and non-exempt beneficiaries, to carry out a full exercise of the double grossing calculation after applying the attribution and allocation rules.

A deed of variation was needed to put things right. Fortunately, the nephew and niece were of full age and Winifred decided to move into sheltered housing after all. It therefore became possible to vary the offending provisions of the will, in particular the quasi-life interest, such that the remaining gift of £30,000 fell entirely within the nil rate band. Grossing up of that gift could be avoided.

These principles should be borne in mind when considering the form of will which creates a nil rate band discretionary will trust of which the beneficiaries will often be the surviving spouse, children and grandchildren. Unless there is a *specific* gift of property that qualifies for APR or BPR, the relief in respect of that property will be spread across the whole of the estate including the residue, which may perhaps be exempt as passing to the surviving spouse. It is not enough for the personal representatives merely to appropriate certain assets to the discretionary trust as that will not satisfy the requirement of *s 39A* for a specific gift unless there is, in the circumstances, no choice as to the assets to be appropriated. An alternative, and more flexible, will could create a discretionary trust not merely of the nil rate band but over residue. The trustees could then make the necessary appointments (at least three months after the death but less than two years thereafter) which would constitute specific gifts for the purposes of *s 39A*.

ROUTING THROUGH NON-EXEMPT BENEFICIARIES

15.15 Frequently, will planning is geared towards skipping a generation in order, perhaps through the medium of a discretionary trust, to pass property down to grandchildren. There are advantages to the grandchildren in that, whilst income (generally in excess of £1,000) is taxed in the hands of the trustees at 38.1/45% when distributed to the beneficiaries who have no other income, that tax may be recovered by them. On the other hand, there is one circumstance in which exactly the opposite tactic is helpful: this concerns woodland.

Example 15.11: Woodland inheritance

Consider the testator owns woodlands which, owing to the age and nature of the trees, are now very valuable. There may be a relative who, although elderly, might nevertheless survive the testator. If the woodlands pass under the will to the elderly relative, there is perhaps little chance of a disposal during the lifetime of that relative. An election might then be made to leave the value of the timber out of account when fixing the tax on the death of the original testator, on the basis that the tax would be chargeable on any later disposal but, since the elderly relative would (it is hoped) not dispose of the timber, that relative's death would frank the IHT charge on the first death.

In an ideal world tax-planning would be made on the basis of a combination of deaths 'at the right time and in the right order'. Explaining the scheme and putting in place safeguards to ensure that the elderly relative does not sell the timber may be quite difficult.

In the typical situation where the testator has 'cashed in' the value of his woodlands and has only recently replanted them the value of timber at the time of the testator's death may be quite low because the trees have not yet reached maturity. In these circumstances, the tax charge may be small and it will not be worth making an election for woodlands relief since there is a risk that, when the deferred tax comes to be payable the woodlands may have increased considerably in value. Similarly, it will not be worth making the election for woodlands relief where the transfer of the woodlands is exempt by for example being left to a spouse or to charity. The practitioner must look at the position in the round and compare the tax deferral on the death of the will beneficiary with the cash-flow advantages.

GIFTS OF BUSINESSES

15.16 Often, a simple tax-planning scheme will perhaps involve the gift of non APR/BPR qualifying assets to the surviving spouse and assets that do qualify for such relief to other beneficiaries, perhaps in the next generation. A difficulty can arise where there is a specific legacy of an interest in a business, particularly an underwriting business at Lloyd's and it was noted at **5.22** that there is some question as to whether there can be succession for the purposes of BPR to the business of a (Lloyd's) Name, in that underwriting is in some respects personal to the Name. It is, however, clear that executors may continue the underwriting business so, whilst the rules as to succession in *IHTA 1984, ss 108* and *109* may not apply, the business may be the subject matter of a gift.

Consider the legacy gift to be of 'all my Lloyd's business' – there can be a distinction between what is defined as the business itself and the value that will attract relief, ie between the value on the 'commercial' basis and on the 'relief' basis as discussed in **Chapter 5**. However, this is not only a problem specific to Lloyd's Names consider that there is the same difficulty in relation to the gift of a farmhouse which enjoys premium valuation by virtue of its location. Nevertheless, the will draftsman must take care to define with precision what the legatee will receive and this aspect will be particularly important where there are tensions between the beneficiaries as is so often the case in a death scenario. It may be difficult to decide what assets pass under a gift of 'such of my property as shall be eligible for business property relief' or of 'such interest in my [farmhouse] as shall be eligible for agricultural property relief' and the executors will have to decide with some precision what value attracts relief. Where there is a gift of a business an early transfer to the beneficiary will be essential to ensure that the business can be run efficiently but the executor may need to take from the beneficiary some undertaking or other security for the return of value if there has been, with hindsight, an error in the calculation of relief or the basis on which it should be claimed – see *Hardcastle (executors of Vernede deceased) v IRC* [2000] [STC] 532 (SCD) in **Chapter 5** and HMRC's review of that decision which they consider not to be of general application.

One concern with the decision in *Hardcastle* might be that the ordinary liabilities incurred in the running of the business would not go to reduce the value of the business for BPR purposes, and that itself might be a general principle. It would seem, although the matter is yet to be clarified, that it is these wider implications of *Hardcastle* that HMRC do not accept.

THE MEANING OF 'SPOUSE'

15.17 The law does not always adapt to changing social circumstances and this is evidenced in the challenge mounted in *Holland (Executor of Holland Deceased) v IRC* SpC 350, [2003] STI Issue 2. The appellant had not been married to the deceased, but had co-habitated with him for 31 years until his death on 17 April 2000. She claimed that the property that passed to her on his death should be exempt from IHT under *s 18*. She argued that the word 'spouse' in that section was not restricted to those who were legally married but included those who lived together as husband and wife. In the alternative, she argued that her rights to respect for private and family life and to peaceful enjoyment of her possessions were infringed by discrimination based on status. That claim was under the *Human Rights Act 1998, s 3*, which required *IHTA 1984, s 18* to be read in a way which would remove that discrimination.

The Special Commissioner disagreed. The meaning of a word must accord with the understanding of the ordinary man using the word in its popular sense at the relevant time. Whilst the meaning of the word 'family' had changed over time, the meaning of the word 'spouse' had not. There was authority for that both in *Dyson Holdings Limited v Fox* [1976] QB 503 and *Fitzpatrick v Sterling Housing Association Limited* [2001] 1 AC 27 (HL).

Chapter 16

APR and BPR: the new regime

QUICK GUIDE

- The IHT regime for business and agricultural property has remained relatively stable for several years now but the critical question for the practitioner advising on IHT planning is whether the present rates can safely be relied upon for the next few years.
- If they can, **16.6** shows why it may be prudent to retain assets in the hands of the older generation.
- If the present rates are not to be relied upon, **16.4** and **16.5** put the arguments for taking action now.

WHAT NEW REGIME?

16.1 Apart from the 2017/18 introduction of the residence nil rate band (RNRB/TRNRB), the majority of changes to IHT have been targeted anti-avoidance measures rather than full reappraisals of the tax structure. Any restriction of APR or BPR to achieve redistribution of wealth could have not only a detrimental IHT impact but also an economic impact on continuing investment within the UK: those who have great fortunes tend not to invest them in bank current accounts, which post-recession and in the current BREXIT era have delivered fairly poor returns, but rather in land, especially farmland, and in businesses. It has also been seen that while the former 50% income tax rate (withdrawn 6 April 2013) for high incomes and trusts was very popular with the majority of taxpayers that tax rate did not achieve the increased tax revenues expected but in fact quite the reverse.

Excluded property trusts

16.2 The excluded property trust which falls within *IHTA 1984, s 6* has long since been the focus of HMRC attention not least since such settlements where there is a settlor interest circumvent the application of the gift with reservation provisions. Another issue in this area, which was primarily concerned with income tax but which had a knock-on effect for capital tax planning, is the extent to which a shadow director may be subject to a charge under *ITEPA 2003, Ch 5* in respect of living accommodation. That was one of the issues raised in *R v Dimsey* and *R v Allen* [1999] STC 846 before the Court of Appeal which in turn led to some tightening up of CGT in relation to offshore trusts.

However, measures introduced in *F (No2) A 2017* have considerably changed the playing field in the offshore arena by carrying through draconian changes first announced in 2015. These comprise:

- Reducing the qualifying period for deemed domicile to 15 out of 20 years UK tax residence.
- Revival of UK domicile of origin for 'returning non domicilliaries'.
- Extending the deemed domicile status to both income tax and CGT.
- Bringing UK residential property held in any form of foreign ownership within the scope of IHT.

The effect and detail of these changes are outside the scope of this book and reference should be made to *Inheritance Tax 2017/18* (Bloomsbury Press) and *Trust and Estates 2017/18* (Bloomsbury Press).

Reform

16.3 Any UK government will have a direct focus on encouraging, nurturing and developing industry which promises to deliver a financial benefit to the UK economy. In the past this has led to a fiscally friendly approach to industry investment with reliefs for risk-taking by investing in business or acting as a 'business angel' made more generous. However, it must be acknowledged that for some years now the farming industry has been undergoing very serious financial difficulties through a combination of European competiveness, trade restrictions, currency fluctuations, political policy and BREXIT. Could these factors prompt a major reform or realignment of APR and BPR? There are a number of obvious targets for reform:

- *The first* might be a return to some form of 'working farmer' relief. In the view of the popular voter, there may well be a distinction between the reliefs which should be accorded to a family who draw all of their income from farming, and those who do not, eg a prosperous accountant who wisely invests part of his capital in a farm which he operates through a contracting agreement.
- *A second* might be making access to APR conditional upon the agreement of the landowner to provide much greater rights of access for the public than at present. In other words, APR would have to be 'earned' by

positively encouraging access, including suitable provision for parking, provision of stiles, signposts and the like.

- *A third* might be the withdrawal of APR for the pure land investor.
- *A fourth* might be the withdrawal of BPR for the passive investor in unquoted trading companies.
- *A fifth* might be a return to a more tiered rate of percentage relief.
- *A sixth* could be relief linked to environmental/green energy compliance.

USE OF 100% RELIEFS WHILE THEY LAST

16.4 It is a fact that the present regime is complicated but benign and there are occasions when it is more tax effective to retain assets qualifying for the relief rather than to transfer them. However, as a general approach where the assets are substantial and donees can be trusted to look after them, there is much merit in encouraging large transfers of property subject to 100% relief. Where a gift in which there is no reserved benefit is made early enough, such that the transferor survives by seven years, all the complications of clawback can be avoided.

Interesting situations arise where planning opportunities are currently available which could easily be lost by changes to the rates of relief.

Example 16.1: Cedric – control shareholding

Cedric established a successful business and a few years ago, his company listed on the London Stock Exchange. The capital is now £15 million, divided into 7.5 million ordinary shares (which are unquoted) and an equal number of 'A' ordinary, which are quoted – both share classes carry equal votes. Cedric holds all the unquoted ordinary shares (7.5 million) and 2 million of the 7.5 million quoted shares.

Both holdings qualify for BPR at 100%, the unquoted shares falling within *IHTA 1984, s 105(1)(bb)* and the quoted shares falling within *s 105(1)(cc)* – the latter is in point since by virtue of Cedric's holding of unquoted shares, the quoted shares give him control and thus attract BPR at 50%.

Cedric wants to establish a charitable trust funded by a gift of some of the unquoted shares. For IHT purposes that would be an exempt transfer and, for CGT purposes would not give rise to a chargeable gain by virtue of *TCGA 1992, s 257*. However, dependent on the number of unquoted shares gifted, this could cause Cedric's retained shareholding to fall below a control holding, resulting in the loss of the 50% BPR on the quoted shares.

Subject to company and commercial considerations, such as the difference in value between the ordinary and the 'A' ordinary shares, a gift of the quoted shares to the charity would leave in Cedric's hands assets all of which still qualify for BPR at 100%.

LOCKING INTO THE 100% RATE EVEN ON FAILED PETS

16.5 A feature of the legislation which may help the tax planner in respect of deaths and other chargeable events occurring after 9 March 1992 when APR became available for certain categories of property at 100% (*F (No 2) A 1992, s 73* and *Sch 14*) – see **Chapters 3** and **6**. The rates apply both to a PET which fails and to the top-up charge on death in respect of a lifetime chargeable transfer (*Sch 14, para 9(1)*). For the purposes of *IHTA 1984, ss 113A* and *113B*, relating to the clawback provisions and the savings provisions for clawback in respect of replacement property, it is to be assumed that:

- the post *F (No 2) A 1992* rates of relief came into effect at the time that the transfer was made; and
- in relation to the charge that would otherwise have applied to top up the tax from the lifetime rate to the death rate, the portion of the value transferred which would have qualified for relief at the new rates does so qualify.

Where shares qualified for BPR within *s 105(1)(b)*, it was possible to disregard the amendments in deciding whether or not the shares were affected by *s 113A(3A)*, ie shares which were quoted at the time of the chargeable transfer or were unquoted shares or securities within *s 105(1)(b)* or *(bb)* and stayed unquoted throughout the period from the date of the chargeable transfer to death of the transferor.

A similar provision in *FA 1996* at *s 184(6)* increased the relief in relation to transfers of value made after 5 April 1996 whilst *s 184(6)(b)(ii)* made it clear that the new rates also had effect for the purposes of any charge to tax by reason of any event occurring after 5 April 1996 in relation to transfers of value made before that date. The effect of the rule on a PET which fails is therefore that 100% APR or BPR is available, provided the conditions are observed which will save the PET from clawback and regardless of changes that might take place to reduce that relief.

Whilst it may therefore be argued that, on the present law, it is possible to 'lock into' the 100% rate, it may be over-optimistic to suggest that such an action would render the transaction immune from future legislative change. It would no doubt be argued, in representations after the draft legislation was published, that such changes were retrospective and threatened transactions which had been entered into in reliance on the earlier law. A government with a strong majority that is committed to simplifying the law might disregard such arguments on the ground that, in relation to PETs which at the date of enactment had not yet failed, there was still the possibility that the transferor might survive the gift by seven years. Both *F (No 2) A 1992* and *FA 1996* introduced provisions as part of a trend not only towards simplification but towards increasing the reliefs: new legislation to withdraw the reliefs could easily adopt a much harsher line.

The moral must be, therefore, that practitioners should use the reliefs for as long as they last, but recognise that the present provisions for 'lock in' are as vulnerable as the reliefs themselves.

The effect of a PET that is relieved at 100% is very similar to that of a CLT and that can be seen from the following comparison Table of the treatment of lifetime gifts:

Event	*PET Treatment*	*CLT Treatment*
1. Immediate charge to IHT?	No: potentially exempt	No: value reduced to nil
2. Death within seven years of transfer: assets retained by donee	No IHT: full APR	No IHT: full APR
3. Death within seven years of transfer where clawback applies	IHT charge (subject to nil rate band)	
(a) as to the transfer itself	On the full amount	On the full amount
(b) as to aggregation with the later transfers, eg on death	Full value is aggregated	Value is nil, so an advantage in making the transfer by CT
4. Transferor lives seven years from gift	No IHT: transfer has become exempt	No IHT: falls out of cumulation

LOSS OF CGT FREE UPLIFT ON DEATH

16.6 The existence of the free CGT uplift on death is incredibly valuable as a feature of tax planning, but it does force the taxpayer to consider whether they should 'volunteer' to pay CGT at the time of a lifetime transfer in order to save a greater IHT charge on death. For example, the owner of freehold investment property does not qualify for APR nor BPR nor for holdover relief under *TCGA 1992, s 165*. Any gift of the property may trigger a charge to CGT save a gift to a non-settlor interested relevant property trust, where holdover is available under *TCGA 1992, s 260*. One potential and therefore speculative legislative change not directly affecting APR and BPR, might be the withdrawal of the right to holdover under *s 260*, in the absence of a positive charge to IHT, ie chargeable value in excess of the nil rate band. A more targeted amendment might provide a mere set-off of one tax against the other, so as to ensure a 'tax take' of up to 40% of the funds transferred. That would be exceptionally harsh but would affect non-business property and, as such, might be doctrinally acceptable to a cash-strapped government committed to redistribution of wealth.

For the time being, the tax practitioner must, in trying to mitigate the burden of IHT, bear in mind that a gift of business assets subject to holdover under *s 165* will still leave the donee liable to pay the tax on the eventual gain realised.

Example 16.2: Sally – failure to advise fully?

Sally, an elderly widow whose late husband had used his nil rate band, was concerned about the burden of IHT on her estate. She owned shares in the family trading company that qualified for BPR at 100% and which stood at a significant gain over acquisition value. She wanted to encourage her daughter Lucy to become more involved as manager of the company, so consulted her solicitor Jack about transferring some of the shares to Lucy but in a tax efficient manner.

Jack advised that BPR should be available and put the share transfer in hand, remembering (at the last moment, just before Sally signed the forms) to advise Sally that there might be CGT implications and that it might be possible to hold over the gain on the shares which was duly done. He also acted for Lucy and warned her about the clawback provisions of *IHTA 1984, s 113A* but Lucy regarded that as less of a risk as she considered the shares to be a long-term investment.

Sally unexpectedly died a few months after the gift. Retested under the clawback rules, the shares, still held by Lucy, continued to meet the BPR conditions such that no IHT was in point. However, the CGT tax-free uplift that would otherwise have been available on Sally's death, had been lost. There was, however, no claim in negligence against Jack: he had mentioned CGT at the time, and Sally seemed to be in good health, and likely to live for some time, when the transfer took place.

THE LAST RESORT: LIFE ASSURANCE

16.7 Life assurance, but at an acceptable cost, can be a considerable help to the tax planner.

There are (see **Chapter 4** at Example 4.3) specific occasions when, for technical reasons, APR may not be available for a time and some form of life assurance in these circumstances is no more than prudent management. Similarly, **Chapter 4**, Example 4.8 (relating to successive transfers) and Examples 4.5 and 4.6 (relating to inherited ownership), show various conditions which must be complied with. Insurance might be appropriate to cover the risk of a death at a time of non-compliance. Exactly the same position may arise in relation to clawback where, through outside forces, a sale takes place of property which had qualified for BPR and, for commercial reasons, it is difficult immediately to reinvest the funds in qualifying property so as to preserve the situation from clawback. The use of insurance was considered in *Swain Mason v Mills & Reeve* [2011] EWHC 410 (Ch), although only in a limited context.

There is one other specific use for life assurance which is relevant to all business entrepreneurs, but perhaps to farmers most of all, and this concerns provision for those members of the family who do not work on the farm or in the business. Farmers will seldom admit to having enjoyed a good year but, when there is a trading surplus, the farmer will far more readily apply the funds in the purchase of land and machinery than in making provision for his non-farming children. If the tax adviser can persuade the farmer to set aside insurance premiums each year for a policy written specifically for the benefit of the non-farming relative, the situation of the farmer late in life, and of his children after his death, may be infinitely easier than would otherwise have been the case. Similarly, where the inheritance structure is such that the farming son inherits the live and dead stock and a share only in the land, leaving him with the obligation of buying out his brothers and sisters at market value, life assurance taken out by a son/daughter may provide them with at least part of the funds that they need to raise in order to carry on farming after their father has died.

Chapter 17

Heritage property: what qualifies, undertakings and access

QUICK GUIDE

- There are six categories of heritage property:
  - pre-eminent chattels that may be of national, scientific, historic or artistic interest;
  - pre-eminent collections that may be of national, scientific, historic or artistic interest;
  - land of outstanding scenic, historic or scientific interest;
  - buildings for the preservation of which steps should be taken by reason of its outstanding historic or architectural interest;
  - amenity land essential for the protection and character of those buildings;
  - any object historically associated with those buildings.
- There are broadly four tests (see **17.4** et seq):
  - history and national life;
  - artistic interest;
  - educational importance; and
  - historic association.
- Public access is fundamental and, in relation to any applications, will be considered before qualitative vetting of the property itself (see **17.16**).
- Publicity is also fundamental to the scheme, through the register of conditionally exempt works of art and the other measures of disclosure described at **17.22** onwards.
- There are three strands to be considered in relation to qualifying heritage property:
  - heritage property relief;

- maintenance funds;
- heritage property in lieu.

- Although this book is principally concerned with IHT, it should be noted that there are parallel reliefs from CGT in respect of heritage property which, for completeness, are considered where relevant to the discussion of IHT.

OVERVIEW

17.1 Unlike BPR and APR (and subject to the 'cultural gifts scheme' mentioned below) the relief for heritage property is not absolute but takes the form of deferral. However, unlike woodland relief which also operates on a deferral basis, the relief for heritage property may be one of deferral for an indefinite time. The regime for heritage property allows the owner (or successive owners) to ring-fence that property from exposure to IHT for so long as certain agreed undertakings in relation to that property are complied with. Failure to adhere to that strict compliance requirement will result in an operational clawback of the exemption and render it exposed to an immediate IHT charge.

Running alongside to operate as a further aid to support heritage property, the transfer of assets to an approved trust fund established for the maintenance of a qualifying heritage building, will also be ring-fenced from exposure to IHT for as long as the funds are applied for the approved purpose.

A final feature of the heritage property regime is that in appropriate circumstances and on negotiated terms which can be favourable to the owner, heritage property may be used in satisfaction of a liability to tax.

In few other areas of tax law is the purpose of the relief so clearly focused – simply put, there is a trade-off between the burden of tax on the owner of heritage property and the twin requirements to preserve the asset and provide access to that property for the general public on the other. That trade-off is therefore reliant upon the owner not only assuming but also carrying out those obligations. As with all taxes, this creates a tension between the desire to save the tax and the desire to minimise public intrusion into personal privacy. When first introduced the regime was fairly benign but this is no longer the case.

This work is mainly concerned with reliefs from IHT but in this and the following chapters there are brief references to the scheme of relief introduced by *FA 2012, s 49* and *Sch 14,* by which taxes other than IHT may be reduced or satisfied by the gift of cultural property to the nation under the 'Cultural Gifts Scheme'.

FINDING THE LAW

17.2 The statutory code is mainly to be found in *IHTA 1984, ss 30–35A* but there is relatively little modern case law and few published technical works such that for most practitioners the subject remains somewhat shrouded in mystery. HMRC have provided substantial guidance in 'Capital Taxation and the National Heritage' (CTNH) which provides an excellent point of reference and it is strongly recommended that this comprehensive (but non-statutory) document should be consulted by any practitioner dealing with or simply interested in this specialised area of law.

HERITAGE PROPERTY: QUALIFICATION

17.3 *IHTA 1984, s 31* allows designation of property by the Treasury but on 25 July 1985, responsibility for conditionally exempt transfers, whether under the *Act* or under earlier legislation, was transferred to HMRC.

Section 31(1) identifies six categories of heritage property:

(*a*) objects which are pre-eminent for national, scientific, historic or artistic interest ('pre-eminent chattels');

(*aa*) collections or groups of objects which taken as a whole are pre-eminent for national, scientific, historic or artistic interest ('pre-eminent collections');

(*b*) land of outstanding scenic or historic or scientific interest ('outstanding land');

(*c*) a building for the preservation of which special steps should be taken because of its outstanding historic or architectural interest ('outstanding buildings');

(*d*) land essential for the protection of the character and amenities of a building as above ('amenity land'); and

(*e*) any object historically associated with a building as above ('associated objects').

By *s 31(1A)*, the quality of heritage property is judged by reference to the circumstances that exist after the death of the transferor. This provision was enacted with reference to transfers of value made after 17 March 1986 by *FA 1986, s 101(3)* and *Sch 19, para 8*.

The quality of heritage property: rules after 30 July 1998

17.4 The scheme for conditional exemption was substantially overhauled by *FA 1998* and claims for objects made on or after 31 July 1998 will only qualify for exemption if they meet these higher standards and are either:

- of pre-eminent national, scientific, historic or artistic interest, taking into account any significant association that the property might have with a particular place; or
- historically associated with a building which is itself of outstanding historical or architectural interest.

For this purpose 'pre-eminence' is interpreted to mean that the property is of 'museum quality', ie that a national or local authority, university or independent museum would regard the property as a pre-eminent addition to its existing collection. It is here that the question of location may be relevant, in that the association of the property or group of items with a particular place may enable property to be pre-eminent by reference only to that place but not perhaps to any other. However, the term 'pre-eminent' is not defined in statute nor is there any decided case but CTNH para 4.2 states:

> 'We aim for a practical, common sense view of what constitutes a pre imminent collection or group, guided as necessary by the acceptance in lieu (AIL) tax panel…'

The condition of the article will be highly relevant but unfortunately, practitioners cannot obtain an official opinion or indication in advance, albeit there is a right of appeal to the Special Commissioner. There is common reference to items in this category as 'chattels', which the bulk of them are, but strictly the term is 'objects' with its wider definition and can include such items as fixtures: eg fireplaces, staircases etc (see CTNH para 4.1).

In the case of land, it must be of national importance and recognised as such; thus land not already classified as being within a National Park, SSSI or part of the Heritage Coast etc will likely fail to qualify.

Buildings, other than those located in Scotland, are likely to be listed Grade II* or higher if they are to qualify. In Scotland such matters are referred by HMRC to Historic Scotland and the categories are different with the highest level of grading set at Category A (see Scottish Historic Environment Policy (SHEP) December 2011). HMRC would appear to accept that Category A equates to 'outstanding' but there is little formal authority for that view.

Amenity land need not meet such a stringent test as it merely supports the main property and will take its qualification from that.

Acceptance in lieu

17.5 HMRC have long had the power, with the agreement of the Secretary of State, to accept property in satisfaction of tax itself. This is considered in detail at **Chapter 23**. The administration of that power, under *IHTA 1984, s 230*, has developed standards for the property so accepted that they are now applied in interpreting *s 31(1)*. Guidance was originally within the domain of the Museums and Galleries Commission (MGC), but in April 2012 responsibility passed to the Arts Council England (ACE).

Practitioners should consider the 'Acceptance in Lieu Guidance Notes' available on the Arts Council England website (http://www.artscouncil.org.uk/sites/default/files/download-file/Acceptance_in_Lieu_guidance.pdf) which, although geared to acceptance in lieu, is also relevant to applications for conditional exemption. As this guidance explains, the Acceptance in Lieu (AIL) Panel will make a recommendation to the Secretary of State. However, this is not a 'one way bet'; rather the panel seek the right price and will, on occasion, recommend a higher price for public acquisition than the owner demanded. This guidance concerns pre-eminent chattels rather than outstanding land.

There are four broad hallmarks of pre-eminence applied by the AIL Panel – the subject matter, which must be in good condition, should be:

- historic; or
- artistic; or
- scientific; or
- of local significance, either individually or collectively, or associated with a building in public ownership.

The AIL Panel will gather advice when considering whether property is 'pre-eminent' and will consult widely where the object has local rather than national significance.

Test 1: History and national life

The object must have an especially close association with 'our' history and national life. For this purpose 'national interest' includes interest within any part of the UK (s *230(5)*) but that does not mean that the work must necessarily be British – gifts from foreign sovereigns or foreign governments may count, as will objects that have been acquired abroad in circumstances closely associated with the history of the UK. Note in this connection that any application should include full provenance, in particular details of ownership between 1933 and 1945. The property might be associated with a particular part of the UK or with the development of one of its institutions or industries. Some property may well be below national importance but may qualify through local connections.

Test 2: Artistic interest

Some property is clearly of such quality that any national museum or gallery would be glad to accept it. The category here includes not only famous works but those items which may not be pre-eminent in a national museum but would pass that test from the standpoint of smaller galleries, such as those maintained by local authorities, universities or independent galleries. The test of quality is, theoretically at least, to be judged separately from any undertaking (see below) as to public access.

The AIL Panel are keen to promote the enjoyment of fine works of art, not just in the major cities but at a more local level and whilst the majority of applications come through the main auction houses, the AIL Report Panel is trying to encourage smaller estates to apply. The 2015/16 Report lists 36 acquisitions at a value of £47.2 million which include: the 'lost portrait' of Bonnie Prince Charlie by Allan Ramsay (at time of publication this is currently on view at the Scottish National Portrait Gallery); the magnificent Reynolds portrait of the fifth Earl of Carlisle from Castle Howard (where it will return under an *in situ* loan arrangement with the Tate); one of Van Dyck's finest female portraits, *Portrait of Olivia, Mrs Endymion Porter*, formerly in the collections of the Dukes of Northumberland at Syon House, and a topographical painting of Oxford by Turner; a sheet of drawings by the Renaissance Old Master Benozzo Gozzoli; one of the largest collections of peep-shows in the world; the first two cultural gifts from a company; and the first gift from a living artist. A number of works by important British artists who have died in recent years were also accepted to include two pivotal sculptures by the sculptor and painter William Turnbull and further works from the estate of Lucian Freud, comprising the artist's sketchbooks, drawings, letters and an unfinished self-portrait.

Test 3: Educational importance

These may be important for the study of, for example, a particular scientific development and this category can also include objects which form part of a series or collection and which need not be in one place. It may be a 'virtual collection', in that its components may be dispersed across the whole of the UK. Here, the test is somewhat similar to the 'twelfth chair' criterion, ie is the group of objects impaired if one particular item is removed from it? Does the lack of a particular object affect the unity of the collection or of the series?

Test 4: Historic association

This will typically include manuscripts, works of art, furniture or other items that are pre-eminent by virtue of the specific contribution that they make to the understanding of an outstanding historic building. Examples are of furniture specially commissioned for a particular house, or paintings commissioned for a house, or a group of paintings which have an association with a particular location. Note, in particular, that, under this head, value is not a determining factor and items of modest worth that reflect the taste of the relevant time, may still qualify for exemption.

Applications should be submitted to HM Revenue & Customs, Inheritance Tax, Heritage Team, Ferrers House, PO Box 38, Castle Meadow Road, Nottingham NG2 1BB and applicants are encouraged to send a copy to Acquisitions, Exports, Loans and Collections Unit, Arts Council England, 21 Bloomsbury Street, London WC1B 3HF who will also provide any further advice.

Land and buildings: England

17.6 There is no statutory definition of 'outstanding scenic or historic or architectural interest' in *s 31(1)(b)*. As a matter of practice the issue will be treated like the procedure for listing under the *Planning (Listed Buildings and Conservation Areas) Act 1990*. The general listing principles applied by the Secretary of State are published by the Department for Culture, Media and Sport and for buildings in England the hierarchy is:

- Grade I: exceptional interest;
- Grade II*: particularly important buildings of more than special interest; and
- Grade II: special interest, warranting every effort to preserve them.

In grading buildings, regard is had to architectural interest and to historic interest; but historic interest implies some illustration in the building to warrant statutory protection; so a mere visit by an important person would not on its own confer interest. Land within Grades I and II* is, prima facie 'outstanding' (see CTNH para 6.1). Several factors influence the decision including:

- age and rarity (all before 1700 if retaining original features; many before 1840; the best since 1840; the very best since 1946; only outstanding buildings, under threat, if less than 30 years old);
- aesthetic merit (a key factor, but technological innovation may count);
- selectivity (if there are many examples, only the best are selected);
- national interest (the best regional buildings; local importance);
- state of repair (not a major issue).

HMRC will consult with English Heritage, the Northern Ireland Environment Agency, Historic Scotland, CADW and the Historic Buildings Council.

Land and buildings: Scotland

17.7 Similar principles apply, though the nomenclature is different, thus:

- Category A: buildings of national or international importance, either architectural or historic, or fine little-altered examples of some particular period style or type;
- Category B: buildings or regional or more than local importance, or major examples of some particular period, style or building type;
- Category C(S): buildings of local importance; lesser examples of any period, style or building type, as originally constructed or altered; and simple, traditional buildings which group well with others.

Land within Grade A is prima facie outstanding (see CTNH para 6.1). Similar factors apply to the assessment in Scotland as in England (see Scottish Historic Environment Policy (SHEP) Annex 2) with careful notes taken of: the interior design; the plan form; technological excellence or innovation, material or design quality; the setting; and regional variations. To achieve close historical association with a person or event, the building 'should reflect the person or event and not merely be a witness to them'.

Associated objects

17.8 This is a difficult category of property which has undergone several reviews to refine and define the precise qualifying criteria. The requirement under *s 31(1)(a)* for chattels to be 'pre-eminent' does not apply under *s 31(1)(e)* to associated objects, which need not be chattels and could at the extremes even include fixtures.

Whatever their form it is always necessary for such objects to have a close association with the building and to make a significant contribution to it, either alone or as a group; thus for example, consider a collection of paintings usually displayed in the National Portrait Gallery but which were returned to Hampton Court on loan to form part of the exhibition 'The Tudors'. They briefly returned to precisely the positions on the walls at Hampton Court that they once occupied, giving us a brief (and, it must be said, cluttered) impression of how they once looked. That kind of connection – if they were returned permanently – would eventually satisfy the linkage required by *s 31(1)(e)*.

It is not enough that an object is contemporary with the building rather it must (see CTNH para 6.5) contribute to or aid appreciation of the building or its history. However, the length of that connection does matter and although long ownership does not guarantee IHT relief, if an item has been associated with the building for less than 50 years it will likely fail the test. An exception to this rule might, as HMRC suggest, be a drawing or picture relating to the development of the building itself, or showing how it looked, perhaps when used as a hospital during wartime. There should be a 'demonstrable historical link' between the object and the building but mere additions to a collection, or new works commissioned for the place, or replacements might not qualify, even though they might enhance an existing collection.

It should be noted that *s 31(1)(c)* requires the 'host' building to be of historic or of architectural interest but not necessarily both. HMRC admit (see CTNH, Sch 15) that objects might qualify which had 'a direct relevance to the appreciation of the history of the building in a wider sense, as a place to live and work, or a place where significant events took place'.

HMRC's approach addressing the issues of 'Historical', 'Contribution' and 'Discontinuous Association' is set out in some detail in CTNH Appendix 15.

Preliminary approach

17.9 For most practitioners, the first port of call for advice is likely to be the HMRC website, which has much relevant material or by direct contact with HMRC on Tel: 03000 580058. If the application is in respect of a chattel, HMRC will refer it to ACE and CTNH Appendix 8 contains a model letter from HMRC requesting information in relation to items that qualify in their own right with adapted model letters for land and associated objects: these will provide the practitioner with a useful checklist for taking instructions.

Public access and publicity

17.10 As will be seen from a consideration of the procedure for application for conditional exemption and the events that can cause that exemption to be lost (see **Chapter 18**), it is fundamental to the heritage property scheme that there be an agreement between the owner of the designated heritage property and the public, whereby in return for the tax 'forgone' the public gain rights of access with that right of access governed by detailed undertakings. *IHTA 1984, s 35A* provides a mechanism for the variation of undertaking as to access already given.

Difficulties have arisen in the past because whilst taxpayers were keen to enjoy exemption from taxation there was less enthusiasm for the disturbance or intrusion caused by the access granted to the general public and as a consequence in some instances that access was curtailed or made more difficult. However, this is now more strictly policed and the threat of loss of the conditional exemption for non-observance of the access agreement has served as good deterrent.

NEW UNDERTAKINGS

Chattels

17.11 *IHTA 1984, s 31(2)* sets out the undertakings to be given in return for conditional exemption for a pre-eminent chattel or pre-eminent collection. The undertakings are as follows:

- the property will be kept permanently in the UK;
- the property will not leave the UK temporarily except for a purpose, and a period, approved by HMRC;
- steps must be taken as agreed between HMRC and the person giving the undertaking for the preservation of the property; and
- steps must also be taken as above to secure reasonable access to the public.

Forms of undertaking are set out in CTNH Appendix 8: conditional exemption (iii) for chattels and conditional exemption (iv) for land and buildings. Appendix 8 also shows how the successful entry might look on HMRC website.

Certain documents may be excluded from an undertaking under a power given to HMRC in *s 31(3)* where those documents contain information which ought to be treated as confidential, either for personal or other reasons.

Land and associated chattels

17.12 Where the heritage property is outstanding land, buildings or an associated chattel, the undertakings required are as follows:

- in the case of outstanding land, for its maintenance and the preservation of its character;
- in the case of any other property, for its maintenance, repair and preservation and, if it qualifies as an associated chattel, an undertaking to keep it associated with the building concerned; and
- in either case, for reasonable access.

In practice, the undertaking will incorporate a detailed plan for the management of the land, negotiated between the taxpayer (or his agent) and the agency representing HMRC. Importantly, access does not mean free right to roam over the entire site rather it will commonly be limited to existing and new permissive rights of way, with appropriate extension to suit the needs of (horse and cycle) riders.

Amenity land

17.13 Where land qualifies as heritage property, not in its own right but because it protects the character and amenities of a heritage building, a further undertaking is required that 'specified steps' will be taken for the maintenance, repair and preservation of the land and for access. Those 'specified steps' will be agreed between HMRC and the person giving the undertaking, and will be set out in the undertaking (see *s 31(4A)*). Separate undertakings may be required where the relevant property is held by different people or people acting in different capacities.

The rules as to amenity land apply to property falling within the definition in *s 31(4C)*. The property affected may be either:

- a building for the protection of whose character and amenities the land is essential in the opinion of the Treasury (*s 31(4C)(a)*); or
- any other area of land which serves that purpose and which lies between the amenity land and the building itself or is, in the opinion of the Treasury, physically closely connected with the relevant land or building (*s 31(4C)(b)*).

This amenity land (also called '*section 31(4A)* land') is the subject of its own particular code of undertakings. Where *s 31(4A)* requires an undertaking for the maintenance, repair, preservation and access of property, that undertaking is required even though some other effective undertaking may exist for its maintenance, repair, preservation and access (see *s 31(4D)*). Any undertaking given pursuant to *s 31(4A)* is given with respect to the relevant land, and it is for the person who seeks exemption of the amenity land by designation to secure that any undertaking necessary under *s 31(4A)* is given. The 'specified steps' in *s 31(4A)* must ensure that access to the amenity land is not confined 'by appointment only' to access.

It is the nature of exemption of amenity land that it rests on exemption of the principal property, which may well be in separate ownership. This gives rise to complicated inter-property and inter-taxpayer arrangements to secure exemption for all parties and HMRC consider the breadth, depth and extent of 'supportive undertakings' in CTNH para 6.7.

Publicity of the terms of undertakings

17.14 The undertaking that accompanies an application for conditional exemption will include provisions to secure reasonable access to the public. It may also include matters such as the publication of the terms of the undertaking itself and any other information relating to the property which, but for *s 31(4FB)*, would be treated as confidential. However, *s 31(4FB)* is itself subject to the 'privacy' qualification of *s 31(3)*.

PUBLIC ACCESS UNDER THE 'NEW' RULES

17.15 The 'by appointment only' system of public access gave considerable control to the heritage property owner and had the result that assets might gain conditional exemption even though only one or two members of the public might actually exercise their right to view. This is not what was intended by the legislation and it is no longer acceptable that public access can be restricted in this way. Accordingly, access 'by appointment' is no longer acceptable for post 1998 undertakings whilst the heritage property owner that was subject to the 'by appointment' condition agreed prior to 1998 must provide some measure of direct open access. Where the heritage property is particularly delicate, for example where it consists of ancient manuscripts at risk of deterioration through human touch or light, it may be possible to negotiate restricted open access combined with access by appointment.

What amounts to reasonable access?

17.16 *IHTA 1984, s 31(4FA)* requires the access to the public that is agreed between the owner and HMRC to be 'reasonable'. What is reasonable will, in the case of any particular asset, be what has been agreed and is set out in the undertaking. The measure of open access that HMRC will seek is set out in CTNH at:

- para 4.5 and Appendix 1A for pre-eminent works of art;
- para 5.6 for land; and
- para 6.13 for buildings, amenity land and associated objects.

The level of access will be determined by:

- the nature of the asset;
- the type of asset; and
- the preservation and maintenance needs of the asset.

Originally, HMRC set out what they considered to be the *minimum* number of days of access per year but it has long since been recognised that this 'once size fits all approach' is not appropriate in all cases and CTNH para 6.13 indicates that flexibility may be built in, stating: 'There is, therefore, no hard and fast rule.' Even so there must be a serious intention to provide good access.

Owners of works of art fall into three broad categories:

- existing exhibitors;
- major private collectors; and
- minor private collectors.

Existing exhibitors

These are claimants whose works of art are already on public display such that access is likely satisfied, and the claim can be referred to ACE for 'pre-eminence' vetting (see CTNH Appendix 1A, para 8).

Major private collectors

These claimants have several items which are not yet displayed. Access proposals should be well formulated by the time the claim is made.

Minor private collectors

Access may be a problem as HMRC will want the items to be displayed permanently or regularly in a public building. The owner must therefore 'line up' a museum, historic house or gallery that is willing to 'house' the item but not every such institution wants items on periodic loan. The alternative is '*in situ*' access – isolation of the site is a relevant factor; thus, for example, less will be demanded in respect of a windswept Scottish island accessible only at low tide than in respect of a mansion ten minutes' drive from a motorway. The requirements that apply to works of art (where access has been negotiated since 31 July 1998) is set out in CTNH Appendix 1A.

HMRC wish to promote open access through European Heritage Open Days. These are organised by the Heritage Open Days National Partnership but occur only rarely and will not by themselves satisfy even the minimum requirement, since HMRC would generally expect a minimum of one month per year.

One possibility is a 'heritage buddy' scheme, to display valuables at the premises of an associate whose house is open to the public where there could be some sharing of overhead costs. Items need not be on permanent display and for example, historically associated objects might be stored in the 'heritage house' in bookcases or in suitable furniture and unless exempted in the higher category, *viz* pre-eminent chattels, such storage may be acceptable. However, problems can arise where the owner has no control over the display building and for that reason default provisions should be included in the undertaking.

The '*HMRC Trust & Estates Newsletter*' dated September 2017, reminded those who have claimed relief for heritage buildings of the requirements for public access and indicated that future agreements will be much stricter both in the terms agreed and the policing thereof.

Access: the chattel, not the owner

17.17 The personal circumstance of the owner is irrelevant rather it is access to the asset that is all important. The key principles, set out in CTNH Appendix 1A and illustrated by pertinent examples, are as follows:

- display may be of part of a set, for example one place setting of silver, crockery etc rather than all 12;
- archives may still benefit from the 'by appointment' rules, see the addendum to CTNH Appendix 1A. They should be catalogued and a copy deposited with the National Archives;
- owners of manuscripts or drawings should be willing to display part, for example one page, for short periods;
- it is good enough to see the spine of a book on a shelf, with more access by appointment; and
- the 'Heineken' rule, namely that 'below stairs' items may qualify and if there is good reason not to allow access to the part of the house where they are kept special (well publicised) tours should give access.

The following example illustrates the current law:

Example 17.1: Roger (1) – Bateman Collector

Roger established two valuable collections: The first was of silver emanating from the renowned Bateman family workshop. The second collection, comprised a number of original records and manuscripts of the Bateman family which provide detailed information on the workshop and go well beyond what collectors could find in specialist works on the subject.

The expense of maintaining the collections prevented Roger from living in a particularly grand style and thus both collections were stored in a fairly modest private house. Roger died in May 2010, and both collections (together with his entire estate) were left to his long-term companion Mary who made a successful application for conditional exemption. Old age and increasing infirmity now make it difficult for Mary to continue allow the public to visit the house all year round to view the collection and she therefore seeks to restrict access to five days each year.

This would not be acceptable in respect of the collection of silver and unless Mary could agree an acceptable minimum period of access, conditional exemption would be refused (see examples (c) and (f) in the Access Notes). However, with regard to the Bateman workshop records, Mary may be able to argue that frequent handling may damage them – she could negotiate restricted access by offering to have the collection properly catalogued and lending them to the local county record office.

Example 17.2: Official example (1)

CTNH Appendix 1A para 13(a), uses the example of a Scottish property on whose walls are hung various portraits of family and other notable Scottish leading figures. The property was previously open to the public for 25 to 30 days per year attracting thousands of visitors per year. The arrangement falls to be renewed and the proposal is to open for only 25 days per year in the summer months with good advance publicity. That is satisfactory, provided that access is possible by appointment at other times and that items not on the visitor route are available to special tours or on request.

Example 17.3: Official example (2)

CTNH Appendix 1A para 13(d) uses an example which, although fictitious, nonetheless has a ring of truth about it:

D lives with a relative and has provisionally agreed with two local authority art galleries that they would display her pictures for one month each year – the National Trust will take one picture on 'permanent loan', and a local auctioneer will display one picture in the foreseeable future. Whilst some of the plan is good enough, the arrangement with the auctioneer needs clarification, and a 'back up' facility needs to be put in place for the picture to go to a public gallery for a month a year or three months every three years.

Example 17.4: Official example (3)

CTNH Appendix 1A para 13(h) contains a detailed but imaginative and flexible approach to the problem:

H has inherited some pre-eminent works of art but lives in a modest home wholly unsuitable for opening to the public. He is willing to arrange loans to museums or galleries of his choosing and even to have open access at home but is unwilling to part with the works of art for a month each year to a public collection.

Only part of the requirements for public access would be satisfied but access at H's home in these circumstances would suit neither the owner nor the public – H would need to consider the security and practical issues in providing access without prior appointment. The public interest too would be better served if access should be provided in public galleries or this would be the preference. HMRC would look for H to agree to undertake a proactive role in seeking out suitable exhibitions with access at his home required to supplement this, both with and without appointment. Once agreed HMRC would need to monitor the viability of the arrangements put in place on a year-by-year basis and if so satisfied, would continue the arrangement.

Means of access

17.18 Heritage chattels will seldom benefit from being constantly moved from one location to another and the owner must make a decision as to where the chattel will be kept and displayed. HMRC have suggested six broad options:

- the owner's residence;
- the place where the owner normally keeps the chattel;
- a museum or gallery open to the public;
- a public building such as the local Record Office;
- Heritage Open Days organised by the Heritage Open Days National Partnership; or
- a regional or touring exhibition.

These options are far from exclusive, and HMRC are open to alternative but realistic suggestions by the owner. Where an object is provided on loan for display in the public collection of a national institution, local authority or university, it will usually count as open access, provided that arrangements are in place for viewing when the item is not on public display. It is considered, this will also be the case even where the article forms part of the reserve collection of the national institution but beware as the fact that the article is not permanently on display may give some indication that it is not of 'pre-eminent' quality.

Damage to chattels

17.19 Whilst there is a government indemnity scheme intended to provide cover against loss or damage in respect of objects loaned, the heritage property owner should have sufficient personal resources such that, in the event of loss or damage, he can restore the heritage property.

Other access details: charges, limited numbers of visitors and guided tours

17.20 The heritage property owner is not prevented from charging the public for access but that charge must be reasonable in relation to the public. Subject to agreement with HMRC, the owner may limit the number of visitors at any one time and may restrict the open access to guided tours. The owner may (and likely will) forbid photography of the items but the owner may not restrict access to pre-eminent chattels by appointment on grounds that the owner can neither find a suitable venue for open access nor afford to provide one. There must be a measure of open access as a condition of securing the exemption but that access may not be 'virtual'. Access may be on pre-booked tours, perhaps for quite small numbers of visitors and if visitors, having booked, fail to visit, that will nonetheless count as access.

Venue for provision of access

17.21 This must be specified in the undertaking, and will be obvious where the chattels are historically associated with a building or are located in a private building which is open to the public. Equally, where items are on long-term loan to public collections, the gallery or museum premises will be the nominated building. There may be situations where it is not practicable to specify the location of the chattel and in this situation the owner and HMRC must agree. The principle behind undertakings under the rules is that the owner must agree where the articles will be displayed and publicise it well in advance of the specified period, so that the public are aware of when and where the heritage property may be seen. Such an arrangement may be combined with access by appointment.

Example 17.5: Roger (2) – restricted access

After negotiation with HMRC, Mary agrees to loan the collection of Bateman silver to her local art gallery which is happy to set aside one small cabinet for its display. There is no particular connection between the Bateman family and the locality, but Roger himself was well known, and many local people are curious to see the collection that had driven him to a lifetime of parsimony. Mary secures an immediate financial advantage, in being relieved of the not inconsiderable cost of insuring the silver as part of the contents of the home.

Mary deposits a catalogue of the Bateman records with the local reference library and agrees that, for two months a year, the Bateman workshop records themselves should be deposited at the museum to provide open access and that, for the rest of the year, the public may see the records at Mary's home by appointment only.

PUBLICITY

17.22 The level of publicity that is appropriate will in part depend on the nature of the heritage asset. For example, in February 1987 the then Chancellor of the Exchequer, Norman Lamont, in reply to a parliamentary question, set out the norm for property other than chattels by indicating that publicity would normally include some or all of the following five procedures:

- Owner to inform the British Tourist Authority (VisitBritain) or, where appropriate, the Scottish Tourist Board (VisitScotland) and the Highlands and Islands Enterprise (HIE), of the opening arrangements and of subsequent changes.
- Owner to advertise the opening arrangements in one or more suitable publications with national circulation.
- Owner to display a notice outside the property giving details of the opening arrangements.
- Owner to agree that the advisory body or bodies (and its/their agents), who confirmed that the relevant property was eligible for relief and with whom the owner had negotiated detailed management agreements, might divulge the access arrangements to anyone who enquired about them.
- Owner to agree to such other publicity as the advisory body or bodies considered to be appropriate. That could include displaying a notice in some public place in the locality – for example, the local post office (in the days when they still existed), the local library, the local tourist office or town hall – or in the newsletter of a local preservation society.

It is acknowledged that the management agreement would normally also provide scope for other agreements between the owner and the advisory body.

The register of conditionally exempt works of art

17.23 The starting point for any member of the public seriously wishing to inspect heritage property is the HMRC website accessed at http://www.hmrc.gov.uk/gds/heritage/lbsearch.htm which is pleasantly set out and easy to search, whether by location or by type of article. This is the register of conditionally exempt works of art available for viewing by appointment with the owners.

What the owner need not disclose to the public

17.24 An owner of a heritage chattel need not disclose to the public everything contained in the undertaking negotiated with HMRC and personal information, not relevant to public access, may be excluded, ie the name of the owner, the value of the asset and the private address of the owner (except where it is relevant for public access, perhaps because the house is itself heritage property or because that house is also the venue for access to other exempt items). In this context, it will be possible to conceal the usual location of an exempt chattel, either for security or other reasons, except where the chattel is in a building that is or will be open to the public. Whilst the value of the asset, as already noted, need not be disclosed, nor indeed the amount of tax that has been deferred, other details of the asset itself must be shown. Specifically, the owner is expected to disclose, through the website, certain basic information, according to the nature of the asset. Examples of entries appear in CTNH.

Disclosure: pre-eminent chattels and pre-eminent collections

17.25 The website must show the following:

- a brief description of each object;
- the full address of the building and the opening times (where the chattel is in a building that is open to the public); and
- where the chattel is not in a building open to the public, the periods when open access will be available, and the name, address and telephone number of the person who will give details of actual opening days and venues. That person might be the owner or might be an agent.

Disclosure: outstanding buildings and amenity land

17.26 The website must show the full address of the building and the opening times.

Disclosure: land

17.27 The requirements are twofold:

- where the land is – ie the name of the estate and the nearest town or village; and
- the name, address and telephone number of the person who can give further details on request.

In addition to the HMRC list referred to at **17.23** the owner will normally be required to give details of access to local tourist offices and, in the case of buildings, to advertise in one of the annual Heritage Guides.

Footnote: foreign-owned works of art

17.28 A work of art that is normally kept overseas can become liable to IHT on the death of its owner merely because it happens to be in the UK at the event date. This is recognised in *s 5(1)(b)* (formerly ESC F7) which operates to waive the IHT liability thereon if property comprising a foreign work of art has been brought to the UK solely for public exhibition or for cleaning or for restoration. Similarly, if the work of art is owned by a discretionary trust, or is otherwise comprised in settled property in which there is no qualifying interest in possession, the charge to tax that would otherwise arise under *IHTA 1984, s 64* will also be waived.

Cultural Gifts scheme

17.29 The Cultural Gifts scheme is essentially a facility to satisfy tax demands rather than a form of relief. The taxes concerned are income tax, CGT and corporation tax. The scheme is described, with other payment options, at **23.11.**

Chapter 18

Heritage property: application for exemption and variation of undertakings

QUICK GUIDE

- A claim must be made within two years of the chargeable event. Where the circumstances warrant it, a protective or provisional claim should be made on the basis set out at **18.4**.
- The full claim should contain the information mentioned at **18.5**.
- The most important single issue, namely the variation of undertakings under *IHTA 1984, s 35A*, is discussed at **18.10**.

THE OLD LAW

18.1 It is useful to have an understanding of the pre *FA 1998* provisions. In that era there was no specified time limit for making an exemption application which might be claimed in respect of chattels, outstanding land, buildings of outstanding interest and their amenity land and associated chattels (as noted in **Chapter 17**, prior to *FA 1998* the definition of those assets that could benefit from exemption was capable of wide interpretation).

An undertaking was required for the preservation of chattels which involved: keeping them permanently in the UK, to maintaining them, preserving their character, securing reasonable public access and publicising the availability of that access. That undertaking, once agreed with HMRC, could not later be altered or modified. The obligation to publicise the availability of access was qualified but the owner did not have to disclose any information that might be considered confidential. Where a new undertaking was given to continue an existing exemption, it was only necessary to repeat the terms of the previous undertaking.

As previously seen in **Chapter 17**, under the old law there could be quite considerable restriction to public access, and great use was made of the 'by prior appointment' limitation.

REQUIREMENTS UNDER THE NEW LAW: THE TIMING OF THE CLAIM

18.2 For transfers made during lifetime or on death on or after 17 March 1998, a claim to conditional exemption must be made no more than two years after the date of that transfer (*IHTA 1984, s 30(3BA)*). That having been said, *s 30(3BA)* provides the possibility of an extension by stating 'or (in either case) within such longer period as the Board may allow'. Such a late claim will be considered, but HMRC will exercise their discretion to admit the claim only in exceptional circumstances and there is no right of appeal against that decision. Specifically, the failure to make the claim on time is not in itself an acceptable reason to allow a later claim.

Even though the ability to make claims for conditional exemption is a more specialised area of law than the ability to vary dispositions within two years of death, it must now become standard practice, on taking instructions in connection with the administration of an estate, to consider whether there might be any heritage property in respect of which a claim, albeit provisional, should be made.

Reference to HMRC discretion to extend the two-year time limit is made in CTNH para 3.5. However, it is also made clear that merely overlooking the need to make a claim will not be a good enough reason, nor the failure to claim some other relief such as APR or BPR, nor the delays attendant upon putting in place a post-death variation. HMRC would normally expect such matters as the entitlement to APR or BPR to be resolved within two years of the death/transfer. There is also the 'don't waste our time' warning in CTNH Appendix 1A para 12 indicating that HMRC will discontinue negotiations and will seek to charge tax instead, where:

- more than six months have elapsed from receiving the exemption claim, but even rudimentary access proposals have not been settled; or
- more than six months have elapsed since agreement that the item was pre-eminent, yet detailed access had not been agreed.

There is the facility available of making a protective claim where a practitioner believes that he will not know within the two-year period whether or not APR or BPR applies and this is discussed at **18.4**. However, this should not encourage the making of a 'frivolous' claim, merely to delay collection of IHT and such an action could and likely would attract penalties as could the making of a claim to APR or to BPR that cannot, in truth, be supported by the facts of the case.

Heritage property in discretionary trusts

18.3 *IHTA 1984, s 79* sets out the procedure for relevant property trusts. Where heritage property is included in the trust and there has been a conditionally exempt transfer, either on or before the occasion on which it becomes comprised in the settlement, the ten-year charge that would otherwise be imposed under *s 64* will not take effect on any ten-year anniversary that falls before a charge is triggered by a chargeable event in respect of that property. The nature of that charge is described more fully in **Chapter 20**. Where the property is comprised in the trust and there has been neither a charge nor a disposal:

- the property is designated as heritage property;
- the appropriate undertaking has been given by the 'appropriate person'; and
- the property is 'relevant property' within the code for the taxation of discretionary trusts.

Prior to *F (No 2) A 2015*, not only must the claim to exemption have been made before the date of the ten-year charge, it must have been made long enough before the date of that charge for the claim to be processed and for the heritage property to be designated as such before the date on which the ten-year charge will otherwise fall due. As part of the trust simplification process, this time constricted rule was amended by *F (No 2) A 2015* and carried into *IHTA 1984, s 79(3A)* which now permits trustees up to two years post the date of the ten-yearly charge in which to claim the exemption.

Heritage maintenance funds are considered in **Chapter 21**. A claim in respect of a heritage fund is referred to in *IHTA 1984, Sch 4, para 1*, and the wording of *para 1(2)* is guidance that the claim should be made in advance of setting up the settlement.

In no other situations can a claim be made in advance of the taxable event.

Protective or provisional claims

18.4 A claim to exemption may be protective or provisional but either way, HMRC expect the claimant to pursue the matter 'without undue delay'. If the claim is followed up within a reasonable time frame, HMRC may charge tax on the basis that the claim has been abandoned. Such action will only be carried out after proper notice has been given to the taxpayer (see **18.2**).

CTNH para 3.5 makes it clear that a formal or protective claim must include the following elements:

- a statement that the exemption is being claimed, specifying the event to which the claim relates;
- confirmation that there is no present intention to sell the asset or assets;

- particulars of each asset or group of assets covered by the claim; and
- a brief statement why each asset is considered to qualify for exemption, including confirmation that proposals will be made for public access.

In addition, particulars of any previous exemption should be supplied, showing the date of the death or event, official reference, tax rate applicable, and details of the asset.

THE CLAIM ITSELF

18.5 Interestingly, there is no prescribed claim form but an application will, however, be met with a request for certain prescribed information, as was noted in **Chapter 17**. CTNH Appendix 8 sets out HMRC guidance in the form of model letters appropriate to the type of property involved. For works of art, the requirements are as follows:

- full descriptive valuation of each asset, including a statement showing how it meets the pre-eminent standard;
- contact details of the person to contact to inspect the objects, location of each asset and confirmation of access to inspectors;
- previous tax history, ie particulars of any previous occasion of conditional exemption, whether from estate duty, capital transfer tax, IHT or CGT – in each case, any reference from HMRC should be quoted;

  the owner's name, address, telephone number and NI number;
- the owner's proposals as to maintenance and preservation of the asset;
- the proposals for providing public access to the asset and publicising such access;
- confirmation of authority to HMRC to disclose or publicise information and undertakings relating to the exemption;
- confirmation that there is at present no intention to sell the asset;
- three colour photographs, of professional standard, of each item, together if possible with a jpg of at least 300dpi sent by email;
- confirmation that there is at present no intention to sell the asset;
- confirmation that the owner will not take the asset out of the UK without prior approval from HMRC;
- whether the object is to be offered in lieu of tax;
- for items within a group, a statement of what constitutes the group and why the group is pre-eminent.

Where the claim is in respect of land or of amenity land or of historically associated objects the letter will set out similar requirements, thus:

- contact details of the person to contact to inspect the property;
- previous tax history, ie particulars of any previous occasion of conditional exemption, whether from estate duty, capital transfer tax (CTT), IHT or CGT – in each case, any reference from HMRC should be quoted;
- the owner's name, address, telephone number and contact details;
- the owner's proposals as to maintenance and preservation of the property;
- the proposals for providing public access to the property(which in relation to an outstanding building should include open access on at least 25/28 days per year during the summer months, with at least ten days falling on weekends and the spring and summer bank holidays) and for publicising such access;
- confirmation of authority to HMRC to disclose or publicise information and undertakings relating to the exemption;
- confirmation that there is at present no intention to sell the property;
- confirmation that the owner will not, without HMRC approval, remove historically associated objects from the property;
- six copies of a map at 1:10,000 scale showing the boundary of the property, any scheduled monuments (with status), other buildings for which exemption is claimed, rights of way, listed sites, nature reserves, woodlands on the Inventory of Natural England, other land in the same ownership, essential amenity land; and land either let or otherwise burdened;
- recent photographs of the exteriors of the buildings and of the principal rooms;
- for associated chattels, confirmation that that association with the building concerned goes back at least 50 years; a justification of the claim, confirmation that the chattels will stay in that building; and will stay in the place specified in that building as mentioned in the claim.

In practice, other information will also be relevant to support the claim. It may not be possible for the claimant to gather all the information required within the two-year period and in that case, the recommendation is to make the claim as fully as possible, indicating when the balance of the information will be available.

It is important to note that for the purposes of the two-year time limit within *IHTA 1984, s 30(3BA)* the claim is not valid until all the information has been received and that claim will not be processed by HMRC until it is complete in all respects. Above all, the claimant must show commitment to providing public access.

No intention to sell

18.6 HMRC consider that a new claim, ie one in respect of which no previous undertaking has been given, cannot succeed if there is any intention to sell the heritage property and for that reason suitable assurances must be given that no sale is in prospect. Some such condition may be implied from the requirements of *s 31(2)(b)* as to the undertakings that are to be given. However, when the claim is instead for the renewal of old undertakings in practice HMRC adopt a slightly less harsh attitude.

As will be seen in **Chapter 19,** heritage property may be subject to very high historic rates of estate duty. It might therefore seem to be good planning to seek a fresh undertaking at current rates but only to break it later and in effect substitute the new rate of tax for the old. It would appear that this is not caught under GAAR but this has yet to be tested.

Triggering a chargeable transfer on which to hang a claim

18.7 A feature of the heritage property code just noted is that property may be burdened with an estate duty liability that may be clawed back. To prevent this, it may be helpful to achieve a transfer on death under the IHT rules, which will 'wash' the old estate duty liability, and to then apply for exemption on death. However, where the death triggers a transfer that is exempt, ie by transfer to the surviving spouse, there can be no application for exemption.

The solution is a deed of variation redirecting the heritage property to a non-exempt beneficiary. That enables an application for conditional exemption to be made.

Example 18.1: Harry – successive deaths

Harry died in 1942 and the rate of estate duty payable on his estate was 60%. Harry left his daughter, Joyce, an antique chest on which Joyce obtained conditional exemption. Joyce died 18 months ago in December 2015 leaving her estate, which included the chest then worth £60,000, to her husband Cyril for life, with the remainder to her son Richard.

Joyce's estate is administered by Cyril's old friend and solicitor, Jack, who does not consider that any particular action is needed with regard to the chest since no IHT is due on the estate by reason of the spousal exemption.

Richard is no lover of antiques and thinks that, in a few years' time, the old chest will translate into a source to meet the school fees for his children in four or five years' time. He consults Cerys who advises that a sale of the chest would trigger a significant tax charge. For example, if the trustees were now, with Cyril's consent as tenant for life, to sell the piece for its present June 2017 value of £100,000, there would be a 20% CGT charge on the £40,000 gain over its probate value and 60% estate duty charge on the proceeds.

Cerys advised that a more tax efficient solution would be to vary Joyce's will to transfer the chest to Richard noting that such action must be put in place before the second anniversary of Joyce's death in December 2017. Richard, can give the necessary undertaking in order to maintain the conditional exemption and as he is not yet in need of the funds for school fees he can honestly say he does not immediately intend to sell.

On the eventual sale of the chest, the tax charge will be by reference to Joyce's estate. Estate duty will no longer be in point and the IHT charge will be at 40%, a saving of 20%. Alternatively, Richard might consider a private treaty sale to a heritage body.

Claims processing

18.8 Although HMRC have specialist teams to advise on chattels, they will in most cases discuss with specialist advisers the proposals for the maintenance and preservation of the assets or management, where land is concerned, and, in particular, will discuss with relevant authorities the proposals for the provision and publicity of public access. Obtaining that advice is not a quick process and dependent on the item involved, it may take up to a year or more. A report will then be prepared setting out the terms of the undertaking and a draft of that report will be sent to the claimant for agreement. Only after that agreement is reached will HMRC formally designate the assets under *IHTA 1984, s 31*. The procedure is similar in respect to that of an approved heritage maintenance fund, although in that case HMRC must also be satisfied as to the terms of the trust documents, that the assets to be transferred to the fund are appropriate, and that the amount is suitable to cover the projected maintenance expenditure for which the fund is set up.

Supervision of owners' compliance with undertakings

18.9 Whilst designation of the assets under *s 31* provides access to the conditional exemption it does not signal the end of the claim and HMRC will take steps to ensure that owners comply with the undertakings given. This takes the form of a three-point staged process: a five-year programme; an annual programme; and spot checks.

- During the five-year programme, advisers to HMRC inspect land and buildings – they ask the taxpayer whether there have been any disposals of exempt assets and require confirmation that all undertakings have been, and are being, observed.
- The annual programme requires owners of outstanding land or outstanding buildings (with amenity land and associated chattels) to make an annual report to the advisers. The report must detail the maintenance and preservation of the assets and the provision of public access. HMRC will check that these annual reports are made. The programme that does not fall into any particular time scale is that which requires owners or their personal representatives to inform HMRC of any disposals of exempt assets, including any transfer occurring on death, since such a disposal is a breach of the undertaking.
- Spot checks are, as the name suggests, random inspections made by HMRC of heritage chattels to ensure compliance with undertakings.

Variation of undertakings

18.10 *IHTA 1984, s 35A* (introduced by *FA 1998*) is a powerful provision. It enables an existing undertaking to be varied from time to time by agreement between HMRC and the person who is bound by the undertaking, and covers any undertaking whether given under:

- *s 30* – the general section relating to conditional transfers;
- *s 32* – an undertaking given in respect of a chargeable event (see *s 32(5)(b), (5AA)*, as described in **Chapter 19**); or
- *s 32A* – the 'associated properties' rule.

In itself, there is nothing sinister in the ability to vary an undertaking by agreement, however, the controversy is in the form of the award of power to the First-tier Tribunal to vary the undertaking. The conditions specified under *s 35A(2)* are that:

- HMRC have made a proposal to vary an undertaking and have put it to the person who is bound by the existing undertaking;
- the person bound by the undertaking has not, within six months of that proposal, agreed to the variation; and
- it is just reasonable, in all the circumstances, for the variation to be made.

The tribunal may direct that the existing undertaking is to have effect, from a date that it specifies, as if the person bound by the undertaking had actually agreed to the HMRC proposals. The tribunal, when making the ruling, must specify a date for the variation to come into force that is not less than 60 days from the date of the decision but the direction of the tribunal will not take effect if, before the date specified in that direction, the taxpayer and HMRC can agree different terms (see CTNH para 3.16).

That time limit does not, however, apply to relief from CGT (as to which, see below). Equally, this requirement does not affect the tax that is chargeable on a transfer of value that occurs before 18 March 1986 in respect of a death or other event occurring after 17 March 1986. This follows from the general transitional provision in *FA 1986, Sch 19, para 40(1)*, that nothing in the amendments to the *1984 Act* brought about by *FA 1986* is to affect the tax chargeable on a transfer of value that occurs before 18 March 1986.

MANAGING CHANGE: THE TRANSITION FROM THE PRE FA 1998 REGIME TO THE CURRENT REGIME

18.11 Following the introduction of *FA1998* which paved the way for a much more robust administration and implementation of the conditional exemption, HMRC entered into correspondence with existing heritage owners to negotiate open access to conditionally exempt heritage objects. Whilst HMRC were keen to reach agreement the owners were reminded that HMRC could take the matter to a Special Commissioner and it was made clear that by appointment only' access should continue only 'in exceptional circumstances i.e. where there are overriding conservation reasons why the heritage object should not be placed on open display'.

The correspondence highlighted the considerations that will always be relevant in relation to heritage property. For example, where an object may not be suitable for a national collection, an appropriate period of access might be in the range 5–25 days, but 'clearly the nearer to national collection standard [the] object is, the more days' open access we shall expect [the owner] to offer'. In other words, the intention is that conditional exemption will now be granted only in respect of high-quality heritage property and, given that high quality, a substantial measure of open access will be required.

There was evidence in *Undertakings of A and B* [2004] SpC 00439 (see **18.15**) that very little use had been made of the 'by appointment' viewing arrangements and there had also been few loans of chattels to public exhibitions. The tax deferred under the exemptions, taking the two owners together, was approx. £33 million. The point does not appear from the report of the case but at that time (September 2004) unpaid inheritance tax attracted interest at 4%, putting the cost to the general public of the deferral relief in those particular cases at approx. £1.3 million in a full year. Given the sums involved, it is understandable that HMRC sought a change to the undertaking.

There was reference in those cases to the burden on the estate owner of agreeing to open access, and it was recognised that the proposals under review by the Special Commissioner would impose substantial additional obligations, costs and disadvantages on the owners without the benefit of any compensation. The Special Commissioner observed 'Parliament no doubt saw the burdens, financial or otherwise, placed on the owner as a *quid pro quo* for continuing deferral of IHT'. He concluded that part of his enquiry by holding that, 'I do not regard the absence of compensation to be a feature of the *1998 Act* regime that necessarily requires me to conclude that [varying the undertaking] will inevitably and always be unjust and unreasonable.'

Museum quality properties

18.12 Under the pre *FA 1998* regime, an item could qualify as heritage property even though it was not 'pre-eminent', either because it was historically associated with an eligible building or because it was of 'museum' quality albeit yet not outstanding. In this chapter, property of this kind is referred to as 'museum quality property'.

Museum quality property can be transferred without loss of an existing exemption and the new owner may replace the undertaking that had been given by the previous owner, but that does not mean that there will be no tax to pay on the transfer itself. Unless the transfer is exempt from tax for some other reason, perhaps because it is being transferred to the spouse of the owner or to a charity, the effect of retaining the exemption will be only to avoid a charge, described below as the 'clawback' charge that would otherwise apply on the transfer of heritage property without undertaking. The new owner may claim new conditional exemption; but, if the article is of museum quality only, there is of course the strong possibility that it will fail the new 'pre-eminent' test.

On the transfer of heritage property under the current rules, the undertaking to be given by the new owner will have to meet different tests according to whether the existing exemption relates to estate duty, to CTT or to IHT. In summary, where the existing exemption relates to estate duty, the new undertaking will reflect the relevant provisions of the law for estate duty and will not be subject to the new requirements as to public access. However, where the existing exemption is in respect of CTT or IHT, the new undertaking must comply with rules in force at the time the new undertaking is given, with the result that open access will be required for the future, even where it was not previously given.

Sales of heritage property

18.13 Where the transition from the old regime to the new causes loss of conditional exemption (for whatever reason) the owner may want to consider a sale of the property and this may be particularly relevant to property which is of museum quality only. However, there is no guarantee that such property will be accepted in satisfaction of tax – the Secretary of State has the power to decline to accept even property which is of pre-eminent quality, so the owner of museum quality property may face an uphill struggle. The fact that the property once qualified for relief under the pre *FA 1998* regime is absolutely no guarantee that it can now be used to satisfy a tax liability. For use in satisfaction of tax, the chattel must qualify under the new rules, which will normally involve the pre-eminent quality test.

Review and variation of undertakings

Pre FA 1998 undertakings

18.14 Two categories of pre-existing undertaking are exempt from the provisions of *FA 1998* (see *Sch 25, para 10*). These are:

(*a*) undertakings that were given when estate duty was in point; and

(*b*) undertakings given under *FA 1975, s 31(2)*.

This second category (*b*) concerned the exemption from CTT of chattels transferred on a death occurring before 7 April 1976. The regime applicable before 7 April 1976 is set out in **Appendix 5** at the end of this book.

Post FA 1998 undertakings

All post *FA 1998* undertakings may be subject to review and there are no circumstances under which a current undertaking, ie one given on or after 31 July 1998, may not be varied. The variation will be by agreement or imposed with the consent of the tribunal. More important, perhaps, are changes of pre FA 1998 undertakings that fall outside the exclusions mentioned above. The change sought must relate either to the preservation and maintenance of the asset or to the provision of reasonable public access and the chief change will be the securing of open access rather than access by appointment only; although of equal importance will be the publicity arrangements, the requirements for which are more stringent than hitherto. Usually, an undertaking will be left undisturbed for at least five years.

Undertakings of A and B

18.15 The *FA 1998* regime for seeking variation did not proceed quickly not least since it was considered that the government was bound by old undertakings and had no right, by the enactment of *IHTA 1984, s 35A*, to renege on those agreements. This led to the test case of *Re an application to vary the undertakings of A; Re an application to vary the undertakings of B* [2005] STC

(SCD) 103 and any practitioner faced with a proposal to vary an undertaking should read the report in full.

The case decided two issues:

- whether it is just and reasonable in all the circumstances to require the proposed variation of an undertaking to be made; and
- if so, whether a direction to that effect should be made.

In brief, the Commissioner concluded that *s 35A* did not interfere with the rights of owners so as to produce substantial injustice nor violate human rights, but that, in the particular circumstances of the two cases, no direction should be made to give effect to the proposed variation. The decision has not since been challenged.

A was born in the 1950s and lived with his wife and family at A house, a Grade I listed house set in a large park close to a city. On the death of the previous owner, conditional exemption was granted for some 150 items, or pairs or sets of items, which had been valued by Christies at £20.89 million. In view of the grant of conditional exemption, those values were not examined but, if correct, the IHT deferred was approximately £8 million. A could have claimed that A house itself was heritage property, but instead paid IHT on the house and grounds, so no public access obligations arose in respect of the house itself. In considering whether or not to claim conditional exemption, A judged that the practical and administrative burden of access 'would be manageable and would not, as he saw it, interfere unduly with his family life'. Only 13 days after the designation of the 150 items for conditional exemption, the budget day press release of 17 March 1998 announced the proposed changes to the exemption regime.

B, born in the 1960s, was the owner of B house, where he lived with his wife and family. B house is Grade I listed, set in a large park, also Grade I listed, which are at the heart of a substantial estate which includes a number of important historical sites. On the death of the previous owner, B had to decide whether to claim conditional exemption in respect of the house itself as heritage property. In response to B's claim HMRC stipulated a minimum of 28 open days per year. B considered that would cause him and his family undue inconvenience and disturbance, so he withdrew the heritage claim in respect of the house and parkland, preferring to pay the tax. Christies advised B that nearly 780 chattels could qualify for heritage relief, so B applied for conditional exemption.

HMRC considered the insurance and security implications of extended access and the implications of the *Disability Discrimination Act 1995*. They visited other heritage sites to see how many visitors might be attracted to A house and to B house if the undertakings were to be reviewed. They chose fewer numbers of days of access than they would require in respect of new claims for conditional exemption to reflect the transitional and quasi-retrospective nature of the powers given them by *s 35A*. In relation to owner A, they took account of the tax at stake and of the fact that A had been admitting tours to the family home over the years. They also noted that a room in A house had occasionally been used for functions to raise funds for charities. They took expert advice as to the quality of the items to be included as 'Part 1 objects', ie being of outstanding artistic and/or historic interest. Their proposal identified 49 Part 1 objects for owner A. The local museum and art gallery was interested in exhibiting some of the Part 1 objects, although the Special Commissioner later observed that 'the local museum and gallery's curator had been enthusiastic about taking a few great paintings from A house but had had to back-pedal once the costs and implications were realised'.

HMRC approached B house in the same manner. The estate within which B house stood had the benefit of conditional exemption but very little use had been made of the 'by appointment' viewing arrangements and there had been few loans of chattels to public exhibitions. HMRC received expert advice that the chattels were of pre-eminent quality and lendable to museums as an alternative to admitting the public to B house itself. They identified 51 Part 1 objects.

All of this preparatory work was very relevant to the consideration by the Special Commissioner of the procedural fairness with which HMRC had acted. A house had been in the family for centuries and contained important items recorded on A's database, most of the paintings being family portraits. The other items were rugs, carpets, tapestries, china, furnishings, books and sculptures spread throughout the house. Keeping the collection was 'a constant and burdensome commitment'. The pictures had been in the family for between 250 and 450 years. Since 1998, there had been from one to five visitors to A house per year. A had lent items to museums and galleries. Significantly, special interest groups visited A house. Here, the numbers were larger: as many as 215 individuals in 2002.

Following his tour of A house, the Special Commissioner summarised his impressions:

> 'It is a family house. There are no facilities for the disabled. Were the public to be admitted and the numerous stealable objects (other than the Part 1 objects) to be placed out of sight or in some other way concealed, an extensive removal task would have to be undertaken. There is no obvious place to stow them away. Paintings could be removed and transferred to more accessible places but this would be at the risk of damage to the paintings on panels and to all the frames of the portraits. The great townscapes in the long hall could be moved, but this would leave unsightly marks. Because the children occupy the only downstairs rooms where there are no Part 1 objects, it would not be practicable to lock the stealable objects away in

> those rooms on open days. All in all, the reorganization of the household furnishings and paintings to meet the open access proposals would, on each occasion, take a great deal of time…'.

The circumstances at B house were slightly different. Numerous items, great and small, of significant value, were spread throughout the house, including many exempt carpets and smaller rugs, all of which were delicate and there were many tapestries. As with taxpayer A, there was no suggestion that any of the chattels had been bought for their potential as investments. As already noted, after negotiation, B pursued a claim for exemption in relation to the estate but not the house, on which he paid IHT. B had received threatening letters in the past which he took seriously and he was concerned that, if B house were open to the public, there would be an increased risk to himself, his family, the house and its contents – that increased risk was corroborated by the county police force. The collection of sculpture was in a separate building; the family used every room in B house save one; there were private possessions and photographs everywhere; most of the portraits were in fine frames that might be damaged if not moved with professional care.

Previous viewing

18.16 At house B, there had been little use of the 'by appointment' viewing arrangement since the grant of conditional exemption. On one day in 2000, there were visits by two special interest groups. There was evidence from B that, if it had been suggested that open access would be required, he would have considered his application for exemption in a different light, and would have considered selling certain of the chattels to pay the bulk of the tax on the rest of them. B was concerned about the personal security of his wife and children, which had been increased by recent publicised thefts from open houses. Opening the house would force on him a complete review and upgrade of security measures. It would increase wear and tear on the fittings and soft furnishings and there were no normal visitor facilities, such as toilets, whilst the facilities relating to access for the disabled were limited.

Were the proposals for variation of the undertakings just and reasonable in all the circumstances?

The Special Commissioner held that he was required to act fairly and to decide whether the expectations of the owners, in giving the existing undertakings, were such that it would be an abuse of power to disappoint those expectations. He must then also decide whether there is an overriding public interest that would justify the disappointment of those expectations. He concluded that HMRC were not the creators of the new access and publicity policy; rather HMRC were doing what Parliament had told them to do. He therefore considered that the argument of the owners was misconceived and that the only matter he must decide was whether, in all the circumstances, the proposals for variation of the undertakings were just and reasonable.

It was agreed that the Part 1 objects were of outstanding quality and most of them satisfied the new 'pre-eminent' quality test. The owners argued that the proposals to vary the undertakings ('the Proposals') were not justified simply because few people had taken advantage of viewing the chattels by appointment. He concluded that there was nothing inherently wrong with the 'by appointment' viewing system. He reviewed the public interest considerations in the context of any greater exposure to crime that might be involved. He concluded that the Proposals would expose the taxpayers to greater security risks, which was a negative factor in the balance; and that open access would involve extra security requirements that would make viewing the chattels a less rewarding experience; but that that was only a minor point and did not displace any overriding public interest in making the chattels available for public viewing.

Is it unreasonable to force upon owners proposals which they will find unacceptable?

This issue is likely to be the most important for the future. The taxpayers argued that the Proposals would bring taxpayers generally to the 'breaking point' at which they would prefer not to accept the demands for public access but instead opt to pay the tax involved, probably by selling part of the collection. Thus, it was argued, the public interest suffers in two ways: there is a loss to the integrity of the collection; and a loss of any existing public access to the entire collection. On the evidence, the Special Commissioner was not satisfied that the breaking point had been reached in either case.

There is no compensation to the owner

It was argued that the lack of compensation to the owner who was given an existing undertaking, but was affected by the Proposals, was interference with his rights without compensation. HMRC disagreed: the owner was compensated in each case. The specific requirements under consideration were that the owner must grant access on specified days between 1 April and 31 October without prior appointment. They must notify HMRC of open days by 31 October in the previous year. They must arrange conducted tours to cover the rooms in which the Part 1 objects were on display, a minimum of four tours per day in the first year. They must keep all the exempt items in an appropriate room if normally located there before the opening.

The alternative option was to lend every one of the Part 1 objects to a public gallery or museum for a continuous period of at least three months in every 39 months, notifying HMRC of the arrangements three months before the loan. They must: allow publicity on the HMRC website; provide a copy of the undertaking to the public on reasonable request; provide reasonable details to local tourist information centres; and provide images of the Part 1 objects and notification of the availability of those objects for loan to certain specified galleries.

Cost

The taxpayers would probably incur additional costs and suffer disruption as a result of all these requirements. Specifically, there was evidence that the cost of providing barrier posts and rope throughout the houses would be £47,000 at A house and £36,000 at B house. There would be a labour cost in preparing the houses for open access; the price of druggets/matting, etc was put at £12,540 at A house and £6,000 at B house.

Security and risk

There was evidence as to the security risk (and, in particular, to a handful of travelling families, of whom it was said that they are 'violent towards each other, [but] not generally violent to members of the public although they do threaten violence … they would leave [the owner's] wives and children alone because this was part of their "code"'). Those houses already had security systems and good security staff and alarm systems. Evidence was heard of arrangements made by the National Trust. The Commissioner was satisfied that variation of the undertakings would expose the owners, their families and their households to additional security risks. Whilst those risks might not be as acute as the taxpayers feared, 'their reasonable apprehensions must be taken into account in assessing the reasonableness of the Proposals'.

Retrospectivity

The Special Commissioner accepted that *s 35A* did involve an element of retrospectivity but that that was intended by Parliament; and that no taxpayer can have a legitimate expectation that is immune to the possibility of subsequent statutory alteration.

Double taxation

The taxpayers argued that, in certain circumstances, the sale of a chattel to fund IHT payable on a breach of an undertaking that had been required by the variation would give rise to a double tax charge. However, the Special Commissioner was not convinced by the logic of the comparison given in argument. There was no necessary connection between the sale by a taxpayer of a particular asset and the breach of the IHT undertaking that had been given in relation to it. The possibility of two tax charges was 'not a consideration of any compelling weight, to be taken into account in applying the statutory test of whether it is just and reasonable to make the direction under *s 35A*'.

Intrusion

Another specific issue was the extent to which the impact on the family life of the taxpayer and enjoyment of his house should count against the interest of the public in access to the heritage property. After a review of the evidence, the Commissioner concluded that the Proposals were:

> 'onerous and intrusive. The time required to prepare the house for open access viewing, to remove and store away vulnerable chattels on the viewing route and to lay out druggets and roped barriers will, I expect, subtract from the days when the house is suitable for normal family life. Overall the proposals will involve a real disruption to the lives of the families.'

The loan alternative

The Commissioner found several problems with this alternative arrangement. Unless the owner can arrange to lend every one of the Part 1 objects, there will be intrusion and risk of much the same nature as if the open access option were to be adopted. More specifically, one large painting at B house would be difficult to move and to place in a public gallery. The paintings on wooden panels at A house and at B house could be damaged through movement to a new location. HMRC argued that purpose-built homes for those items could be constructed, but it was accepted that the maintenance costs would be great.

Whilst museums and galleries expressed enthusiasm for exhibiting items, there were difficulties. A curator had given thought to issues such as safe transport, invigilation, security and insurance of paintings that might be borrowed from A house. The curator thought that they could be contextualised by information given on their historic and social backgrounds in ways that would allow many members of the public to appreciate them. A Museums Officer for the county in which B house was situated was interested in various paintings and portraits for loans for between two and five years. Within that time frame, the very large investment that would be made in security, transport and hanging would become worthwhile. However, a sizeable amount of lottery funding would be required for the venue that the officer had in mind for the B house paintings. The venue would need upgrading.

However, it had to be accepted that, at the moment, the public facilities were quite inadequate for the proposed loans. Both in relation to A and B, the relevant authorities had had to withdraw from the idea of loans once the cost implications were revealed. The Commissioner therefore concluded that the museum/gallery option could not be regarded as a reasonable condition for continued conditional exemption of the chattels.

Taking all the circumstances into account, the Commissioner considered that the burdens placed on each owner would so outweigh the benefit to the public as to make it neither just nor reasonable for him to direct that the variation proposals should take effect. He therefore did not make any direction requiring the existing undertakings to be varied. The case was very fully argued and aired many points that affect the general law.

Appeal

18.17 The current legislation provides no right of appeal from the decision of the Special Commissioner/tribunal. Such right of appeal, if any, as may exist will be under the general law in respect of judicial review of the Special Commissioner/tribunal ruling.

Preserving earlier exemption: replacement undertakings

18.18 The regime for 'clawback' of exemption is considered in **Chapter 19**, but it is appropriate here to consider the kind of undertaking that will be effective not to shelter an existing transfer from tax, but to preserve an earlier exemption. An undertaking given for that purpose is in this book called a 'replacement undertaking'. A replacement undertaking given on or after 31 July 1998 must comply with the rules on undertakings that were in force on the date the replacement undertaking is given (except, as already noted, where the rules concerning estate duty apply). Effectively, this means that the new owner is bound by the new rules relating to the maintenance and preservation of the property, public access, publicity about the exemption and the rules as to review and variation. As already noted, a transfer of 'museum quality' chattels may not attract conditional exemption under the new rules, but a replacement undertaking may save those chattels from the clawback charge. This is a fairly narrow field: it is understood that HMRC receive about one new case per month. That said, there should be no delay by the practitioner (see CTNH para 3.15).

Does the system work?

18.19 Anecdotal evidence suggests that, for those who accept the full limitations of public access, the current regime does work by operating to preserve for the nation assets that might otherwise be sold and pass into foreign ownership – claimants now better understand their obligations. As stated in CTNH Appendix 1A para 24, 'Access is for the general public, not just for specialists and enthusiasts.' There is significant flexibility in the arrangements set out in CTNA Appendix 1A.

Perhaps the only improvement might be some 'democratisation' of the process by which provincial practitioners and their clients, owners of smaller estates, become more closely involved. At present, nearly all the work centres on the leading auction houses, which may lead to a concentration of fine objects in London, whose public places are already well endowed with fine things. The scope for close examination of fine heritage objects is more difficult in economically deprived areas, where perhaps their civilising effect might be much appreciated.

Chapter 19

Heritage property: loss of the exemption

QUICK GUIDE

- A breach of undertaking, or death, or a disposal will trigger a 'recapture charge' to claw back the relief previously given. In relation to a composite holding of heritage property, sale or breach of undertaking in respect of any constituent part can endanger relief on the whole. The rules are complicated, and the flow charts at **19.4** may help the practitioner through the maze.
- Before calculating the charge to tax, 'the relevant person' must be defined by applying the rules in **19.10**. It is then necessary to establish whether it will be the lifetime rate or the death rate that is to apply, using the rules at **19.10** onwards. Finally, the event must be placed in its context. There is guidance in the examples at **19.7**.
- Loss of the exemption may trigger an extra penalty, in that the previous exemption from ATED may be lost: see **19.22**.

THE GENERAL RULE

19.1 The legislation (see *IHTA 1984, s 32(1)*) uses the expression 'chargeable event' to describe the circumstances that trigger the 'recapture charge' or clawing back of the relief that had previously been given. Where the transfer of the property was a PET, the chargeable event will occur on or after the death of the transferor.

Breach of undertaking under s 32

19.2 The failure, in any material respect, to observe and abide by the terms of the undertaking given in respect of heritage property is a chargeable event but they may be extenuating circumstances which prevent that failure from being classed as a breach: for example, during the widespread outbreak of foot and mouth disease some years ago now, the question arose as to whether owners of heritage land would be in breach of access undertakings if their property fell into an exclusion zone which prevented access. In Tax Bulletin special edition 'Foot and Mouth Disease', issued on 9 May 2001, HMRC stated that they would not regard it as a breach of the undertaking for the owner to decide, after judging the risks, that access or visits should be suspended. Where there was a commitment to allow visits on a number of pre-arranged days in each year, HMRC would not insist, for tax compliance purposes, that the days missed should be made up later in the year.

Failure to provide access will often result in a Breach. CTNH Appendix 1A Part III paras 18–26 describe good practice where access is by prior appointment and as a general practice the owner should facilitate access, not frustrate it. The approach from the public will arise from a search of the HMRC website and ideally, arrangements should be made there and then for the viewing. Specifically:

- Appointments should be made as soon as requested, rather than the agent asking for a telephone request to be re-made in writing.
- Inspection need not be at the place where the item is normally kept.
- Agents should not deter visitors: the member of the public should normally be able to name the day and a choice of viewing dates should be offered, to include three weekdays and two weekend days in the next four weeks.
- This is public access, for recreation as much as for scholarship, so owners should not refuse unless the enquirer is clearly a time-waster or poses a real security risk. Proof of identity may be required but not personal references. CCTV may be used within legal restrictions.
- Numbers might be a problem: it is acceptable to say 'limited space: groups of no more than five people'.
- Owners should not discriminate against a potential visitor since access is for the public, not just specialists.
- A reasonable charge may be made but not to cover third-party costs, such as viewing at a separate location.
- It is acceptable to say 'don't touch' or don't photograph'.
- Owners must ensure that their agent is helpful and does not deter the public. Such action would amount to denial of access and HMRC will investigate if notified of the problem.
- Articles may be moved, for example for renovation or short-term loan; if so, the owner should notify the public where the article is and when it will be back.

- Holidays should be accommodated: a member of the public may be able to visit only because holidaying in the area, which should be allowed for. Make arrangements in advance and keep them unless absolutely impossible.

Where a member of the public notifies HMRC via the website that access has been frustrated or denied, the assertion will be examined very carefully. If there has been a breach of undertaking, HMRC will seek its remedy and if that cannot be done, or if this is not the owner's first offence, HMRC will claim tax on the current value of the item concerned. That could be the entire contents of a house, but more likely only the article that the member of the public wanted to see. If the owner was just obstructive, it might be the whole collection.

Example 19.1: Enthusiasts together

A silver enthusiast noted from the HMRC website a collection of household silverware. He happened to know the owner's agent personally, so rang him and was without delay put in touch with the owner, a fellow member of The Silver Society. The owner invited him to lunch and to see and handle the artefacts, which were of the finest quality, and to chat about a shared interest. It was a privilege, rather than the exercise of a taxpayer's right, to be welcomed into the home in that way. All the owner asked in return was a letter acknowledging that the visit had taken place, as evidence for his 'access' file, which the enthusiast gladly sent soon afterwards.

Death or disposal

19.3 The chargeable event occurs where the person beneficially entitled to heritage property dies (see *s 32(3)(a)*) or if the heritage property is disposed of by any means (*s 32(3)(b)*). However, this rule is subject to qualification and thus there will be no chargeable event, whether arising by death or by disposal, if there is a transfer of the heritage property:

(i) by

- the personal representatives of the deceased; or
- the trustee of settled property; or
- the person next becoming entitled to settled property;

(ii) to

- a heritage body; or
- HMRC in satisfaction of tax.

Section 32(4) provides that a death or disposal of the property after transfer to a heritage body is not a chargeable event with regard to that heritage property unless there has, after the disposal, been a further conditionally exempt transfer of it. However, it is not immediately clear how there can again be a conditionally exempt transfer of heritage property after it has been accepted by HMRC in satisfaction of tax or after it has been sold to a heritage body. Neither the legislation nor established commentaries on the legislation are any help in imagining how this part of *s 32(4)* comes into play and CTNH para 7.3 is somewhat reticent on the point.

Neither death nor a disposal of the heritage property is a chargeable event where further conditions are satisfied namely:

- where the transfer, which would otherwise be a chargeable event, is itself conditionally exempt (see *s 33(5)(a)*).
- where a person whom HMRC consider to be 'appropriate' gives an undertaking under *s 31*, or where the property is amenity land within *s 31(1)(d)* and appropriate undertakings are given under *s 31 s 32(5AA)*.

The composite heritage estate: 'associated properties' under s 32A

19.4 Separate heritage property assets may commonly be held by several different persons, often members of the same family, or it may be held partly by family members and partly by trustees, for example, a house, fine contents and amenity land (ie land within *s 31(1)(d)* which is essential to the protection of the character and amenities of the house). These separate entities are 'associated properties' for the purposes of *s 32A(1)* which provision operates with the aim of holding associated properties together, and makes the exemption granted in respect of any element of the composite property conditional upon observance of the conditions and undertakings in respect of the whole. The workings of *s 32A* are complicated, being a set of rules adapted to different circumstances with such rules illustrated in the series of flow charts that appear below.

IHTA 1984, s 32A: Associated properties

Note: lettered boxes are referred to in later flow charts:

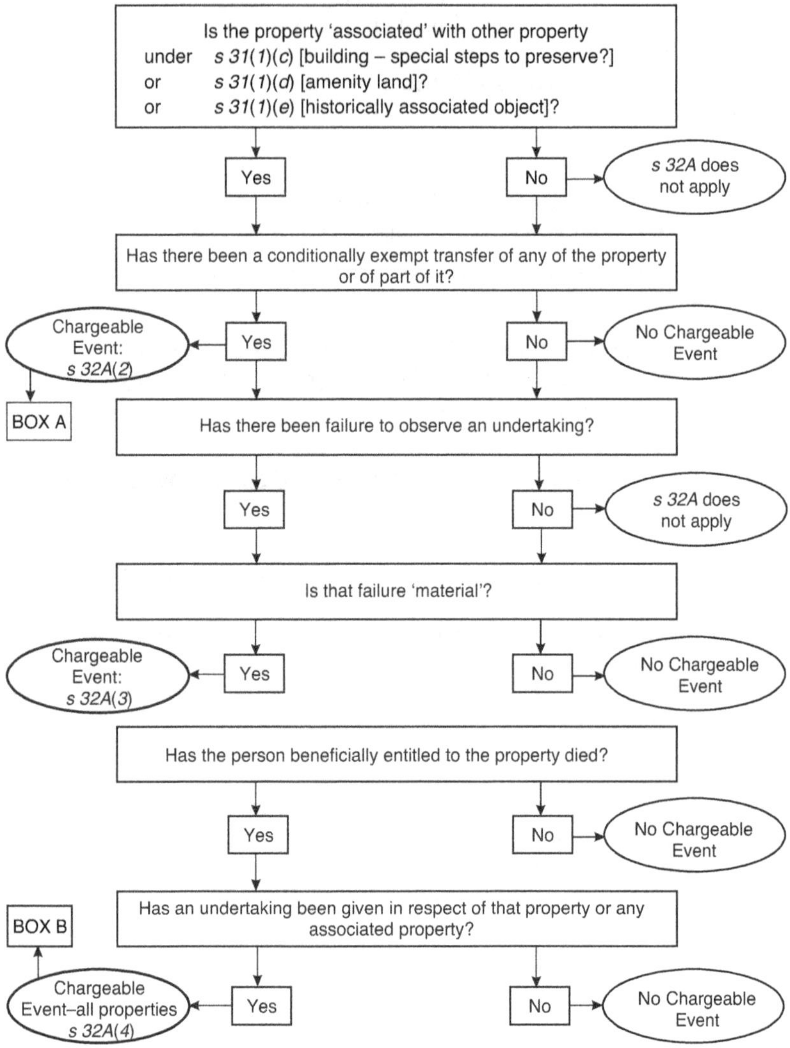

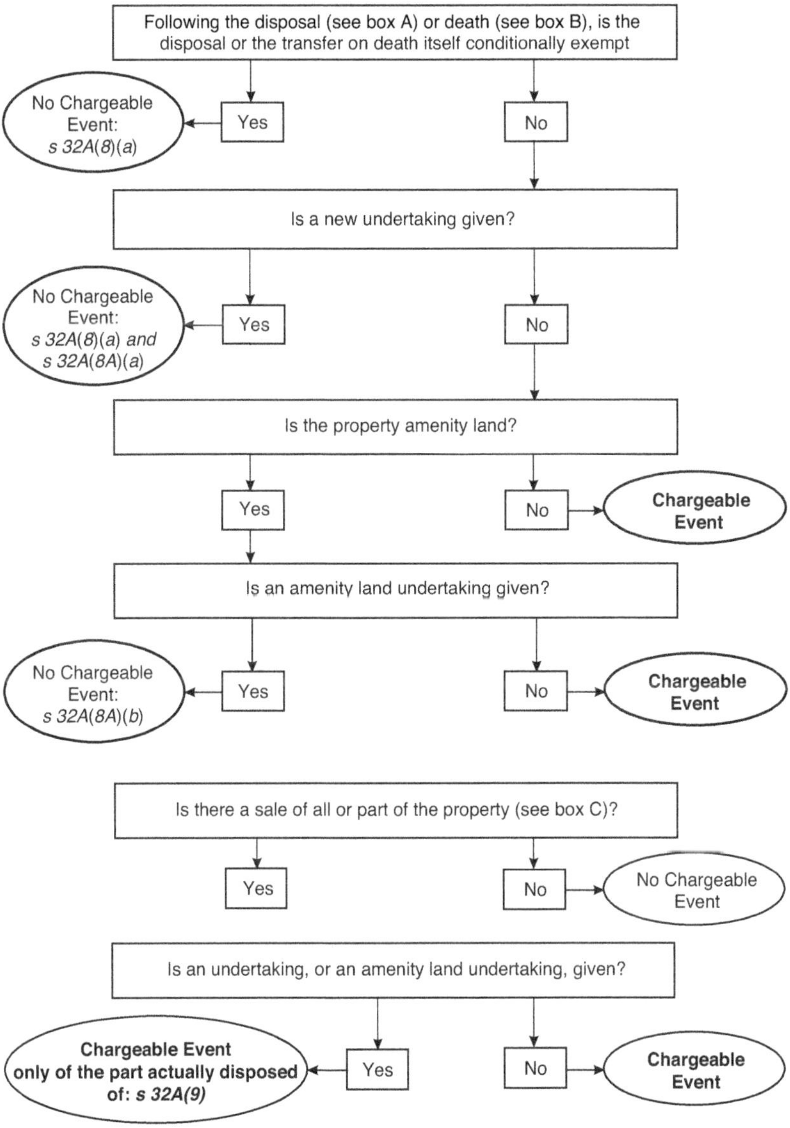
Following the disposal (see box A) or death (see box B), is the disposal or the transfer on death itself conditionally exempt
Yes
No
No Chargeable Event: s 32A(8)(a)
Is a new undertaking given?
Yes
No
No Chargeable Event: s 32A(8)(a) and s 32A(8A)(a)
Is the property amenity land?
Yes
No
Chargeable Event
Is an amenity land undertaking given?
Yes
No
No Chargeable Event: s 32A(8A)(b)
Chargeable Event
Is there a sale of all or part of the property (see box C)?
Yes
No
No Chargeable Event
Is an undertaking, or an amenity land undertaking, given?
Yes
No
Chargeable Event only of the part actually disposed of: s 32A(9)
Chargeable Event

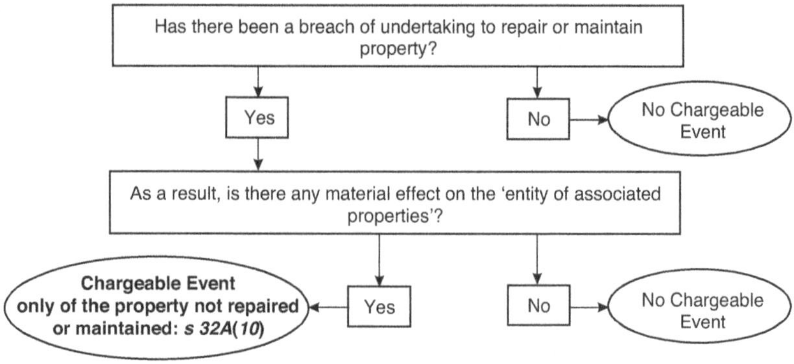

Selling (only) the family silver

19.5 The potential charge under *s 32A* can extend to the whole of the property in respect of which exemption was granted, even though the bulk of it has been retained (but see the *de minimis* exception under *s 32A(10)* described below and illustrated in the flow chart above). The chargeable event can also occur in respect of the breach of an undertaking in respect of any part of the composite estate.

The death of a person entitled to associated property, or its disposal by sale or gift or otherwise, triggers a chargeable event in respect of all the associated properties. This is subject to an exception where there is a disposal of the property:

(i) by

- the personal representative of the deceased; or
- the trustees; or
- the next person entitled to the property;

(ii) to

- a heritage body; or
- HMRC in satisfaction of tax.

This part disposal affects all the associated properties, unless appropriate undertakings are given in respect of the associated properties that are not disposed of. This means there is the opportunity of refreshing the undertakings given in respect of associated property, even where it is not the actual article in respect of which the undertaking is given that has been disposed of, but instead some other associated item. Generally speaking, this will not be an onerous burden for the owners of heritage property because the kind of undertaking that is given, for example with regard to chattels, is likely to provide for a measure of open access where the chattels qualify for conditional exemption, perhaps not in their own right as being pre-eminent, but by virtue of the connection with the house that qualifies for exemption.

Nevertheless, from an administrative point of view, the disposal of one of the elements of a composite heritage estate will cause considerable administrative work in the refreshing of all the undertakings given in respect of all the associated properties that are not disposed of.

'RELEVANT DISPOSAL'

19.6 *IHTA 1984, s 32A(7)* is concerned with the death of a person entitled to heritage property or part of heritage property following a 'relevant disposal'. For this purpose, a relevant disposal is one that complies with *s 32A(5)(a)* or *(b)*, ie a sale by private treaty to a heritage body or disposal in satisfaction of tax. That death is not a chargeable event with respect to the property concerned unless there has been a conditionally exempt transfer of that property after the relevant disposal. In the same way, the disposal is not in itself a chargeable event unless there has again been a conditionally exempt transfer. Given that the disposal, to qualify under *s 32A(5)*, is to a heritage body or in satisfaction of tax, it is not immediately clear how the circumstances envisaged under *s 32A(7)* could arise, nor is there any official guidance on the point.

Section 32A(8) and *(8A)* provide that the death of a person entitled to property, or its disposal otherwise than by sale, is not a chargeable event if the transfer itself is conditionally exempt or if the *s 32A(8A)* conditions are satisfied, ie the giving of appropriate undertakings. This is illustrated in the flow charts above.

Part disposals: effect limited to the item sold

19.7 The impact of *s 32A*, if rigorously applied, would restrict the proper management of heritage estates since any disposal of any part of the estate could cause a clawback of all of the relief granted, but this is not always a problem.

Members of the National Trust will know that, on more than one occasion, property may have been frozen in time with assets left undisturbed for many years following the death of their owner. A good example is Mr Straw's House at 7 Blyth Grove, Worksop, the interior of which has remained unchanged since the 1930s, with contemporary wallpaper, Victorian furniture and household objects. Occasionally, this situation can give a quite remarkable snapshot of life and times long gone. Equally, however, many fine estates that qualify as heritage property are living entities for which time does not stand still; consider for example trees planted at the time of Capability Brown which no longer enhance a view but obscure it. A property may contain artefacts that are of intrinsic value, but which cannot be publicly displayed because of shortage of space. Land on the edge of an estate may be taken for road widening.

To address these issues, *s 32A(9)* allows for undertakings to be given in respect of property so as to preserve the estate as a whole from the chargeable event by providing that the disposal is a chargeable event only with respect to the part that is actually disposed of. HMRC also have the power to apply a *de minimis* provision where it appears to them that the heritage entity as a whole has not been materially affected by the happening of what would otherwise be a chargeable event, such as the disposal of one of the associated properties. Where so satisfied, HMRC may direct that the disposal or breach of undertaking is a chargeable event only in relation to the property so disposed of, or in respect of which the undertaking has been broken.

However, these provisions do not really allow 'dynamic' collecting (though see below).

Example 19.2: Josiah – a living collection

Josiah's fine collection of silver passed on his death to his daughter Anastasia, who cherished it and gave undertakings to secure heritage relief which have been rigorously observed. However, she wanted to weed out a few pieces that were not well hallmarked and replace them with better examples of the work of the same goldsmiths to enhance the collection; however, such action would trigger a breach of the undertaking. Only if Anastasia can afford to buy the new items mainly out of new money, and meanwhile pay the tax on the items sold, will she be able to carry on the collecting that her father established. There is no 'rollover' relief except in cases of loss (see below).

The outcome in Example 19.2 might be different if Anastasia could exchange an item in her collection for the other property, swapping spoons for forks for example. CTNH para 7.2 comments that, though an exchange of property is a disposal, and therefore a chargeable event, a fresh undertaking could be given in respect of the replacement.

Loss or destruction of heritage property

19.8 The general rule used to be that the loss or destruction of heritage property was not an occasion for an IHT charge, even where insurance monies had been received, but this would not apply if the loss arose out of a clear breach of an undertaking. However, where for example a building has not been completely destroyed, any sale of it other than to an approved heritage body will be a chargeable event.

Where items are missing, HMRC make the aggressive assumption that their disposal took place at some time since the item was last clearly seen 'where it should have been'. A mid point in time is assumed with the market value of the item agreed as at that date. No allowance is made for CGT and interest on the tax charged runs from the assumed time.

CGT rule

For CGT, the rules are different, since the loss of an asset is normally treated as a disposal giving rise to a chargeable gain where any compensation received is greater than the acquisition cost (see *TCGA 1992, s 23*). If the destruction of the building is irreparable and all the insurance money is laid out in acquiring a replacement building on other land, then the old building and the new building are regarded as an asset which is separate from the land, and the old building is treated as lost or destroyed.

Replacement of chattels, etc

HMRC guidance in respect of heritage property indicates that there is some flexibility in the interpretation of the replacement facility; thus in the case of loss of furniture, purchase of other furniture to serve the same function in relation to the heritage property would be acceptable or if works of art are lost, purchase (in so far as that might be possible) of other similar works of art would be acceptable. Applying the rules in *TCGA 1992, s 23*, and the 'functional' test set out above, heritage property could in theory be replaced by a modern building which would not be suitable for designation as heritage property in its own right but that would not be relevant for CGT purposes.

Associated objects

As is nearly always the case, a separate code applies to property which, though not heritage property itself, nonetheless supports it. CTNH Appendix 7 para 4 states, if a building is totally destroyed by fire, most, if not all, of the historically associated objects are likely to have gone up in flames with it thereby leaving essential amenity land potentially affected. Where the destruction does not arise from a breach of an undertaking, the receipt of insurance monies does not trigger an IHT charge on the supporting property and provided the original undertakings continue to be observed, the ancillary property is unaffected. A fine estate may yet exist without the palace that used to grace it: see, for example, Clumber Park, or Lady Jane Grey's home at Bradgate Park, Leicestershire.

Breach of undertaking accompanied by destruction

If, for any other reason, there is a breach of undertaking, the IHT charge will apply in the normal way. The loss of the principal heritage property may naturally mean that the ancillary land, not being heritage property in its own right, loses its claim to heritage relief. It follows that by passage of time a further occasion of charge may arise when the undertaking in respect of the land falls to be renewed and the conditions no longer apply for re-designation as heritage property. Similarly, for CGT there is no immediate charge on that amenity land until loss of conditional exemption is triggered through other circumstances.

The destruction of heritage property can naturally affect maintenance funds, which are considered in **Chapter 21**. It is likely that, with no heritage property to maintain, the Direction by the Treasury under *IHTA 1984, Sch 4* that supported the maintenance fund will be withdrawn. At that point, the trustees must consider how they apply the fund, since application of the fund to endow another property or resettlement of new maintenance trusts would avoid the charge.

The charge to tax

Taxable value

19.9 This is, frankly, difficult material but there is guidance in CTNH Appendix 2. The value on which tax is charged is that established at the time of the chargeable event, and where the item is sold at auction that 'value' is the net proceeds, not the hammer price: *Tyser v Attorney General* [1938] Ch 426. However, where there has been improvement since the last grant of conditional exemption the value of that improvement will be ignored for this purpose.

Example 19.3: Overlooked masterpiece

When clearing her late mother's country house prior to sale, Jennifer put to one side a fine oil painting in respect of which conditional exemption had previously been given on the basis that it was a fine representation (albeit in poor condition) of family property that was involved in a famous historic incident. That fact aside, the attribution had been merely 'Venetian School'. Jennifer had the work valued for probate purposes.

Jennifer gave an undertaking to preserve conditional exemption. She then spent money with experts who, after paint analysis and detailed consideration of style, materials and technique, were satisfied that the work dated from 1730 and really was a Canaletto. Encouraged by that, Jennifer incurred significant expense in having the painting restored which served to increase its value. Jennifer now wants to sell, triggering a recapture charge, but is worried about the taxable value.

CTNH Appendix 2 para 3 states: 'The extent to which the value for tax is attributable to clearly identifiable capital additions or improvements to the property made by its owner since we last accorded CE will in practice be omitted from the value for tax when a chargeable event occurs.'

Arguably, all that has happened is that a painting's true identity has now been recognised and carries the correct value applied to a Canaletto; but realistically that would never have happened without incurring the cost of research and restoration. The taxable value should either be merely for a work of the Venetian School; or if for a Canaletto, then heavily discounted to allow for the costs that Jennifer has incurred, and for a margin for being bold enough to incur it.

Relevant person

19.10 *IHTA 1984, s 33* employs the expression 'the relevant person', which term is defined in *s 33(5)* in relation to three scenarios:

Scenario 1: simple case

Where there has been only one conditionally exempt transfer of the property before the chargeable event, the relevant person is the person who made the transfer.

Scenario 2: two or more conditional transfers, one of which was made more than 30 years ago

In this more complex case, it is necessary to consider the ownership of the property in the period of 30 years leading up to the chargeable event:

- Where the most recent transfer of the property was more than 30 years ago, the person who made that transfer is the relevant person.
- Where there have been several transfers, of which one had taken place in the last 30 years, that transferor is the relevant person.

Scenario 3: two or more conditional transfers all within the last 30 years

Where there have been two or more transfers within the last 30 years, HMRC may select the relevant person and there is no right of appeal. This prevents avoidance of tax by channelling heritage property through, perhaps, members of the family who do not have taxable estates to ensure that, in the event of a chargeable event, the charged tax would be as low as possible.

These rules contain traps for executors and trustees, who may have to retain sums to pay tax clawed back on breach of undertaking without knowing how much may be involved. Practitioners facing this issue may wish to read *Re Bedford, Russell v Bedford* [1960] 3 All ER 756 and *Re Scott* [1916] 2 Ch 268.

19.11 Where an undertaking has been broken or, on the death of the owner, no new undertaking is given, liability for the tax falls on the person who would have been entitled to the proceeds had the property then been sold.

If the chargeable event is a disposal, whether by sale or gift, the person liable for the tax is the one by whom or for whose benefit the property was disposed of.

THE RATE OF IHT APPLIED: LIFETIME TRANSFERS

19.12 Where the relevant person is alive at the time of a chargeable event, lifetime rates apply. This could occur where either:

- there was a lifetime transfer made before 18 March 1986; or
- an immediately chargeable transfer was made after 17 March 1986.

(Note: It is not possible to claim conditional exemption in respect of PETs until they fail by reason of the death of the transferor within seven years and there is no need for conditional exemption in respect of a PET that becomes fully exempt).

Where the lifetime rates apply, the value of the property attracting the charge is added to the cumulative 'clock' of chargeable lifetime transfers made by the relevant person in the seven years immediately preceding the chargeable event. If PETs made by the *relevant person later* become chargeable on his or her death no adjustment is made to the clawback charge (*s33(2A)*).

No allowances

19.13 Whilst the nil rate band is available to shelter the charge to tax, access to any unused annual exemptions or APR or BPR is denied – the IHT rate is that in force at the time of the chargeable event but that rate may not benefit from taper relief.

Transfers on death

19.14 The rules are more complicated where the relevant person is dead (see CTNH Appendix 2 para 7):

(1) Where the conditionally exempt transfer was made in the lifetime of the relevant person, the tax is charged at lifetime rate with the amount on which the charge is levied treated as if it were the top slice of the estate on death. This is also the case if that transfer was also a gift with reservation (*s 33(2)*).

(2) Where the conditionally exempt transfer was made on the death of the relevant person, the tax is charged at the death rate with the amount on which the charge is levied treated as if it were the top slice of the estate on death.

(3) Where the transfer was by way of settlement of which the relevant person was the settlor, the lifetime rate will be used if the settlement was created during lifetime; otherwise the death rate will be used.

(Note: Where the chargeable event occurs after 17 March 1986 but the death was before that date, the lifetime rates that apply are those in force after 17 March 1986).

Example 19.4: George – sale of lifetime gift

In March 1992, George (who always used his annual exemptions against gifts made on 6 April each year) gifted his library of rare manuscripts worth £350,000 to his grandson Stephen. George died in June 1994, and under his will left the remainder of his estate, valued at £200,000, to his niece Mary.

To avoid an IHT charge on the failed PET, Stephen lodged the collection with Cardiff University, claiming conditional exemption. The tax charge on the estate was as follows:

	£
George's estate at death	200,000
Nil rate band (1994/95)	(150,000)
Chargeable estate	50,000
IHT thereon at 40% – payable by Mary	20,000

Had Stephen not claimed the conditional exemption, the tax charge on the failed PET and on the estate would have been as follows:

	£
George's estate at death	200,000
Value of failed PET	350,000
	550,000
Nil rate band (1994/95)	(150,000)
	400,000
IHT on PET (40% on £200,000*) – payable by Stephen	80,000
IHT on the estate (40% on £200,000) – payable by Mary	80,000

* *The failed PET has first claim on the nil rate band thus £350,000 – £150,000*

Many years later, Stephen's need for cash resulted in the return to him of the manuscripts, which he sold at auction in June 2017 realising net proceeds after CGT of £425,000 – the disposal of the manuscripts is a chargeable event. The IHT charge on £425,000 at the lifetime rates applicable in 2017/18 is calculated as follows:

	£
George's estate at death	200,000
Sale proceeds	425,000
	625,000
Nil rate band (2017/18)	(325,000)
Chargeable amount all attributable to the sale proceeds	300,000
IHT at 20% payable by Stephen	60,000

No further tax (nor refund available) is payable on Mary's inheritance.

INCREASING RATES OF IHT

19.15 Since 15 March 1988, both the lifetime and death rate of IHT above the nil rate band has been at a flat rather than graduated rate – thus 20% for the former and 40% for the latter (excepting where level of charitable legacies on death meets the condition for the reduced 36%). However, the effective rate of IHT may increases progressively dependent on the estate and cumulative total of recognised chargeable events.

A chargeable event affecting conditionally exempt property will attract a lower charge before the death of the relevant person and before any substantial lifetime transfers by the relevant person. The difficulty for the owner of conditionally exempt property is in knowing who actually is the relevant person at any one time, since the way the rules work, as described at **19.10** above, demonstrates that the identity of that person changes according to a retrospective test.

Example 19.5: Fiona – cumulative sales

Fiona, died in May 2004 leaving an estate valued at £200,000. Fiona had made the following lifetime gifts on which conditional exemption was claimed at that time:

(*a*) a collection of early 20th century Lalique glassware, worth £25,000, to Jason who sold it all in September 2005 for £40,000;

(*b*) a bronze statue worth £35,000 to Sharon, who sold it in June 2006 for £50,000;

(*c*) scientific instruments used in an 1850 geographical survey worth £30,000, to Simon and sold by him in May 2007 for £90,000; and

(*d*) a painting worth £40,000, to Tom, who is considering its sale in March 2018. It is likely to fetch £400,000.

Jason's sale in 2005 is in the band £200,000–£240,000 (estate value + sale proceeds) at a time when the nil rate band stood at £275,000, so nil IHT is payable (£245,000 – £275,000).

Sharon's sale is in the band £240,000–£290,000 (£200,000 + (£40,000 + £50,000)). By June 2006, the nil rate band was £285,000, so £5,000 (£290,000 – £285,000) of the proceeds attract an IHT charge of £2,000 (£5,000 at 40%).

Simon's sale is in the band £290,000–£380,000 (£200,000 + (£40,000 + £50,000 + £90,000)) In May 2010, the nil rate band was £325,000 so £55,000 (380,000 – £325,000) of the proceeds attract an IHT charge of £22,000 (£55,000 at 40%).

Should Tom proceed with his proposed sale in March 2018, the proceeds will attract the full 40% charge as the nil rate band, which has not increased 6 April 2009, has already been fully exhausted.

SET-OFF AND CREDIT

19.16 The circumstances that trigger a chargeable event may also give rise to a chargeable transfer. When that happens, the IHT on the chargeable transfer is allowed as a credit against the IHT payable in respect of the chargeable event but this will not apply to a PET. Where a chargeable transfer is not itself a chargeable event, the tax paid in respect of the chargeable transfer stands as a credit that may be used by way of set-off against the IHT liability on the next chargeable event to occur. The credit may be set off either against IHT or against its predecessor CTT. The set-off may apply even where the conditional exemption arose under the estate duty rule, or under the rules for relief before 7 April 1976 or those that applied thereafter.

Example 19.6: Lucy – tax credit

Lucy had made no lifetime transfers. On her death in February 1992, her estate comprised the value of her house and cash totalling £250,000, plus a collection of rare china worth £100,000. The china was left to her sister Margaret, who claimed conditional exemption on it.

Margaret, who had never married and had not made any lifetime transfers, died in December 2017 when the china was worth £125,000. Margaret's other estate assets totalled £500,000 which all passed to her cousin Fred. He had no wish to retain the old china and did not claim conditional exemption.

Margaret's death is a chargeable event in relation to the china with the tax charge by reference to Lucy's estate calculated as follows:

	£
Lucy's estate	250,000
Value of china (December 2017)	125,000
	375,000
Less nil rate band (2017/18)	(325,000)
	50,000
The china, valued at £125,000, is treated as the top slice of Lucy's estate with tax charged thereon at 40%	20,000

However, there is also a chargeable transfer on Margaret's death calculated as follows:

	£
Value of china	125,000
Other estate assets	500,000
	625,000
Less nil rate band (2017/18)	(325,000)
Chargeable	300,000
Total tax charge at 40%	120,000

Of this £120,000, ⅕ thereof (£125,000/£625,000), is attributable to the china, a tax charge of £24,000 (£120,000 × ⅕). That sum is allowed as a credit against the £20,000 charge in respect of Lucy's estate, reducing it to nil.

The set-off can apply in a different way too – for example consider a PET of property that is already conditionally exempt. At the time of the making of the PET, the conditions and the undertaking may be honoured, with the result that there is no chargeable event. If, however, the PET fails or a chargeable event occurs after the PET and before the death of the transferor, chargeable event tax becomes payable although, at that time, it is not known that the PET will fail. When the PET does in fact fail, the tax on the chargeable event is allowed as a credit against the tax on the failed PET.

Example 19.7: Henry – credit against later charge

Henry managed a family business situated in premises which are part of the national industrial heritage. He died in May 2004, leaving his daughter Sylvia an estate worth £350,000, including the heritage site worth £200,000. Conditional exemption is claimed by Sylvia, who for a time continues both to retain the site and to run the business.

In November 2012, Sylvia, who had utilised her nil rate band and annual exemptions, gave the site and the business to her son Sean (the PET). By then, the business was worth £100,000 and the site was worth £300,000. Sean did not give the undertakings nor claim the exemption; instead, he sold the business and the site for £375,000 in August 2013.

Sylvia died in December 2017 causing the PET to Sean to fail.

There is a chargeable event triggered by the PET by Sylvia to Sean in relation to the previous conditional exempt death transfer by Henry to Sylvia:

	£
Henry's estate excluding the site	150,000
Value of the site at time of gift to Sean (November 2012)	300,000
	450,000
Nil rate band at time of the chargeable event (2012/13)	(325,000)
Balance all attributable to the site*	125,000
'Event charge' @ 40%:	50,000

* *Representing the top slice of the 'estate'*

The gift by Sylvia to Sean was a PET when made, but it becomes chargeable by virtue of Sylvia's death within seven years. Clawback applies to negate any BPR that might have reduced the values, because Sean has sold everything before Sylvia's death. The position is as follows:

	£
Failed PETs:	
Business	100,000
Heritage site	300,000
	400,000
Less nil rate band (2017/18)	(325,000)
Chargeable at 40%	75,000
Attributable:	30,000
Business (¼)	(7,500)
Heritage site (¾)	22,500
Less 'event charge' credit	(50,000)
IHT now payable on the failed PET	NIL

THE 'CLOCK' OF CHARGEABLE TRANSFERS

19.17 There is adjustment of the cumulative total of chargeable transfers of the person who made the last conditionally exempt transfer before the chargeable event, if that person is still alive. The hypothesis is that:

- the transferor, whether or they are the 'relevant person',
- has made a transfer of value of the amount chargeable on the chargeable event,
- at the time of the transfer,
- where that amount is added to their own cumulative total of transfers of value,
- with the result that the rate of tax chargeable on all later transfers made by the transferor may therefore, be increased.

The legislation is contained in *IHTA 1984, s 34(1)* and *(2)*, as varied by the particular circumstances set out in *sub-s (3)*, which relate to settlements where:

- the person who made the last conditionally exempt transfer of the property before the relevant event is *not* the relevant person in relation to that event;
- the property is, or has been at any time within the last five years, comprised in a settlement;

- the transfer was made within the 30 years immediately preceding the event; and
- the transferor has made a conditionally exempt transfer of the property within those 30 years.

Where all of these four conditions are satisfied, it is the settlor who is treated as having made a transfer of the value that is chargeable on the chargeable event. The value in respect of which the chargeable event is triggered is added to the cumulative total of the settlor. The effect is to increase the rate of IHT chargeable on later transfers by that settlor.

However, where the relevant person is *not* the last person to make a conditionally exempt transfer of the property, the calculation of the clawback charge will not affect cumulation of tax on his own estate (although the amount may affect the clawback charge). Equally, previous chargeable, non-exempted, charges are unaffected: there is no extra charge on such transfers.

The old rules in relation to heritage property

19.18 It is in the nature of heritage property that the history not only of the property but of tax transactions relating to it can span a lengthy time frame. For that reason, it may be important for the practitioner to have an awareness of the rules that applied under estate duty and CTT regimes. For a full account, the reader is directed towards specialist works such as *Foster Inheritance Tax* and *Tolley's Administration of Estates* – a summary of the relevant rules may also be found at **Appendix 5** to this book.

Transferable nil rate band

19.19 The introduction of the transferable nil rate band by *IHTA 1984, ss 8A–8C* can cause complications. As already noted, for both heritage relief and woodland relief, it may be necessary to look back to an earlier death to establish the tax charge and it follows that the availability of the nil rate band can affect that tax charge. *Section 8C* takes account of this, where a heritage or woodland clawback charge applies.

Clawback of heritage relief whilst surviving spouse still alive

19.20 If the event triggering the charge happens before the death of the surviving spouse, it is necessary to recalculate the available (or used) nil rate band. Legislation provides the following:

- first, find the nil rate band for the first spouse to die, defined in *s 8A(4)* as 'NRBMD';

- next, find the current nil rate band, ie that in force at the time of the event triggering the charge. This is defined by *s 8C(2)* as 'NRBME';
- next, establish 'E'. This is the excess of the nil rate band over the chargeable transfer at the first death; effectively, the unused nil rate band; and
- finally (for this stage of the computation), calculate 'TA': this is the amount on which the clawback is charged.

Then apply the fraction

$$\left(\frac{\text{E}}{\text{NRBMD}} - \frac{\text{TA}}{\text{NRBME}}\right) \times 100$$

to discover the percentage of the nil rate band in respect of which a claim may be made.

Example 19.8: Mary Brown – transferable nil rate band

Mary Brown's rare vintage car, worth £100,000, was the subject of a conditional exemption at the time of her death in November 2007. Her son had the car re-sprayed in May 2009, which destroyed the essential character of the vehicle and was in breach of the undertaking, triggering a chargeable event. Mary Brown had made chargeable lifetime transfers of £25,000 and left her estate, apart from the car, to her husband William, whom she had married late in life.

(i) Facts:

Nil rate band at November 2007 (NRBMD) = £300,000

Nil rate band at May 2009 (NRBME) = £325,000

Cum. chargeable transfer (M) = £25,000

Conditional Exemption (TA) = £100,000

(ii) Unused nil rate band, November 2007 calculation:

Applying the formula in *s 8A(2)* and *(4)*

(E/NRBMD – TA/NRBME) × 100

(E) = (£300,000 – £25,000) = £275,000

Computation:

$$\left(\frac{\pounds 275{,}000}{\pounds 300{,}000} - \frac{\pounds 100{,}000}{\pounds 325{,}000}\right) \times 100 = 60.8975\% \text{ of the nil rate band}$$

If there is more than one breach of undertaking, or where there are several clawback charges, the nil rate band that may be transferred is reduced by the proportion of the nil rate band clawed back by all the deferred charges, ie by the aggregate of TA/NRBME in respect of each triggering event (see *s 8C(3)*).

Clawback of heritage relief after the second death

19.21 If a deferred charge is triggered after the nil rate band has been transferred, *s 8C(4)* reduces the nil rate band of the first spouse to die (*s 8C(5)*).

The nil rate band of the first to die of the spouses or civil partners is first adjusted by applying *Sch 2* (the uprating provisions that give the benefit of any reduction in the tax that applies because the nil rate band has been increased over time). That uprated nil rate band is then potentially both increased and decreased:

- the increase can apply where the first spouse to die might have more than one nil rate band available, perhaps being a widow or widower;
- the decrease is the amount of any increase in that band by virtue of the nil rate band transfer rules.

The language of the legislation is rather convoluted but the following example may serve as an illustration:

Example 19.9: Interaction of clawback and transferable nil rate band

Sir James's first wife was Peralia who became UK domiciled in 1975 but died in March 1991 leaving her estate to her husband – she had not used her nil rate band. In 1997, Sir James married his second wife Sally and, on his death in November 2009, he left her his estate, having made only one lifetime chargeable transfer of £50,000 (a failed PET).

The family seat qualified as heritage property, but a sale of amenity land for £200,000, to meet Lady Sally's debts after her death in September 2023, triggered a clawback charge under the 'associated property' provisions of *s 32A*.

The nil rate band on Peralia's death in 1990/91 was £128,000 and in November 2009 the nil rate band had increased to £325,000, so the total nil rate band available to Sir James's executor was £600,000 (£325,000 – £50,000) + £325,000). The nil rate band at Sally's death in 2023 is, say, £375,000.

The reduction under *s 8C(5)* is of the increase in the nil rate band from Sir James's death to that of Sally of £50,000 (£375,000 – £325,000). Thus, Sir James's executor has available a reduced (double) nil rate band of £350,000 (£600,000 – (£50,000 – £200,000)).

Note that, if a Scottish law claim to 'legitim' is made after a claim to transfer unused nil rate band following the death of the second parent, HMRC may adjust the claim in respect of the unused nil rate band accordingly (see *s 147(10)* and IHTM43041).

Sting in the tail: ATED liability on top of recapture charge

19.22 *FA 2013, Pt 3*, introduced the Annual Tax on Enveloped Dwellings (ATED) but *FA 2013, s 155(2)* treats the value of such a dwelling, pending any chargeable event under *s 32*, as zero where:

- all or part of the dwelling is designated under *s 31*;
- an undertaking has been given under *s 30*; and
- as a result, a transfer of the dwelling is exempt as heritage property.

There is a similar provision under for settled property (*s 155(3)* and *(4))* where the taxable value is taken to be zero on any day if no chargeable event under *s 32* has occurred from the conditionally exempt occasion to the beginning of the next day.

Thus a chargeable event may have much wider and possibly more urgent effect than just the IHT charge.

Compliance

19.23 Compliance generally is discussed later in **Chapter 22**. Where conditional exemption is lost, the owner must deliver an account in form IHT100 within six months of the end of the month in which the 'loss event' arose and accounts filed late will trigger penalties.

Chapter 20

Heritage property: the regime for relevant property trusts

QUICK GUIDE

- This chapter is only concerned with relevant property trusts only, namely discretionary trusts and any new lifetime trust created post 21 March 2006 (excepting privileged trusts).
- As may be expected, the combination of rules for relevant property trusts and for heritage property makes for some fine distinctions and difficult computations. If 'clawback' arises, it is necessary to identify the 'relevant person' (except where the choice lies with HMRC), and apply the computation rules in **20.7** whilst taking note of the anti-avoidance rule described at **20.8**. The current form IHT100 is a considerable improvement on its predecessor and the calculation guide will assist.
- The thrust of the trust reforms contained in *FA 2006* have resulted in fewer life tenants who hold a qualifying life interest for IHT purposes, but those that do are treated in the same manner as the owner of the property.

TREATMENT OF RELEVANT PROPERTY TRUSTS: RULES APPLICABLE AFTER 8 MARCH 1982

20.1 *IHTA 1984, s 78* (which only relates to events happening after 8 March 1982, regardless of the regime under which the original conditional exemption was granted) sets out the criteria leading to conditional exemption from an exit charge which would otherwise arise in relation to a relevant property trust. The conditions are that:

- the property has been comprised in the settlement throughout the six years ending with the transfer of it, or with any other event that would otherwise have given rise to an exit charge; and
- the property is designated as qualifying heritage property by HMRC under the *s 31* regime; and

- the appropriate *s 31* undertaking is given by such person as HMRC consider to be appropriate in the circumstances; or
- in the case of amenity land falling within *s 31(1)(d)*, the appropriate undertaking for amenity land is given with regard to the amenity land by such person(s) as HMRC consider appropriate.

Detailed HMRC Guidance is set out in CTNH Appendix 2 at:

- para 15, which deals with qualifying interest in possession trusts.; and
- paras 16–21, which deal with the relevant property trust tax regime.

As was noted in **Chapter 18**, any claim under *s 78(1)* must be made within two years after the date of the transfer or other event (or exceptionally within such longer period as HMRC allow). The regime for relevant property trusts links directly to the requirements of *ss 32, 32A, 33(1), (3)–(7)* and *35(2)* and the charging provisions under *ss 32* and *32A*. This replicates the following requirements in respect of designation and of undertakings on or after 31 July 1998:

- chattels must be pre-eminent;
- undertakings must ensure that there is a measure of open public access;
- undertakings must satisfy the new requirements for greater publicity;
- information about conditionally exempt property must satisfy the new higher standards;
- the new rules as to varying undertakings apply; and
- the rules in *s 35A* apply directly to property in a relevant property trust so as to affect undertakings that were given before the coming into force of *FA 1998*.

Whenever conditional exemption is granted it must be recognised that there is the possibility of a tax charge (quaintly described in CTNH as a 'ripened' charge) at some later point. However, there is a mechanism to credit tax against concurrent and future claims which serves to mitigate the effect of the clawback charge (*s 33(7)*).

CLAWBACK CHARGE ON RELEVANT PROPERTY TRUSTS

20.2 As with absolute ownership, unless conditionally exempt, the occasion of certain events will give rise to an IHT charge where:

- an undertaking is broken in some material respect; or

- the heritage property is sold; or
- the heritage property is disposed of otherwise than by sale, including on an occasion which triggers an exit charge under the normal regime for taxing relevant property trusts.

Calculating the tax charge

20.3 Relevant property trusts have their own regime for calculating the IHT charge and this is dependent on the identity of the settlor as the relevant person and the timing of the chargeable event thus:

- If the chargeable event occurs before the first ten year anniversary of the date of the trust creation to fall after the heritage property became comprised within the trust, and the 'relevant person' is the settlor, the rate of tax is 30% of what it would otherwise be (*s 78(4)(a)*).
- If the chargeable event occurs after the trust's first ten-year anniversary but before the second ten-year anniversary after the heritage property became comprised within the trust, and the 'relevant person' is the settlor, the rate of tax is 60% of what it would otherwise have been (*s 78(4)(b)*).

The rule observed previously, relating to the reinstatement of the cumulative total of transfers or 'clock' of the transferor, does not apply to trustees under this regime.

THE 'RELEVANT PERSON' AND SPECIAL RULES

20.4 IHTA 1984 s 78(3) provides that the settlor of a relevant property trust will be the relevant person but if there is more than one settlor, HMRC may choose which of them is to be the relevant person for this purpose.

The rate of tax depends upon whether the settlor (identified as the relevant person) is alive or dead at the time of the chargeable event and that in turn affects the rate of tax:

- If the selected settlor is alive, he is treated as having made a chargeable transfer which will require reference to be made to his cumulative 'clock' (*33(1)(b)(i)*).
- If the selected settlor is dead, having died after 13 March 1975, the deemed transfer is added to and treated as the top slice of his estate at death but yet the tax rate will depend on whether the actual settlement was created in lifetime or on death (*33(1)(b)(ii)*).

- If the selected settlor is dead, having died before 13 March 1975, the rate of tax to be applied is the rate that would have applied to the value of the heritage property at the time of the chargeable event (*IHTA 1984, s 78(5)*):
  - if the settlor had died when the chargeable event occurred;
  - the value transferred on his death was equal to the amount on which estate duty was chargeable when he in fact died; and
  - the value of the property now being charged had been added to the estate on death and formed the top slice of it.

Example 20.1: Nellie – charges on selling silver

In August 1983, Michael created a discretionary trust into which he made a chargeable cash transfer of £100,000.

In January 1989, he transferred to the trust a collection of silver worth £40,000 – his cumulative 'clock' of transfers in the previous seven years (including the initial cash transfer of £100,000) totalled £135,000.

The trustees claimed and received conditional exemption on the silver at that time.

In February 1995, finding the obligations of insurance and protection burdensome, they appointed the silver *in specie* to a beneficiary, Nellie, who also claimed the benefit of the conditional exemption. The collection was by this time worth £50,000.

In January 2000, Michael died leaving an estate of £200,000. His cumulative total of lifetime transfers made in previous seven years was £150,000.

In March 2000, Nellie, who had only kept the silver so as not to hurt Michael's feelings, sold it all for £75,000.

The IHT charges are as follows:

Event	*IHT charge on the silver*	*Reason*
January 1989 – on transfer into trust	Nil	Conditional exemption applied
August 1993 – the first ten-year anniversary	Nil	Conditional exemption applied
February 1995 – appointment to Nellie	Nil	The trustees had held the silver for six years so conditional exemption applied

Event	*IHT charge on the silver*	*Reason*
March 2000 – on sale by Nellie	60% of the lifetime rate	This was the second decade of the trust and the gift of the silver was made in Michael's (as settlor) lifetime even though he had died by the time of the chargeable event

	£
Sale proceeds	75,000
Michael's 'clock'	150,000
Michael's estate	200,000
Upper end of range	425,000
Transfer in the range £350,000–£425,000:	
£75,000 at 20%	15,000
Charge is 60% of £15,000	9,000

Exemption from the ten-year periodic charge

20.5 IHTA 1984 *s 79* provides an exemption from the periodic charge in respect of property comprised within the trust where the property was subject to a conditional exemption either on or before it became comprised within the trust.

Exemption will also apply where the disposal of the trust property qualified for relief from CGT under *TCGA 1992, s 258(4)* (the regime for relief of heritage property from CGT) and there has been no chargeable event in relation to it since that exempt disposal.

Two exemption regimes

20.6 There is slightly different treatment within the relevant property trust according to whether the exemption is granted under IHT or CGT.

- In relation to property conditionally exempt from IHT, that exemption continues until there is a chargeable event.
- In relation to property exempted for CGT, that exemption continues until the property is disposed of (as defined under the CGT regime) or there is a breach of the undertaking.

Either way, at the trust's next ten-year anniversary post IHT exemption loss, the standard IHT regime for taxing relevant property trusts will apply and, at that point, the cumulative total applicable to the trust will be used unaffected by exemption on any previous ten-year anniversaries.

The exemption from the ten-year anniversary charge does not automatically apply if there has been neither a conditionally exempt transfer nor a CGT relieved disposal in favour of the trust on the transfer into the settlement or before it. However, where the usual undertakings are given, and designation is given by HMRC, such qualifying property can be relieved from the forthcoming ten-year charge. Prior to 18 November 2015, such a claim for exemption and designation had to be made sufficiently in advance of the ten-year anniversary to ensure its acceptance by HMRC before the anniversary date. However, *F(No2)A 2015* relaxed the provision extending the clearance window to up to two years post the ten-year anniversary date (note: CTNH Appendix 2 has not yet been updated to reflect this change).

A charge to tax arises on the first chargeable event that would occur under the general regime but for conditional exemption. That charge, otherwise applied by virtue of *s 79(4)*, will not arise if, between the date of the settlement and the date of the notional chargeable event, there has been an occasion which is conditionally exempt. This is because, in those circumstances, the normal rules for taxing relevant property trusts would give rise to a charge. There would be a real event giving rise to a full charge.

CALCULATING THE RATE OF IHT ON NEW HERITAGE PROPERTY IN A RELEVANT PROPERTY TRUST

20.7 The HMRC guidance is set out in CTNH Appendix 2 para 21.

Step 1 – The appropriate 'start date' for the period to which the rates will apply is the later of:

- 13 March 1975
- the commencement of the settlement
- the last ten-year anniversary of the trust that occurred before the day on which the heritage property came to the trust.

Step 2 – The 'relevant period' runs from the appropriate date established at Step 1 to the date of the event that triggers the charge.

Step 3 – That 'relevant period' is divided into complete successive quarters (of a year) and the rate is charged which over a period of 50 years, reaches a maximum of 30%. It is illustrated (in whole years) in the table below and is applied as follows:

- first 40 quarters: 0.25%
- next 40 quarters: 0.2%
- next 40 quarters: 0.15%
- next 40 quarters: 0.1%
- next 40 quarters: 0.05%.

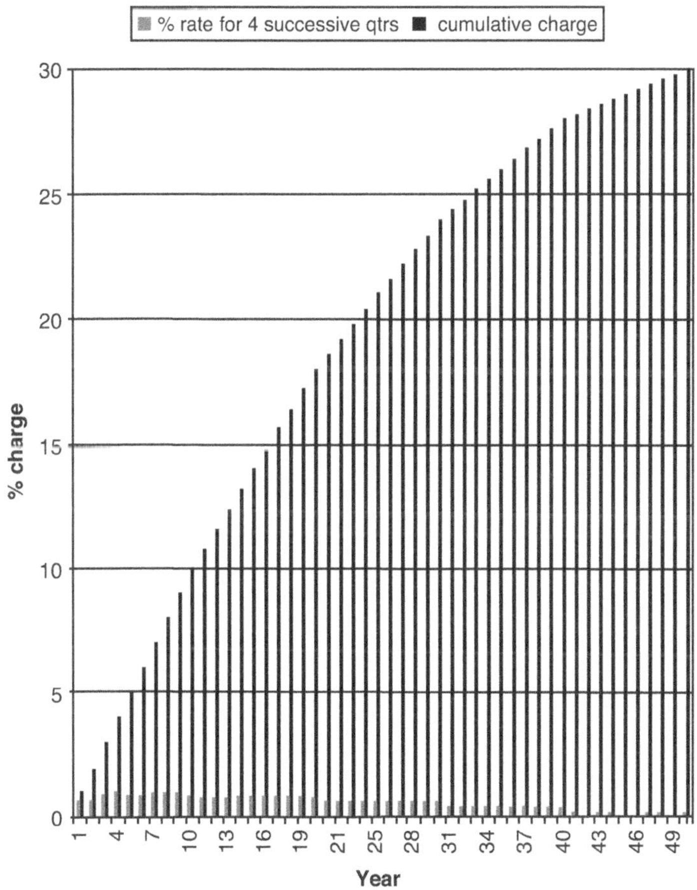

The charge is levied on the value at the time of the event, but subject to the 'market value' rule which provides for substitution of the sale price where the sale was not at full market value (*s 79(5A)*).

In relation to relevant property trusts, *s 207(3)* provides that the persons liable for tax charged under *s 79(3)* are:

- the trustees of the settlement; and
- any person for whose benefit any of the property, or income from it, was applied at or after the time of the event triggering the clawback charge.

Anti-avoidance: prevention of the purchase of heritage property to keep down periodic charges

20.8 *Section 79(9)* sets out the position for calculating the ten-year anniversary charge tax levied under the general rule of *s 64*. The value of any consideration given for heritage property is calculated using a statutory fiction with the price paid for the heritage property treated as if it were a distribution triggering an exit charge. This has a cumulation effect when calculating the next ten-year charge on any other property within the trust but subsequent ten-year charges will not be affected unless, of course, more heritage property has been purchased in the meantime.

Chapter 21

Maintenance funds

QUICK GUIDE

- Relief is available on funds set aside for the maintenance of heritage property and the various reliefs are catalogued at **21.2**. The trust must be:
  - of an acceptable wording;
  - with acceptable trustees;
  - holding acceptable property;
  - for the maintenance of suitable heritage property.
- Once established and authorised, rules apply to supervise the fund and its accounts. A flow chart at **21.12** lists the progress of the procedure.
- Anti-avoidance rules abound, including those as to 'excess value' and 'purchase' noted at **21.16**. Further anti avoidance rules are summarised at **21.18** and illustrated thereafter.
- Relief is clawed back where the fund is used for unacceptable purposes, such as:
  - the release of capital;
  - dispositions by trustees;
  - the loss of heritage status in respect of the property for which the fund was established; or
  - the withdrawal of the Treasury Direction: see **21.24**.
- Where funds were previously in a relevant property trust, a somewhat simpler calculation applies: see **21.28**.
- There is a CGT regime for maintenance funds; for completeness, it is outlined at **21.32**.

OUTLINE OF THE SCHEME

21.1 More often than not it is not the original expense of the outlay of the asset but the upkeep that costs – this is particularly so with certain heritage property and has inevitably led to the loss or destruction or dilapidation of many fine properties. That very loss to the national heritage is the focus for the funds described in this chapter as this part of the legislation is designed to encourage the settlement of funds that are to be used towards the cost of upkeep of qualifying heritage property. The IHT code for maintenance funds contains strict provisions to curb abuse, leading to complexity which will, in itself, be enough to deter many taxpayers from setting up such funds. However, assets can eventually be returned to the settlor or applied for non-heritage purposes albeit subject to an appropriate tax charge. The scope of the exemption has been enlarged so that maintenance funds may be used to benefit a wide range of heritage property.

The main relieving provisions in *IHTA 1984* that trigger an examination of the code relating to maintenance funds are:

- *s 27* (the general provision in respect of exempt transfers to maintenance funds);
- *s 57A* (inserted into the regime for taxing interest in possession trusts);
- *s 58* (defines 'relevant property' for the IHT regime for discretionary trusts, the definition of which specifically excludes property passing to maintenance funds); and
- *s 77*, introduces the code for maintenance funds in *Sch 4*.

Exemption from several taxes is granted to a trust fund set up for the maintenance, repair or preservation of heritage property, and for the provision of reasonable public access to it. As a niche subject there is very little case law on heritage property, but there is substantial official HMRC guidance in CTNH at chapter 8.

Summary of the relief

21.2 Listed below are the main points of the relief:

- IHT is not charged when property is put into the fund: see *s 27, s 57(5)* and the combination of *s 65* and *Sch 4, para 16(1)*.
- The charge to CGT which might otherwise arise when property is transferred into the fund may be deferred.
- The fund is not subject to a ten-year anniversary charge.

- At the option of the trustees, the fund is not treated as settlor interested for income tax purposes but it does suffer tax at the rate applicable to trusts.
- At the option of the trustees, cash paid from the fund and applied for heritage purposes is not treated, as might otherwise be the case, as the income of a person by virtue of that person's occupation of the heritage property.
- IHT is not charged where capital is taken from the fund for heritage purposes, including resettlement into a new maintenance fund.
- CGT deferral may be claimed where assets are transferred from the maintenance fund to another trust, including a maintenance fund or to an individual.
- Stamp duty land tax liability is restricted to the single fixed conveyance duty.

'As you were' tax provisions apply where any part of the maintenance fund is used for a purpose not connected with heritage. In particular, under these provisions, depending on who receives the money, there may be charges to IHT. There will be an income tax charge at up to 30% on any of the fund income that has been taxed only at the rate applicable to trusts and which has not been applied for heritage purposes. Normal CGT rules apply, such that holdover relief under *TCGA 1992, s 260* will be available in respect of a transfer from a relevant property trust. Stamp duty land tax can be payable in certain circumstances.

CONDITIONS FOR THE RELIEF

21.3 The legislation is mainly to be found in *IHTA 1984, Sch 4*, and the conditions for the relief are set out in *paras 2–4*. Where those conditions are satisfied and a claim has been made, HMRC give a 'Direction' under *para 1* but the direction may also be given in advance of the creation of the fund, where HMRC are satisfied that they would be obliged to give a 'direction' once the property was so comprised. Transitional relief is allowed under *para 1(3)* in respect of property comprised in a settlement as a result of transfer of value that was exempt under *FA 1976, s 84* and made before *FA 1982, s 94* came into force.

The conditions relate to:

- the wording of the trusts of the maintenance fund;
- the trustees of the fund;
- the property in the fund; and
- the property to be maintained.

The first condition: the trust instrument

21.4 Drafting the maintenance fund trust is critical to the success of the application and it should be appreciated that it may take time to arrive at a form that is acceptable both to the party wishing to set up the fund and to the 'other side', as represented by HMRC. The following areas should be addressed:

- definitions;
- retention or sale of trust fund;
- power of accumulation;
- maintenance trusts;
- default trusts of income for qualifying charities;
- powers of appointment and ultimate default trusts;
- power of investment;
- powers in relation to real property;
- power to borrow;
- trustee charging clause;
- power to enter into transactions with the trustees;
- power to appoint nominees;
- appointment of new trustees;
- provision of accounts; and
- stamp duty certificate.

CTNH para 8.5(c)(iii) sets out HMRC requirements and observations as to the trustees' powers.

The six-year rule

Schedule 4, para 3 sets out the terms of a trust that will be acceptable for a maintenance fund.

The first is 'the six-year rule' (see *para 3(1)(a)*) which states that none of the property held on the trusts can at any time in the period of six years, beginning with the date on which that property came to the maintenance fund, be applied for a 'non-heritage purpose'. Acceptable purposes for the use of the property are basically:

- the maintenance of certain categories of heritage property; or

- the making of that property available to a heritage body or qualifying charity.

More specifically, *para 3(1)(a)(i)* identifies a number of uses which are acceptable for property in a maintenance fund. These uses are all in relation to the heritage property for the maintenance of which the fund has been established.

In this chapter, such heritage property is called 'qualifying property', the name given to it by *para 3(2)*, and the funds or assets set aside for the maintenance of qualifying property are called the 'trust property'. The categories of heritage property that may benefit from a maintenance fund are, land, buildings, amenity land and historically associated objects; but critically not works of art (see CTNH para 8.3).

Paragraph 3(4) contains another definition of a 'qualifying charity'. Such a charity is one that exists wholly or mainly for the maintaining, repairing or preserving for the public benefit of either buildings of historic or architectural interest, land of scenic, historic or scientific interest, or objects of national, scientific, historic or artistic interest. 'National interest' includes interest within any part of the UK. The acceptable uses for the trust property are, in relation to qualifying property:

- maintenance, repair or preservation of qualifying property;
- provision for public access to qualifying property;
- maintenance, repair or preservation of the trust property;
- improvement of the trust property where such improvement is reasonable, having regard to the purposes of the trusts of the maintenance fund; and
- paying the expenses of trustees in relation to the trust property.

Alternative use of the income of the fund under para 3(1)(a)(i)

The second limb of *para 3(1)(a)* is the alternative provision where the trust property is not used for the maintenance etc of the qualifying property. This second limb will be appropriate where the qualifying property is, for example, sold by allowing for the application of surplus income of the maintenance fund to a qualifying charity or to a heritage body as defined in *Sch 3*.

Alternative use of the income or capital of the fund under para 3(1)(b)

The second main condition that must be included in the trusts of an acceptable fund is that none of the property can, within the period of six years from the establishment of the trust or within the lifetime of the settlor if he dies less than six years after the making of the trust, pass otherwise than to a heritage body or qualifying charity.

There is a rider to this rule which applies where a maintenance fund is established following an interest in possession (irrespective of whether a qualifying interest or no) in a trust. This particular situation is covered by *para 15A*, which came into force in respect of events after 16 March 1987 and is considered at **21.17** below. Where that code applies, the effect of *para 3(5A)* is to apply a further condition to *para 3(1)(b)* namely:

- none of the property can:
  - on ceasing to be held on the maintenance fund at any time in the six-year period; or
  - if the tenant for life dies within that period, at any time before the death of the tenant for life;
- devolve otherwise than to a heritage body or a qualifying charity.

As to income, the rule is absolute. That income, unless accumulated, must be used only for the qualifying purposes set out at the beginning of this section. That rule applies not merely for the first six years but for the entire duration of the fund. Any maintenance fund trust must, by its drafting, satisfy this rule.

Qualifications to the first condition

21.5 The rules just set out are varied where property was previously comprised in another settlement and is transferred to the maintenance fund under circumstances that qualify for exemption under the rules in *Sch 4*. The condition as to income still applies, namely that it may be applied only for the maintenance purposes set out above. In these circumstances, it is assumed that the six-year period will already have expired at some time in relation to the previous settlement but where that period has not expired, the six-year period begins not with the date on which the property comes to the present, maintenance fund, but with the date on which it was comprised in the earlier settlement. These adjustments to the general rule are contained in *Sch 4, para 4*.

Maintenance of heritage property

21.6 A distinction is drawn between:

- repairs to the qualifying property; and
- alterations.

Essentially, the fund may be used only for repairs; thus the provision of a new roof for a house would generally qualify, but the acquisition of new property would not. By way of illustration of running repairs and maintenance costs that will qualify, see the guidance in CTNH para 8.7 which may be summarised as follows.

Buildings

- preserving stonework and fabric;
- internal redecoration;
- general maintenance;
- heating costs where necessary to maintain or preserve the building;
- heating costs where necessary to preserve qualifying contents; and
- heating costs where necessary for the purpose of public access.

Qualifying land

- bracken control;
- scrub clearance;
- woodland management;
- replacement of mature trees, for example in parkland;
- fencing;
- maintenance of hedges, hedge-banks and walls; and
- upkeep of ditches and drainage.

Gardens

- the cost of seeds and plants, materials and implements;
- the repair and heating costs of greenhouses; and
- wages for gardeners.

Historically associated objects

- cleaning and restoration of pictures;
- repairs to tapestries, curtains and carpets; and
- repairs to clocks, furniture and similar objects.

Public access requirements

21.7 The appropriate expenditure may be either capital or income in nature. The main illustrative categories of appropriate expenditure are as follows:

- provision and maintenance of car parks;
- provision and maintenance of public lavatories;
- wages of guides and attendants;
- special fire precautions;
- lighting, heating and security costs where necessary for public access; and
- maintenance of roads and drives where necessary for public access.

The property within the maintenance fund itself will often need management; for example, it may consist of a landed estate that supports a stately home. All expenses necessary for the proper and 'prudent' management of the assets of the fund will qualify and the trustees may spend either capital or income on repairs and maintenance of the property. In a similar way, the property in the heritage fund may be improved so as to generate a higher income which could involve the purchase of extra farmland or other expenditure which might assist good management of the trust assets. All of this is, however, subject to the qualification that these improvements must be 'reasonable having regard to the purposes of the trust'. The prime requirement is the maintenance of the qualifying property and improvement of the property in the fund must take second place. HMRC view on this and on 'character and amount appropriate' is set out in CTNH para 8 5(v).

Expenses and other items

21.8 Professional fees, any rates and taxes in respect of the capital or income of the fund and certain other expenses are properly payable from the fund. In particular, insurance of the qualifying property itself is a proper expense, as is the cost of public liability insurance in relation to public access.

The second condition: the trustees

Status

21.9 By *Sch 4, para 2(1)(b)*, the trustees must be approved by HMRC and they must include either a trust corporation or a solicitor or an accountant or a member of such other professional body as HMRC may allow or recognise.

Residence

The trustees must be resident in the UK at the time that HMRC make the 'direction', with 'resident' defined for this purpose as being satisfied where the general administration of the trusts is ordinarily carried on in the UK, and the trustees or a majority of them are resident in the UK (*Sch 4, para 2(2)*). Where there is more than one class of trustees, a majority of each such class must be resident in the UK. The residency of a trust corporation is determined as for the purposes of corporation tax, involving the test of where the central management and control of the trust corporation resides.

The third condition: the content of the maintenance fund

21.10 The third principal condition to be satisfied is that the property transferred, or to be transferred, to the maintenance fund is 'of a character and amount appropriate for the purposes of those trusts'. The official guidance in CTNH para 8 5(v) states that assets which provide little or no income are, on the face of it, unsuitable for a maintenance fund, particularly if these assets themselves require money to be spent on them for their maintenance. Equally, the amount of property transferred to the fund should not be excessive in relation to the likely burden of expenditure on the fund. The fund need not be established all at once and it is possible for a small initial fund to be established, provided that it is clear that further assets will be added to the maintenance fund later.

Further condition: anti-avoidance 'charity' rules

21.11 A further condition applies before transfer to a maintenance fund can attract relief but this is not specific to such funds in isolation; rather it applies to several forms of transfer involving gifts to charities. Despite public perception, transfers of value to charities are not always exempt (see *s 23(2)*) and this is so where the transfer:

- takes effect only on the termination of some interest or period after the transfer of value; or
- is dependent on some condition which is not satisfied within 12 months after the transfer; or
- is defeasible.

(For this purpose, a disposition which has not in fact been defeated on the first anniversary of the transfer, and is not defeasible after that time, is treated as not being defeasible.)

Furthermore, in the context of charities, exemption shall not apply (see *s 23(3)*) in relation to property carved out of other property where:

- the interest so carved out is less than the full interest of the donor to charity; or
- the interest so carved out is given for a limited period only.

Section 27(2) carries across the application of both *s 23(2)* and *s 23(3)* to maintenance funds and also imports *s 24(4)* in relation to gifts made in similar circumstances to political parties.

It is here that CTNH para 8.4 is helpful in summarising the effect of *s 27(2)*. The tax reliefs for a maintenance fund will not apply where the transfer into an existing fund results in the trustees acquiring a reversionary interest only by purchase. Equally, the tax reliefs will not apply where property is transferred to the maintenance fund in exchange for a reversionary interest which does not form part of the estate of the person who acquires it.

PROCEDURAL POINTS

21.12 The application for exemption should be made to the designated property section of:

HMRC Capital Taxes, Heritage Section, Ferrers House, PO Box 38, Castle Meadow Road, Nottingham, NG2 1BP.

It must be made within two years of the transfer although, by *IHTA 1984, s 27(1)(a)* HMRC may exercise their discretion to extend (see CTNH para 8.4). Where the application is made after the maintenance fund has been set up, HMRC must give the 'direction' if satisfied that the conditions mentioned in this chapter have been satisfied. Note, however, that by authority of *Sch 4, para 2(1)(a)(ii),* when a claim to relief is made HMRC will test whether the proposed fund will do what is intended and specifically the value settled must be realistic (which HMRC interpret to be at least £10,000). Where the property has not yet been settled, HMRC have a discretion as to whether or not to give a 'direction'.

Satisfying the condition as to timing may present personal representatives with significant problems since in the absence of a 'pilot' trust, it will take time to agree the terms of the trust instrument so as to ensure compliance with the relevant conditions. Alternatively, the personal representatives may hold property on discretionary will trusts and may be keen to exercise their discretion within two years of the date of death, so that the action of passing of assets into the maintenance fund is written back to date of death. Equally, there may be issues involving preparation and agreement of a deed of variation, again within the two-year time limit.

In such a situation, the practical advice is to contact the Heritage Section of HMRC and discuss possible solutions. In that way if the personal representatives do find that they are running out of time, an application to HMRC to exercise its discretion can be considered. The flow chart below, outlining the procedure, may perhaps be used as a checklist.

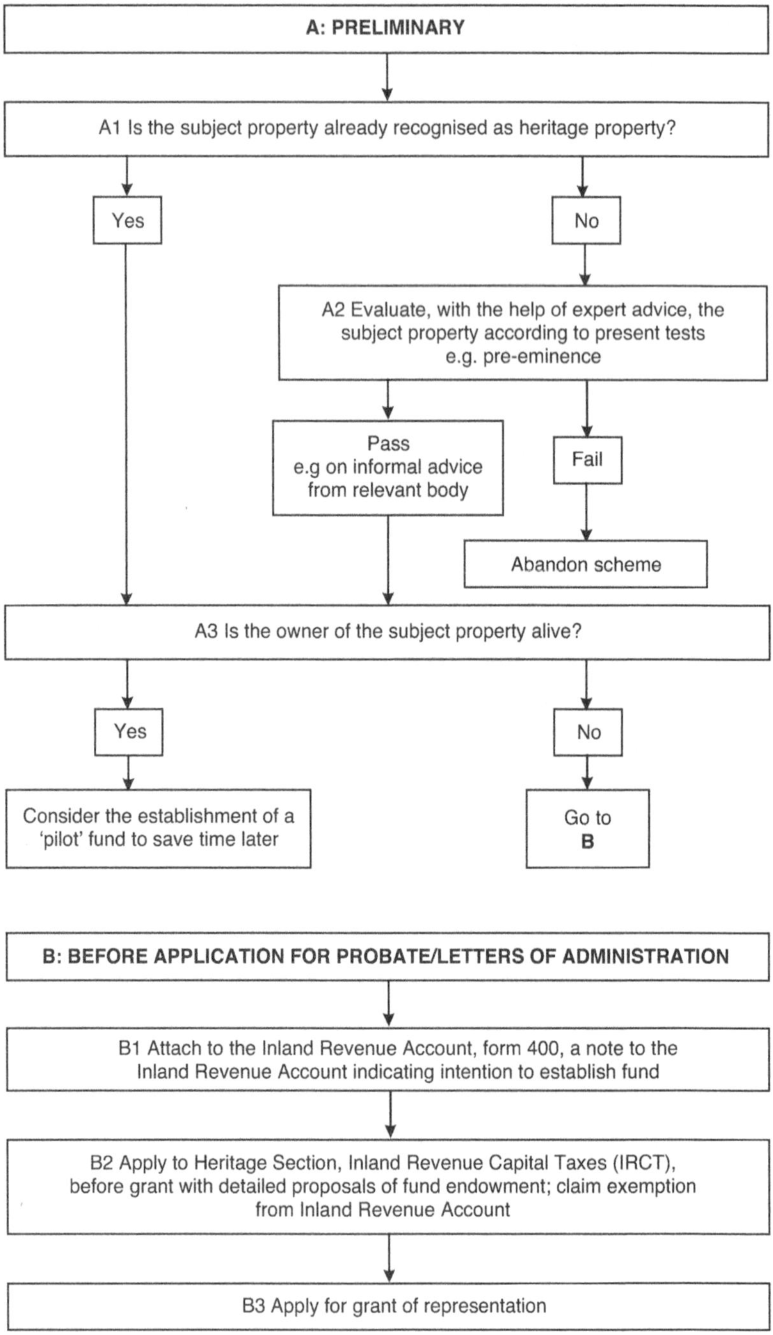
A: PRELIMINARY
A1 Is the subject property already recognised as heritage property?
Yes
No
A2 Evaluate, with the help of expert advice, the subject property according to present tests e.g. pre-eminence
Pass e.g on informal advice from relevant body
Fail
Abandon scheme
A3 Is the owner of the subject property alive?
Yes
No
Consider the establishment of a 'pilot' fund to save time later
Go to B
B: BEFORE APPLICATION FOR PROBATE/LETTERS OF ADMINISTRATION
B1 Attach to the Inland Revenue Account, form 400, a note to the Inland Revenue Account indicating intention to establish fund
B2 Apply to Heritage Section, Inland Revenue Capital Taxes (IRCT), before grant with detailed proposals of fund endowment; claim exemption from Inland Revenue Account
B3 Apply for grant of representation

C: THE APPLICATION ITSELF

↓

C1 Make more formal application for relief and ask for the claim to be admitted. Details as in the next boxes.

↓

C1(a) The subject property details:

- here set out the details
-
-

↓

C1(b) Maintenance programme

(a) recent expenditure on the subject property
(b) forecast expenditure over the next three years

↓

C1(c) Fund details: capital aspects

- assets to be held in the fund
- running costs of above assets on same basis as maintenance programme

↓

C1(d) Fund details: revenue aspects

- income of each part of the fund
- projected administration costs

↓

C1(e) Details of proposed trustees

↓

C1(f) Draft trust instrument, to specify:

- here set out the main provisions, for example, based on those referred to at **21.4** above
-
-

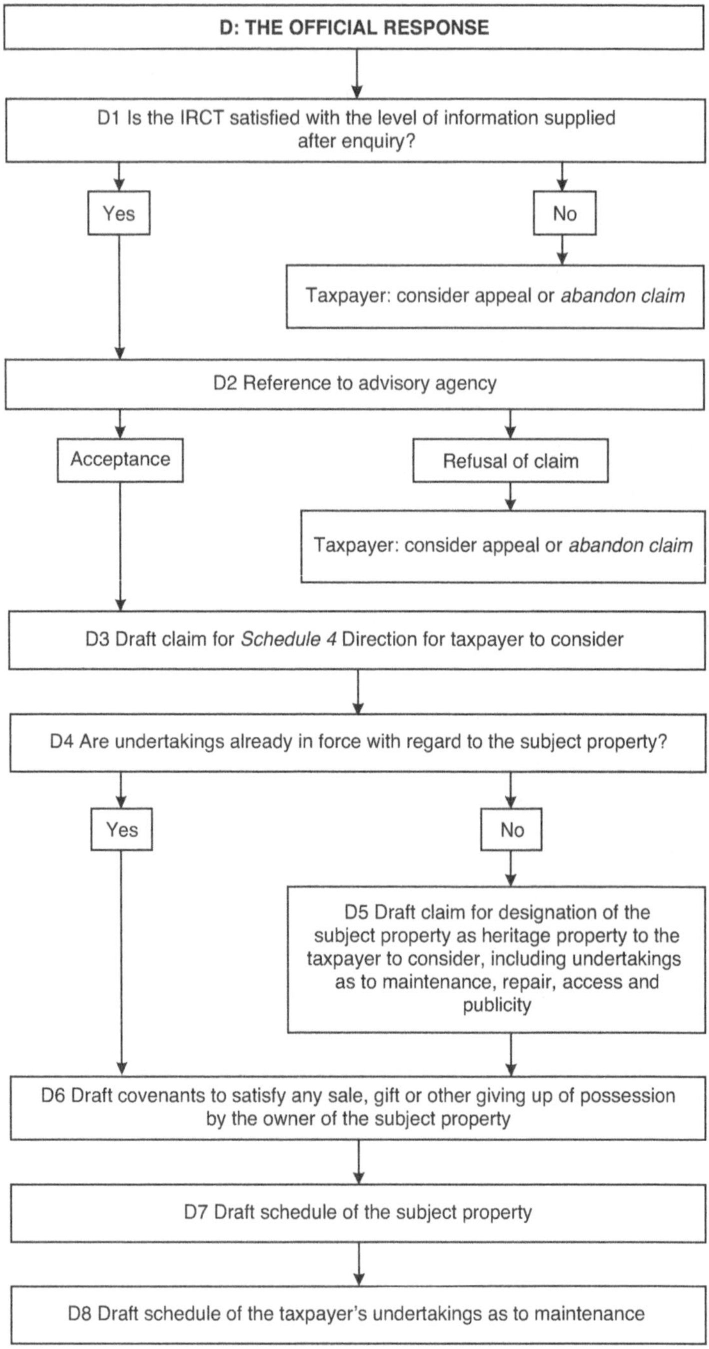
D: THE OFFICIAL RESPONSE
D1 Is the IRCT satisfied with the level of information supplied after enquiry?
Yes
No
Taxpayer: consider appeal or abandon claim
D2 Reference to advisory agency
Acceptance
Refusal of claim
Taxpayer: consider appeal or abandon claim
D3 Draft claim for Schedule 4 Direction for taxpayer to consider
D4 Are undertakings already in force with regard to the subject property?
Yes
No
D5 Draft claim for designation of the subject property as heritage property to the taxpayer to consider, including undertakings as to maintenance, repair, access and publicity
D6 Draft covenants to satisfy any sale, gift or other giving up of possession by the owner of the subject property
D7 Draft schedule of the subject property
D8 Draft schedule of the taxpayer's undertakings as to maintenance

E: NEGOTIATION BETWEEN IRCT AND TAXPAYER

F: COMPLETION BY THE TAXPAYER

↓

F1 Taxpayer completes trust instrument for the fund

↓

F2 Taxpayer (owner of subject property) completes undertakings

↓

F3 Trustees of the maintenance fund complete undertakings

G. CONFIRMATION BY THE REVENUE

↓

G1 IRCT issue letter of confirmation

- that the subject property is 'qualifying property' within *Sch 4 para 3* (*2*)
- of a Direction under *Sch 4 para 1* (*1*)
- that the maintenance fund is exempt IHT under *IHTA 1984, s 27* (*1*)

Supervision of the fund

21.13 Once the trust is established and a 'direction' has been made, HMRC will still have certain powers of supervision. HMRC may withdraw that 'direction' if they consider that the trustees are exercising their powers in a manner that is inconsistent with the purposes of the maintenance fund. CTNH para 8.10 states that, in these circumstances, HMRC will normally give the trustees an opportunity to put matters right before exercising the authority under *Sch 4, para 5* to withdraw the benefit of the tax relief on the fund itself. In the event that the 'direction' is withdrawn in this way, the tax charges, set out later in this chapter, would apply.

Accounts

21.14 HMRC may revoke the 'direction' if such revocation is warranted by circumstances affecting any property or its administration. HMRC can call for the accounts of the maintenance fund and information concerning the fund.

Where the income of the fund exceeds £25,000 gross, HMRC will expect independent examination and production of annual accounts with a balance sheet and a statement of income and expenditure, plus an annual report on the activities of the fund describing what the trustees have done during the year – this is necessary not least since, as will be seen below, a charge to tax may arise when property is withdrawn from the fund for non-heritage purposes. It is necessary to show how the income of the fund had been taxed, what part of the fund was accumulation of income, and what part of the income had been applied for maintenance of the heritage property. By *Sch 4, para 7*, HMRC are given all the rights and powers of a beneficiary in relation to the enforcement of the trusts of the maintenance fund, and in relation to the appointment, removal and retirement of trustees.

Where the income of the fund is less than £25,000 the requirements are less onerous – see CTNH para 8.10. In neither case is the fund required to instruct an external professional audit, though an audit clearly counts as examination.

Entry of property to the fund

21.15 Property may become comprised in a maintenance fund in a number of ways:

- by straightforward transfer exempted under *s 27*, for example by an individual or a company; or
- by a transfer from a qualifying interest in possession trust (see below), where the exemption will arise under *s 57* or *s 57A* (in respect of transfers after 16 March 1987); or
- where the property has been 'relevant property within a discretionary trust' before being transferred to the maintenance fund, in which case exemption arises through the operation of *s 65* and *Sch 4, para 16(1)*.

For this purpose, a qualifying interest in possession trust is defined by *s 57A(1A)* to include any one of the following:

- an interest established prior to 22 March 2006;
- an IPDI;
- a disabled person's trust; or
- a transitional serial interest; or
- a trust coming within the narrow category of '*s 5(1B)* interests' (see below).

As to *s 5(1B)* interests, the amendment was introduced by *FA 2010* where the person became entitled to the interest on or after 9 December 2009. The issue here is of countering avoidance schemes where a person became entitled to an interest in possession by virtue of a 'no benefit' transaction, ie one that was not a transfer of value because, under *s 10*, it was not intended to confer gratuitous benefit, ie a purchased life interest illustrated as follows:

- A had made the settlement on B for life, remainder to C and D;
- B then bought the trust fund from A;
- so neither A nor B made a transfer of value because it was a commercial transaction, B however enjoyed the interest in possession but for the anti-avoidance provisions of *s 53*; and
- C or D bought their interests for full value.

Such arrangements were intended to move property into a relevant property trust without suffering the IHT 'entry' charge as a chargeable lifetime transfer of value. The amendment in *FA 2010* operated to apply the relevant property regime to such an interest in possession where a person became entitled by virtue of a commercial transaction.

The transfer to the maintenance fund will be exempt in so far as the value transferred becomes comprised in the fund at, or immediately after, transfer and there is in force a 'direction' in respect of that maintenance fund either at or after the transfer. As was noted in the previous section, the anti-avoidance provisions that relate to exempt transfers to a charity apply here with equal force. Specifically, where property transferred to the fund is a qualifying interest in possession in property comprised in a settlement, that settlement must come to an end in relation to the property transferred when it moves to the maintenance fund (see *s 56(3)(a)*).

Contributions to maintenance funds from discretionary trusts: anti-avoidance rules

'Excess value'

21.16 The transfer of funds from a relevant property trust to a maintenance fund would normally attract the exit charge, because the funds then cease to be relevant property, but relief from that charge will apply subject to qualifications in *Sch 4, para 16*. The first difficulty concerns what is known as 'excess value', as determined by *para 16(2)* – it is possible for the transfer, and even the tax that would have been charged on the funds before APR or BPR and before grossing up, to be greater than the value of the property itself that reaches the fund (after deduction of any money received by the trustees of the relevant property trust in respect of the property). This test is applied immediately after the transfer of the property to the fund as illustrated below:

Example 21.1: Myers Discretionary Trust – transfer between trusts

The Myers Discretionary Trust holds 72% of the issued share capital of Myers Engineering Ltd, an unquoted investment company whose net asset value is £1.2 million. The trust holding is valued at £734,400.

Myers Hall Preservation Trust has been established for the protection and maintenance of heritage property in North London. The trustees of the Myers Discretionary Trust transfer a 20% holding in the family company to the trustees of the Myers Hall Preservation Trust: that holding is valued at £60,000. The 52% retained holding is valued at £499,200.

The fall in value of the Myers Discretionary Trust fund is £235,200 (£734,400 – £499,200). The value received by the Hall trustees is £60,000. The difference, £175,200 is prima facie taxable.

On one view of *Sch 4, para 16(2)*, there is a reduction in value of the property transferred to the fund, viz (20/72 × 734,400), ie 204,000 – 60,000 = 144,000. Applying *s 65*, under general principles, the loss in value is greater, ie £175,200 as calculated above. Either way, the sum would have to be grossed up.

However, should BPR apply, thus in this example should the company be a trading company, there would be no liability.

'Purchase rule'

There is an anti-avoidance provision which denies exemption from the tax charge on transfer to a maintenance fund and this applies where the fund trustees buy an interest from a relevant property trust and pay for that interest in money or money's worth. This provision is also extended to apply where the trustees of the maintenance fund distribute property to a person for a consideration in money or money's worth.

Equally, the exemption is denied where, within 30 days of property leaving the relevant property trust, the individual who acquired that property for money or money's worth transfers that property to the maintenance fund. This could also apply where the transferor became entitled to the property through transactions which involved purchase, for example where the trustees of a relevant property trust had no power to appoint property to a maintenance fund they could appoint it to an individual, who would in turn transfer it to a maintenance fund.

It is in this context that *Raikes v Lygon* [1988] 1 All ER 884 is relevant. This concerned an application to the court to decide whether the trustees could transfer property to a maintenance fund to maintain a historic property which was itself part of the trust. It was acknowledged that, by *Settled Land Act 1925, s 64*, the trustees could enter into transactions which were for the benefit of the beneficiaries or of the settled land. However, the difficulty for the trustees was that in order for the maintenance fund to comply with the statutory requirements that have already been described in this book, it would have to include a charity or national museum as default beneficiaries if the primary purposes of the trust were to be exhausted (see **21.4** above). Such beneficiaries were not beneficiaries of the existing trust.

The tax advantages of the proposed action were put to the court, which held that those advantages themselves justified the proposed transfer to a maintenance fund – the transfer was for the benefit of the settled land itself and of the existing beneficiaries. The charity and museums were introduced only as part of the price of obtaining the tax advantage, a consideration that did not invalidate the proposal as an exercise by the trustees of their power.

Property entering a maintenance fund from an interest in possession trust

21.17 The original legislation was restrictive and for deaths after 16 March 1987, *FA 1987* inserted *IHTA 1984, s 57A* to provide relief, subject to anti-avoidance provisions, where:

- immediately before their death the deceased was the life tenant (with a qualifying interest) of trust property; and
- that property enters a maintenance fund within two years of the death.

For this purpose, it does not matter whether the maintenance fund is the same settlement as that in which the life tenant had a qualifying interest (see *s 57A(1)(a), (b)*). Where the conditions of *s 57A(1)* are satisfied, the exemption for property entering a maintenance fund applies as if the property had entered the maintenance fund on the death of the life tenant even though there may have been a delay of up to two years before the actual transfer of the property. A disposition or other event occurring between the date of the death and the date on which the property enters the maintenance fund is not treated as a transfer of value, nor is it an occasion for a charge to IHT.

By *s 57A(3)*,the two year window is extended by 12 months where court proceedings (by reason of an application under the *Variation of Trusts Act 1958)* are necessary before the property can move from the interest in possession trust to the maintenance fund. Practitioners may feel that the extension of only 12 months is not particularly generous.

ANTI-AVOIDANCE RULES

21.18 There are further anti-avoidance provisions that apply to transfers to maintenance funds apply:

- relief will not be allowed (see *s 57A(4)*) in respect of a conditional or defeasible transfer to a maintenance fund; nor
- where the interest transferred to the fund is no more than a life interest in trust property; nor
- where the trustees buy an interest under a settlement which contained the property immediately before the death of the tenant for life; nor
- where the property entering the maintenance fund is purchased or has at any time since the death of the tenant for life been purchased.

There is also a further restriction on the relief: certain property, such as minority shareholdings in unquoted companies, may have more than one value, according to the context in which it is held and this was the case with Myers Engineering Ltd in **Example 21.1** (the operation of the related property rules in *s 161* is but one example of this). If the value of the property transferred to the maintenance fund, when seen in isolation, is lower than the value of that same asset when considered as part of the deemed transfer on death of the life tenant, relief is restricted to the lower value.

Example 21.2: Thomson Fabrication Ltd – limited relief

Thomson Fabrication Ltd was formerly a trading company but as a result of dwindling trade during the recession years in 2017 it is now an investment company which holds only a portfolio of industrial units (see *Moore v IRC* [1995] STC (SCD) 5) such that BPR is no longer available. The net asset value of the company is £1.76 million.

Edwin Thomson holds 40% of the shares, his wife Sally 20%, with the remainder held by the trustees of the Thomson 1999 Trust. Edwin's shares on their own would be subject to a discount of 30% but when considered with Sally's holding by applying the related property rules of *s 161(2)(b)*, a discount of only 20% is appropriate, valuing his holding at £563,200.

Edwin transfers 25% of his shareholding (representing 10% of the capital) to a maintenance fund. In the hands of the fund's trustees, the shareholding, after applying an 80% discount, is worth £35,200. However, the fall in value of Edwin's estate is much greater calculated on the value of his original holding as part of a 60% holding (when taken with Sally's shares) to part of a 50% holding.

Relief will be restricted to the lower value in the hands of the Trustees of £35,200.

The final anti-avoidance provision is *s 57A(6)*, which merely provides, for completeness, that a person is treated as buying the property if he becomes entitled to it as a result of transactions which include:

- any disposition for any consideration;
- either of that or other property; and
- whether to that person or to some other person.

The provision can be likened to the operation of *s 142* (deeds of variation) where again any consideration for the variation denies access to the favourable tax treatment sought.

Schedule 4, para 15A contains provisions to bring the treatment of maintenance funds that arise following a qualifying interest in possession trust into line with other maintenance funds.

PROPERTY LEAVING THE MAINTENANCE FUND

21.19 The favourable tax treatment of maintenance funds was summarised at the beginning of this chapter and in essence states that provided funds are applied for the purposes of the protection of heritage property, those tax advantages can be preserved. Where, however, the trustees no longer apply the fund to maintain heritage property and take action to return it to the settlor or to release it to the other beneficiaries of the trusts that were established for the maintenance fund, a charge arises – particulars of which are set out at *IHTA 1984, Sch 4, para 8*. HMRC guidance is set out in CTNH Appendix 5.

It may, however, be helpful first to consider the consequences of the 'proper' use of heritage funds, ie use for the purpose of which the funds were established. Application of the funds in this way secures exemptions from charge which are provided for by *Sch 4, paras 9* and *10*. Essentially, the exemptions from charge are these:

- reverter to settlor;
- use for heritage purposes;
- transfer to a new maintenance fund; and
- transfer to a suitable purpose, namely charity, a political party, a heritage body or another non-profit making body.

Reverter to settlor

21.20 There is no charge to tax where property leaves a maintenance fund and again becomes property to which the settlor is beneficially entitled or to which his wife or her husband (or civil partner in either case) is so entitled or, where the settlor has died in the two years preceding the withdrawal of property from the maintenance fund, the settlor's widow or widower is entitled to (*Sch 4, para 10(1)*). There are qualifications to this rule: the 'excess value' rule and the 'purchase' rule, both outlined at **21.16** above.

The specific legislation covering these points is to be found in *Sch 4, para 10(2), (4)* and *(5)* but further restrictions apply:

(1) The exemption from charge is not available unless the person who becomes beneficially entitled to the property is, at the time that he receives the property, domiciled in the UK (see *para 10(8)*).

(2) Further, it will be remembered that the relief in respect of maintenance funds is available only where the six-year rule applies, as described in **21.4** above. To prevent circumvention of that rule, the tax charge on 'non-heritage' use of property in a maintenance fund is denied where that property came to that fund, either directly or indirectly, from a discretionary trust or another maintenance fund.

Use for heritage purposes

21.21 The use by the trustees of the property in the maintenance fund for heritage purposes will not trigger any charge. Specifically, the purposes that qualify for this treatment are those set out in *para 3(1)(a)(i)* or *(ii)*, which were examined at **21.4** above, ie broadly, use for maintenance, repair or preservation, securing access in respect of the property, or transfer to a heritage body or qualifying charity. HMRC interpretation of purposes which are an acceptable use of maintenance funds is set out in CTNH para 8.7.

Transfer to another maintenance fund

21.22 The transfer of property from one maintenance fund to another does not trigger any clawback of relief, provided that the property reaches the new fund in normal circumstances within 30 days – this is confirmed in CTNH para 8 6(c). That time limit is extended to two years where the settlor has died and his death would, but for the transfer to the new heritage fund, be an occasion of charge. As before, this exception is not available where the 'purchase rule' applies, and the exception is subject to the 'excess value' restriction already noted at **21.16** above.

Other acceptable uses

21.23 *IHTA 1984, s 76* contains a general relief, which applies equally to a maintenance fund, in respect of property that comes to be held for charitable purposes namely:

- property held for charitable purposes only;
- the property of a political party qualifying for exemption under *s 27*; and
- the property of a '*Schedule 3*' body, ie one of the heritage bodies identified in *IHTA 1984, Sch 3*.

Section 26A applies to transfers of value made after 17 March 1986 – it provides that such a transfer that would fall to be a PET is regarded as exempt to the extent that the value transferred by it has been designated as heritage property or could be so designated and which has been disposed of by private treaty sale to a *Schedule 3* body or has been transferred to a *Schedule 3* body otherwise than by sale or has been accepted in satisfaction of tax under *s 230*. These procedures are described in more detail in **Chapter 23**.

Relief under *s 76* is subject to the 'excess value' restriction and to the 'purchase' rule, both noted at **21.16** above.

Events giving rise to clawback of relief

21.24 There are several events that can give rise to the loss of relief that arose when a 'direction' was given in respect of a maintenance fund under *Sch 4, para 1* and the occasions of charge where the exemptions mentioned above do not apply are as follows:

- payment of capital to anyone other than the persons, or for the purposes, mentioned above;
- dispositions by the trustees, including depreciatory transactions or omissions;
- loss of heritage status in respect of the subject property; and
- withdrawal of the Treasury Direction.

Each of these sections is examined in turn below.

Release of capital

Schedule 4, para 8(2)(a) imposes a charge where settled property leaves the maintenance fund without being applied for the 'heritage' purposes for which the fund was established, ie maintenance, repair or preservation of the 'subject' property (*para 3(1)(a)*). The value attracting the charge is calculated on the standard 'before and after' drop in value principle of the maintenance fund and by the same token if the tax is paid out of the assets that remain in the fund, the value must be grossed up. The tax charge is at 'the appropriate rate' calculated as indicated below.

Transactions by the trustees

If the trustees of a maintenance fund pursue an action (or deliberately fail to exercise a right) which serves to reduce the value of the property in the fund, an IHT charge arises under *para 8(2)* unless:

- the property was applied for a heritage purpose; or
- it was a commercial transaction and thus was not a transfer of value because it was not a disposition intended to confer gratuitous benefit and is, therefore, exempt under *s 10*; or
- it was the grant of an agricultural tenancy for full value falling within the exceptions to transfers of value in *s 16*.

Loss of heritage status

The 'raison d'être' of a maintenance fund is to support heritage property and where there is a loss of heritage status of the property that is the subject matter of the maintenance fund the benefits secured will fall away. Accordingly, when the subject property no longer qualifies for conditional exemption by failure to comply with *Sch 4, para 3* such event will trigger the IHT charge on the maintenance fund itself.

Withdrawal of Treasury Direction

The withdrawal of the 'direction' under exercise of the power in *Sch 4, para 5* automatically triggers an IHT charge on the maintenance fund.

The rate of tax: general cases

21.25 *Schedule 4, para 8(4)* defines the tax rate as being whatever results from a consideration of *paras 11–15* (there is a special code under *para 11* for property that came to a maintenance fund from a relevant property trust, considered at **21.28** below but this section of the book is concerned with the rate of charge that will apply in all other circumstances). It is the higher of two rates where the first rate is that determined by *Sch 4, para 13*, and the second rate is that determined by *para 14* – HMRC guidance is set out in CTNH Appendix 5 para 5.

The first or 'para 13' rate

This is the rate that was described at and illustrated in the table shown in **Chapter 20** and repeated below:

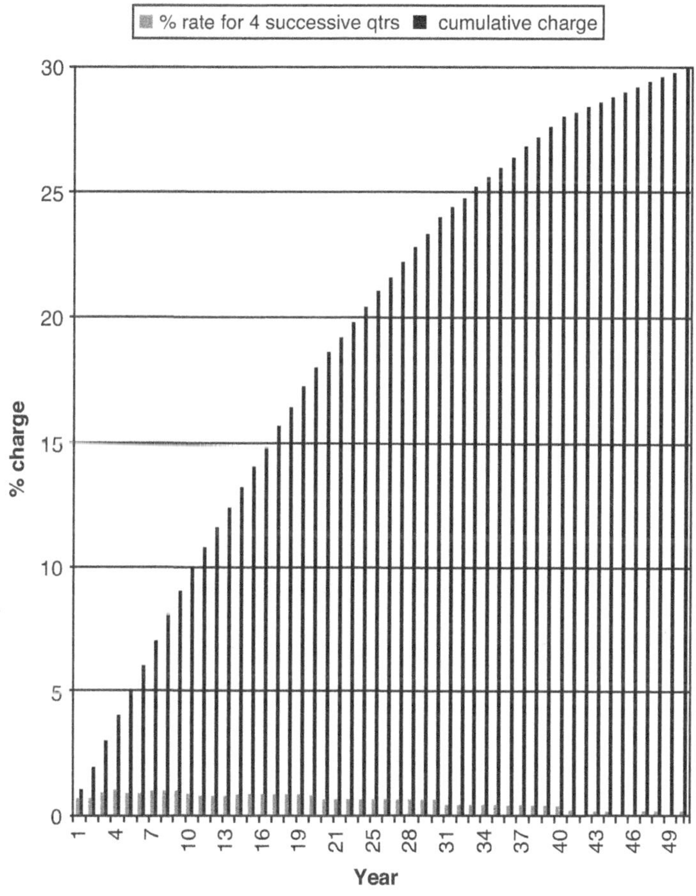

It can be seen that the rate of charge is relatively high in the early years and, although the cumulative rate increases over time, the increments get smaller in the later years and where the time charged relates to the period starting from the day on which the property first entered the maintenance fund regime to the day before the event that gives rise to the charge (*para 13(2)*). That period is varied where there has been a previous taxable event where the time-related charge runs from the date of the previous taxable event, disregarding the period from entry to the fund to that taxable event (see *para 13(3)*). However, that rule itself is subject to an exception in respect of an 'excess value' charge under *para 9(4)* (see *para 13(4)*).

The second or 'para 14' rate

This rate is determined with reference to the tax position of the settlor or the life tenant and pivots on whether or not that person is alive at the time of the charge (even if the death of the settlor occurs within seven years of the occasion of charge on the maintenance fund). *Paragraph 14(1)* imposes a charge:

- at the effective rate at which tax would be charged on the relevant amount;
- if that were the value transferred by a chargeable transfer;
- made by him on the date of the present charge to tax;
- in accordance with the lifetime rates in *s 7*.

The effective rate is more easily calculated than the rates that apply to periodic and exit charges in respect of relevant property trusts and is simply the tax chargeable as a percentage of the amount on which it is charged (see *para 14(8)*).

Example 21.3: Hubert – adoption of the higher tax rate

Hubert, who had made no previous chargeable transfers, established the Huntsman Preservation Trust in June 2009, transferring to it £500,000 for the upkeep of Grade 1 listed kennels and other property on the Huntsman estate.

Not all citizens admire either the architecture of the buildings or the purpose for which they were used and in 2016 vandalism caused major destruction of the kennels. The trustees, considering the buildings to be beyond repair, applied £200,000 in June 2017 for the support of field sports generally – this is not a charitable purpose.

The rate of charge applied on:

The para 13 rate:

Complete quarters in the period from June 2009 to June 2017:	32
Table rate:	0.25%
Para 13 rate (0.25% × 32):	8%
Potential charge (£200,000 at 8%):	£16,000

The para 14 rate:

Event date:	Within the first ten years of trust creation
Effective rate:	Rate on creation of the trust
Rate on creation of the trust:	0% (as value within the nil rate band)
Para 14 rate:	0%
Potential charge (£200,000 at 0%):	£0

The *para 13* rate at 8% is higher and will be applied.

Schedule 4, para 15A imports charging rules in respect of maintenance funds that follow qualifying interest in possession trusts. Thus, the reference to the settlor in *para 14(1)–(3)* is re-interpreted as being a reference to the life tenant, and there are appropriate amendments to the remainder of *para 14* wherever necessary.

Death of life tenant

21.26 Where the settlor is dead and the chargeable event occurs on the death of the life tenant (by reason of their qualifying interest), the rate of charge will again depend on the particular circumstances surrounding the creation of the fund. The full 'death' rate of tax will apply if the fund was created on death but if the fund was created during the settlor's lifetime then even if the settlor has since died, and/or that death occurred within seven years of the creation of the fund, the lifetime rate is used.

Having found the rate of tax to be applied, it is then necessary to determine the effective rate. This is the rate at which tax would have been charged if the amount now taxable had been the top slice of the estate on death where under the *para 14(2)* and *(9)* such slice is charged at:

- the lifetime rates in respect of property settled by a lifetime transfer; or
- the death rates if settled on death.

If the death of the settlor occurred before 13 March 1975, the IHT charge is calculated as the top slice of the estate on which estate duty had been charged (see *para 14(3)*). However, that reference to the amount on which estate duty was chargeable only fixes the size of estate with which the charge is concerned this is the effective rate which would have been charged in accordance with *s 7*. *Schedule 2, para 6* stipulates that the appropriate rate is the most recent reduction in force where:

- tax is chargeable under the 'clawback' provisions now being considered; and
- the charge arises after any reduction in the general rates of IHT,

Schedule 4, para 14(4) and *(5)* address the situation where property in a maintenance fund was previously comprised in another maintenance fund whereby HMRC have an option to calculate the tax by reference to:

- the settlor of the present fund; or
- the settlor of the previous fund; or
- the life tenant of either fund.

Note: Paragraph 14(6) and *(7)* create a form of cumulation of charge. Where property in a maintenance fund has been withdrawn on some previous occasion, giving rise to a charge to tax under these clawback provisions, any further clawback event within seven years (see *para 14(6)(a)*) triggers aggregation of the amount that was charged on the first occasion with the present clawback charge.

Example 21.4: Sebastian – withdrawal of funds

In December 2002, Sebastian transferred a majority shareholding in his residential property management company worth £600,000 to a maintenance fund – this was the only transfer made during his lifetime.

In June 2010, the trustees released part of the holding to Sebastian's daughter, Rachel. The value of their remaining holding, which fell below the control threshold, was diminished by £400,000 but the market value of the shares transferred was £150,000.

In May 2013, Sebastian died leaving an estate worth £625,000.

In January 2017, the trustees released the remainder of the shares, now worth £1.1 million, to Sebastian's son, Jonathan.

	£
1. Tax charge on the event in June 2010 –	
Value lost by the trust	400,000
Value received by Rachel	(150,000)
Excess value to be charged	250,000

	£
Paragraph 13 calculation:	
01.12.02–16.06.10 = 34 qtrs	
34 × 0.25% = 8.5%	
£250,000 at 8.5% =	21,250
Paragraph 14 calculation:	
£250,000 in the bracket £0–250,000: within nil rate band= 0%	
£250,000 at 0%	NIL
Actual charge:	
Charge greater of £21,250 or £0 =	21,250

2. Tax charge on the event in 2017 –

Paragraph 13 calculation:	
Value of transfer	1,100,000
01.12.02–01.01 2017 = 56 qtrs	
(40 × 0.25%) + (16 × 0.2%) = 13.20%	
£1,100,000 at 13.20% =	145,200
***Paragraph 14* calculation:**	
Transfer by Sebastian –	
lifetime transfer	600,000
estate at death	625,000
value transferred	1,100,000
Total	2,325,000
Lifetime settlement by Sebastian using lifetime rates:	
nil rate band, May 2013	325,000
value transferred from fund treated as the top slice of capital falling within the bracket £1,225,000–£2,325,000	1,700,000
tax rate	20%
Para 14 rate	20%
Actual charge:	
The *para 14* rate at 20% is higher than the *para 13* rate at 13.20%	
Tax charge (£1,1000,000 @ 20%)	220,000

** The lifetime rate applies as the settlement was created in the settlor's lifetime*

(Note: the seven-year period was originally longer; in respect of events that occurred before 17 March 1986, the period was ten years).

Special property reliefs

21.27 The general clawback charge operates as a charge on a transfer of value and to the extent a transfer is exempt under the general provisions it will similarly escapes the effect of clawback. In the same way, APR and BPR may be available in appropriate circumstances to shelter the transfer of property from a maintenance fund. Perhaps, not infrequently, agricultural land may have been introduced to a maintenance fund some years ago when the reliefs were less generous than under the current regime and neither would it be unusual for a heritage house to be supported by agricultural land. The normal tests will apply for APR and HMRC guidance suggests that BPR will be available 'if the trustees are themselves running a business', which would seem to be a somewhat restrictive interpretation of BPR.

Calculation of the clawback charge where property was previously comprised in a relevant property trust

21.28 The clawback charge that arises on a maintenance fund in respect of property that came to it from a relevant property trust is refreshingly simple. It is a time-related charge on the scale set out in *Sch 4, para 11(2)*, replicated at **21.25** above where the starting point for the time charge is determined is the latest of the following dates:

- 13 March 1975;
- the last ten-year anniversary of the 'originating' trust; and
- the day, or the most recent day where there is more than one, on which the property that is now subject to the clawback became subject to a discretion in the originating trust before it ceased to be subject to that discretion (or, where the property has slipped in and out of discretion, where it last ceased to be subject to a discretion).

Practical difficulties may arise where the records of the originating trust are incomplete or lack detail, or where it is not immediately apparent which assets were subject to a discretion and which were, for example, comprised in a separate settled fund for a beneficiary in respect of which there was a qualifying interest in possession. The fear here is that the cost of research and of computing the tax charge may at least equal if not eclipse the charge itself.

The period in respect of which the clawback charge applies runs from the relevant date (see above) to the day before the clawback event. It has already been noted that there can be a period during which property is not in a maintenance fund but comes to be comprised in one and this may be up to two years following the death of the settlor or the life tenant, or up to 30 days in any other situation. This is a reference to 'the permitted period' identified in *para 9(2)* where that permitted period is counted as part of the period for the calculation of clawback (see *para 11(4)*).

THE IMPACT ON MAINTENANCE FUNDS OF OTHER TAXES

Income tax

'One estate'

21.29 The special rule 'one estate' that applied to landed estates was repealed on 1 April 2001.

Income tax treatment of maintenance funds

21.30 The main income tax code for maintenance funds is *ITA 2007, Part 9, Chapter 10, ss 507–517*, formerly *ICTA 1988, ss 690–694*.

Under *ITA 2007, s 508*, the trustees may elect for income not to be treated as that of the settlor (see *ITA 2007, s 508(3)*) with the result that the income is taxed at the rate applicable to trusts. Given that the trust rate is currently 45% or 38.1% (2017/18) whereas many individuals are taxed at a lower rates of 40% and less (2017/18) the election carries little benefit. Where no election is made the income arising that is not used for maintenance of the heritage property is treated as income of the settlor by virtue of their retained interest in, or occupation of, the heritage property.

An election under *ITA 2007, s 508* is by notice in the form approved by HMRC and must be made on or before the first anniversary of the normal self-assessment filing date by 31 January for the year to which they relate. Split year treatment is allowed under *ITA 2007, s 509* where, for example, the 'direction' under *IHTA 1984, Sch 4* comes to an end during the year.

A different rule applies where income from a maintenance fund is treated as income of the settlor and is used to reimburse that settlor for expenditure that he has incurred for heritage purposes. If the expenditure is deductible from the profits of a trade or UK property business carried on by the settlor or would be so deductible, but for the reimbursement, *ITA 2007, s 511* avoids double taxation by providing that that income is not treated as reducing the heritage expenditure and is not treated as income of the settlor except for the purposes of the maintenance fund code. In the rare but complicated situation, where a trust comprises both settled property which is part of a maintenance fund and other property, those funds are treated as separate settlements for the purposes of this code.

The benefit of an income tax election continues only for as long as those funds are used for heritage purposes. If property in a maintenance fund reverts to the settlor or is applied other than for heritage purposes, the benefit of the election is lost – a charge is levied on the income of the fund that has not been used for heritage purposes. The legislation, as rewritten in *ITA 2007, s 512*, sets out four main situations where this may happen:

- property used otherwise than for property maintenance – Case A;
- property not used for property maintenance, nor accumulated, nor paid to a heritage body – also Case A;
- property leaves the settlement but does not pass to a heritage body – Case B;
- heritage direction no longer applies – Case C; and
- property leaves the settlement and goes to a heritage body – Case D; or
- property leaves the settlement and an interest in the heritage fund itself is acquired by a heritage body – also Case D.

Thus, even where income is applied by payment to a heritage body or other qualifying body, it cannot escape the charge, levied under *ITA 2007, s 512*, where the heritage or other qualifying body has purchased an interest in the fund, unless that purchase is from a heritage body. The charge is on all of the income in 'the relevant period' (see *ITA 2007, s 513*), namely since the creation of the settlement or since the previous charge under this legislation.

The rate actually applied by *ITA 2007, s 515* is equivalent to the additional rate of income tax for the year of assessment (currently 45% for 2017/18) but reduced by the trust rate for that year (also currently 45%), and so for the moment is nil. The tax charge, when it is a positive rate, is in addition to any other tax charged on the fund income under any other provision.

Stamp duty and stamp duty land tax

21.31 *FA 1980, s 98* provides that stamp duty is not to be chargeable on any document relating to property which comes out of a settlement if, as a result of that property becoming comprised in another settlement, there is no IHT charge on the property. The likely situations that will qualify for this relief are the '30-day' resettlement or the 'two year on death' situations, noted at **21.28**

above. Where part only of the property goes into the new maintenance fund, stamp duty on the remainder of the property is unaffected.

An exemption for stamp duty land tax is available under *FA 2003, s 69* where the property is sold to:

- The Historic Buildings and Monuments Commission for England (Historic England);
- The National Endowment for Science, Technology and the Arts (NESTA);
- The British Museum;
- The National Heritage Memorial Fund (NHMF); or
- The National History Museum.

Capital gains tax

21.32 *TCGA 1992, s 260(2)(b)(iii)* allows CGT holdover on transfer of property to a maintenance fund even where property passes to a maintenance fund from a qualifying interest in possession. Similarly, it can also apply where the '30-day' or 'two-year on death' situations arise, noted at **21.28** above.

It is considered that holdover relief is available in cases of direct transfer between separate settlements which are maintenance funds.

The trap which triggered where the settlor retained an interest under the trust resulting in the gains of the trustees being assessed on the settlor has been removed with the repeal of *TCGA 1992, s 77* by *FA 2008* with effect for 2008/09 and later years. The following *hypothetical example* illustrates the problem: readers will need to study it if they share the writer's fear that rates for CGT may change over the years, and that we may again see *TCGA 1992, s 77* or something similar.

Example 21.5: Hector – hypothetical example

In August 2018, Hector informed HMRC that he intended to set up a maintenance fund for the family estate and to transfer to it some of the estate cottages. He hoped to reduce the value of his estate for IHT but setting up the trust took time, and it was not put in place until July 2019. In August 2021, Hector informed HMRC that he intended to add more property to the trust but HMRC advised him that a Treasury direction would be needed under *Sch 4, para 1*.

The cottages were transferred into the maintenance fund in October 2021 but soon after were sold by the trustees. Hector was assessed to tax on the gain and paid it. Work on the trust annual report a year later reminded the trustees that there was no Treasury direction in place, although that did arrive in October 2022, whereupon the trustees made the election under (what was then) *ICTA 1988, s 691*.

HMRC rejected the election in so far as the gain was concerned: there was no 'direction' in place at the time the gain was realised and thus no ability to make an election. Hector claimed reimbursement of the CGT under *TCGA 1992, s 78*.

Chapter 22

Compliance

> QUICK GUIDE
>
> - As part of the drive to enforce and strengthen IHT compliance HMRC routinely impose stringent penalties for failure to fully comply with the statutory requirements. The penalties for non-compliance are set out at **22.11** to **22.17**.
> - Executors and those advising them must take all appropriate steps to accurately establish the size of the estate and the quantum of lifetime gifts made within the seven years preceding death.
> - Estimates are acceptable only within certain limits, noted at **22.5**. Values should be fair and not artificially depressed.
> - Nearly all IHT forms have, within the last few years, been overhauled. The most relevant forms are considered at **22.18** onwards.

CHARGEABLE LIFETIME TRANSFERS (CLT)

22.1 Practitioners often assume that the reporting of a CLT is entirely dependent on whether any IHT falls due for payment and if death occurs within seven years of the transfer, it is mistakenly considered sufficient to only include particulars of the settlement in the HMRC IHT400 account at that stage. This is wholly incorrect. *IHTA 1984, s 216(1)(a)* requires 'every person' who is liable as a transferor for tax on the value transferred by a chargeable transfer, 'or would be so liable if tax were chargeable on that value', to deliver an account unless the requirement to deliver such account is specifically exempted. The current regulations for the present purposes are now:

- the *Inheritance Tax (Delivery of Accounts) (Excepted Estates) Regulations 2004, SI 2004/2543*;
- the *Inheritance Tax (Delivery of Accounts) (Excepted Transfers and Excepted Terminations) Regulations 2008, SI 2008/605*; and
- the *Inheritance Tax (Delivery of Accounts) (Excepted Settlements) Regulations 2008, SI 2008/606*.

22.1 *Compliance*

Compliance with these regulations is now enforced by a raft of legislation, which has arisen in part from the amalgamation of the former HM Customs & Excise and Inland Revenue into a single HMRC body. It seeks to rationalise HMRC powers so that the same team of officers can enforce compliance across all taxes. The main legislation is:

- *FA 2007, Sch 24*: Penalties;
- *FA 2008, Sch 36*: Information and Inspection Powers;
- *FA 2008, Sch 37*: Record Keeping;
- *FA 2008, Sch 38*: Disclosure of Tax Avoidance Schemes;
- *FA 2008, Sch 39*: Time Limits;
- *FA 2008, Sch 40*: Penalties;
- *FA 2012, Sch 38*: Tax agents: Dishonest Conduct;
- *FA 2009, s 96 and Sch 48*: Information and inspection powers.

Under the *Inheritance Tax (Delivery of Accounts) (Excepted Settlements) Regulations 2008, SI 2008/606,* which is effective for chargeable events occurring on or after 6 April 2007, there is no requirement to file form IHT100 where:

- a trust is set up by a UK-domiciled settlor,
- with UK trustees,
- where there are no related settlements, and
- the value of the notional aggregate transfer does not exceed 80% of the nil rate band (£325,000 until 6 April 2021), ie currently £260,000. (Note that the absence of any positive tax liability is irrelevant for this purpose.)

However, practitioners are reminded that that figure of notional aggregate includes:

- the value of the relevant property trust assets at the ten-year anniversary;
- the historical value of non-relevant property in the trust for chargeable events occurring before 18 November 2015;
- the value of related settlements value;
- chargeable transfers by the settlor in the seven years before creating this trust (or later substitution under *s 67(s)*); and
- capital distributions made in the last ten years.

Since the trust may contain assets of uncertain value, such as land or unquoted shares, practitioners will often err on the side of caution by filing IHT100 even where it might not be needed. HMRC guidance is set out in IHTM 06100 et seq.

This work is concerned only with the details of compliance as it affects APR, BPR, and the woodland and heritage reliefs, but for a more general review of the legislation, reference should be made to *Ray & McLaughlin's Practical Inheritance Tax Planning* 15th edition (Bloomsbury Professional), or to the *Core Tax Annuals*, especially *Inheritance Tax 2017/18* (Bloomsbury Professional).

DUTY OF THE PERSONAL REPRESENTATIVE

22.2 Completion of the HMRC IHT400 account has always given rise to problems but HMRC guidance is available in the form of published step by step narratives supplemented by a toolkit.

The HMRC guidance is practical, for example, in urging personal representatives not to rush the form completion but to take sufficient time and care to complete the forms accurately whilst still adhering to the filing deadline. The guidance was updated in July 2017.

The 'HMRC Inheritance Tax toolkit' is intended to 'cover risks seen in a majority of forms IHT400 received'. However, the toolkit introduction makes it clear that 'risks arising in connection with claims to BPR, APR and reliefs associated with works of art and other national heritage assets are specifically not dealt with'. The toolkit was updated in April 2017.

In *Re Sir Malcolm Arnold Deceased, Day v Harris and Others* (2010) the first defendant (a chartered tax adviser) felt it to be his duty as executor to make searching enquiries into lifetime transactions carried out by his fellow executor, the plaintiff, who had been attorney for the deceased. The defendant reported on transactions in the last years of the deceased life and qualified the IHT400 in part to reflect his researches. The plaintiff 's application for the removal of the first defendant as executor by reason of partiality towards other beneficiaries and for spending too much time and cost on the compliance aspects of the estate, failed.

The personal representative should approach preparation of form IHT400 with the prospect of an HMRC review in mind. IHTM09032 urges the Inspector to 'build up a picture of the deceased or transferor' and directs the Inspector to consider the taxpayer's lifestyle and financial affairs – if the form does not disclose what might be expected, the Inspector should make searches and check other information sources to identify any potential errors, omissions, under-valuations or tax avoidance or tax evasion.

HMRC guidance

22.3 It should be recognised that the HMRC manuals, which are in the form of instructions to its own officers, are not wholly impartial but do form a clear and useful starting point for anyone unsure of the accepted attitude to particular tax issues; it would not be advisable to blatantly disregard the guidance. There is a great deal of information on the HMRC website: apart from the text of the IHT Manual there is the 'Guide to completing your Inheritance Tax account (IHT400 notes)' which provides specific help in completing the various supporting schedules – the guide was updated on 28 June 2017.

As demonstrated in *Re Sir Malcolm Arnold Deceased, Day v Harris and Others* (2010), the appropriate level of enquiry will depend on the facts and on the knowledge by the personal representative of the financial affairs of the deceased in his lifetime. A thorough search should be made of the deceased's papers, such as the records kept for self-assessment and copy tax returns. If gifts come to light even after the personal representatives have made reasonable enquiries, those particulars should be supplied to HMRC and where there is prompt disclosure, the only 'penalty' suffered is likely to be that the correct tax is paid and interest for late payment. Where there is evidence of a more serious nature, a clearance certificate, if one has previously been issued, may be set aside, and HMRC can then pursue additional tax and penalties.

Property details

22.4 *IHTA 1984, s 216(3)* specifies the details of the property to be shown in the return and it is subject to the rules set out in *s 216(3A)* and *(3B)* discussed below. Where the personal representatives take out a full grant, they must show all property which formed part of the estate of the deceased immediately before his death, other than the property which would not form part of the estate except for the gift with reservation provisions.

The requirement to show 'all property which formed part of the estate of the deceased immediately before his death' was examined in *Daffodil (administrator of Daffodil Deceased) v IRC* SpC 311, [2002] STI Issue 20. The wonderfully named Mr and Mrs Daffodil had owned a bungalow as tenants in common. The former died intestate in 1994 and although the latter could have applied for a grant of representation in her husband's estate, she did not do so and continued to live in the bungalow until her death in February 2000. Her son, as administrator of his mother's estate, obtained a grant of letters of administration in August 2000. The value of the entire estate was something over £320,000, including her own (but not inherited) share of the bungalow at £95,000 and the IHT liability was duly paid.

The son then wanted to sell the bungalow, but his lawyer advised him that his father's share had not passed to his mother by survivorship on death, because ownership had been as tenants in common, and that the son must therefore apply for a grant of letters of administration of his father's estate to complete the title – this he duly did. A notice of determination was issued by HMRC stating that the interest of Mrs Daffodil in the estate of her late husband formed part of Mrs Daffodil's estate for IHT. The son argued that, because his late mother had not taken out a grant of letters of administration to the estate of his late father, his mother was not entitled to property that had belonged to his father at the date of her death.

The Special Commissioners agreed with HMRC that at the date of her death, Mrs Daffodil had the right to require the whole of the estate of Mr Daffodil to be transferred to her. That right was 'property' within *s 272*. The half share of the bungalow owned by Mr Daffodil was therefore part of the estate of Mrs Daffodil.

A similar issue arose in *Thompson (Thompson's Executor) v CIR* SpC 429. A widow had died in 1971, leaving her house to her daughter who, in turn, died in 2002. It appeared that the daughter had never signed any document accepting title to the house. On appeal from a notice of determination charging IHT on the value of the house, the Special Commissioner held that the daughter was the sole owner of the house, despite her inaction, and it was part of her estate for IHT.

In *G Silber (Lerner's PR) v HMRC* [2012] TC02369 the issue was whether an amount paid to a company was a loan or a gift. Mr Lerner was a shareholder in a company whose records showed that he had lent it £107,210 more than seven years earlier. Following his death HMRC issued Notice of Determination that this loan was part of his estate. The executor appealed, saying it was a gift, but the First-tier Tribunal, having read the company accounts, accepted them as *prima facie* evidence of a loan, not a gift: that was confirmed by a balance sheet drawn up after the death.

The following should be regarded as basic components of a starter checklist:

- Where there are family trusts, establish exactly who owns what, and for whose benefit.
- Check for evidence of partnerships, of tenancies, licences and informal arrangements.
- Where is the administration of a farm really carried on?
- What efforts have the directors of a trading company made to reinvest 'surplus' cash?
- Has partnership land been registered in the names of the partners?
- Do the accounts reflect what is happening now, or are they, in effect, out of date?
- How are freeholds and improvements shown?

On death, completion of the forms demands answers, with evidence, to such questions. A 'lifetime audit' can throw up many points where the position could be improved.

The use of estimates

22.5 *Section 216(3A)* is the statutory authority for the inclusion in the IHT account of estimates. This is often an issue where the deceased owned a business or an interest in one and it is sufficient to make use of a provisional estimate of the value, provided that:

- the return contains a statement to the effect that an estimate has been used; and
- the personal representatives undertake to deliver a further account as soon as the value has been ascertained.

Section 216(3B) is the authority for HMRC to control the use of estimates. HMRC expect personal representatives to make a thorough search of the deceased's papers and whilst there is no absolute duty to make enquiries of any particular person, HMRC consider it reasonable to expect the personal representatives to make an approach to any professional advisers who may have dealt with the affairs of the deceased, such as solicitors, accountants, stockbrokers, financial advisers and banks.

Robertson v IRC

22.6 The use of estimates was considered by the Special Commissioners in *Robertson v IRC* [2002] STI 19. Mrs Stanley had died on 10 October 1999, owning both a house and contents in Scotland and a property in England. The executor, a solicitor, Mr Robertson, considered that a valuation of the Scottish house and contents was required but he had no direct knowledge of the property in England, save a letter indicating that it was occupied by longstanding tenants at a rent of £260 per annum, and a photograph showing what appeared to be a small cottage. What Mr Robertson was not aware of was that the English property was set in five acres of grounds.

Mr Robertson promptly submitted to HMRC an inventory of Mrs Stanley's estate, showing an estimated value of the Scottish house of £60,000, its contents at £5,000, and the English cottage at £50,000 – the IHT of approx. £400,000 was paid at that time and confirmation was issued in due course. Mr Robertson later received a valuation of the English property in the sum of £315,000 and promptly submitted a corrective inventory to HMRC settling the additional IHT due of approximately £120,000 at the same time. HMRC considered that the disparity between the original estimates and the later figures indicated that Mr Robertson had not fulfilled his obligation under *s 216(3)* to make the fullest enquiries that were reasonably practicable in the circumstances. HMRC imposed a penalty of £9,000 against which Mr Robertson appealed.

The Special Commissioner held that Mr Robertson had made the appropriate statutory statement and was not negligent in delivering an incorrect account. An executor was required to make the fullest enquiries that were reasonably practical in the circumstances – the executor had been to the home of the deceased and had submitted estimates but disclosed them for what they were. In the particular circumstances, it had been prudent to go ahead on the basis of estimates only as that enabled the executor to obtain confirmation rapidly, so as to be able to sell the house in Scotland. Mr Robertson had followed acceptable legal practice by complying with his common law duties as an executor and was not liable to any penalty.

A chorus of articles in the professional press supported Mr Robertson's actions and this perhaps prompted HMRC to issue a newsletter in May 2002 on the subject. It noted that the particular form that Mr Robertson had used, Cap A3, did not contain an express declaration to the effect that, where estimates are supplied, the personal representative undertakes to provide exact valuations as soon as practicable. That form has now been replaced by IHT400, which incorporates a declaration along the lines suggested by the Special Commissioner.

HMRC have indicated a willingness to help personal representatives fulfil their obligations without incurring penalties but there is an expectation that the exact value of property should be given, and not merely an estimate. If there is a clear need to obtain a grant urgently, personal representatives may feel they need to submit an estimated account of the value of a particular item but even in such circumstances, the personal representatives must make the fullest enquiries that are reasonably practicable. The estimate itself should be as accurate as possible and a personal representative who employs a professional to value the property should ensure that the estimate is a reasonable one.

HMRC do not accept the oft seen practice of 'special' valuations for probate, for example that a house is worth £200,000 'but for probate the value is £175,000'. If the valuation reports a lower figure 'for probate', HMRC may seek penalties from the valuer for assisting in the furnishing of an incorrect return.

Robertson v IRC (No 2)

22.7 In *Robertson v IRC (No 2)* [2002] STI 25, Mr Robertson pressed home that advantage of his earlier success and claimed costs. Costs, HMRC argued, were not allowable (see *Special Commissioners (Jurisdiction and Procedure) Regulations 1994, SI 1994/1811, reg 21*) unless the Commissioners were satisfied that a party to the proceedings had acted 'wholly unreasonably' in connection with the hearing.

The Special Commissioner in awarding costs, held that neither the evidence presented by HMRC, nor the agreed statement of facts showed any rational basis for bringing the proceedings against Mr Robertson. HMRC had not addressed the issue of negligent furnishing of an account, or the issue of what was reasonably practicable. What Mr Robertson did was in accordance with standard practice and was not negligent; rather he had made the fullest enquiries that were reasonably practicable in the circumstances.

Where it seems likely the asset will qualify for APR or BPR at 100%, personal representatives are often reluctant to go to the expense of a full valuation as no tax is at stake and the value of the asset as returned for IHT will not be 'ascertained' within the meaning of *TCGA 1992, s 274*. However, HMRC in their Newsletter of April 2004 indicated that under no circumstances should nominal or ill-considered values be used, since if it later transpires that relief is not due, penalties could arise if a substantial uplift in value is appropriate.

VALUATION: COMPLIANCE ISSUES

22.8 HMRC perceive valuation as a key area of risk and that is understandable. Penalty issues can and do arise where an incorrect valuation included in the account results in an understated IHT liability and there has been a lack of reasonable care in that valuation. By contrast HMRC have indicated that, if instructions for the valuation of a property have been given on the correct basis, any uplift in value subsequently agreed is 'unlikely' to attract a penalty. The 'correct basis' is defined as:

> 'a hypothetical sale in the open market under normal market conditions and marketed properly with no discounts for a quick sale or the time of year etc.'

To show that 'reasonable care' had been demonstrated, the executor should obtain three valuations from different estate agents or a professional (ie Royal Institution of Chartered Surveyors (RICS)) valuation if a definitive valuation was necessary. HMRC confirmed in 2011 their previous stance, which is to consider whether 'reasonable care' was taken by asking:

- Was professional advice sought?
- Were instructions given on the correct basis?
- Was the valuer's attention drawn to particular features of the property (eg development potential)?
- Was anything unusual about the valuation questioned?

Cairns v HMRC

22.9 In *Cairns v Revenue & Customs* [2009] UKFTT 00008 (TC), HMRC had imposed a penalty on Mr Cairns, a solicitor acting as a personal representative. The penalty related to a valuation of £400,000 in respect of the deceased's residence, as returned on form IHT200 (the forerunner to the IHT400). This was a valuation by chartered surveyors in January 2004, which was stated to be an 'arbitrary figure pending investigations as to costs involved in upgrading' and it was heavily qualified due to the poor state of the property. Mr Cairns was uncertain of the property's value, but considered that the existing valuation was sufficient in the meantime. The District Valuer subsequently valued the property at £600,000 as at the date of death, which was also the amount for which the property was sold. The Special Commissioner was asked to consider whether Mr Cairns submitted an incorrect IHT account, and whether he had acted negligently.

The Special Commissioner held: 'the mere failure to obtain another valuation when it has not been established that a second valuation would have led to a different figure being inserted in the statutory form does not constitute negligent delivery of an incorrect account…On the evidence before me, even if it were concluded that an incorrect account was delivered or furnished, it is simply not possible to conclude that it was negligently delivered or furnished except in one minor respect'. The 'minor matter' referred to concerned the fact that the valuation obtained had been heavily qualified, and was a provisional estimate. Mr Cairns had not disclosed this in the IHT account and that omission to do so was a careless error. However, the Commissioner added that 'it was minor, technical and of no consequence whatsoever'.

The Commissioner concluded that there had been a 'narrow, technical failure'. The account was incorrect and the sum of £400,000 should have been described as a provisional estimate; but whilst that failure was negligent, it was held to be a 'failure of the merest technicality'. Mr Cairns was held to have acted 'perfectly sensibly and reasonably throughout'. The Commissioner added that, even if he had been wrong to dismiss it, he would have reduced the penalty to a nominal amount or recommended that it be so reduced.

That case pre-dates the new penalty regime, but nevertheless may offer some comfort to the personal representatives of a deceased person's estate on the circumstances in which penalties can be imposed for an incorrect IHT return. However, it also provides a warning as to the degree of disclosure required in respect of provisional valuations and estimates, to avoid an accusation by HMRC of careless behaviour. In general, provision of good, detailed information in support of values will often achieve earlier agreement with HMRC.

PERSONAL APPLICATIONS FOR GRANTS OF REPRESENTATION

22.10 It is increasingly common to find that application has been made for a grant of representation in a substantial estate without the use of a solicitor.

Commonly, this will be a 'personal application' where family members are confident of their ability or where an executor, not being a solicitor, has the relevant experience, perhaps as a member of the Society of Trust and Estate Practitioners (STEP). All professional agents assisting with preparation of HMRC accounts are expected to make 'reasonable yet searching enquiries', so that the person actually signing the declaration in the account is making a true and complete return. All the information known to professional agents at the time of delivery of the account should be reflected in that account.

PENALTIES: THE CURRENT REGIME

22.11 For a full account of the penalty regime, the reader is referred to *Ray & McLaughlin's Practical Inheritance Tax Planning* 14th edition, (Bloomsbury Professional) or *Inheritance Tax 2017/18* (Bloomsbury Press) or *Trust & Estates 2017/18* (Bloomsbury Press). Penalties, which are in addition to tax and interest, may be incurred for failing to deliver an account, make a return or comply with a notice seeking information. Fraud or negligence in the provision of information or accounts and returns may also be penalised. *FA 2007* applied, in *Sch 24*, the penalty regime for taxes other than IHT. *FA 2008, Sch 40* took that further. The changes apply with effect for all chargeable events occurring on or after 1 April 2009.

A penalty of £100 applies for failure to deliver an account on time, and is now routinely imposed (*IHTA 1984, s 245(2)*). A further penalty of not more than £60 per day may be claimed, though this is much less common. This is per day of default running from the day on which the failure to deliver the account has been declared as a default, either by a court or by the Special Commissioners. The daily charge runs up to the day on which the account is delivered. Usually, HMRC will give notice of an action to seek penalties so the taxpayer has one last chance to comply. If an account is delivered more than six months after the due date but before HMRC have taken proceedings before the tribunal the penalty is increased to £200. If the account is delivered more than 12 months after the due date a penalty of up to £3,000 may be charged.

The penalty is not imposed where the taxpayer can show reasonable excuse. The procedure involves the simple completion of a form setting out the circumstances, but the penalty will be charged unless the taxpayer delivers the account without unreasonable delay after the excuse has ceased: see *s 245(7)*. The mere fact that the estate is complicated is not itself a factor: there must be some other element.

A person not liable for the tax (eg an agent or adviser) who fraudulently or negligently supplies any incorrect document or information in connection with a chargeable transfer is liable to a penalty of up to £3,000. A person who assists in or induces the supply of any account, information or document he knows to be incorrect is liable to a penalty of up to £3,000 (see *s 247(3), (4)*).

A person may have supplied an account, information or document without fraud or negligence but later discover that it is 'materially' incorrect (ie where the tax in issue is £1,000 or more). Such a return is treated as having been negligently supplied unless the error is remedied without unreasonable delay. If another person (eg an agent or adviser) supplied the account, information or document and the person liable to the tax discovers that it is incorrect, he must inform the Board of the error without unreasonable delay. Otherwise, he is liable to a penalty as if he personally had negligently supplied the incorrect account, information or document (*s 248*).

A tax agent engages in 'dishonest conduct', see *FA 2012, Sch 38 para 3(1)*, where he or she does something dishonest, whilst in the course of acting as a tax agent with a view to bringing about a loss of tax revenue. It matters not that the agent was instructed so to act; nor whether any tax was actually lost. Penalties range from £5,000 to £50,000 plus 'naming and shaming'.

Overview of the penalty provisions

22.12 Whereas other aspects of the modernisation of penalties do not specifically concern IHT, that tax is included in *FA 2008, ss 122–123* and *Sch 40*. The intention of the new legislation is to address five areas of concern:

- incorrect returns;
- failure to notify a new taxable activity;
- late filing and late payment;
- failure to keep records, and powers to seek information; and
- other regulatory failures.

There is evidence of fewer applications for grants of representation by lawyers, whilst the number of personal applications is steady. There also seems to be some discrepancy between the number of trusts notified to HMRC and the number of family businesses in the UK, in respect of which it is often said that part is held in family trusts.

FA 2007, Sch 24 is amended and extended by *FA 2008, Sch 40* so as to include:

- accounts under *IHTA 1984, ss 216* and *217*;
- information or documents required by *s 256*; and
- statements or declarations in connection with deductions, exemptions or reliefs.

In IHT situations, there is often shared responsibility for information gathering, most notably in relation to lifetime gifts. The new rules apply to errors on returns for tax events and periods beginning on or after 1 April 2009, where the filing date is 1 April 2010 or later. However, there is no penalty in the case of an 'innocent error' (ie broadly where a person has taken reasonable care in completing the return, and has taken reasonable steps to notify errors if appropriate). The deceased's personal representatives will have taken reasonable care where they:

- follow the guidance provided about filling in forms such as the IHT400 and IHT205/207;
- make suitable enquiries of asset holders and other people (as suggested in the guidance) to establish the extent of the deceased's estate;
- ensure correct instructions are given to valuers when valuing assets;
- seek advice about anything they are unsure of;
- follow up inconsistencies in information they receive from asset holders, valuers and other people; and
- identify any estimated values included on the form.

Where an agent is acting, HMRC expect the personal representatives to review the form before signing it and to question anything that does not accord with what they know about the deceased. Simply signing an account completed by an agent is not regarded as taking reasonable care although the use of the toolkit might. If IHT is payable other than on death, HMRC expect the transferor (or trustees) to deliver a full and complete return of the transaction concerned, and to have sought professional advice as necessary.

Levels of penalty

22.13 There are three broad degrees of culpability:

- The lowest level of culpability, described as 'careless', is where the taxpayer fails to take reasonable care in completing the return.
- The 'middle' level of culpability is 'deliberate but not concealed', where the return is wrong and results from the deliberate action of the taxpayer, but the taxpayer makes no arrangements to conceal the inaccuracy.
- Finally, the most serious level of wrongdoing is that which is 'deliberate and concealed', where the taxpayer has deliberately sent in a wrong return and deliberately tries to conceal the parts of the return that are wrong, for example by submitting false evidence in support of false figures.

The following maximum levels were established by *FA 2007, Sch 24*, but their scope was extended by *FA 2008*:

- The 30% rate: 30% of the potential lost revenue, where the taxpayer was careless.
- The 70% rate: 70%, where the action of the taxpayer was deliberate but not concealed.
- The 100% rate: 100%, where the action of the taxpayer was deliberate and concealed. This can apply in two circumstances. Under *FA 2007, Sch 24, para 4(1)(c)*, the full penalty could apply to 'deliberate and concealed action'. That penalty is extended to cover 'third party' acts. The collection of tax lost through third-party inaccuracy is extended to include any inaccuracy resulting from the supply of false information or the withholding of information.

Careless inaccuracies are corrected before deliberate inaccuracies; and deliberate but not concealed inaccuracies are corrected before deliberate and concealed inaccuracies. In calculating the lost tax, account is taken of any overstatement in any document given by the taxpayer that relates to the same tax period.

The IHT in issue on death can greatly exceed what an ordinary family person would be likely to incur in respect of any other tax or on any other occasion. Preparation of IHT400 should fully reflect the IHT in issue and the professional risks involved.

There is facility in *FA 2007, Sch 24, para 11* for 'special reduction' of a penalty if HMRC think it right because of 'special circumstances', but that does not include inability to pay, nor the fact that the increased liability of one taxpayer may reduce the liability of another. There is a 100% cap on penalties (see *FA 2007, Sch 24, para 12(4)*) to cover the situation where a penalty could be charged against both the taxpayer and a third party.

Penalties will not be claimed from a deceased person who had not been compliant, but that in no way exonerates delinquent personal representatives. The penalty regime in *FA 2007, Sch 24* did not previously apply to IHT, so specialists in that area of practice may not have become familiar with the scheme of that schedule. In a situation where there are various errors, such as could easily apply in the administration of an estate, careless inaccuracies are corrected before deliberate inaccuracies; and deliberate but not concealed inaccuracies are corrected before deliberate and concealed inaccuracies. In calculating the lost tax, account is taken of any overstatement in any document given by the taxpayer that relates to the same tax period.

The case of *Timothy Clayton Hutchings v HMRC* [2015] UKFTT 9 (TC) is a stark lesson in the manner in which the penalties operate and it highlights HMRC's significantly tougher approach. In that case, the taxpayer deliberately withheld information relating to a substantial gift of an offshore bank account and as the disclosure was prompted, a hefty penalty equal to 50% of the IHT involved was imposed on the taxpayer.

Mitigation of 30% penalties

22.14 Where the disclosure is prompted, a careless error, attracting a penalty of 30%, may be reduced, but not below 15%. However, the mechanism of valuation in deceased estates relies on disclosure which, in turn, includes answering the question whether the property is to be offered for sale. The value of the property is routinely referred to the Valuation Office. If it is known that a property is to be sold within a reasonably short time after the death, then it is quite common for the gross sale proceeds to be taken as the value at the date of death.

Mitigation of 70% penalties

22.15 Where the action of the taxpayer was deliberate but not concealed, the starting point for penalties is 70%, but it may be reduced under *FA 2007, Sch 24, para 10(3)* to a minimum of 20%. The mitigation will depend on the 'quality' of the disclosure, to include timing, nature and extent. Under *FA 2007, Sch 24, para 10(4)*, a penalty that is otherwise chargeable at 70% cannot be reduced below 35% if the disclosure is prompted.

Mitigation of 100% penalties

22.16 Similar rules apply to errors that fall into the 100% regime, but they are harsher. For unprompted disclosure, the minimum penalty is 30%; but, where the disclosure arises only after HMRC have raised questions, the penalty cannot be less than 50%.

THE CRIMINAL OFFENCE OF TAX FRAUD

22.17 *FA 2000, s 144* introduced a specific criminal offence punishable by up to seven years in prison. It is aimed at tax fraud but applies only to the fraudulent evasion of income tax and not to other taxes. It can be tried either summarily or before a jury. The essence of the new defence is dishonesty, such conduct to be judged by what is considered dishonest by the ordinary standards of reasonable and honest people. There is no need for any positive act of deception. Refraining from notifying a chargeability to tax, which could easily happen in the administration of an estate, whether by a lay personal representative or, it must be said, by a professional, could come within the offence.

The distinction between avoidance and evasion has not, in the opinion at least of HMRC, been changed by the introduction of the new offence. If a scheme has a sound technical basis, and if there is no concealment of the true facts, a criminal offence has probably not been committed.

FORMS AND PROCEDURES

22.18 Good practice starts with careful preparation. Most APR/BPR cases before the First-tier Tribunal turn on the particular facts. Good work, therefore, starts with the forms that accompany the returns and apart from IHT400 the most relevant of these for the purposes of this book are:

- IHT405: Houses, land, buildings and interests in land;
- IHTD36: Land, buildings and interests in land (lifetime chargeable transfers);
- IHT412: Unlisted stocks and shares, and control holdings;
- IHT413: Business and partnership interests and assets;
- IHTD38: Business relief, business or partnership assets (lifetime chargeable transfers);
- IHT414: Agricultural relief;
- IHTD37: Agricultural relief (lifetime chargeable transfers);
- IHT419: Debts owed by the deceased; and
- IHT420: National Heritage assets – Conditional exemption and offers in lieu of tax.

Agricultural relief

22.19 Usually, form IHT405 or D36 (lifetime transfers) will be required in respect of any land held with the agricultural value claimed entered in column F or E as appropriate. It is frequently if not routinely claimed by HMRC that the agricultural value is lower than open market value perhaps by an average of 30% but this rule of thumb approach should never be accepted at face value and certainly in the case of bare land, should be fiercely resisted (see **3.1**). The factors that may affect value are more fully described in IHT405 than in D36, which is an older form.

To claim the relief, form IHT414 or D37 will also be needed. The IHT400 Notes, 'Guide to completing your Inheritance Tax account', Schedule IHT414 explain the detail expected when filling in these forms. In particular, details are needed of the actual agricultural activities carried on, whether by the owner or another. That involves finding any letting agreements, or other agreements as to the use of the land, if not occupied by the owner.

Farmhouses and farm cottages are a major source of contention and the form should be completed with care and sufficient detail to justify the claim.

Business relief

22.20 Often, form IHT412 (unquoted shares) will be relevant with IHT413 or D38 pertinent to the claim for relief. The forms require the taxpayer or agent to distinguish an interest in the business separately from a mere asset in the business as the tax treatment differs. The notes in the Guide are far too brief to help anyone to fill in the form satisfactorily without further research but IHT413 asks for three years' accounts: these will normally be only what is already available, not accounts prepared to the date of death or specially prepared for the return. These forms take time to complete and demand both care and attention: the questions are short but the answers, or the research to produce them, may be very demanding.

Woodland relief

22.21 There is no separate form to claim this relief; rather the claim should be made on form IHT405, column F, or on D36, column E.

Heritage relief

22.22 Use form IHT420 and Schedule IHT420 in the Guide. This form differs from the others in that it is used not only to claim relief when submitting a return in form IHT400 but also when negotiating an offer in lieu of tax, as described in the next chapter. The Guide is brief, and good on that account, but many taxpayers will need to do some research or seek specialist advice before completing the return.

DISCLOSURE OF TAX AVOIDANCE SCHEMES (DOTAS)

22.23 In January 2011 the *Inheritance Tax Avoidance Schemes (Prescribed Description of Arrangements) Regulations SI 2011/170* were published. The effect is that from 6 April 2011 transfers into trusts are within the DOTAS regime for IHT. A useful flowchart to determine whether a particular scheme is subject to DOTAS has been published by HMRC in their guidance notes (see www.hmrc.gov.uk/aiu/dotas.pdf). To determine whether DOTAS applies HMRC have published a four-stage test in their guidance:

- **Test 1**: Are there arrangements or proposals for arrangements which result in property becoming 'relevant property'?
- **Test 2**: Are those arrangements or proposals for arrangements such that they enable a 'relevant property charge advantage?
- **Test 3**: Is the tax advantage a main benefit of the arrangements?
- **Test 4**: Is the arrangement on the list of grandfathered schemes and schemes that are not within the regulations. If so it may have to be disclosed. The list of grandfathered schemes can be found in the guidance.

The main aim of the extension of the disclosure rules is to restrict disclosure to those schemes which are new or innovative. This is achieved by exempting from disclosure those schemes which are the same or substantially the same as arrangements made available before 6 April 2011.

Legislation was included in *FA 2014* to impose advance penalties on taxpayers who entered into a tax avoidance scheme. In such cases where the tax avoidance scheme involved is notifiable to HMRC under DOTAS, or is counteracted under the GAAR, HMRC have (from 17 July 2014) the power to issue a Notice to Pay. This will require the taxpayer to pay the tax that HMRC consider to be due, even though the taxpayer has appealed against the assessment and does not believe that he will owe any tax. Controversially this also applies to those who entered into tax schemes many years ago where the dispute has not yet been settled.

It was further announced that users of defeated avoidance schemes will be required to disclose their use to HMRC and pay any tax in dispute upfront. Penalties will apply for non-compliance.

FA 2014 included measures against so-called high-risk promoters of tax avoidance schemes awarding HMRC two new powers. The first enables HMRC to issue a promoter a 'conduct notice', demanding full details about a particular tax planning product and its intermediaries and users. The second allows HMRC to issue 'monitoring notices' to selected promoters to provide details of all their products, intermediaries and users within five days of being asked to do so. 'Monitoring notices' will require the approval of the First-tier Tribunal.

On 31 July 2014 HMRC issued a consultation document 'Strengthening the Tax Avoidance Disclosure Regimes' and a summary of responses was later published in December 2014. The 2014 Autumn Statement contained a proposal for the introduction of a new threshold condition to ensure that promoters do not use the DOTAS regime as a way of obtaining advance clearance by repeatedly tweaking schemes and re-submitting them for consideration. The new threshold condition would consider how often a promoter contacts HMRC to gain approval of a scheme.

Specifically aimed at IHT, which hitherto had been largely unscathed by the DOTAS rules, it is intended that the disclosure requirements will be more keenly focused on those IHT avoidance schemes which operate outside the normal planning routines involving available reliefs and exemptions. On 2 September 2015 HMRC issued fresh Guidance Disclosure of Tax Avoidance Schemes which brings readers up to date with the changes introduced in *FA 2014* and *FA 2015*. The latter changes include:

- increasing the penalty for persons using schemes which do not comply with reporting requirements;
- protecting persons who voluntarily provide information about potential failures to comply with DOTAS;
- requiring promotors to notify HMRC if their details or those of one of their schemes change;
- enabling HMRC to publish information about schemes and promoters that it receives under the DOTAS rules;
- giving HMRC powers to obtain details of users of undisclosed schemes from introducers; and
- creating a new power requiring additional information to be sent to clients.

Following responses made to the consultation on its proposed draft changes to the hallmark regulations, which closed in September 2015, HMRC made a number of amendments which changes became effective from 23 February 2016. In April 2016 HMRC issued a further consultation document which provided welcome examples of ordinary IHT planning which could secure a tax advantage but which HMRC considered to be nonetheless outside the DOTAS hallmark because such tax planning was neither contrived nor abnormal. These included:

- straightforward outright gifts;
- lifetime transfers into flexible or discretionary trusts;
- investment into assets that qualify for IHT reliefs;
- arrangements that are within a statutory exemption, ie paying full consideration for the continued use of a gifted asset;
- certain routine insurance-based arrangements, ie loan trusts.

Chapter 23

Payment of IHT

QUICK GUIDE

- Unless the instalment basis is available and so elected, the IHT on an estate must be paid before obtaining the grant of probate. For unrepresented taxpayers, the funding of that tax has been made easier by the procedure outlined in **23.4**.
- Certain categories of illiquid assets qualify for an instalment facility outlined at **23.5**.
- As will be seen at **23.8**, care should be taken to observe the rules relating to instalment. Even greater care should be taken when releasing assets from an estate before all the tax has been paid.
- The scheme by which heritage property may be used in satisfaction of IHT is considered in detail at **23.11** onwards, noting especially the discount arrangements or 'douceur' at **23.12**.
- *FA 2012* introduced a new facility whereby heritage property might be given to the nation to satisfy taxes other than IHT. The new rules are outlined at **23.18**.

THE GENERAL RULES

23.1 Subject to a contrary intention in the will, the IHT for which the personal representatives are liable is treated as part of the administration expenses of the estate (see *IHTA 1984, s 211(1)*). By the operation of *Administration of Estates Act 1925, s 34(3)* and *Sch 1, para 8*, tax is chargeable against the capital assets of the estate in the order set out in *AEA 1925, Sch 1*. *IHTA 1984, s 211* limits the applicability of this rule to the value of property in the UK which:

- vests in the personal representatives of the deceased; and
- was not, immediately before the death, comprised in a settlement.

Where the amount of IHT paid by the personal representatives on the value transferred by a chargeable transfer made on that death (rather than during lifetime) does not fall to be borne as part of the testamentary and administration expenses of the estate within the rules in *s 211(1)*, that IHT should be repaid to them by the person in whom that property is vested and to which the value of the tax is attributable. This will give the executor a right of action against the donee for recovery.

'Free of tax' legacies

23.2 Clarity of intention should be encouraged and thus where it is intended that some specific gifts should bear their own tax and others not, the will should make this clear. In some cases the law will dictate that the gift must bear its own tax as in the case of a legacy of foreign property but this can lead to a situation where a pecuniary legacy is payable and the assets out of which it is to be paid include both UK and foreign property. Foreign property will bear a proportion of the IHT unless the will expresses that the foreign property be given free of tax.

Many of the cases in this area are of some antiquity, examining phrases such as 'free of estate duty' and 'free of CTT' but *IHTA 1984, Sch 6, para 1* effectively updates all such references to earlier capital taxes on death so as to apply to IHT. Such a clause will, however, normally relate only to the liability to tax on the death of the testator as demonstrated in *Re Pattersons Will Trusts, Lawson v Payn* [1963] 1 All ER 114 which held that a gift of property free of tax to A for life, with remainder to B, will relate in normal circumstances only to tax payable on the death of the testator and not to any charge to tax on the death of A or on the transfer arising by disposal by A of his interest in his lifetime. *Re Owers* [1940] 4 All ER 225 established the rule that an instruction that residue should carry testamentary expenses does not include payment of tax on foreign property.

Payment of tax on failed PETs

23.3 *IHTA 1984, s 199(1)* places the liability for settling the 'top up' charge on:

- the transferee;
- any person in whom the property is vested (whether or not beneficially);
- any person beneficially entitled to an interest in possession; and
- any person for whose benefit the property or income is supplied where the property has become settled.

The transferee is the first person liable but if the transferee does not pay, the personal representatives may become liable under *s 204*. This naturally has caused practitioners some concern, which was discussed between the Law

Society and HMRC, who indicated that they will not normally assess IHT on personal representatives, provided that:

- the executors have made the fullest enquiries that are reasonably practicable to discover the existence of lifetime transfers;
- the executors have done all they can to make full disclosure of lifetime transfers to HMRC;
- the executors or other personal representatives have obtained a clearance certificate; and
- the estate has been distributed before the existence of the chargeable transfer comes to light.

A special rule applies under *s 208* for woodlands, in that the person liable for tax chargeable under *s 126* in relation to a disposal of timber is the person who is entitled to the proceeds of sale (or would be entitled to those proceeds if the disposal were a sale).

Due date for payment

23.4 The normal payment date is six months from the end of the month in which the transfer takes place and in relation to failed PETs, this is six months after the month in which the death occurred (see *s 226(3A)*). Special rules apply where a lifetime transfer is made after 5 April and before October in any year where tax on transfers falling within that period is due (see *s 226(1)*) on 30 April of the following year. *FA 2009, s 105* aligned the interest rates for IHT with those for other taxes with effect from 21 July 2009.

To ease the difficulty and cost for executors of funding IHT before grant of probate, HMRC introduced the direct payment scheme, by which the financial institution that holds funds for the deceased is asked to pay the tax direct to HMRC. The scheme came into effect on 31 March 2003 but is voluntary, so the practitioner should check with the relevant financial institution whether they participate in the scheme, which applies only to accounts in the sole name of the deceased. If the deceased had several accounts with the same bank or building society, including loan or credit card accounts, it is likely that only the net capital amount will be available to pay tax.

The procedure is simple. First, obtain an IHT reference for the estate, either online through the HMRC website or using form IHT422. Then download form IHT423 from the HMRC website, fill it in and send it, not to HMRC, but to the institution holding the money. A separate form IHT423 must be completed for each financial institution from which the personal representatives wish to transfer money to pay the tax.

Payment of tax by instalments

23.5 *IHTA 1984, s 227* provides that, on an election by the person paying the tax, IHT due on qualifying property may be paid by ten equal yearly instalments. 'Qualifying property' means:

- any land, anywhere;
- '*s 228*' shares or securities (see below); and
- a business or an interest in a business.

By *s 229*, the instalment option is available in respect of woodlands under the charge in *s 129*. The tax may be paid by ten equal yearly instalments, the first being payable six months after the end of the month in which the transfer within *s 129* is made.

The conditions of eligibility for the instalment option are:

- the transfer is made on death; or
- the tax is being paid by the person who benefits from the transfer; or
- the transfer comes within the settled property regime and either the tax is borne by the beneficiary or the property stays in the settlement.

At first sight this would seem to exclude IHT due on failed PETs but actually the instalment option is available, see *IHTA 1984, s 227(1C)*, where:

- the property was owned by the transferee from the date of the transfer to the death of the transferor (or if earlier of a transferee); or
- the charge arises under clawback of APR or BPR following the death of the transferor.

Where the property consists of unquoted shares or unquoted securities, further conditions apply under *s 228(3A)*, ie that the shares or securities remained unquoted throughout the period beginning with the date of the transfer and ending with the death of the transferor (or the earlier death of the transferee).

'Section 228 shares'

23.6 *IHTA 1984, s 228* applies to four categories of shares or securities of a company, namely:

- shares giving control;
- unquoted shares which attract (together with tax due on other instalment assets for which the same person is liable) not less than 20% of the tax chargeable by that person in the same capacity;

- unquoted shares where the tax cannot be paid without undue hardship; and
- other shares neither giving control nor unquoted but where the value of the transfer is over £20,000 and the shares are at least 10% of the nominal value of all the shares in the company at that time, or they are ordinary shares and their nominal value is at least 10% of the nominal value of all the ordinary shares of the company at that time.

'Ordinary shares' means (see *s 228(4)*) shares carrying a right to dividends not restricted to those at fixed rate, or shares with a right to conversion to shares carrying such a right.

Interest on instalments

23.7 Interest is added to each instalment, running from the date at which the instalment is payable unless the instalment is attributable to shares, securities, business, interest in a business, land/buildings qualifying for APR or woodlands (*IHTA 1984, s 234(2)*). The rule as to interest on instalments does not apply to dealing in investing companies, unless they are also holding companies or market-makers within *s 234(3)(b)* or *(c)*. 'Market maker' is defined in *s 234(4)*.

Termination of the right to pay tax by instalments

23.8 An instalment election once made is not binding on the taxpayer and the tax involved together with interest thereon may be settled in full at any time without penalty. However, if the property in respect of which an instalment option has been made is later sold, the instalments basis is forfeit and the tax unpaid (plus accrued interest where applicable) becomes payable immediately (see *s 227(4)*). Where part only has been sold, a proportionate part of the tax is payable but where the sale takes place before the date on which the first instalment is payable, tax is payable on that first instalment date.

There are two bear-traps within *s 227(4)*:

(1) It states that 'the tax for the time being unpaid, with interest to the time of payment, may be paid at any time' but HMRC insist that there must be no 'half election'. Accordingly, if the executors elect to pay tax by instalments and subsequently wish to make an advance payment on account that advance payment has, in theory at least, the effect of causing the instalment basis to be forfeit such that all the remaining instalments become due. HMRC have subtly changed their stance in recent times and there is some indication, although not consistent, that a partial repayment in excess of the standard instalment would not affect the ability to continue with the instalments basis.

(2) Where a person (say) the executor has paid tax in one lump sum which might have otherwise been paid by instalments and he is entitled to recover that tax from the beneficiary, that beneficiary is entitled to refund the tax to the executor on the same instalment basis as had been available to (but not elected for by) the executor (*s 213).*

This very issue was demonstrated in *Howarth's Executors v IRC* [1997] STC (SCD) 162. The will of the deceased was proved by the son, his wife and an employee (the lawyer) of the firm of solicitors who dealt with the administration of the estate. The executors elected to pay the tax (then CTT) of £34,380 by instalments. At the request of the family, the lawyer agreed that the assets of the estate might be transferred out of the names of the executors and into the names of the beneficiaries.

The beneficiaries made the payments of instalments but the son was then made bankrupt at a time when the tax outstanding plus interest amounted to £8,084. HMRC served a notice of determination on each of the executors, who appealed but none appeared at the appeal and the lawyer was absent due to ill health. On his behalf, it was argued that HMRC should have registered their charge (arising under *s 237*) against the property and that registration would have protected the interests of the lawyer when the land was sold. The lawyer had not been told by HMRC that there had been any problem with payment of instalments until almost two years after he had retired owing to ill health (he suffered from high blood pressure and diabetes). Attending court would have caused a further deterioration in his health, as indeed would enforcement proceedings to recover from him the outstanding tax that the family should have paid. Not being a beneficiary, he had received nothing from the estate.

The Special Commissioners held that a charge in favour of HMRC in relation to unpaid tax and interest was created under *s 237(1)*. The property itself ceased to be subject to that charge when sold, 'but the property for the time being representing it shall be subject to it', according to *s 238(1)*. The HMRC charge thus moved from the property of the deceased to the proceeds of sale and there was no need for HMRC to register the charge.

There was some sympathy for the lawyer, but in his capacity as an executor, he had agreed to pay the tax by instalments and had agreed to the transfer of the assets out of the names of the executors and into the names of the beneficiaries. CTT was a personal liability of each executor or personal representative and it was not easy to shed that responsibility, as had been noted in *IRC v Stannard* [1984] STC 245.

The HMRC charge

23.9 Where tax charged on the value transferred by a chargeable transfer is unpaid, a charge is imposed by *s 237(1)* in favour of HMRC on any property to the value of which the value transferred is wholly or partly attributable (see *Howarth's Executors v IRC* at **23.8**). The charge is also imposed on any property comprised in a settlement where the chargeable transfer is the making of the settlement or is made under the settled property regime.

Section 237(3), as originally enacted, excluded from the HMRC charge personal or moveable property situated in the UK which was beneficially owned by the deceased immediately before his death and vested in his personal representatives. The definition of 'personal property' includes undivided shares in land held on trust for sale, whether statutory or not, but by virtue of *FA 1999*, excludes leaseholds.

Section 237(3A) applied to failed PETs. Property (or an interest in it) disposed of to a purchaser before the death of the transferor is not itself subject to the HMRC charge but the following is:

- property which has been disposed of by any other means; and
- property which at the death of the transferor represents either the original property or any interest in it.

Section 237(3C) imposes a charge in favour of HMRC in respect of unpaid tax where such tax relates to:

- chargeable events relating to conditionally exempt transfers (*s 32*);
- the charge in relation to associated properties (*s 32A*);
- exemption from the ten-year charge *s 79(3)*);
- the charge to tax where property leaves maintenance funds (*Sch 4, para 8*); or
- the charge on the failure of conditional exemption in respect of objects and buildings (*Sch 1, para 5* or *Sch 3, para 5*).

The charge under *s 237(3C)* will apply except where the event giving rise to the charge was a disposal to a purchaser of the property or object in question and, in such a case, on any property for the time being representing that property or object.

PURCHASE OF OWN SHARES BY AN UNQUOTED TRADING COMPANY

23.10 Sometimes, families who run an incorporated businesses plough back a considerable amount of the profit and BPR encourages retention of liquidity within the family company, subject only to the rules as to excepted assets considered in **Chapter 7**. The general rule is that any payment by an unquoted trading company for the purchase of its own shares will be treated by *Corporation Tax Act 2010* (*CTA 2010*), *s 1000* as a taxable distribution for corporation tax, save for those exceptions where the strict requirements of *Chapter 3* of *CTA 2010*, in particular *s 1033,* are met.

During the lifetimes of the shareholders, the exceptions most commonly encountered are those where the transaction is for the benefit of the trade of the company and is not for the avoidance of tax. Specifically, however, *CTA 2010, s 1033(3)* allows the purchase of own shares by an unquoted trading company *not* to be treated as a distribution where:

- the proceeds of that purchase are all, or nearly all, used to pay IHT; and
- the transaction is completed and the tax paid within two years of the death.

The rule is itself subject to an exception, which extends its application. Whilst, in general, the proceeds of the sale of shares must be applied in paying IHT, those proceeds may also be applied in paying CGT arising on the disposal of the shares to the company.

This relief may be relied on, in IHT cases, only where it is shown that the IHT could not be paid by any other means without undue hardship. *CTA 2010, s 1044* allows for advance clearance of payments under this code. The application must be in writing and must set out particulars of the relevant transactions (*CTA 2010, s 1045(1)*) and it is more likely to succeed if it demonstrates the hardship on which the application is based. HMRC have 30 days in which to respond (*CTA 2010, s 1045(5)*) but there is no right of appeal against that decision.

PUBLIC PURCHASE OF HERITAGE PROPERTY BY PRIVATE TREATY SALES: OUTLINE

23.11 On the sale in the open market of property that has been the subject of conditional exemption, that exemption, whether for estate duty, CTT or IHT, is forfeit. It is also quite likely that the sale will give rise to a gain on which CGT is chargeable. Although it is common practice for auctioneers to charge a 'buyer's premium', which may have the result of reducing the other charge that is made against the seller on the hammer price, the sale charges may be considerable but will be allowable against the gain for computation of CGT.

Items are reviewed by the Arts Council England (ACE) which will refer them to the Acceptance in Lieu Panel (AIL). The AIL Panel seeks to establish the true market value of the item and, in 10% to 15% of cases, will actually recommend a higher figure than was proposed by the owner. Given the limited funds currently available to museums, the AIL scheme is very important to them.

A sale by private treaty to a '*Sch 3*' body, ie a heritage body within *IHTA 1984, Sch 3,* does not trigger a forfeit of the conditional exemption, nor does it give rise to a CGT charge. That tax exemption comes, however, at a price in that the vendor is asked to accept a lower price than he would get in the open market. The sweetener or the 'douceur', to encourage the owner of heritage property to sell to the nation, is an administrative arrangement which is intended to benefit both parties.

There is a secondary saving. As previously explained, the clawback of exemption from IHT or its predecessors can involve not only a charge in relation to the event that triggers the clawback but also a tax charge found by reaching back to the cumulative total of transfers made by the last person in respect of whom conditional exemption was given. Thus, the clawback charge may affect not only the value of the heritage property itself, but also the value of other family wealth. Sale by private treaty to the public leaves the calculation of tax on other family property undisturbed and it can also benefit the current cum. lifetime 'clock' of a person who has settled heritage property.

IHTA 1984, s 230(1) refers to 'the application of any person liable to pay tax or interest'. This suggests not merely that only such a person may make the application, but that an application may be made only in respect of property in which the applicant has a beneficial interest or which he holds as personal representative of the deceased former owner. Thus, a beneficiary liable to pay tax might offer property that he already owns or might offer property coming to him by an inheritance, which was the trigger for the liability to tax. What such a person, presumably, may not do is offer other property which he holds in some other capacity, as, for example, the trustee of a family trust. Equally, an executor can offer property in the estate that he administers but cannot, according to HMRC, offer his personal property unless he is not only an executor but also a beneficiary in the estate.

Acceptance in lieu of tax is available (see CTNH Chapter 11 para 11.12) in respect of pre-eminent objects, land, and of objects kept in a building that has been accepted in lieu of tax or belongs to the nation, or is looked after by the state, for example English Heritage or another *Sch 3* body.

'Douceur'

23.12 The calculations assume agreement to the value of the heritage property on the principles described earlier in this book. A calculation is made of the net of tax value, after taking account of CGT on the sale itself and clawback of IHT (or its predecessors) but costs of sale are not taken into account in this notional tax calculation even though, as noted above, they may be considerable. The 'douceur' is then applied to the net result and the net of tax value is enhanced. The starting figure for 'douceur' is 25% of the combined notional tax figure, although that percentage may be increased on low-value objects or reduced on high-value objects. It is always reduced on land, generally to 10% but a douceur is not allowed on amenity land: for this point (CTNH chapter 11 para11.18).

The purchaser must be aware of the full facts and the taxpayer should authorise HMRC to disclose their potential liability to tax. When the figures are agreed, the buyer settles the special price, ie the net of tax market value, plus the 'douceur'.

Example 23.1: HMRC illustration updated (1) – Appendix 11

The illustration given in CTNH Appendix 11 (originally set in July 2010 (2010/11) but updated for this book to July 2017 (2017/18)) is a useful starting point for understanding the mechanism. The example assumes that a work of art is likely to fetch £300,000 (gross) at auction, that such a sale would realise a notional gain of £275,000, chargeable at 20% and resulting in a notional CGT liability of £55,000. The example also assumes that there would be an IHT charge at 40% on the market value reduced by the CGT charge.

The clawback charge is calculated as follows:

	£
Sale price	300,000
Less CGT on assumed gain	(55,000)
	245,000
40% assumed tax charge	(98,000)
Net value after combined notional tax	147,000
Douceur at 25% of notional tax (£55,000 + £98,000)	38,250
Special price	185,250

Example 23.2: HMRC illustration updated (2) – Appendix 12

Applying similar principles, the illustration set out in CTNH Appendix 12 (again updated to 2017/18) is in respect of land, showing the lower level of douceur:

	£
Assumed market value of the land	300,000
Assumed CGT at 20% on assumed gain	(55,000)
	245,000
Assumed clawback of IHT at 40% on assumed net value (£300,000 – £55,000)	(98,000)
Net value after combined notional tax:	147,000
Douceur: at 10% of notional tax (£55,000 + £98,000):	15,300
Special price:	162,300

The 'other side' of each of these calculations is that HMRC write off the total notional tax of £153,000 (£55,000 + £98,000). The public end up paying the special price: thus, it is easily seen why the purchasing body must know 'the full story'. Whilst, in the CTNH examples, it simplifies the calculations, in that the expenses of sale are ignored, many owners and their advisers will in effect rework the calculations to take into account expenses, not least because there will be some expense for professional advice sought to negotiate a private treaty sale to the public.

Stamp duty and stamp duty land tax

23.13 No duty is payable on any transfer of property accepted in satisfaction of tax (*National Heritage Act 1980, s 11*). The acquisition of an interest that is chargeable to stamp duty land tax by a charity is exempt from stamp duty land tax under *FA 2003, s 68* and *Sch 8*. Acquisition by certain heritage bodies is exempt under *s 69*, as mentioned at **21.31**.

Value added tax

23.14 It is possible, but perhaps unlikely, that property offered in lieu of tax could be subject to a VAT charge on disposal but this could apply to property that had been exhibited in a house open to the public, or property that had been used for some business purpose. The impact of VAT on works of art is governed by *VATA 1994, Sch 9, Group 11*. The disposal of property in satisfaction of tax under *s 32(4)* or *s 32A(5)* or *(7)* is not chargeable (see *Group 11, Item 3*). The corresponding item for CGT is *Item 4*, with additional *Items 1* and *2* respectively for estate duty and claims under *IHTA 1984, Sch 5*. Transfer of property to the state is not treated as a 'supply'.

Interest and the valuation date option

23.15 The value of heritage property can and does fluctuate. Since 17 March 1987, the basis for acceptance of property in satisfaction of tax is the value on the date of offer and not the value at the date the property is actually accepted and in times of inflation, this rule could work against the owner. On the other hand, interest on the tax in issue ceases to run from the date of the offer rather than from the date of acceptance, even though, in effect, the owner may have had the use and enjoyment of the property meanwhile.

This rule is not absolute and by SP 6/87, there is a facility to opt for revaluation as at the date on which the property is accepted. Where that option is taken, interest on the unpaid tax runs until acceptance. By para 5 of SP 6/87, where the 'offer date' option remains open and is chosen by the owner, interest on the tax to be satisfied by the item will cease to accrue from that date rather than from the date of acceptance.

The date of completion of the offer of chattels and their acceptance will be important to the issue of risk as such risk remains with the owner of the chattel until the Secretary of State has nominated the institution that will take charge of the item, either on a temporary or on a permanent basis, and that institution has collected the object. It is therefore important that any insurance that the owner of the item maintains should be kept in force until the item is delivered in this way. Different conditions should apply where the item is already on loan to a public museum and in such a case, the owner should try to negotiate that risk passes at an earlier stage, ie when the 'memorandum of acceptance' has been agreed and exchanged between the parties. The risk in respect of land passes only on completion of the assurance by way of conveyance or transfer to the body nominated to receive it.

'Supermarket' heritage offer

23.16 A problem arises if an offer of property fails when that tax becomes payable, plus interest, even though a new offer may later succeed. Where an owner holds many items of heritage property, together worth far more than the liability to tax that must be satisfied, the owner may 'set his stall out' by making a multiple offer. This allows HMRC to self-select the items they most want and the owner reduces his risk of exposure to an interest charge.

Offers 'in situ'

23.17 In 1997, the Department for Culture, Media and Sport indicated changes to the previous scheme for offers of property '*in situ*'. In essence, such an offer is made by the owner on the basis that the chattel, which must be pre-eminent and not merely associated, remains at the private property where it has, in effect, always been. The ownership of the chattel passes to the state and is allocated to a public museum. A loan agreement is set up between the owner and the museum.

Present procedure is that the owner to whom the offer is made must find a museum that will take ownership of the article and lend it back to them before the *in situ* offer is submitted to HMRC. In outline, the agreement will cover at least the following issues:

- *Public access*: the building where the item remains must be open to the public. The minimum access likely to be required is 100 days per year.
- *Preservation*: had the item been in a museum, there would have been in place environmental controls and measuring devices. If it is to remain in the present property, standards of environmental control must be appropriate. This will be fixed by the ACE and by the museum which becomes the owner of the chattel. The owner must pay.
- *Future conservation*: the museum that becomes the owner of the chattel will be responsible for these arrangements.
- *Security*: again, the ACE will advise. The owner must stand the initial cost of the security arrangements, but later upgrades will be met by the museum that owns the chattel.
- *Insurance*: the museum to which the chattel is allocated may seek a government indemnity for up to 99% of the 'special price', as determined in accordance with the principles discussed earlier in this chapter. The owner covers the other 1%.
- *Future owners*: an *in situ* agreement runs for the lifetime of the owner of the chattel who puts it into effect, but is terminated if that person leaves the property at which the chattel is kept. Under the latest scheme for agreements of this kind, any attempt to achieve automatic succession as between one member of the family and the next is excluded. A successive occupier of the heritage property may indicate that he or she wishes to enter into a new *in situ* agreement; but, if the museum owning the chattel is not happy with that, new arrangements must be made. Such matters

should be referred to arbitration, as indeed should any provision that allows the museum to raise the issue of whether it is still appropriate for the chattel to remain at the property once 21 years have elapsed from the date of the original agreement.

- *Exhibition*: the museum must have the right to remove the chattel occasionally for short periods of public exhibition or for conservation purposes. Equally, if the former owner of the chattel no longer observes the terms of the agreement, the museum must have power to remove it.

Gifts to the nation: the Cultural Gifts Scheme

23.18 The aim of measures introduced by *FA 2012, s 49* and *Sch 14*, was 'to stimulate lifetime giving by encouraging taxpayers to donate pre-eminent objects, or collections of objects, to the nation'. What follows is only an introduction to the Cultural Gifts Scheme. Objects may be lent or given to appropriate institutions so that public access is achieved. The scheme specifically excludes trustees and personal representatives: see *Sch 14, para 2(2),* for whom the existing arrangements may be good enough. The relevant taxes are income tax, CGT and, for companies corporation tax (other than in passing, the taxation of companies is not, generally, dealt with in this work). An individual donor who provides items that are accepted under the scheme is granted a reduction in tax liabilities, which may be spread over up to five years (not available for companies) and that reduction is 30% (20% for companies) of the value of the object as agreed.

There is an overall budgetary limit for the scheme of £30 million, to include acceptance in lieu for IHT already described (the old scheme used to cost about £14 million pa).

23.19 A 'qualifying gift' is made where:

- the taxpayer offers pre-eminent property to the nation;
- the taxpayer owns the property outright, not jointly;
- the offer fits the scheme;
- the offer is registered under the scheme;
- the offer is accepted, in whole or in part; and
- the gift is made pursuant to the offer.

Timing is important: *Sch 14, para 1(3)* defines the offer registration date as that on which the offer was indeed registered, but that is only part-way through the process and is no guarantee of a successful conclusion. However, if the process is successful, that date defines the earliest tax year for which a tax reduction can apply.

23.20 Once the offer has been registered, experts will advise on pre-eminence and value. The annual budget is tested and if the recommendation is accepted and the relevant Minister agrees, the taxpayer is notified and has 30 days within which to agree.

There are ten quite complicated rules to apply the tax reduction in the year of registration and four following years. The available reduction is allocated by the taxpayer to years and extinguishes or reduces the liability for that year. The taxpayer might get it wrong, by allocating part to a future year when his liability turned out to be less than forecast: that is just bad luck, and part of the reduction may simply be wasted. The official response to this apparent injustice was that they could not monitor variable take-up of the reduction, since that would make budgeting impossible. Similarly, death intervening before full use of the reduction may waste the saving. There is no proposal to allow credit the unused balance against taxes on death.

Appendix 1

Statutory Instruments

THE INHERITANCE TAX (MARKET MAKERS) REGULATIONS 1992, SI 1992 NO 3181

Interpretation

(2) In these regulations 'subsection (7)' and 'subsection (4)' means subsections (7) of section 105 and subsection (4) of section 234 respectively of the Inheritance Tax Act 1984.

Modification of Subsection (7) and Subsection (4)

(4) Subsection (7) and subsection (4) shall have effect as if—

- (a) the reference to The Stock Exchange in paragraph (a) of each of those subsections were a reference to either of The Stock Exchange and LIFFE (Administration and Management) (both being recognised investment exchanges within the meaning of the *Financial Services Act 1986*), and
- (b) references to the Council of The Stock Exchange in paragraph (b) of each of those subsections were a reference to the investment exchange concerned.

Appendix 2

Extra-statutory Concessions

F15 WOODLANDS

Finance Act 1986, Sch 19, para 46 denies potentially exempt transfer treatment for Inheritance Tax purposes to all property comprised in a single transfer any part of which, however small, is woodlands subject to a deferred Estate Duty charge. By concession the scope of this paragraph will henceforth be restricted solely to that part of the value transferred which is attributable to the woodlands which are the subject of the deferred charge.

F16 AGRICULTURAL PROPERTY AND FARM COTTAGES

On a transfer of agricultural property which includes a cottage occupied by a retired farm employee or their widow(er), the condition in *IHTA 1984, ss 117* and *169* concerning occupation for agricultural purposes is regarded as satisfied with respect to the cottage if either

- the occupier is a statutorily protected tenant, or
- the occupation is under release granted to the farm employee for his/her life and that of any surviving spouse as part of the employee's contract of employment by the landlord for agricultural purposes.

F17 RELIEF FOR AGRICULTURAL PROPERTY

On a transfer of tenanted agricultural land, the condition in *IHTA 1984, s 116(2)(a)* is regarded as satisfied where the transferor's interest in property either

- carries a right to vacant possession within 24 months of the date of the transfer, or
- is, notwithstanding the terms of the tenancy, valued at an amount broadly equivalent to the vacant possession value of the property.

Appendix 3

HMRC Statements

STATEMENTS OF PRACTICE

SP 12/80 Business Relief 'Buy and Sell' Agreements

The Board understands that it is sometimes the practice for partners or shareholder directors of companies to enter into an agreement (known as a 'Buy and Sell' Agreement) whereby, in the event of the death before retirement of one of them, the deceased's personal representatives are obliged to sell and the survivors are obliged to purchase the deceased's business interest of shares, funds for the purpose being frequently provided by means of appropriate life assurance policies.

In the Board's view such an agreement, requiring as it does a sale and a purchase and not merely conferring an option to sell or buy, is a binding contract for sale within *section 113* of *IHTA 1984*. As a result the inheritance tax business relief will not be due on the business interest or shares. (*Section 113* of *IHTA 1984* provides that where only property would be relevant business property for the purpose of business relief in relation to a transfer of value but a binding a contract for its sale has been entered into at the time of the transfer, it is not relevant business property in relation to that transfer.)

SP 6/87 Acceptance of property in lieu of Inheritance Tax, Capital Transfer Tax and Estate Duty (8 April 1987)

This statement is largely superseded by Capital Taxation and the National Heritage, to which frequent reference is made in the main text.

REVENUE INTERPRETATIONS

RI 95 (December 1994) Inheritance Tax — business and agricultural relief

The Inheritance Tax (IHT) legislation provides relief for transfers of agricultural property and for business property. We have been asked for our views on the availability of relief—

- where agricultural property is replaced by business property (or vice versa) shortly before the owner's death; and
- on the donor's death, where the donee of a potentially exempt transfer of agricultural property has sold it and reinvested the proceeds in a non-agricultural business (or vice versa).

A 'potentially exempt transfer' (PET) is a lifetime transfer which only becomes chargeable to IHT if the donor dies within seven years of the transfer.

All statutory references in this article are to *Inheritance Tax Act (IHTA) 1984.*

IHT business and agricultural relief reduces the value of the relevant business property, or the agricultural value of agricultural property, by either 50 or 100 per cent. The rate of relief depends on the nature of the property and interest held.

The qualifying conditions for the relief include requirements of a minimum period of ownership and, in the case of agricultural property, of occupation of the property for agricultural purposes immediately before the transfer. If, and to the extent that, the same property may qualify for relief to both agricultural property and business property, *s 114* prevents double relief.

There are also rules which allow for the sale and replacement of qualifying property. The replacement is qualifying property only if it, and the original qualifying property, have together been owned (and in the case of agricultural property, occupied) for a combined minimum period.

In the Revenue's view, where agricultural property which is a farming business is replaced by non-agricultural business property, the period of ownership of the original property will be relevant for applying the minimum ownership condition to the replacement property. Business property relief will be available on the replacement if all the conditions for that relief are satisfied. Where non-agricultural business property is replaced by a farming business, if the latter is not eligible for agricultural property relief, *s 114(1)* does not exclude business property relief if conditions for that relief are satisfied.

There could be cases where, for example, agricultural land is not part of a farming business, so any replacement could only qualify for business property relief if it satisfied the minimum ownership conditions in its own right. However, our experience suggests such cases are likely to be exceptional.

Where the donee of a PET of a farming business sells the business, and replaces it with a non-agricultural business, the effect of *s 124A(1)* is to deny agricultural property relief on the value transferred by the PET. Consequently, *s 114(1)* does not exclude Business Property Relief if the conditions for that relief are satisfied; and in the reverse situation, the farming business acquired by the donee can be 'relevant business property' for the purposes of *s 113B(3)(c).*

RI 110 (April 1995) Inheritance Tax — Valuation of assets at the date of death

Where the value of an asset is ascertained for Inheritance Tax (IHT) purposes on the owner's death, this is also taken as the beneficiary's acquisition value for Capital Gains Tax (CGT) purposes. The Revenue have been asked to say whether they will ascertain the value of the estate assets using IHT principles, where either:

- the asset is wholly exempt or relieved from IHT; or no IHT is payable on the deceased's estate,
- in order to provide a value for any other Revenue purpose, in particular the CGT acquisition value.

The value of an asset for IHT purposes is usually the price it would realise if sold on the open market. In certain circumstances special rules may apply to give a different value. For example, under the related property provisions of the *IHT 1984, s 161*, property held jointly by husband and wife is treated as a single unit in arriving at the value of their respective interests.

IHT is charged on the assets of a person's estate on death if their value together with the value of any chargeable lifetime gifts exceeds the IHT 'threshold' £154,000 for deaths and other chargeable events occurring on or after 6th April 1995. There are various exemptions and reliefs. These include the exemption for assets given to a surviving spouse and up to 100 per cent relief for agricultural business property.

If an asset is wholly exempt of relief from IHT, neither the personal representative of the deceased nor the Revenue can require the value of that asset to be ascertained for IHT purposes.

Where it is evident that any possible increase or decrease in the value of the chargeable assets of the estate, as included in an Inland Revenue account, will leave the total value of the estate below the IHT threshold, it will not be necessary to ascertain the value of all of the individual assets for IHT purposes. In some cases, particularly where the estate is close to the threshold, values may be considered but not necessarily 'ascertained'.

For example, the value included in the Inland Revenue Account for a holding of shares in an unquoted company might appear to the Revenue's Shares Valuation Division (SVD) to be too high. In this situation, as no IHT is at stake, SVD is unlikely to negotiate an ascertained value for IHT. On the other hand, if the value included seems too low, SVD may negotiate an ascertained value if the likely amount of IHT at stake warrants this.

If the value of an asset is not ascertained for IHT, the normal rules of *TCGA 1992, s 272* will apply to determine the CGT acquisition value of the beneficiary.

The Revenue have also been asked how they will approach the valuation of a holding of shares in a unquoted company where not all of the company's assets qualify for IHT Business Property Relief, so that the shares are not wholly relieved. Again, SVD's approach will depend very much upon whether in any event IHT is payable and, if so, the amount of tax involved. SVD is unlikely to negotiate an ascertained value for the holding if very little or no tax is at stake.

RI 121 (August 1995) Inheritance Tax — relief for untenanted agricultural land

Inheritance Tax — relief for tenanted agricultural land — FA 1995, section 155

FA 1995, s 155 increased the rate of Inheritance Tax relief for transfers of tenanted agricultural land from 50 per cent to 100 per cent. The full relief applies to transfers, made on or after 1st September 1995, of agricultural land which is let on a tenancy starting on or after that date. The purpose of full relief is, as indicated in the Revenue Press Release of 27th January 1995, to boost the government's reforms of the law on agricultural tenancies, now contained in the *Agricultural Tenancies Act 1995. ATA 1995* applies to England and Wales and not to Northern Ireland or Scotland.

Application of Inheritance Tax for Relief

The Revenue have received enquiries asking whether the increased relief will apply only where the tenancy in question is within the provisions of *ATA 1995*.

The relief will apply to all agricultural tenancies, throughout the UK, starting on or after 1st September 1995, provided that all the statutory conditions for relief are met. In particular, a tenancy starting on or after that date by reason of statutory succession to an existing tenancy is not excluded from the full relief.

RI 210 Bookmakers' Pitches and capital gains tax

(Extract)

A pitch will also be 'property to which (the deceased) is beneficially entitled', *IHTA 1984, s 5(1)*, and where it formed part of the deceased's business its value may qualify for inheritance tax business property relief at 100% if the statutory conditions are met, *IHTA 1984, ss 103–114*.

Appendix 4

HMRC and other press releases

This appendix contains only a selection of the most pertinent HMRC releases. It is strongly recommended that practitioners consult HMRC website on a regular basis since increasingly it is only there, following changes to the HMRC Manuals, that changes in HMRC approach are notified.

2 NOVEMBER 1983 HANSARD

Blood Stock Industry (taxation)

Mr Latham asked the Chancellor of the Exchequer on what authority the Capital Taxes Office indicated that horse owners and breeders will not be permitted 50[now 100]% agricultural relief on capital transfer tax and also, in certain circumstances, not be permitted 50[now 100]% business relief; and if he will instruct the office to reverse this decision.

Mr Moore: Under the provisions of the Capital Transfer Tax, agricultural relief is given for agricultural property occupied for agricultural purposes. Business relief is given if the business is carried on for gain. Whether any particular property qualifies under these provisions is a matter of the proper interpretation of the law in relation to the fact of the particular case. If the taxpayer does not accept the view of the Inland Revenue he may appeal.

19 SEPTEMBER 1984 ICAEW

Memorandum TR557: CTT: 'buy and sell agreements'

[This memorandum is of considerable length and has not been set out. It is referred to at **6.14** and **14.11**.]

21 APRIL 1986 LAW SOCIETY'S GAZETTE

Assets used for life tenant's business: Business Property Relief

Under the Code in *IHT 1984, sections 103* to *114*. Business Property Relief is given at either [50]% or [100]% by adjusting the valuation of 'relevant business property' as defined in *section 105*. In *Featherstonehaugh v IRC* [1984] STC 261, it was decided that relief at what is now the [100]% rate was available for land on which a sole trader was life tenant and on which he had carried on a farming business prior to the transfer on his death.

The Inland Revenue's understanding of this decision is that, where there is a transfer of value of a life tenant's business or interest in a business (including assets of which he was life tenant which were used in that business), the case falls within *section 105(1)(a)* and the [100]% relief is available. Where, by contrast, the transfer of value is only of any land, building, machinery or plant used wholly or mainly for the purposes of a business carried on by the life tenant and in which he had an interest in possession under a settlement, the relief is only available at the rate of [50]% if the transfer takes place in circumstances in which the business itself is not being disposed of.

Thus, if land in a settlement has been used for the purpose of the life tenant's business, the 100% relief is only available if the transfer of value is one of his business as a whole (or an interest in it), including the property in which he has a life interest.

19 FEBRUARY 1987 INLAND REVENUE LETTER

Gifts with reservation: directors' remuneration

You are concerned about the possible effects of the provisions concerning gifts with reservation in a case where a company director or employee makes a gift of shares in a company, but stays in the company as a director or employee in return for a salary or other benefits that are no greater than what might reasonably be provided under an arm's-length deal between unconnected parties. The application or otherwise of these provisions to any given arrangement would, as you know, be a matter for determination in the light of the relevant facts and law of the date of the donor's death.

That said, I am pleased to tell you that the Revenue would accept that the continuation of reasonable commercial arrangements in the form of remuneration and other benefits for the donor's services to the business entered into prior to the gift would not, by itself, amount to a reservation provided that the benefits were in no way linked to or affected by the gift. For completeness, I might add that if in your example the gift of the shares was into trust and the settlor was also a trustee of that trust and a director or employee of the company, the fact that the terms of the trust empowered trustees to retain directors' fees would not, of itself, amount to a reservation of benefit. In this context too, the Revenue will need to be satisfied that the remuneration and other benefits were on a reasonable commercial basis. What is 'reasonable' would of course depend on the facts of the case.

Where, however, as part of the overall transaction, including the gift, new arrangements are made, it will be necessary to examine all the relevant facts to determine whether the remuneration package amounts to a reservation of benefit to the donor by contract or otherwise.

5 MARCH 1987 INLAND REVENUE LETTER

Gifts with reservation: Director's remuneration

I confirm that a downward adjustment made at the time of the gift to a director's pre-existing remuneration to bring the arrangement into line with the level no higher than would be commercially justifiable will not be regarded as giving rise to a reservation of benefit.

18 MAY 1987 INLAND REVENUE LETTER

IHT: Gifts with reservation

I am now able to write to you about the points concerning the provisions on gifts with reservation.

It does not seem realistic to think in terms of precise and comprehensive guidance on how the gifts with reservation provisions will be interpreted and applied since so much will turn on the particular facts of individual cases. However, as the provisions are similar to those adopted for estate duty, the relevant estate duty case law and practice provide a helpful guide to the interpretation and application of the IHT legislation. That said, may I turn to your specific concerns.

Gifts of Land

(1) Consistent with the assurance given last year by the Minister of State and Standing Committee G (Hansard, 10 June 1986, col 425) the estate duty practice on the treatment of gifts involving a share in a house where the gifted property is occupied by all the joint owners including the donor will apply. The donor's retention of a share in the property will not by itself amount to a reservation. If, and for so long as, all the joint owners remain in occupation, the donor's will not be treated as a reservation provided the gift is itself unconditional and there is no collateral benefit to the donor. The payment by the donee of the donor's share of the running costs, for example, might be such a benefit. An arrangement

will not necessarily be jeopardised merely because it involves a gift of an unequal share in the house.

(2) In other cases the donor's occupation or enjoyment of the gifted land will only be disregarded if the occupation is for full consideration in money or monies worth as provided in *FA 1986, Sch 20 para 6(1)(a)* or if it is by way of reasonable 'care and maintenance' provision within *para 6(1)(b)*. Whether an arrangement is for full consideration will of course depend on the precise facts. But among the attributes of an acceptable arrangement would be the existence of a bargain negotiated at arm's-length by parties who were independently advised and which followed the normal criteria in force at the time it was negotiated.

(3) …

Gifts involving family businesses or farms

(4) A gift involving a family business or farm will not necessarily amount to a gift with reservation merely because the donor remains in the business, perhaps as a director or partner. For example, where the gift is of shares of a company, the continuation of reasonable commercial arrangements in the form of remuneration for the donor's ongoing services to the company entered into before the gift will not of itself amount to a reservation provided the remuneration is in no way linked to or beneficially affected by the gift. Note in this connection the Inland Revenue letters of 19 February and 5 March 1987, both reproduced in Appendix 5 at 20.1.6 and 20.1.7. Similar considerations will apply in the case where the gift is into trust which empower the trustees, who may be the donor, to retain director's fees etc for his own benefit: see the letter of 19 February 1987.

(5) The 'Munro' principle will also be relevant in determining the tax treatment of gifts affecting family farms where the donor and the donee continue to farm the land in pursuance of arrangements entered into prior to and independently of the gift. In cases where this principle does not apply, the test of 'full consideration' for the purposes of *para 6(1)(a)* will need to be satisfied with regard to the donor's occupation of the land. In applying that test we shall take account of all the circumstances surrounding the arrangement including the sharing of profits and losses, the donor and the donee's interests in the land, and their respective commitment and expertise.

Gifts of chattels

(6) …

Settlor's retention of reversion

(7) …

17 MARCH 1988 HANSARD

IHT and Family companies

Viscount Mackintosh of Halifax asked whether, given their stated aim of creating a climate where family companies can flourish, it is their intention that clearance under [*ICTA 1988, s 219(2)*] will in all cases be refused on the death of a controlling shareholder where sufficient dividends can be voted by the executors to enable the Inheritance Tax liability to be met by instalments on the basis that hardship would not arise as the executors control the company and thus control the flow of dividends.

Lord Brabazon of Tara: I understand that there have been very few cases of the kind in question. However, where the company has surplus funds sufficient to discharge the inheritance tax liability, the Revenue take the view that there would be no hardship since the liability could be met by dividend payments from the company.

11 SEPTEMBER 1991

Inheritance Tax Act 1984, section 98(1) — deferred shares

Following recent legal advice, the Revenue's interpretation of *IHTA 1984, section 98* has changed.

Until now the Capital Taxes Offices have taken the view that when deferred shares come to rank equally with another class of shares there would be no alteration in the rights of the shares within the meaning of *IHTA 1984, section 98(1)(b)*, but there would be an alteration in the company's share capital within the meaning of *IHTA 1984, section 98(1)(a)*.

The Board of Inland Revenue has now been advised that an alteration of rights, within the meaning of *IHTA 1984, section 98(1)(b)*, occurs when deferred shares come to rank equally with another class of shares. Accordingly, claims for Inheritance Tax will be raised where deferred shares, issued after 5 August 1991, subsequently come to rank equally, or become merged, with shares of another class.

18 MARCH 1992

Valuation of agricultural tenancies

[This records an agreed summary of a meeting between representatives of the Law Society of England and Wales, the Law Society of Scotland and the Inland

Revenue held on 3 February 1992 to discuss the specific issue of valuation of agricultural tenancies. The note is of some length and is not set out here: it appears in *Tolleys Yellow Tax Handbook 2013–2014 and earlier editions.*

FEBRUARY 1993 INLAND REVENUE TECHNICAL BULLETIN

Milk Quota

For inheritance tax purposes, where agricultural land, or an interest in agricultural land, is valued and the valuation of the land reflects the benefit of milk quota, agricultural relief is given on that value.

Where milk quota is valued separately, it will normally constitute an asset used in the business, within *IHTA 1984, s 110*, so that business relief under *IHTA 1984, s 105(1)(a)* may be available.

Other press releases in this area that may be of interest to the practitioner but which are not set out here, many being freely available on HMRC website.

Business relief:

- Extension of non-statutory clearances service for HMRC business customers: 1 May 2008

Heritage property:

- Leasehold Enfranchisement of conditionally exempt property: 7 May 1993
- 'Heritage property: *IHTA 1984, s 30*: December 1995
- The Register of conditionally exempt works of art: 16 January 1996

Compliance:

- Simpler payment scheme for IHT: 27 March 2003

Appendix 5

Heritage property: the old rules

These are retained, not merely as an academic museum piece, but to provide the context for those cases that fall under an earlier set of rules, and which can still be relevant in this field. Note also the changes introduced by *FA 2012, Sch14* referred to at the end of this appendix.

DEATHS BEFORE 7 APRIL 1976

Conditional exemption from the charge to capital transfer tax (CTT) could be available in respect of heritage property so as to reduce the charge on a death before 7 April 1976. The Treasury could designate heritage property. Whilst the tests for qualification as heritage property fell into the same categories as they do now, there was no requirement that chattels be 'pre-eminent'. Conditional exemption was available in respect of property which either belonged to a person immediately before his death or was comprised in a settlement in which the deceased had an interest in possession. Property comprised within the discretionary trust did not qualify. The value of heritage property was excluded from the value deemed to be transferred on the death of the deceased.

Undertakings had to be given, but they were in many respects less demanding than the undertakings now required, as has been noted in **Chapters 17** and **18**. The Treasury had discretion to exclude confidential material from an undertaking. The early rules in respect of heritage property included the association of properties where one was a building of outstanding historic or architectural interest and the other was land or historically associated objects in relation to the building. On the breach of an understanding, capital transfer tax became chargeable. Where that breach occurred after 31 December 1984, the charge was made under *IHTA 1984, Sch 5*. As with the present code for heritage property, relief was available in respect of CGT, although that would apply only in respect of lifetime disposals.

The charge to capital transfer tax

A breach of an undertaking in respect of a death before 7 April 1976 triggered a charge to capital transfer tax on the previously exempted property. If the property was a building, or land adjoining it that was essential for the preservation of its character, or an object historically associated with it, a charge to CTT also arose in respect of that other associated property. The charge arose under *FA 1975, s 32* on chattels or under *FA 1975, s 34* on land or buildings. For a charge arising after 31 December 1984, the appropriate legislation is *IHTA 1984, Sch 5.*

The person who would be entitled to receive the proceeds of sale (or income from the property) at the time of the chargeable event for capital transfer tax was the person liable for that tax. There was, under the old rules as now, an escape from the charge to capital transfer tax where the property was sold to a heritage body or, if not sold, transferred to a person who gave a fresh undertaking. In respect of those chargeable events, as now, acceptance by the Revenue of property in lieu of tax was not treated as a disposal of the property and did not, therefore, trigger a charge. Theft or loss or destruction of heritage property without the consent of the owner did not trigger a charge to capital transfer tax, as long as the terms of the undertaking had been complied with.

It might happen that heritage property was acquired by a heritage body, not in satisfaction of tax but by purchase at auction, perhaps where agreement could not be reached between the owner and the heritage body as to its value. Where that happened, capital transfer tax was chargeable on the disposal. Where the property concerned had been exempted on more than one death before 7 April 1976, the charge to tax was by reference to the last death: see *IHTA 1984, Sch 5, paras 1(ii)* and *3(ii).*

Under the old rules, as now, a *de minimis* provision applied where the failure to comply with an undertaking did not materially affect the heritage property as a whole. That provision (see *Sch 5, para 3(iii)*) allowed for a charge to capital transfer tax on only the particular part of the property affected by the undertaking.

APPLICATION OF THE NEW RULES

The new rules may apply in certain circumstances, and will apply in others. The situation affected is one where tax would become payable under the old estate duty rules or the capital transfer tax rules but, before that tax becomes payable, there has been a conditionally exempt transfer of the same property after 6 April 1976 under the new regime that applies with effect from that date. Where this happens, HMRC can elect to charge under the new rules or the old if there has been no conditionally exempt transfer of property on death. Where, on the other hand, there has been a conditionally exempt transfer on death, capital transfer tax becomes chargeable only under the new rules.

Administration

In the event of a breach of undertaking, sale or gift, a person who is liable for the capital transfer tax that then arises should make a return within six months of the end of the month in which the event occurred. HMRC may withdraw the exemption by issuing a notice of determination. Such a notice, as with other notices of determination, may be challenged by an appeal to the Special Commissioners.

Calculation of the charge: event within three years of death

In calculating the charge to capital transfer tax, it is first essential to determine whether the occasion of charge occurs within three years after a death or at some later time. The charge arises through disposal of the property or failure to observe an undertaking. Where the event occurred within three years of death, the capital transfer tax, otherwise chargeable by reference to that death, was recalculated as if the exempted heritage property had been part of the estate. This rule can apply only in respect of deaths occurring before 6 April 1979. As a result, surely, most of the relevant estates will now have been fully administered.

Once the capital transfer tax has been recalculated, part of it is apportioned, by value, to the main estate and part of it to the heritage property. A sale, or a breach of undertaking such as a disposal, can trigger a liability to CGT. Here, there is a curious 'one way set-off' rule. The liability to CGT may be allowed against the capital transfer tax, but there is no allowance for the capital transfer tax liability against any liability to CGT.

Computations: chargeable event more than three years after the death

Once three years have elapsed from the death, the calculation of capital transfer tax due on the estate as at that death is not disturbed by the happening of a chargeable event. Instead, there is a recalculation of the amount of the capital transfer tax that would have been chargeable according to the rules set out below. This is charged on the property itself. The assumptions made are that:

- no exemption applied to the property; and
- the value of the property at the time of death was its value at the time of the chargeable event; or, where the property has been sold, the value of the property at the time of death was the proceeds of sale.

This causes a recalculation of the total capital transfer tax that would have been payable on the estate. A proportion is then charged with respect to the heritage property by reference to the value of the asset, calculated as above, as a proportion of the value of the estate as recalculated.

For this purpose, the figure used is that for the sale only where it is the true value, ie where:

- the sale was not intended to confer any gratuitous benefit on any person; and was either
- at arm's length between unconnected persons; or
- such as might be expected if it had been at arm's length.

For this, see *IHTA 1984, Sch 5, paras 2(2)(b), 4(b)* and *6*. Where those conditions do not apply, the open market value is used in substitution for sale price. This calculation applies only to the charge on the property that is sold; the tax on the estate itself is not changed.

Net proceeds, not gross

Fortunately (and somewhat exceptionally) the charge to tax is on the net proceeds after deducting any costs of sale and after deducting any CGT chargeable by reason of the sale. The authority for the first of these propositions is an old one, *Tyser v A-G* [1938] Ch 426, naturally a case on estate duty. For the second rule, the authority is *TCGA 1992, s 258(8)*.

Recalculation for each event

This recalculation in respect of chargeable events that occur more than three years after the death is applied separately with regard to each sale that takes place. The effect is not cumulative. For instance, where the property sold is one of several objects which did not constitute a set, the sum must be completed separately each time. It may be that this even applies where two separate items, not constituting a set, are sold from the same collection at the same time.

Special rules for sets of objects

The rule is different where:

- the items did form a set at the time of death; and
- two or more of them qualified for conditional exemption; and
- two or more of the components of the set attract a charge to capital transfer tax through the happening of chargeable events at different times.

IHTA 1984, Sch 5 is applied as if these events occurred when they first originated. If that was within three years of the death, the tax chargeable on the aggregate value of the transfer on death must be adjusted for each of the later events. If the first chargeable event occurred more than three years after the death, there is a re-adjustment of the charge in respect of each object as a result of the chargeable event on each later component of the set (see *Sch 5, para 2(3)*).

However, by *Sch 5, para 2(4)*, this tiresome rule will not apply where:

- there are two or more chargeable events;
- by disposals to different persons,

and it can be shown that:

- those persons do not act in concert; and
- are not connected with each other.

ESTATE DUTY

The statutory authority for conditional exemption from estate duty was *FA 1930, s 40*. Those provisions continue to operate under *IHTA 1984, Sch 6, para 4*, where property which qualified under the old estate duty rules remains subject to conditional exemption and, for example, has still not been sold or, which perhaps is increasingly rare, has not since been the subject of a transfer, either under the rules for capital transfer tax or those for IHT.

Breach of undertaking: estate duty cases

The scheme of *FA 1930, s 40* was similar to the present regime in many respects. Thus, the breach of an undertaking or the disposal of property triggered a chargeable event for estate duty. Again, the market value rule was applied in respect of a chargeable event by way of sale that was not at arm's length. By *IHTA 1984, Sch 6, para 4(3)*, the estate duty charged is a credit against the IHT.

As noted before, with use of the expression 'Russian roulette', this is an instance where HMRC are presented by the legislation with a wide choice. Where there is a disposal of conditionally exempt property triggering a charge under *FA 1930, s 40*, HMRC may charge tax either under that section, or under *IHTA 1984, s 32* or *32A*. Where, however, death has intervened and conditional exemption has been allowed on that death, HMRC are restricted to such charge as they may make under *IHTA 1984, s 32* or *32A* and may not make any estate duty charge.

Where it can still be applied, the estate duty is charged at the rate which applied to the estate of the last person to die in relation to whose death relief was obtained; that, at least, is the apparent effect of *FA 1930, s 40*. By *FA 1969, s 39*, if the death was after 15 April 1969, the proceeds of sale, or value at the time of the breach of undertaking, of the heritage property is aggregated with the estate, and the rate recalculated in the way described in **Chapter 19**.

LOSS OF TAPER RELIEF ON GIFTS

If the heritage property had been subject to a lifetime gift, and relief had been given in respect of it, it becomes necessary to refer to the Inland Revenue press release of 3 May 1984 for assistance in the calculation. It was noted here that, by *FA 1960, s 64*, taper relief could, in general cases, be applied to property given away within a certain period before the death. However, HMRC will not allow *s 64* taper relief to reduce the charge that arises on the chargeable event. The charge must apply to the full value or sale proceeds.

GIFTS TO THE NATION

FA 2012, Sch 14, amended the Estate Duty rules in *FA1930, s 40* with effect from 1 April 2012 where a gift is made under the scheme introduced by *FA 2012*. Where that gift would trigger a recapture charge under the Estate Duty rules *Sch 14* limits the charge to the amount by which the effective rate of the recapture charge, but for *Sch 14*, would exceed 40%. That relates to the tax after set off of credit as provided by *IHTA 1984 s 33(7),* see **Chapter 20,** but repayments are barred by *Sch 14, para 33(5).*

Index

[*All references are to paragraph number*]

Accounts
maintenance funds, and, 21.14
Agricultural land or pasture
definition 2.2
Agricultural property
definition 2.1
extra statutory concession, Appendix 2
Agricultural property relief (APR)
Agricultural Holdings Act tenancies, 3.28
agricultural land or pasture, 2.2
'agricultural property' 2.1
agricultural purposes, 3.17
agricultural value, 3.1
'agriculture', 2.2
amount, 12.4
areas of concern
controlled companies, 4.9
Gladstone v Bowers agreements, 4.13
grazing agreements, 4.13
joint ownership, 4.2
land sold within four years after death, 4.26
no contract for sale, 4.1
partnerships, 4.3
replacement property, 4.6–4.8, 4.12
sale of related property within three years after death, 4.27
sale of shares within 1q2 months of death, 4.25
successions, 4.10
successive transfers, 4.11
trusts, 4.4–4.5
vacant possession of land held out outside a partnership, 4.28
valuation of agricultural property, 4.14–4.24
background, 1.2
buildings used for intensive rearing of livestock or fish
ancillary occupation, 2.5
generally, 2.4
Agricultural property relief (APR) – *contd*
clawback
basic rule, 10.13
business property, and, 10.25
death before clawback event, 10.13
disregard of relief, 10.15–10.18
effect, 10.17
exception, 10.18
flowchart, 10.21
gifts with reservation, 10.23
identification issues, 10.19
introduction, 10.1
'original property', 10.16
partial clawback, 10.14
professional advice, and, 10.24–10.26
relief from, 10.19–10.23
replacement property, 10.19–10.22
surrender of farm tenancies, and, 10.26
'transferee', 10.16
compliance
chargeable lifetime transfers, 22.1
DOTAS, 22.23
duty of personal representative, 22.2–22.7
forms and procedures, 22.18–22.22
penalties, 22.11–22.16
personal applications for grants of representation, 22.10
tax fraud, 22.17
valuation, 22.8–22.9
computation, 1.7
conservation land, 2.27
contract farming, 12.16
controlled companies, 4.9
corporate 'wrapper', 12.11
distinction from BPR, 1.9
diversification into non-agricultural enterprises, 12.20–12.21
DOTAS, 22.23

Agricultural property relief (APR) – *contd*
'double discount cases', 3.7
election, 12.2
excluded property trusts, and, 16.2
exotic grazing, 2.24
extra statutory concession, Appendix 2
family limited partnerships, 12.12–12.15
farm cottages
generally, 2.6
occupation, 3.26
farmhouses
Antrobus (No 1) v CIR, 2.13–2.14
Arnander v HMRC, 2.19
big houses, 2.10
checklist, 2.14
early cases, 2.11
'elephant' test, 2.15
family home, and, 2.20
generally, 2.7
Higginson v CIR, 2.12
historical background, 2.8
historical context, 2.9
HMRC V Hanson, 2.18
illness, and, 2.23
lifestyle farming, 2.13
McKenna Deceased v HMRC, 2.19
not in identical ownership with land, 2.22
Rosser v IRC, 2.17
shared ownership of land, 2.22
Starke v IRC, 2.16
substantial houses, 2.21
tests, 2.8–2.11
farming companies, 2.28
farming trusts, 3.5–3.6
forms and procedures, 22.19
Gladstone v Bowers agreements, 4.13
grant of agricultural tenancy, 12.17–12.18
grazing agreements, 4.13
history, 1.2
'hobby farming', 12.10
holiday lettings, 12.21
horses, 2.24
intensive rearing of livestock or fish
ancillary occupation, 2.5
generally, 2.4

Agricultural property relief (APR) – *contd*
interaction with other reliefs
generally, 1.6
instalment relief, 1.8
joint ownership, 4.2
land sold within four years after death, 4.26
life assurance, and, 16.7
milk quota, 2.26
no contract for sale, 4.1
non-qualifying property, 2.30
occupation
agricultural purposes, 3.17
Atkinson Deceased v HMRC, 3.24
Basic Farm Payment (BFP), 3.19
contract farming, 3.27
controlled companies, 4.
diversification, 3.18
dwellings, 3.25
farm cottages, 3.26
generally, 3.17–3.27
ill farmers, 3.23
introduction, 3.3 9
joint ownership, 4.2
occupiers, 3.20
partnerships, 3.21, 4.3
proof, 3.22
share farming, 3.27
successions, 4.10
successive transfers, 4.11
test determinative of rate of relief, 3.4
trusts, 4.4–4.5
working farmer relief, 3.9
100% rate
conditions, 3.3–3.4
'double discount cases', 3.7
farm business tenancies, 3.15
farming trusts, and, 3.5–3.6
introduction, 3.2
land let on new tenancies, 3.15
ownership and occupation, 3.17–3.27
renewal of tenancy, 3.16
working farmers, and, 3.7–3.14
overlap with BPR
amount, 12.4
APR more generous than BPR, 12.6
BPR more generous than APR, 12.7

Agricultural property relief (APR) – *contd*
overlap with BPR – *contd*
contract farming, 12.16
corporate 'wrapper', 12.11
diversification by farmers into non-agricultural enterprises, 12.20–12.21
election, 12.2
family limited partnerships, 12.12–12.15
farms, 12.8
grant of agricultural tenancy, 12.17–12.18
historical perspective, 12.1
'hobby farming', 12.10
holiday lettings, 12.21
priority, 12.2
reason for turning hand to farming, 12.10–12.11
replacement assets, 12.5
security of tenure, 12.18
share farming, 12.16
sheltering proceeds of sale of business in agricultural property, 12.9
switching, 12.5
territoriality, 12.3
ownership
agricultural purposes, 3.17
Atkinson Deceased v HMRC, 3.24
Basic Farm Payment (BFP), 3.19
contract farming, 3.27
diversification, 3.18
dwellings, 3.25
farm cottages, 3.26
generally, 3.17–3.27
ill farmers, 3.23
introduction, 3.3
occupiers, 3.20
partnerships, 3.21
proof, 3.22
share farming, 3.27
test determinative of rate of relief, 3.4
working farmer relief, 3.9
partnerships, 4.3
qualifying property
agricultural land or pasture, 2.2
conservation land, 2.27

Agricultural property relief (APR) – *contd*
qualifying property – *contd*
farm cottages, 2.6
farmhouses, 2.7–2.23
farming companies, 2.28
horses, 2.24
intensive rearing of livestock or fish, 2.4–2.5
introduction, 2.1
milk quota, 2.26
set aside land, 2.27
short rotation coppicing, 2.25
territoriality, 2.29
wildlife habitats, 2.27
woodland, 2.3
rates
50%, 3.28
100%, 3.2–3.27
reason for turning hand to farming, 12.10–12.11
reduced relief, 3.28
replacement property
introduction, 4.6
occupation test, 4.7
ownership test, 4.8
share or securities, 4.12
switching, 12.5
role, 1.1
sale of related property within three years after death, 4.27
sale of shares within 12 months of death, 4.25
security of tenure, 12.18
set aside land, 2.27
share farming, 12.16
sheltering proceeds of sale of business in agricultural property, 12.9
short rotation coppicing, 2.25
successions, 4.10
successive transfers, 4.11
switching, 12.5
tax planning
See also **Tax planning**
buy and sell agreements, 14.11–14.12
deathbed planning, 14.28
deductibility of debts, 14.7–14.10
gifts with reservation, 14.3–14.6
'money box', 14.18

Agricultural property relief (APR) – *contd*
tax planning – *contd*
relevant property trusts, 14.2, 14.21–14.27
retirement, 14.14–14.17
share options, 14.13
structures, 14.19–14.20
unfortunate gifts, 14.1
tax shelters, 13.1–13.14
territoriality
generally, 2.29
introduction, 1.5
overlap with BPR, 12.3
trusts, 4.4–4.5
use for intensive rearing of livestock or fish, 2.4
vacant possession of land held out outside a partnership, 4.28
valuation of agricultural property
Alexander v IRC, 4.17
Arkwright v IRC, 4.21
Baird's Executors v IRC, 4.15
general considerations, 4.14
Gray v IRC, 4.16
joint property, 4.21
lotting for sale, 4.18
natural units, 4.18
other aspects, 4.19
planning permission, 4.20
Price v HMRC, 4.22
tenancies, 4.15
valuation of farmhouses, 4.23
wildlife habitats, 2.27
will planning
See also **Will planning**
attribution rules, 15.12
drafting points, 15.10–15.11
gifts of businesses, 15.16
nil rate band, 15.1–15.9
non-exempt beneficiaries, and, 15.15
s 39A, 15.13–15.14
'spouse', 15.17
woodland, 2.3
working farmers, and
companies, and, 3.11
difficulties, 3.10
generally, 3.7
limitation by value or by acreage, 3.12
occupation test, 3.9

Agricultural property relief (APR) – *contd*
working farmers, and – *contd*
restrictions, 3.14
successions, 3.13
'wholly or mainly engaged', 3.8
Agricultural purposes
definition, 3.17
Agricultural value
definition, 3.1
Agriculture
definition, 2.2
AIM companies
business property relief, and, 8.3
Alterations to property
maintenance funds, and, 21.6
Amenity land
See also **Heritage relief**
generally, 17.3
undertakings, 17.13
Animals
business property relief, and, 5.35
Annual exemption
interaction with other reliefs, and, 1.6
Annual tax on enveloped dwellings (ATED)
heritage relief, and, 19.22
Anti-avoidance rules
maintenance funds, and
charitable gifts, 21.11
excess value, 21.16
generally, 21.18
purchase rule, 21.16
relevant property trusts, and, 20.8
heritage relief, and, 19.22
Artists
business property relief, and, 5.11
Associated objects
See also **Heritage relief**
generally, 17.8
introduction, 17.3
Associated operations
planning issues, and, 14.22–14.23
Associated properties
loss of exemption, and, 19.4–19.5
Authors
business property relief, and, 5.11

Basic Farm Payment (BFP)
occupation, and, 3.19

Bookmakers' pitchers
business property relief, and, 5.39
Breach of undertaking
loss of exemption, and, 19.2
Buildings
heritage relief, and
England, in, 17.6
introduction, 17.3
Scotland, in, 17.7
undertakings, 17.12
tax shelters, and, 13.11
Business property relief (BPR)
AIM shares, 8.3
amount, 12.4
animals, 5.35
artists, 5.11
authors, 5.11
background, 1.2
bookmakers' pitchers, 5.39
business or interest in a business
activities constituting a business, 5.4
artists, 5.11
assets supporting the underwriting business, 5.19
authors, 5.11
'business', 5.3
corporate/non-corporate distinction, 5.7
disposal of part of partnership share, 5.15
generally, 5.3
Jersey LLPs, 5.23
limited (liability) partnerships, 5.21–5.24
Lloyd's Name losses, 5.18
Lloyd's underwriters, 5.17–5.19
'mere asset' exclusion, 5.13–5.14
multiple businesses, 5.16
musicians, 5.11
Nelson Dance Settlement v HMRC, 5.14
non-investment businesses, 5.5
painters, 5.11
partnership interest, 5.12
restrictive effect of s 105(3), 5.8
sculptors, 5.11
sole proprietor's business, 5.10
succession, 5.20
unincorporated business, 5.7
woodlands, 5.9
Business property relief (BPR) – *contd*
business or interest in a business – *contd*
wrong kind of business for relief, 5.6
buy and sell agreements, 6.14
capital reconstruction, 6.8
clawback
additional conditions, 10.8–10.9
basic rule, 10.2
computational issues, 10.3–10.4
disregard of relief, 10.7
flotation, after, 10.7
identification issues, 10.10
introduction, 10.1
'original property', 10.9
partial clawback, 10.5–10.7
pre-death loss relief, and, 10.6
professional advice, and, 10.24–10.26
relief from, 10.10–10.12
replacement property, 10.11–10.12
stock exchange listing, and, 10.2
takeovers, and, 10.2
'transferee', 10.9
compliance
chargeable lifetime transfers, 22.1
DOTAS, 22.23
duty of personal representative, 22.2–22.7
forms and procedures, 22.18–22.22
penalties, 22.11–22.16
personal applications for grants of representation, 22.10
tax fraud, 22.17
valuation, 22.8–22.9
computation, 1.7
controlling holdings in quoted company
clearances, 5.31–5.32
'control', 5.29
generally, 5.28
preference shares, 5.30
restriction on relief, 5.29
corporate/non-corporate distinction, 5.7
debt, 7.14
distinction from APR, 1.9
DOTAS, 22.23
election, 12.2

Business property relief (BPR) – *contd*
equipment, 5.37
excepted assets
Barclays Bank Trust Co Ltd v IRC, 7.11
'business', 7.7
corporate structures, and, 7.6
debt, and, 7.14
'group', 7.13
group structures, 7.1
hybrid companies, 7.10
introduction, 7.1
McCall v HMRC, 7.8–7.9
mixed development, 7.10
ownership, 7.2
personal use, 7.2
Salaried Persons Postal Loans Ltd v HMRC, 7.12
test period, 7.1
use, 7.2–7.3
valuation principles, 7.4–7.5
exceptions
excepted assets rule, 7.1–7.14
introduction, 9.1
liquidations, 6.17
market makers, 5.8
reconstructions, 6.17
underlying assets, 6.15
'wholly or mainly holding investments', 9.1–9.23
excluded property trusts, and, 16.2
family limited partnerships, 12.12–12.15
forms and procedures, 22.20
grossing up, 6.3
'group', 7.13
history, 1.2
holdover relief, and, 6.2
holiday lettings, 12.21
hybrid companies, 7.10
intellectual 'tools' and property, 5.36
interaction of reliefs, 1.6
interaction with other reliefs
generally, 6.2
instalment relief, 1.8
introduction, 1.6
quick succession relief, 6.1
introduction, 5.1
land and buildings used in the business
animals, 5.35

Business property relief (BPR) – *contd*
land and buildings used in the business – *contd*
bookmakers' pitchers, 5.39
equipment, 5.37
generally, 5.33
intellectual 'tools' and property, 5.36
machinery and plant, 5.34
royalties, 5.38
land and buildings used in the tenant for life's business
applicability of lower rate of relief, 5.45
'assets used in the business', 5.44
generally, 5.40
history of relief, 5.42
'interest in possession', 5.41
other conditions, 5.46
timing of transfer, 5.43
liabilities, and, 8.16–8.24
life assurance, and, 16.7
limitations, 6.5–6.22
limited liability partnerships
accounts and formalities, 5.22
generally, 5.21
Jersey LLPs, 5.23
transition, 5.24
liquidations, 6.17
Lloyd's underwriters
assets supporting the business, 5.19
generally, 5.17
Names losses, 5.18
machinery and plant, 5.34
making or holding investments, 9.1–9.23
market makers exception, 5.8
'mere asset' exclusion, 5.13–5.14
mixed development, 7.10
musicians, 5.11
no contract for sale, 6.13–6.15
non-investment businesses, 5.5
overlap with APR
amount, 12.4
APR more generous than BPR, 12.6
BPR more generous than APR, 12.7
contract farming, 12.16
corporate 'wrapper', 12.11
diversification by farmers into non-agricultural enterprises, 12.20–12.21

Business property relief (BPR) – *contd*
overlap with APR – *contd*
election, 12.2
family limited partnerships, 12.12–12.15
farms, 12.8
grant of agricultural tenancy, 12.17–12.18
historical perspective, 12.1
'hobby farming', 12.10
holiday lettings, 12.21
introduction, 1.9
priority, 12.2
reason for turning hand to farming, 12.10–12.11
replacement assets, 12.5
security of tenure, 12.18
share farming, 12.16
sheltering proceeds of sale of business in agricultural property, 12.9
switching, 12.5
territoriality, 12.3
ownership
capital reconstruction, 6.8
detailed rules, 6.10–6.11
excepted assets, and, 7.2
extension of period, 6.9
general rule, 6.5
modification of rule, 6.6
replacement property, 6.7
unquoted shares, 6.8
painters, 5.11
partnership interest, 5.12
personal use, 7.2
qualifying property
business or interest in a business, 5.3–5.24
categories, 5.2
controlling holdings in quoted company, 5.28
land and buildings used in the business, 5.33–5.39
land and buildings used in the tenant for life's business, 5.40–5.46
unquoted shares, 5.27–5.32
unquoted securities giving control, 5.25–5.26
quick succession relief, and, 6.1

Business property relief (BPR) – *contd*
'quoted'
generally, 6.18
list of exchanges, 6.21
other trading exchanges, 6.20
recognised stock exchange, 6.19
quoted shares and securities, 8.2
rates
generally, 6.4
grossing up, 6.3
introduction, 6.1–6.2
reconstructions, 6.17
related property, 8.23
relevant business property
business or interest in a business, 5.3–5.24
categories, 5.2
controlling holdings in quoted company, 5.28
land and buildings used in the business, 5.33–5.39
land and buildings used in the tenant for life's business, 5.40–5.46
unquoted shares, 5.27–5.32
unquoted securities giving control, 5.25–5.26
replacement property, 6.7
role, 1.1
royalties, 5.38
sculptors, 5.11
sole proprietor's business, 5.10
Statements of Practice, Appendix 3
succession
generally, 6.16
introduction, 5.20
successive transfers, 6.16
tax planning
See also **Tax planning**
buy and sell agreements, 14.11–14.12
deathbed planning, 14.28
deductibility of debts, 14.7–14.10
gifts with reservation, 14.3–14.6
'money box', 14.18
relevant property trusts, 14.2, 14.21–14.27
retirement, 14.14–14.17
share options, 14.13
structures, 14.19–14.20
unfortunate gifts, 14.1

Business property relief (BPR) – *contd*
tax shelters, 13.1–13.14
territoriality
generally, 12.3
introduction, 1.5
transfers after 6 April 1996, 6.12
treatment of liabilities
general rules, 8.16–8.20
guarantees by business owners, under, 8.22
sole traders, of, 8.21
two-year rule
general rule, 6.5
modification of rule, 6.6
unincorporated business, 5.7
unquoted shares
generally, 5.27–5.32
period of ownership, 6.8
valuation, 8.4–8.13
unquoted securities giving control
'control', 5.26
generally, 5.25
valuation
AIM shares, 8.3
basic principles, 8.1
discounts, 8.10–8.13
excepted assets, and, 7.4–7.5
general rules, 8.14–8.15
introduction, 8.1
loan notes, 8.24
quoted shares and securities, 8.2
related property rules, 8.23
shares, 8.24
treatment of liabilities, 8.16–8.22
unquoted shares, 8.4–8.13
valuation of unquoted shares
basis to be applied, 8.10–8.11
Caton's Administrators v Couch, 8.5
CVC and others v Demarco Almeida, 8.8
Denekamp v Pearce, 8.7
discounting, 8.12–8.13
generally, 8.4
Hawkings-Byass v Sassen, 8.6
minority interests in property, 8.13
Phoenix Office Supplies v Larvin, 8.9
'wholly or mainly holding investments'
Best v HMRC, 9.22

Business property relief (BPR) – *contd*
'wholly or mainly holding investments' – *contd*
building and developing, 9.19
Burkinyoung v IRC, 9.3–9.4
caravan parks, 9.7–9.14
carrying on a business for gain, 9.20
Clark Deceased v HMRC, 9.21
Earl of Balfour Deceased v HMRC, 9.17
Farmer v IRC, 9.16
farming partnership, 9.17
Furness, 9.9
George v CIR, 9.11–9.12
Hall v CIR, 9.7
holiday homes, 9.13–9.14
introduction, 9.1
level of activity, 9.5
livery business, 9.15
Mallender, 9.20
Martin v IRC, 9.2
mixed trade, 9.1
mixed use estate, 9.16
money lending, 9.18
Moore Deceased v IRC, 9.2
overview, 9.23
Pawson v HMRC, 9.14
Pearce Deceased v IRC, 9.8
Phillips Deceased v HMRC, 9.18
Piercy's Executors v HMRC, 9.19
Powell v IRC, 9.8
property development, 9.19
property management, 9.21–9.22
Stedman's Executors v IRC, 9.11
Vigne v HMRC, 9.15
Weston v IRC, 9.10
Zetland Settlement v HMRC, 9.6
will planning
See also **Will planning**
attribution rules, 15.12
drafting points, 15.10–15.11
gifts of businesses, 15.16
nil rate band, 15.1–15.9
non-exempt beneficiaries, and, 15.15
s 39A, 15.13–15.14
'spouse', 15.17
winding-up exclusion, 6.22
woodlands, 5.9

Buy and sell agreements
generally, 6.14
planning issues, 14.11–14.12
Statement of Practice, Appendix 3
'By appointment only' access
generally, 17.1
variation of undertakings, and, 18.11

Capital gains tax
maintenance funds, and, 21.32
Capital reconstruction
business property relief, and, 6.8
Caravan parks
'wholly or mainly holding investments', and, 9.7–9.14
Carrying on a business for gain
'wholly or mainly holding investments', and, 9.20
Chargeable lifetime transfers
compliance, and, 22.1
Charitable gifts
maintenance funds, and, 21.11
Chattels
heritage relief, and
generally, 17.5
introduction, 17.3
undertakings, 17.11
Clawback
advice, and, 10.24
agricultural property relief
basic rule, 10.13
business property, and, 10.25
death before clawback event, 10.13
disregard of relief, 10.15–10.18
effect, 10.17
exception, 10.18
flowchart, 10.21
identification issues, 10.19
introduction, 10.1
'original property', 10.16
partial clawback, 10.14
relief from, 10.19–10.23
replacement property, 10.19–10.22
surrender of farm tenancies, and, 10.26
'transferee', 10.16
business property relief
additional conditions, 10.8–10.9
basic rule, 10.2
computational issues, 10.3–10.4
Clawback – *contd*
business property relief – *contd*
disregard of relief, 10.7
flotation, after, 10.7
identification issues, 10.10
introduction, 10.1
'original property', 10.9
partial clawback, 10.5–10.7
pre-death loss relief, and, 10.6
relief from, 10.10–10.12
replacement property, 10.11–10.12
stock exchange listing, and, 10.2
takeovers, and, 10.2
'transferee', 10.9
discretionary trusts, 10.27
gifts with reservation, 10.23
maintenance funds, 21.24
introduction, 10.1
loopholes, 10.30
professional advice, and, 10.24
relevant property trusts, 20.2
value shifting, 10.29
Compliance
chargeable lifetime transfers, 22.1
duty of personal representative
estimates, 22.5
generally, 22.2
HMRC guidance, 22.3
property details, 22.4
Robertson v IRC, 22.6
Robertson v IRC (No 2), 22.7
forms and procedures
agricultural relief, 22.19
business relief, 22.20
heritage relief, 22.12
introduction, 22.18
woodland relief, 22.21
heritage relief, and, 19.23
penalties
culpability, 22.13
generally, 22.11
levels, 22.13
mitigation, 22.14–22.16
overview of provisions, 22.12
purpose, 22.12
personal applications for grants of representation, 22.10
tax fraud, 22.17
valuation
Cairns v HMRC, 22.9

Compliance – *contd*
valuation – *contd*
generally, 22.8
Conservation land
agricultural property relief, and, 2.27
Contract farming
distinction between APR and BPR, 12.16
occupation, and, 3.27
tax shelters, and, 13.4
Controlled companies
agricultural property relief, and, 4.9
Controlling holdings in quoted company
clearances, 5.31–5.32
'control', 5.29
generally, 5.28
preference shares, 5.30
restriction on relief, 5.29
Conversion into equity
tax shelters, and
cash, 13.10
debt, 13.9
Cultural Gifts Scheme
generally, 23.18–23.20
introduction, 17.29

Death of person
clawback, and, 10.13
loss of exemption, and, 19.3
maintenance funds, and, 21.26
Deductibility of debts
effect of Finance Act 2013, 14.8–14.9
generally, 14.7
restriction on offset, 14.10
Deferral basis
heritage relief, and, 17.1
Destruction of property
loss of exemption, and, 19.8
Disclosure of tax avoidance schemes (DOTAS)
generally, 22.23
Discretionary trusts
clawback, and, 10.27
maintenance funds, and, 21.16
Disposal of property
loss of exemption, and, 19.3
Douceur
public purchase of heritage property by private treaty sales, and, 23.12

Election
distinction between APR and BPR, 12.2
Equipment
business property relief, and, 5.37
Excepted assets rules
Barclays Bank Trust Co Ltd v IRC, 7.11
'business', 7.7
corporate structures, and, 7.6
debt, and, 7.14
'group', 7.13
group structures, 7.1
hybrid companies, 7.10
introduction, 7.1
McCall v HMRC, 7.8–7.9
mixed development, 7.10
ownership, 7.2
personal use, 7.2
Salaried Persons Postal Loans Ltd v HMRC, 7.12
test period, 7.1
use, 7.2–7.3
valuation principles, 7.4–7.5
Excluded property trusts
business property relief, and, 16.2
Exotic grazing
agricultural property relief, and, 2.24
Expenses
maintenance funds, and, 21.8
Extra statutory concessions
texts, Appendix 2

Failed PETs
payment of IHT, and
due date, 23.4
generally, 23.3
HMRC charge, 23.9
Family limited partnerships
distinction between APR and BPR, 12.12–12.15
Farm business tenancies
occupation, and, 3.15
Farm cottages
extra statutory concession, Appendix 2
generally, 2.6
occupation, and, 3.26
Farmhouses
See also **Agricultural property relief**
Antrobus (No 1) v CIR, 2.13–2.14

Farmhouses – *contd*
Arnander v HMRC, 2.19
big houses, 2.10
checklist, 2.14
early cases, 2.11
'elephant' test, 2.15
family home, and, 2.20
generally, 2.7
Higginson v CIR, 2.12
historical background, 2.8
historical context, 2.9
HMRC V Hanson, 2.18
illness, and, 2.23
lifestyle farming, 2.13
McKenna Deceased v HMRC, 2.19
not in identical ownership with land, 2.22
Rosser v IRC, 2.17
shared ownership of land, 2.22
Starke v IRC, 2.16
substantial houses, 2.21
tests, 2.8–2.11
Farming companies
agricultural property relief, and, 2.28
Farming partnerships
'wholly or mainly holding investments', and, 9.17
Farms
contract farming, 13.4
let farms, 13.3
share-farming, 13.4
tenant buying out landlord, 13.5
Foreign owned works of art
generally, 17.28
Forms and procedures
agricultural relief, 22.19
business relief, 22.20
heritage relief, 22.12
introduction, 22.18
woodland relief, 22.21
Fraud
compliance, and, 22.17
'Free of tax' legacies
payment of IHT, and, 23.2
Fully private companies
tax shelters, and, 13.8

General Anti-Abuse Rule (GAAR)
tax shelters, and, 13.2
Gifts to the nation
payment of IHT, and, 23.18–23.20
Gifts with reservation
clawback, and, 10.23
de minimis, 14.4
further substitution rules, 14.6
generally, 14.3
treatment where BPR is available, 14.5
Gladstone v Bowers agreements
agricultural property relief, and, 4.13
Grants of representation
personal applications, and, 22.10
Grazing agreements
agricultural property relief, and, 4.13
Grossing up
business property relief, and, 6.3

Heritage property
See also **Heritage relief**
maintenance funds
accounts, 21.14
anti-avoidance rules, 21.18
charity, and, 21.11
conditions of relief, 21.3–21.11
content, 21.10
contributions from discretionary trusts, 21.16
entry of property, 21.15–21.17
expenses, taxes and fees, 21.8
impact of other taxes, 21.29–21.32
introduction, 21.1
maintenance, 21.6
procedural points, 21.12
property entering, 21.15–21.18
property leaving, 21.19–21.28
public access, 21.7
summary of relief, 21.2
supervision, 21.13
trust instrument, 21.4–21.8
trustees, 21.9
public purchase by private treaty sales
Cultural Gifts Scheme, 23.18–23.20
douceur, 23.12
generally, 23.11
gifts to the nation, 23.18–23.20
in situ offers, 23.17
interest, 23.15
multiple property offer, 23.16
stamp duty (land tax), 23.13
'supermarket' offers, 23.16

Heritage property – *contd*
public purchase by private treaty sales – *contd*
valuation date option, 23.15
value added tax, 23.14
old rules, Appendix 5
relevant property trusts
anti-avoidance rules, 20.8
calculation of rate of tax, 20.7
calculation of tax charge, 20.3
clawback charge, 20.2
conditions for exemption, 20.1
exemption from ten-year periodic charge, 20.5–20.6
general rules, 20.1
prevention of purchase to keep down periodic charges, 20.8
relevant person, 20.4
Heritage relief
See also **Heritage property**
amenity land
generally, 17.3
undertakings, 17.13
annual tax on enveloped dwellings, 19.22
applications for exemption
content, 18.5
discretionary trusts, 18.3
form, 18.5
intention to sell property, and, 18.6
introduction, 18.2
old law, 18.1
processing of claim, 18.8
protective claims, 18.4
provisional claims, 18.4
supervision of compliance with undertakings, 18.9
timing, 18.2–18.4
triggering a chargeable transfer, 18.7
artistic interest, 17.5
associated objects
generally, 17.8
introduction, 17.3
buildings
England, in, 17.6
introduction, 17.3
Scotland, in, 17.7
'by appointment only' access
generally, 17.1
Heritage relief – *contd*
'by appointment only' access – *contd*
variation of undertakings, and, 18.11
case law, 17.2
categories, 17.3
chattels
generally, 17.5
introduction, 17.3
undertakings, 17.11
compliance
chargeable lifetime transfers, 22.1
DOTAS, 22.23
duty of personal representative, 22.2–22.7
forms and procedures, 22.18–22.22
introduction, 19.23
penalties, 22.11–22.16
personal applications for grants of representation, 22.10
tax fraud, 22.17
valuation, 22.8–22.9
deferral basis, 17.1
DOTAS, 22.23
educational importance, 17.5
exempt property, 18.1–18.19
forms and procedures, 22.22
guidance, 17.2
historic association, 17.5
history and national life, 17.5
interaction with other reliefs, and, 1.6
introduction, 17.1–17.2
land and buildings
England, in, 17.6
introduction, 17.3
Scotland, in, 17.7
undertakings, 17.12
loss of exemption
allowances, and, 19.13
associated objects, 19.8
associated properties, 19.4–19.5
ATED, and, 19.22
breach of undertaking, 19.2
charge to tax, 19.9
chargeable events, 19.1–19.5
'clock' of chargeable transfers, 19.17
composite estates, and, 19.4–19.5
death of person beneficially entitled, 19.3

Heritage relief – *contd*
loss of exemption – *contd*
destruction of property, 19.8
disposal of property, 19.3
general rule, 19.1–19.5
increasing rates of tax, 19.15
lifetime transfers, 19.12
living collections, 19.7
loss of property, 19.8
old rules, 19.18
part disposals, 19.7
progressive sales, 19.14
rate of tax applied, 19.12–19.13
replacement of chattels, 19.8
relevant disposal, 19.6–19.9
relevant person, 19.10–19.11
set-off, 19.16
surviving spouse, 19.20
tax credit, 19.16
taxable value, 19.9
transferable nil rate band, 19.19–19.21
transfers on death, 19.14
nature, 17.1
need for, 1.3
new undertakings
amenity land, 17.13
chattels, 17.11
land and associated chattels, 17.12
old rules, Appendix 5
outstanding land and buildings
England, in, 17.6
introduction, 17.3
Scotland, in, 17.7
'pre-eminence'
associated objects, 17.8
chattels, 17.5
generally, 17.4
land and buildings, 17.6–17.7
'Waverley' criteria, 17.5
pre-eminent chattels, 17.3
pre-eminent collections, 17.3
preliminary approach, 17.9
public access
charges, 17.20
guided tours, 17.20
indemnity scheme, 17.19
in situ access, 17.16
introduction, 17.10
means of access, 17.18

Heritage relief – *contd*
public access – *contd*
new rules, under, 17.15–17.21
numbers of visitors, 17.20
other details, 17.20
personal circumstances of owner, and, 17.17
publicity, 17.22–17.
reasonable access, 17.16
undertakings, 17.11–17.13
venue, 17.21
publicity
amenity land, 17.26
details to be disclosed, 17.24–17.27
foreign-owned works of art, 17.28
generally, 17.22
introduction, 17.14
land, 17.27
outstanding buildings, 17.26
pre-eminent chattels and collections, 17.25
register of conditionally exempt works of art, 17.23
qualifying property
introduction, 17.3
quality, 17.4
quality of property, 17.4
statutory basis, 17.2
territoriality, 1.5
undertakings
amenity land, 17.13
chattels, 17.11
land and associated chattels, 17.12
variation of undertakings
'by appointment only' access, 18.11
introduction, 18.10
museum quality property, 18.12
practice, in, 18.19
replacement undertakings, 18.18
review of undertakings, and, 18.14–18.17
sale of heritage property, 18.13
transition to current regime, 18.11
'Waverley' criteria, 17.5
HMRC charge
payment of IHT, and, 23.9
HMRC interpretations
texts, Appendix 3
HMRC press releases
texts, Appendix 4

HMRC Statements of Practice
texts, Appendix 3
Holdover relief
business property relief, and, 6.2
Holiday lettings
distinction between APR and BPR, 12.21
Holiday homes
'wholly or mainly holding investments', and, 9.13–9.14
Horses
agricultural property relief, and, 2.24
Hybrids
diversification by farmers, and, 12.20
limited liability partnerships, and, 5.21
planning points, and, 14.20
relevant business property, and
generally, 7.6
introduction, 7.1
practical approach, 7.10
'wholly or mainly holding investments', and, 9.11

In situ offers
public purchase of heritage property by private treaty sales, and, 23.17
Income tax
maintenance funds, and
general treatment of fund, 21.30
'one estate', 21.29
Inheritance tax
agricultural property relief
agricultural value, 3.1
areas of concern, 4.1–4.28
clawback, 10.1–10.30
computation, 1.7
distinction from BPR, 1.9
history, 1.2
interaction of reliefs, 1.6
occupation, 3.17–3.27
overlap with BPR, 12.1–12.22
qualifying property, 2.1–2.30
rates, 3.2–3.16
reduced relief, 3.28
tax planning, 14.1–14.28
tax shelters, 13.1–13.14
territoriality, 1.5
will planning, 15.1–15.17
business property relief
clawback, 10.1–10.30

Inheritance tax – *contd*
business property relief – *contd*
computation, 1.7
distinction from APR, 1.9
excepted assets, 7.1–7.14
history, 1.2
interaction of reliefs, 1.6
liabilities, and, 8.16–8.24
limitations, 6.5–6.22
making or holding investments, 9.1–9.23
overlap with APR, 12.1–12.22
ownership, 6.5–6.22
qualifying property, 5.1–5.46
rates, 6.1–6.4
tax planning, 14.1–14.28
tax shelters, 13.1–13.14
territoriality, 1.5
valuation, 8.1–8.24
will planning, 15.1–15.17
compliance
chargeable lifetime transfers, 22.1
DOTAS, 22.23
duty of personal representative, 22.2–22.7
forms and procedures, 22.18–22.22
penalties, 22.11–22.16
personal applications for grants of representation, 22.10
tax fraud, 22.17
valuation, 22.8–22.9
DOTAS, 22.23
heritage relief
exempt property, 18.1–18.19
introduction, 17.1–17.2
loss of exemption, 19.1–19.23
maintenance funds, 21.1–21.32
qualifying property, 17.3–17.29
relevant property trusts, and, 20.1–20.8
territoriality, 1.5
interaction of reliefs, 1.6–1.8
payment
due date, 23.4
failed PETs, 23.3
'free of tax' legacies, 23.2
general rules, 23.1–23.9
HMRC charge, 23.9
instalments, by, 23.5–23.8

Inheritance tax – *contd*
payment – *contd*
public purchase of heritage property by private treaty sales, 23.11–23.20
purchase of own shares by unquoted trading company, 23.10
role of relief, 1.1
territoriality, 1.5
woodlands, 11.1–11.18
working farmer relief, 3.7–3.14
Instalment payments of IHT
eligibility conditions, 23.5
generally, 23.5
interaction with other reliefs, and, 1.8
interest, and, 23.7
qualifying property, 23.5
s 228 shares, 23.6
termination of right, 23.8
Insurance
maintenance funds, and, 21.8
Intellectual 'tools' and property
business property relief, and, 5.36
Intensive rearing of livestock or fish
ancillary occupation, 2.5
generally, 2.4
Interest
payment by instalments, and, 23.7
public purchase of heritage property by private treaty sales, and, 23.15
Interest in possession trusts
maintenance funds, and, 21.17
Investment businesses
planning issues, and, 14.19
Investment properties
tax shelters, and, 13.14
Investments
See also **Business property relief**
Best v HMRC, 9.22
building and developing, 9.19
Burkinyoung v IRC, 9.3–9.4
caravan parks, 9.7–9.14
carrying on a business for gain, 9.20
Clark Deceased v HMRC, 9.21
Earl of Balfour Deceased v HMRC, 9.17
Farmer v IRC, 9.16
farming partnership, 9.17
Furness, 9.9
George v CIR, 9.11–9.12
Investments – *contd*
Hall v CIR, 9.7
holiday homes, 9.13–9.14
introduction, 9.1
level of activity, 9.5
livery business, 9.15
Mallender, 9.20
Martin v IRC, 9.2
mixed trade, 9.1
mixed use estate, 9.16
money lending, 9.18
Moore Deceased v IRC, 9.2
overview, 9.23
Pawson v HMRC, 9.14
Pearce Deceased v IRC, 9.8
Phillips Deceased v HMRC, 9.18
Piercy's Executors v HMRC, 9.19
Powell v IRC, 9.8
property development, 9.19
property management, 9.21–9.22
Stedman's Executors v IRC, 9.11
Vigne v HMRC, 9.15
Weston v IRC, 9.10
Zetland Settlement v HMRC, 9.6

Joint ownership
agricultural property relief, and, 4.2

Land and buildings
heritage relief, and
England, in, 17.6
introduction, 17.3
Scotland, in, 17.7
undertakings, 17.12
Life assurance
generally, 16.7
Lifetime transfers
loss of exemption, and, 19.12
Limited liability partnerships
accounts and formalities, 5.22
generally, 5.21
Jersey LLPs, 5.23
transition, 5.24
Liquidations
business property relief, and, 6.17
Listed unquoted companies
tax shelters, and, 13.7
Livery business
'wholly or mainly holding investments', and, 9.15

Lloyd's underwriters
assets supporting the business, 5.19
generally, 5.17
Names losses, 5.18
tax shelters, and, 13.13

Machinery and plant
business property relief, and, 5.34
tax shelters, and, 13.11

Maintenance funds
See also **Heritage property**
accounts, 21.14
alterations to property, 21.6
alternative use of the income (or capital), 21.4
anti-avoidance rules
charitable gifts, 21.11
excess value, 21.16
generally, 21.18
purchase rule, 21.16
applications for exemption, 21.12
capital gains tax, 21.32
charity, and, 21.11
clawback, and, 21.24
conditions for relief
content of the fund, 21.10
introduction, 21.3
property, 21.10
trust instrument, 21.4–21.8
trustees, 21.9
content of the fund, 21.10
contributions from discretionary trusts, 21.16
death of life tenant, 21.26
entry of property
anti-avoidance rules, 21.18
discretionary trusts, from, 21.16
generally, 21.15–21.17
interest in possession trust, from, 21.17
expenses, taxes and fees, 21.8
impact of other taxes
capital gains tax, 21.32
income tax, 21.29–21.30
stamp duty (land) tax, 21.31
income tax
general treatment of fund, 21.30
'one estate', 21.29
insurance, 21.8
interest in possession trusts, and, 21.17

Maintenance funds – *contd*
introduction, 21.1
loss of heritage status, 21.24
maintenance of heritage property, 21.6
procedural points, 21.12
professional fees, 21.8
property entering
anti-avoidance rules, 21.18
discretionary trusts, from, 21.16
generally, 21.15–21.17
interest in possession trust, from, 21.17
property in the fund, 21.10
property leaving
death of life tenant, 21.26
generally, 21.19
loss of heritage status, 21.24
other acceptable uses, 21.23
rate of tax charge, 21.25
release of capital, 21.24
revertor to settler, 21.20
special property reliefs, and, 21.27
transactions by trustees, 21.24
transfer to another maintenance fund, 21.22
use for heritage purposes, 21.21
withdrawal of Treasury Direction, 21.24
property previously in a relevant property trust, 21.28
public access, 21.7
rates and taxes, 21.8
release of capital, 21.24
relevant property trust, and, 21.28
repairs to property, 21.6
residence of trustees, 21.9
six-year rule, 21.4
stamp duty (land) tax, 21.31
statutory basis, 21.1
summary of relief, 21.2
supervision, 21.13
transactions by trustees, 21.24
trust instrument
expenses, taxes and fees, 21.8
generally, 21.4
maintenance of heritage property, 21.6
public access, 21.7
qualifications, 21.5–21.8
trustees, 21.9

Maintenance funds – *contd*
withdrawal of Treasury Direction, 21.24
wording of trust instrument, 21.4–21.8
Market makers exception
generally, 5.8
Regulations, Appendix 1
'Mere asset' exclusion
business property relief, and, 5.13–5.14
Milk quota
agricultural property relief, and, 2.26
Mitigation
30% penalties, 22.14
70% penalties, 22.15
1000% penalties, 22.16
Mixed development
business property relief, and, 7.10
Mixed trade
'wholly or mainly holding investments', and, 9.1
Mixed use estate
'wholly or mainly holding investments', and, 9.16
'Money box'
planning points, and, 14.18
Money lending
'wholly or mainly holding investments', and, 9.18
Musicians
business property relief, and, 5.11

Normal expenditure out of income
interaction with other reliefs, and, 1.6
Nursing homes
tax shelters, and, 13.12

Occupation and ownership
agricultural property relief
agricultural purposes, 3.17
Atkinson Deceased v HMRC, 3.24
Basic Farm Payment (BFP), 3.19
contract farming, 3.27
controlled companies, 4.9
diversification, 3.18
dwellings, 3.25
farm cottages, 3.26
generally, 3.17–3.27
ill farmers, 3.23
introduction, 3.3
Occupation and ownership – *contd*
agricultural property relief – *contd*
joint ownership, 4.2
occupiers, 3.20
partnerships, 3.21, 4.3
proof, 3.22
share farming, 3.27
successions, 4.10
successive transfers, 4.11
test determinative of rate of relief, 3.4
trusts, 4.4–4.5
working farmer relief, 3.9
business property relief
capital reconstruction, 6.8
detailed rules, 6.10–6.11
excepted assets, and, 7.2
extension of period, 6.9
general rule, 6.5
modification of rule, 6.6
replacement property, 6.7
unquoted shares, 6.8

Painters
business property relief, and, 5.11
Partnerships
agricultural property relief, and, 4.3
business property relief, and, 5.12
Payment of IHT
Cultural Gifts Scheme, 23.18–23.20
douceur, 23.12
due date, 23.4
failed PETs, and
due date, 23.4
generally, 23.3
HMRC charge, 23.9
'free of tax' legacies, 23.2
general rules, 23.1–23.9
gifts to the nation, 23.18–23.20
HMRC charge, 23.9
instalments, by
eligibility conditions, 23.5
generally, 23.5
interest, and, 23.7
qualifying property, 23.5
s 228 shares, 23.6
termination of right, 23.8
public purchase of heritage property by private treaty sales
douceur, 23.12

Payment of IHT – *contd*
public purchase of heritage property by private treaty sales – *contd*
generally, 23.11
in situ offers, 23.17
interest, 23.15
multiple property offer, 23.16
stamp duty (land tax), 23.13
'supermarket' offers, 23.16
valuation date option, 23.15
value added tax, 23.14
purchase of own shares by unquoted trading company, 23.10
'top up' charge, and, 23.3
Penalties
culpability, 22.13
generally, 22.11
levels
generally, 22.13
mitigation, 22.14–22.16
mitigation
30% penalties, of, 22.14
70% penalties, of, 22.15
1000% penalties, of, 22.16
overview of provisions, 22.12
purpose, 22.12
Personal applications for grants of representation
compliance, and, 22.10
Personal representatives
compliance duty, and
estimates, 22.5
generally, 22.2
HMRC guidance, 22.3
property details, 22.4
Robertson v IRC, 22.6
Robertson v IRC (No 2), 22.7
Planning points
additional charity relief, 15.8
associated operations, 14.22–14.23
attribution rules, 15.12
buy and sell agreements, 14.11–14.12
deathbed planning, 14.28
deductibility of debts
effect of Finance Act 2013, 14.8–14.9
generally, 14.7
restriction on offset, 14.10
defining the legacy, 15.11
Planning points – *contd*
drafting wills
defining the legacy, 15.11
use of a cap, 15.10
'early' gifts, 14.1
elderly couple, for, 15.9
gifts of businesses, 15.16
gifts with reservation
de minimis, 14.4
further substitution rules, 14.6
generally, 14.3
treatment where BPR is available, 14.5
hybrid companies, 14.20
investment businesses, 14.19
'money box', 14.18
multiple company structure, 14.20
nil rate band
additional charity relief, and, 15.8
difficulties arising under s 41(b), 15.7
double grossing, 15.6
double NRB, 15.2
gift bearing its own tax, 15.4
infinite NRB, 15.1
mixture of legacies with shares of residue, 15.6
non-exempt 'free of tax' legacy, 15.5
non-exempt 'gross' legacy, 15.4
partly exempt transfers, 15.3
simple grossing, 15.5
non-exempt beneficiaries, and, 15.15
relevant property trusts
associated operations, 14.22–14.23
transfer between settlements, 14.24
treatment of relieved property, 14.2
use to secure 100% relief, 14.21–14.27
retirement, 14.14–14.17
s 39A, 15.13–15.14
share options, 14.13
'spouse', 15.17
structures, 14.19–14.20
transfer between settlements, 14.24
'unfortunate' gifts, 14.1
use of a cap, 15.10
Potentially exempt transfers
interaction with other reliefs, and, 1.6

Pre-eminence
heritage relief, and
associated objects, 17.8
chattels, 17.5
generally, 17.4
land and buildings, 17.6–17.7
'Waverley' criteria, 17.5
Property development
'wholly or mainly holding investments', and, 9.19
Property management
'wholly or mainly holding investments', and, 9.21–9.22
Public access
charges, 17.20
guided tours, 17.20
indemnity scheme, 17.19
in situ access, 17.16
introduction, 17.10
maintenance funds, and, 21.7
means of access, 17.18
new rules, under, 17.15–17.21
numbers of visitors, 17.20
other details, 17.20
personal circumstances of owner, and, 17.17
publicity, 17.22–17.
reasonable access, 17.16
undertakings, 17.11–17.13
venue, 17.21
Professional fees
maintenance funds, and, 21.8
Public purchase of heritage property by private treaty sales
douceur, 23.12
generally, 23.11
in situ offers, 23.17
interest, 23.15
multiple property offer, 23.16
stamp duty (land tax), 23.13
'supermarket' offers, 23.16
valuation date option, 23.15
value added tax, 23.14
Purchase of own shares by unquoted trading company
payment of IHT, and, 23.10

Qualifying property
agricultural property relief
agricultural land or pasture, 2.2
Qualifying property – *contd*
agricultural property relief – *contd*
conservation land, 2.27
farm cottages, 2.6
farmhouses, 2.7–2.23
farming companies, 2.28
horses, 2.24
intensive rearing of livestock or fish, 2.4–2.5
introduction, 2.1
milk quota, 2.26
set aside land, 2.27
short rotation coppicing, 2.25
territoriality, 2.29
wildlife habitats, 2.27
woodland, 2.3
business property relief
business or interest in a business, 5.3–5.24
categories, 5.2
controlling holdings in quoted company, 5.28
land and buildings used in the business, 5.33–5.39
land and buildings used in the tenant for life's business, 5.40–5.46
unquoted shares, 5.27–5.32
unquoted securities giving control, 5.25–5.26
Quick succession relief
business property relief, and, 6.1
Quoted companies
controlling holdings
clearances, 5.31–5.32
'control', 5.29
generally, 5.28
preference shares, 5.30
restriction on relief, 5.29
'quoted'
generally, 6.18
list of exchanges, 6.21
other trading exchanges, 6.20
recognised stock exchange, 6.19
valuation of shares, 8.2

Rates and taxes
maintenance funds, and, 21.8
Reconstructions
business property relief, and, 6.17

Related property
business property relief, and, 8.23
Relevant business property
See also **Business property relief**
business or interest in a business
activities constituting a business, 5.4
artists, 5.11
assets supporting the underwriting business, 5.19
authors, 5.11
'business', 5.3
corporate/non-corporate distinction, 5.7
disposal of part of partnership share, 5.15
generally, 5.3
Jersey LLPs, 5.23
limited (liability) partnerships, 5.21–5.24
Lloyd's Name losses, 5.18
Lloyd's underwriters, 5.17–5.19
'mere asset' exclusion, 5.13–5.14
multiple businesses, 5.16
musicians, 5.11
Nelson Dance Settlement v HMRC, 5.14
non-investment businesses, 5.5
painters, 5.11
partnership interest, 5.12
restrictive effect of s 105(3), 5.8
sculptors, 5.11
sole proprietor's business, 5.10
succession, 5.20
unincorporated business, 5.7
woodlands, 5.9
wrong kind of business for relief, 5.6
categories, 5.2
controlling holdings in quoted company
clearances, 5.31–5.32
'control', 5.29
generally, 5.28
preference shares, 5.30
restriction on relief, 5.29
excepted assets
Barclays Bank Trust Co Ltd v IRC, 7.11
'business', 7.7
corporate structures, and, 7.6
Relevant business property – *contd*
excepted assets – *contd*
debt, and, 7.14
'group', 7.13
group structures, 7.1
hybrid companies, 7.10
introduction, 7.1
McCall v HMRC, 7.8–7.9
mixed development, 7.10
ownership, 7.2
personal use, 7.2
Salaried Persons Postal Loans Ltd v HMRC, 7.12
test period, 7.1
use, 7.2–7.3
valuation principles, 7.4–7.5
introduction, 5.2
land and buildings used in the business
animals, 5.35
bookmakers' pitchers, 5.39
equipment, 5.37
generally, 5.33
intellectual 'tools' and property, 5.36
machinery and plant, 5.34
royalties, 5.38
land and buildings used in the tenant for life's business
applicability of lower rate of relief, 5.45
'assets used in the business', 5.44
generally, 5.40
history of relief, 5.42
'interest in possession', 5.41
other conditions, 5.46
timing of transfer, 5.43
unquoted shares, 5.27–5.32
unquoted securities giving control
'control', 5.26
generally, 5.25
Relevant property trusts
heritage property, and
anti-avoidance rules, 20.8
calculation of rate of tax, 20.7
calculation of tax charge, 20.3
clawback charge, 20.2
conditions for exemption, 20.1
exemption from ten-year periodic charge, 20.5–20.6
general rules, 20.1

Relevant property trusts – *contd*
heritage property, and – *contd*
prevention of purchase to keep down periodic charges, 20.8
relevant person, 20.4
maintenance funds, and, 21.28
planning issues, and
associated operations, 14.22–14.23
transfer between settlements, 14.24
treatment of relieved property, 14.2
use to secure 100% relief, 14.21–14.27
Repairs to property
maintenance funds, and, 21.6
Replacement assets
distinction between APR and BPR, 12.5
Replacement property
agricultural property relief
introduction, 4.6
occupation test, 4.7
ownership test, 4.8
share or securities, 4.12
business property relief, 6.7
clawback
agricultural property relief, 10.19–10.22
business property relief, 10.11–10.12
Retirement
planning issues, and, 14.14–14.17
Royalties
business property relief, and, 5.38

Sculptors
business property relief, and, 5.11
Secondary market listings
tax shelters, and, 13.7
Section 228 shares
payment by instalments, and, 23.6
Security of tenure
distinction between APR and BPR, 12.18
Set aside land
agricultural property relief, and, 2.27
Set-off
loss of exemption, and, 19.16
Settlements
planning issues, and, 14.24
Share-farming
distinction between APR and BPR, 12.16
occupation, and, 3.27
tax shelters, and, 13.4
Share options
planning issues, and, 14.13
Shares
valuation, and, 8.24
Sheltering proceeds of sale
distinction between APR and BPR, 12.9
Short rotation coppicing
agricultural property relief, and, 2.25
Six-year rule
maintenance funds, and, 21.4
Small gifts
interaction with other reliefs, and, 1.6
Sole proprietor's business
business property relief, and, 5.10
Stamp duty (land) tax
maintenance funds, and, 21.31
public purchase of heritage property by private treaty sales, and, 23.13
Statements of Practice
texts, Appendix 3
Successions
agricultural property relief, and, 4.10
business property relief, and
generally, 6.16
introduction, 5.20
Successive transfers
agricultural property relief, and, 4.11
Successive transfers
business property relief, and, 6.16
'Supermarket' offers
public purchase of heritage property by private treaty sales, and, 23.16
Switching
distinction between APR and BPR, 12.5

Tax credit
loss of exemption, and, 19.16
Tax fraud
compliance, and, 22.17
Tax planning
See also **Will planning**
associated operations, 14.22–14.23
buy and sell agreements, 14.11–14.12

Tax planning – *contd*
deathbed planning, 14.28
deductibility of debts
effect of Finance Act 2013, 14.8–14.9
generally, 14.7
restriction on offset, 14.10
'early' gifts, 14.1
gifts with reservation
de minimis, 14.4
further substitution rules, 14.6
generally, 14.3
treatment where BPR is available, 14.5
hybrid companies, 14.20
investment businesses, 14.19
'money box', 14.18
multiple company structure, 14.20
relevant property trusts
associated operations, 14.22–14.23
transfer between settlements, 14.24
treatment of relieved property, 14.2
use to secure 100% relief, 14.21–14.27
retirement, 14.14–14.17
share options, 14.13
structures, 14.19–14.20
transfer between settlements, 14.24
'unfortunate' gifts, 14.1
Tax shelters
businesses
buildings or machinery, 13.11
conversion of cash into equity, 13.10
conversion of debt into equity, 13.9
fully private companies, 13.8
listed unquoted companies, 13.7
Lloyd's underwriting, 13.13
nursing homes, 13.12
secondary market listings, 13.7
farms
contract farming, 13.4
let farms, 13.3
share-farming, 13.4
tenant buying out landlord, 13.5
General Anti-Abuse Rule, 13.2
introduction, 13.1
investment properties, 13.14
Taxes
maintenance funds, and, 21.8
Territoriality
agricultural property relief, and, 2.29
distinction between APR and BPR, 12.3
generally, 1.5
Timber
See also **Woodland relief**
generally, 11.3–11.4
'Top up' charge
payment of IHT, and, 23.3
Transferable nil rate band
loss of exemption, and, 19.19–19.21
Trust instruments
maintenance funds, and
expenses, taxes and fees, 21.8
generally, 21.4
maintenance of heritage property, 21.6
public access, 21.7
qualifications, 21.5–21.8
Trustees
maintenance funds, and, 21.9
Trusts
agricultural property relief, and, 4.4–4.5
Two-year rule
business property relief, and
general rule, 6.5
modification of rule, 6.6

Undertakings
amenity land, 17.13
chattels, 17.11
land and associated chattels, 17.12
variation
'by appointment only' access, 18.11
introduction, 18.10
museum quality property, 18.12
practice, in, 18.19
replacement undertakings, 18.18
review of undertakings, and, 18.14–18.17
sale of heritage property, 18.13
transition to current regime, 18.11
Unincorporated businesses
business property relief, and, 5.7
Unquoted companies
securities giving control
'control', 5.26
generally, 5.25

Unquoted companies – *contd*
shares
generally, 5.27–5.32
period of ownership, 6.8
valuation, 8.4–8.13
valuation of shares
basis to be applied, 8.10–8.11
Caton's Administrators v Couch, 8.5
CVC and others v Demarco Almeida, 8.8
Denekamp v Pearce, 8.7
discounting, 8.12–8.13
generally, 8.4
Hawkings-Byass v Sassen, 8.6
minority interests in property, 8.13
Phoenix Office Supplies v Larvin, 8.9

Valuation
agricultural property
agricultural estate, 4.16
general considerations, 4.14
joint property, 4.21
lotting for sale, 4.18
natural units, 4.18
other aspects, 4.19
planning permission, 4.20
tenancies, 4.15
AIM shares, 8.3
Alexander v IRC, 4.17
Arkwright v IRC, 4.21
Baird's Executors v IRC, 4.15
basic principles
Alexander v IRC, 4.17
Antrobus (No 2) v IRC, 4.23
Arkwright v IRC, 4.21
Baird's Executors v IRC, 4.15
general considerations, 4.14
Gray v IRC, 4.16
overview, 8.1
Price v HMRC, 4.22
business property
AIM shares, 8.3
basic principles, 8.1
discounts, 8.10–8.13
excepted assets, and, 7.4–7.5
general rules, 8.14–8.15
introduction, 8.1
loan notes, 8.24
quoted shares and securities, 8.2

Valuation – *contd*
business property – *contd*
related property rules, 8.23
shares, 8.24
treatment of liabilities, 8.16–8.22
unquoted shares, 8.4–8.13
Cairns v HMRC, 22.9
compliance, and
Cairns v HMRC, 22.9
generally, 22.8
farmhouses, 4.23
general considerations, 4.14
Gray v IRC, 4.16
HMRC interpretations, Appendix 3
IRC v Crossman, 4.14
joint property, 4.21–4.22
long lease of flat, 4.17
lotting for sale, 4.18
natural units, 4.18
other aspects, 4.19
planning permission, 4.20
Price v HMRC, 4.22
shares in quoted companies, 8.2
shares in unquoted companies
basis to be applied, 8.10–8.11
Caton's Administrators v Couch, 8.5
CVC and others v Demarco Almeida, 8.8
Denekamp v Pearce, 8.7
discounting, 8.12–8.13
generally, 8.4
Hawkings-Byass v Sassen, 8.6
minority interests in property, 8.13
Phoenix Office Supplies v Larvin, 8.9
tenancies, 4.15

Value added tax
public purchase of heritage property by private treaty sales, and, 23.14

Value shifting
clawback, and, 10.29

Variation of undertakings
heritage relief, and
'by appointment only' access, 18.11
introduction, 18.10
museum quality property, 18.12
practice, in, 18.19
replacement undertakings, 18.18
review of undertakings, and, 18.14–18.17

Variation of undertakings – *contd*
heritage relief, and – *contd*
sale of heritage property, 18.13
transition to current regime, 18.11
introduction, 1.5

'Wholly or mainly holding investments'
See also **Business property relief**
Best v HMRC, 9.22
building and developing, 9.19
Burkinyoung v IRC, 9.3–9.4
caravan parks, 9.7–9.14
carrying on a business for gain, 9.20
Clark Deceased v HMRC, 9.21
Earl of Balfour Deceased v HMRC, 9.17
Farmer v IRC, 9.16
farming partnership, 9.17
Furness, 9.9
George v CIR, 9.11–9.12
Hall v CIR, 9.7
holiday homes, 9.13–9.14
introduction, 9.1
level of activity, 9.5
livery business, 9.15
Mallender, 9.20
Martin v IRC, 9.2
mixed trade, 9.1
mixed use estate, 9.16
money lending, 9.18
Moore Deceased v IRC, 9.2
overview, 9.23
Pawson v HMRC, 9.14
Pearce Deceased v IRC, 9.8
Phillips Deceased v HMRC, 9.18
Piercy's Executors v HMRC, 9.19
Powell v IRC, 9.8
property development, 9.19
property management, 9.21–9.22
Stedman's Executors v IRC, 9.11
Vigne v HMRC, 9.15
Weston v IRC, 9.10
Zetland Settlement v HMRC, 9.6
Wildlife habitats
agricultural property relief, and, 2.27
Will planning
additional charity relief, 15.8
attribution rules, 15.12
defining the legacy, 15.11

Will planning – *contd*
drafting points
defining the legacy, 15.11
use of a cap, 15.10
elderly couple, for, 15.9
gifts of businesses, 15.16
nil rate band
additional charity relief, and, 15.8
difficulties arising under s 41(b), 15.7
double grossing, 15.6
double NRB, 15.2
gift bearing its own tax, 15.4
infinite NRB, 15.1
mixture of legacies with shares of residue, 15.6
non-exempt 'free of tax' legacy, 15.5
non-exempt 'gross' legacy, 15.4
partly exempt transfers, 15.3
simple grossing, 15.5
non-exempt beneficiaries, and, 15.15
s 39A, 15.13–15.14
'spouse', 15.17
use of a cap, 15.10
woodland, 2.3
Winding-up
business property relief, and, 6.22
Woodland relief
agricultural property relief, and, 2.3
business property relief, and, 5.9
charge to tax on disposal
basis, 11.10
credit for tax charged, 11.13
deduction of expenses, 11.12
generally, 11.9
net value, 11.12
rate of tax, 11.11
commercial occupation, 11.1–11.4
compliance
chargeable lifetime transfers, 22.1
DOTAS, 22.23
duty of personal representative, 22.2–22.7
forms and procedures, 22.18–22.22
penalties, 22.11–22.16
personal applications for grants of representation, 22.10
tax fraud, 22.17
valuation, 22.8–22.9

Woodland relief – *contd*
death, on
charge to tax on disposal, 11.9–11.13
50% reduction, 11.14
generally, 11.5
prior claim to BPR, 11.7
prior relief as agricultural property, 11.6
qualifying property, 11.8
DOTAS, 22.23
extra statutory concession, Appendix 2
50% reduction, 11.14
forms and procedures, 22.21
generally, 11.1–11.18
interaction with other reliefs, and, 1.6
liabilities, 11.16–11.18

Woodland relief – *contd*
lifetime gifts, on, 11.15
old law, 11.4
qualifying property, 11.8
tax treatment, 11.1
timber, 11.3–11.4
trade, 11.2
treatment of liabilities, 11.16–11.18

Working farmer relief
companies, and, 3.11
difficulties, 3.10
generally, 3.7
limitation by value or by acreage, 3.12
occupation test, 3.9
restrictions, 3.14
successions, 3.13
'wholly or mainly engaged', 3.8